AAPS Advances in the Pharmaceutical Sciences Series

Volume 58

The AAPS Advances in the Pharmaceutical Sciences Series, published in partnership with the American Association of Pharmaceutical Scientists, is designed to deliver volumes authored by opinion leaders and authorities from around the globe, addressing innovations in drug research and development, and best practice for scientists and industry professionals in the pharma and biotech industries. Indexed in

Reaxys

SCOPUSChemical Abstracts Service (CAS)SCImago

EMBASE

Hannah Batchelor • Klaus Rose
Editors

Pediatric Formulations

A Roadmap

Second Edition

Springer

Editors
Hannah Batchelor
University of Strathclyde
Glascow, UK

Klaus Rose
klausrose Consulting
Pediatric Drug Development and More
Riehen, Switzerland

ISSN 2210-7371 ISSN 2210-738X (electronic)
AAPS Advances in the Pharmaceutical Sciences Series
ISBN 978-3-031-77239-9 ISBN 978-3-031-77240-5 (eBook)
https://doi.org/10.1007/978-3-031-77240-5

Jointly published with American Association of Pharmaceutical Scientists

This Springer imprint is published by the registered company Springer Nature Switzerland AG
The registered company address is: Gewerbestrasse 11, 6330 Cham, Switzerland

Introduction, Purpose of the Book

This is the second edition of the first book that addresses pediatric formulations in the context of drug development. Why was it not published fifty or hundred years earlier? It is not just a "how to formulate" textbook. The subtitle indicates that we are on a road that reflects a new desire and challenge: let minors participate in pharmaceutical progress. Altogether, this is called "pediatric drug development" (PDD). A road with obstacles, barriers, and illusions. On our road, we hope to help you to further walk through the jungle of information relating to pediatric medicines development in academic training, your workplace, in print, Internet, social media, hearsay, and more.

Medicines have a complex background. Today we have many more effective ones compared to a century ago. There are many players: physicians prescribe medicines; hospital pharmacists procure them in the hospital; other pharmacists provide them outside the hospital. Companies develop and market new drugs. After a while, they lose patent protection and are sold as generics. Academic scientists work on the understanding of diseases, pharmacology, chemistry, and more. Pediatric pharmacologists have elucidated a lot of what the child's body does to a given drug, and what drugs do to the child's body. All drugs are regulated by authorities to ensure only quality medicines are approved; to detect side effects; to uncover counterfeits; and more. Drugs are paid for by reimbursement organizations—private, state-owned, or a combination. And there are the patients. All these are just the core players, each with particular perceptions and interests.

The position of children in our society has changed dramatically. Entire industries provide toys, clothing, literature, electronic gizmos, education, entertainment, and more. Civilization is becoming more and more complex. Everybody has an idea what a child is. And how accurate is this idea? Legally and administratively, this is easy. In 1979, the FDA defined "children" as $\leq$16 years old. The European Union (EU) defines children as from birth to <18 years. By and large, the FDA has adopted this position. All early FDA pediatric regulations were predominantly justified by safety considerations. But the human body does not undergo a metamorphosis in the night of the 16th or 18th birthday. It matures during puberty. Once puberty is achieved, the body is physiologically mature, although administratively and legally the person remains a child until they come of age. Drugs treat the body/disease, not the legal status. The process of puberty has accelerated considerably during the twentieth century in all modern societies. To ask for "pediatric" formulations makes sense in babies, infants, and young children, but not in adolescents. Among all medicinal specialties, pediatrics is a very young discipline. The American Academy of Pediatric (AAP) defines its clientele as up to 21 years, older in persons with special needs. Pragmatically and administratively, there is nothing wrong with that.

The need to differentiate between the administrative and the physiological meaning of the term "child" is important in pediatric pharmaceutical development.

Neonates and young infants cannot digest solid food. Their digestive system needs a minimal degree of maturity for that. When drug development began, most drugs were sold as tablets, which excluded babies and infants. Tablets had to be crushed and reconstituted as liquids to administer to such young patients. But these young persons are only a part of the "pediatric population" that is covered by pediatric legislation.

There were many assumptions that took decades to erode. One was that in "children" everything is different. Depression was believed not to exist in minors. Child cancer was believed to be fundamentally different from malignancies in adults. Today, we have a more differentiated view. Depression exists in minors, too. There are many different types of malignancies. Some occur in adults only, some only in children, some in all age groups, and some predominantly in adults but occasionally also in adolescents. Occasional occurrence in adolescents does not make a disease a "pediatric" disease. Another assumption was the belief that children needed to be protected against clinical trials. Yet clinical trials saved the lives of hundreds of thousands of children with cancer.

Developing age-adjusted drug formulations is not only a technical challenge. Otherwise, this would be the 20th edition of this book. The dynamics behind the development of new medicines and of age-adapted formulations are complex. This book gives background information and recommends further reading to those that want to know more.

There are two types of drugs on the market. Generics, i.e., drugs whose patent protection has expired, and patent-protected new ones, developed predominantly by research-based companies and, to some degree, also by academia. For new drugs, laws, both in the USA and the EU, compel companies to consider also appropriate medicinal products for minors during drug development. One major demand of the authorities is to develop age-appropriate formulations. Generics are exempt from mandatory pediatric product development.

A century ago, there were powerful cough suppressants, labeled as "suitable for children and adults." They were so effective that they could kill a child. We will never know how many children did not survive such treatment. These concoctions contained opioids.

Source: www.wellcomecollection.org

If you go to a pharmacy museum, you see what pharmacies sold in the past. Apart from opioids, there was alcohol, creams and ointments, spices, dried fruits, powder of pulverized mummies, bones and other body parts of convicted criminals, dried fish, and more. Most of this merchandize is no longer regarded as medicine today.

A "candy aspirin" product was introduced as an orange-flavored small-dose chewable tablets for children in 1947. This product received "blockbuster" status almost instantly. It was the subject of widespread advertising campaigns that highlighted its pleasant taste and ease of use in children. However, the American Academy of Pediatrics identified an increase in aspirin poisoning in children, which was due to overdose in children. The overdose may have been related to the fact that each company had a different dose in their tablets, and there was no standardized dose for children at that time coupled with the fact that as children liked the taste they sought out the product as it was like candy. This resulted in action to develop labels that stated keep out of the reach of children as well as moves toward developing a child resistant pack for medicines. The safety caps were introduced in 1966 and by the mid-1970s aspirin mortality rates in children had declined. This example demonstrates the need for careful thought when designing product for children.

Where there is a demand, a lot will be sold. A century ago, there were many medicines against tuberculosis, cancer, infection, aging, and other health challenges. They all provided a good income to manufacturers and pharmacists. Yet they were not proven to work.

Modern pharmaceutical treatment evolved with the scientific revolution and with modern industry, which was chemical industry first, pharmaceutical industry later, and is life science or health industry today. Powerful drugs were synthesized. It took two major catastrophes—sulfanilamide elixir in 1936 and thalidomide in 1961—to open the path to modern drug regulation, where the safety and efficacy of any drug must be proven by clinical and other trials. This signaled the advent of the modern label, a shift from claiming whatever the manufacturer wanted to claim toward a document that reflects what has been proven. This led to new terms such as "off-label": the use of drugs outside of its label, i.e., in a therapeutic area or age group for which the drug is not registered. Unlicensed use in pediatrics often involved crushing tablets or opening capsules and transforming the content into a liquid formulation. Since initial US pediatric regulations, most drugs in children were prescribed off-label. The debate about on-label/off-label is keeping US courts busy since decades. For this introduction, we just emphasize that it is necessary to differentiate between regulatory justifications for PDD and clinical outcomes of pediatric studies and other measures.

Where did the demand for PDD originate? All healthcare players emphasize children's health as their one and only focus of attention. We recommend caution. When the FDA introduced a regulation on pediatric use in 1979, it justified this with safety concerns. But in their demands for separate pediatric studies, the American Academy of Pediatrics (AAP), the FDA, and pediatric researchers did not distinguish between the administrative and the physiological meaning of the term "child." Did they simply forget this? Were there misunderstandings? In the first edition of this book, we did not emphasize this difference. But we note that a standard textbook on pediatric clinical pharmacology simply assumes that the widespread off label use of highly potent modern medicines has the potential to make them dangerous. Well, would that have been the case, pediatric oncology would not have become what it is today. It developed "off-label" decades before the term "off-label" even existed.

Medicine is perceived as something that needs to be taken, not enjoyed. For most adults, this works: You have a headache, you ingest a tablet. Your senses tell you that shape, surface, hardness, smell, and taste are unpleasant but you force the tablet down your throat because you know from experience or because you trust the prescriber that it will help. Young children cannot make this informed decision. They do not understand the medicine-disease connection and/or cannot override the reactions triggered by their senses. That said, also many adults are unable to swallow adult dosage forms. Young children need oral forms that resemble food they can ingest: fluid, semifluid, pseudofluid, or quickly resorbable oral forms.

Pediatric legislation was introduced in the USA in 1997 and in the EU in 2006 as a growing gap was claimed by academic scientists and regulatory authorities between the wealth of information available for adult patients and the limited information in minors. PDD covers three major areas: clinical development; pre-clinical safety and toxicology, mostly young animal studies; and the development of pediatric formulations. For young babies, infants, toddlers, and children, the development of pediatric formulations is probably the most reasonable and the least controversial part of PDD. The EU legislation is more ambitious and asks for a pediatric investigation plan (PIP) early in development. One PIP part must address pediatric formulations.

We come back to the question why this book was not published fifty or hundred years earlier. The answer is simple. Nobody would have understood its need. Most drugs that we use routinely today did not yet exist. The increased demand for pediatric formulations is triggered by regulatory requirements discussed in depth in several chapters. Do these demands translate into clinical improvement? Again, there are different positions.

The debate about better medicines for children has also reached the World Health Organisation (WHO), which has launched the program "make medicines child size." Since the first edition of this book, not much more has been heard from it. Furthermore, the requirements of medicines in the developed world sometimes contradict those of the developing countries. Technologically advanced formulations should not only be good but also be affordable. We have refrained from addressing this additional dimension.

This book covers the anatomy and physiology of babies and minors and the technical state of the art where possible. It hints where to find additional inspiration, such as the food industry, gives a suitable background on the regulatory framework, and offers explanations on contradictory issues. Have we covered everything? Certainly not. Our roadmap is hopefully useful and will help you to form your own opinion. Enjoy.

Contents

1 **Defining a Paediatric Patient** 1
Hannah Batchelor

2 **Paediatric Development and Pharmacotherapy** 9
Hannah Batchelor

3 **General Considerations for Pediatric Formulation Development: Setting Target Product Profiles** 33
Daniel E. Schaufelberger

4 **Acceptability of Formulations: Testing Strategies** 47
Vallet Thibault, Bras Da Costa Sabrina, Wagner Leonie, and Ruiz Fabrice

5 **Palatability of Oral Dosage Form: Considerations for Pediatric Patients** 69
Anne-Marie Pensé-Lhéritier and Leonie Wagner-Hattler

6 **Taste Masking Technologies** 85
Kavil Patel, Rand Turki, Samar Issa, and Fang Liu

7 **Oral Liquid Formulations** 123
David Harris and Peter Kühl

8 **Paediatric Solid Formulations** 149
Sejal R. Ranmal, Susan A. Barker, and Catherine Tuleu

9 **Micropellets: A Modern Multiparticulate Technology Platform for Pediatric Medicines** 179
Norbert Pöllinger

10 **Oral Semisolid Formulations** 211
Carsten Timpe

11 Buccal/Sublingual Drug Delivery for the Paediatric Population 227
Darragh Murnane and Marc B. Brown

12 Topical and Transdermal Drug Delivery 239
William J. McAuley and Marc Brown

13 Parenteral 259
Fiona O'Brien and Kamelia Krysiak

14 The Challenges of Paediatric Pulmonary Drug Delivery 281
Darragh Murnane, Laura Urbano, and Marc B. Brown

15 Paediatric Formulations: Nasal, Ocular and Otic Drug Delivery 301
Hannah Batchelor

16 Rectal Drug Delivery 331
Sara Hanning, Trusha Purohit, and Catherine Tuleu

17 Excipients and Active Pharmaceutical Ingredients 343
Eleanor Dakin, Smita Salunke, Hudson Polonini, Peter Kuehl, and Kevin Hughes

18 Clinical Testing in Children 381
Klaus Rose

19 Bridging of Formulations Used During Clinical Development 395
Norbert Pöllinger

20 Modeling and Simulation for Pediatric Formulation Development 413
Hannah Batchelor

21 Paediatric Pharmaceutical Legislation and Its Impact on Adult and Paediatric Drug Development: The EU Regulatory View 423
Siri Wang and Karl-Heinz Huemer

22 Global Pediatric Legislation for Pediatric Medicines in LMIC Regions 437
Hannah Batchelor

23 Impact of the US and EU Pediatric Pharmaceutical Legislation 445
Klaus Rose

24 The Dangerous Business of Predicting the Future 455
Daniel Bar-Shalom, Hannah Batchelor, Linda F. McElhiney, and Klaus Rose

Index 459

About the Editors

Hannah Batchelor is a pharmaceutical scientist who has worked in academia, the NHS, and within pharmaceutical industry. She is currently based at the Strathclyde Institute of Pharmacy and Biomedical Science, University of Strathclyde in Glasgow. She is Director of Research for the Strathclyde Institute of Pharmacy and Biomedical Sciences.

She works on the design and manipulation of medicines to create age-appropriate drug formulations to maximize clinical efficacy in pediatric patients. Her research interests lie in the optimization of drug formulations to maximize their biopharmaceutical performance and acceptability to children. Her research is informed by the views of children, young people, and parents to ensure that the patients are at the center of new developments. She is also the director of the GIBio facility at Strathclyde, which is an advanced dissolution apparatus designed to provide near clinical data to inform formulation development.

She is passionate about pharmaceutical sciences and is motivated to inspire the next generation of scientists.

Klaus Rose is CEO of klausrose Consulting, Switzerland, and advises on pediatric drug development and how to comply with FDA and EMA pediatric requirements. He has studied Latin languages, psychology, and medicine. After postgraduate clinical training in general medicine in Germany and England, he joined the pharmaceutical industry. He has held various positions in R&D and medical affairs, was Global Head Pediatrics Novartis from 2001 to 2005, and was Global Head Pediatrics Roche from 2005 to 2009. After a year with a regulatory consultancy, he established his own business. He is a frequent speaker on international conferences on pediatric drug development, organizes trainings, and publishes on a regular base.

Chapter 1
Defining a Paediatric Patient

Hannah Batchelor

Abstract The phrase "children are not just small adults" is widely used when discussing clinical research in paediatric populations. However, it is important to recognise that a paediatric population is very diverse and often requires subdivision into sub-categories based on age and growth. This chapter explores the definition of a paediatric patient and some of the considerations that underpin further subdivision of the paediatric population.

Keywords Paediatric patient · Definition · Neonate · Infant · Adolescent

1.1 Introduction

A patient can be defined as "a person who requires medical care". In the case of this book we are mainly concerned with a person who requires pharmacological therapy as part of their care as the focus is on medicinal products.

Paediatric medicine is the branch of medicine that deals with the diseases and treatment of children; however, the sub-classification of paediatric populations can be complex.

1.1.1 Regulatory Guidance on the Definition of "Paediatric"

- The United Nations Convention on the Rights of the Child (UNCRC) defines a child as everyone under 18 years of age [1].

H. Batchelor (✉)
Strathclyde Institute of Pharmacy and Biomedical Sciences, University of Strathclyde, Glasgow, Scotland
e-mail: Hannah.batchelor@strath.ac.uk

H. Batchelor, K. Rose (eds.), *Pediatric Formulations*, AAPS Advances in the Pharmaceutical Sciences Series 58, https://doi.org/10.1007/978-3-031-77240-5_1

- The European Medicines Agency (EMA) defines the "paediatric population" as the part of the population aged between birth and 18 years [2].
- In the USA, drug regulation paediatric patients are defined as children younger than the age of 17 [3].
- The International Conference on Harmonisation (ICH) harmonised guidance puts the upper limit of 18 for paediatric populations [4].

1.1.2 *Legal Classification of the Paediatric Population*

The age of majority is the legal classification of adulthood. In the United Kingdom (UK) law, a person's 18th birthday draws the line between childhood and adulthood [5]. The age of 18 is typical for most of the world; however, there are examples of a range of ages from 15 (in Indonesia, Iran and Yemen), 19 in certain territories in Canada and 20 in Japan and Korea [6]. Note that the age of majority can differ from the age at which an individual can vote, has criminal responsibility, can purchase alcohol or tobacco or can consent to sexual intercourse.

1.1.3 *Cognitive Adulthood*

It is well established that the brain, specifically the development of the prefrontal cortex, undergoes a "rewiring" process that is not complete until approximately 25 years of age [7, 8]. The development of the prefrontal cortex is important for complex behavioural performance, as this region of the brain helps accomplish executive brain functions.

1.1.3.1 Age and Consent to Treatment

In the UK the National Health Service (NHS) states that individuals aged 16 or over are entitled to consent to their own treatment, thus suggesting cognitive responsibility for their own healthcare. At this age the individual is presumed to have sufficient capacity to decide on their own medical treatment, unless there's significant evidence to suggest otherwise.

In addition, children under the age of 16 can consent to their own treatment if they're believed to have enough intelligence, competence and understanding to fully appreciate what's involved in their treatment. This is known as being Gillick competent [9].

The concepts of Gillick competency are by healthcare professionals working with children to strike a balance between respecting children's wishes and ensuring their safety. The origins of Gillick competency trace back to a legal case in the

1980s that explored whether doctors should be allowed to provide contraceptive advice or treatment to individuals under 16 years old without parental consent [9].

The assessment of Gillick competency does not follow a predefined set of questions. Professionals must take into account various factors when determining a child's capacity to provide consent. These factors include:

- Age, maturity, and mental capacity of the child: Professionals consider the child's developmental stage, cognitive abilities and emotional maturity.
- Understanding of the issue: The child's comprehension of the matter at hand is crucial, including knowledge about its implications, advantages, disadvantages and potential long-term effects.
- Awareness of risks and consequences: The child's awareness of the risks associated with their decision and the potential consequences that may arise from it.
- Comprehension of advice and information: Professionals evaluate how well the child understands the advice and information provided to them, ensuring that they have a clear grasp of the relevant details.
- Knowledge of alternative options: The child's understanding of any alternative options available to them is considered, as they should be aware of the different choices and possibilities.
- Ability to provide a rationale: Professionals assess the child's capability to explain the reasoning behind their decision-making process, ensuring that their choices are based on sound judgment.

These factors collectively contribute to the assessment of a child's Gillick competency and their capacity to provide informed consent.

Other factors that may also be considered by the healthcare professional include:

- The child's developmental stage, knowledge of illness and experience of healthcare.
- How are decisions usually made in the family and the typical level of autonomy of the child within their family.

It is important that there is a framework of considerations for analysis, a record of observations and discussions and a documented decision when a child is considered to be Gillick competent in terms of their treatment.

1.1.4 Age Classification Within Paediatrics

Any classification of the paediatric population into age categories is to some extent arbitrary. Decisions on how to stratify the paediatric population need to take into consideration developmental biology and pharmacology, and this may not always correlate to age.

Often clinical trials are designed to evaluate a drug in a defined population which can be based on age; however, this can cause issues with small sample sizes due to the age stratification. Furthermore, regulations state that extrapolation of clinical data from older paediatric groups may be feasible for certain drugs treating specific conditions [4].

Different bodies have used different classifications of age categories within paediatric populations which can add to the confusion in terms of paediatric formulation development. A summary of these is presented below.

The Eunice Kennedy Shriver National Institute of Child Health and Human Development in the United States proposed a series of age ranges in 2011 as shown in Table 1.1 [10].

The ICH regulations [4] define fewer categories as shown in Table 1.2, and these are mirrored in the FDA legislation.

If the established absorption and clearance pathways of a medicinal product and the developmental changes in those pathways are well understood, age categories for evaluating pharmacokinetics could be selected based on significant points where such parameters are expected to vary. When considering efficacy, distinct endpoints may be defined for different paediatric age groups, which may help to define stratification. Creating numerous age groups within the paediatric population could unnecessarily escalate the required number of patients in clinical testing and in longer-term clinical studies, paediatric patients may transition from one age category to another, and the clinical study design and statistical plans should proactively incorporate the changing patient counts within specific age categories.

1.1.4.1 Preterm Neonates

Preterm newborn infants cannot be considered as a homogeneous group of patients, as a significant difference exists between a 25-week gestation, 500-gram newborn and a 30-week gestation newborn weighing 1500 grams. Additionally, a distinction must be made between immature and growth-retarded low-birth-weight babies. Several important factors should be taken into account when considering these patients, including: (1) gestational age at birth and age after birth (adjusted age); (2) immaturity of renal and hepatic clearance mechanisms; (3) issues related to protein binding and displacement, particularly bilirubin; (4) penetration of medicinal

Table 1.1 Subdivision of the paediatric population as defined by the Eunice Kennedy Shriver National Institute of Child Health and Human Development in the United States

Stage	Definition
Preterm neonatal	The period at birth when a newborn is born before the full gestational period
Term neonatal	Birth–27 days
Infancy	28 days–12 months
Toddler	13 months–2 years
Early childhood	2–5 years
Middle childhood	6–11 years
Early adolescence	12–18 years
Late adolescence	19–21 years

Table 1.2 Subdivision of the paediatric population as defined by ICH

Stage	Definition
Term neonatal	Birth–27 days
Infancy and toddlers	28 days–2 years
Childhood	2–11 years
Adolescence	12–17 years

products into the central nervous system (CNS); (5) specific neonatal disease states (e.g., respiratory distress syndrome of the newborn, patent ductus arteriosus, primary pulmonary hypertension); (6) unique susceptibilities of preterm newborns (e.g., necrotising enterocolitis, intraventricular haemorrhage, retinopathy of prematurity); (7) rapid and variable maturation of all physiological and pharmacological processes, leading to different dosing regimens with chronic exposure; and (8) transdermal absorption of medicinal products and other chemicals.

1.1.4.2 Term Newborn Infants (0–27 Days)

Although term newborn infants are more developmentally advanced compared to preterm newborn infants, many of the physiological and pharmacological principles discussed relating to preterm neonates also apply to them. However, there are some specific considerations for term infants. The volumes of distribution of medicinal products may differ from those in older paediatric patients due to variations in body water and fat content, as well as a high body-surface-area-to-weight ratio. The blood-brain barrier is not fully mature, which means that medicinal products and endogenous substances like bilirubin can enter the central nervous system (CNS) and potentially cause toxicity. The hepatic and renal clearance mechanisms in term infants are still immature and rapidly changing, necessitating potential dose adjustments during the first few weeks of life. Limited clearance in these patients can lead to an increased susceptibility to toxic effects of medicinal products, as seen in examples such as chloramphenicol grey baby syndrome. However, term newborn infants may be less prone to certain types of adverse effects, such as aminoglycoside nephrotoxicity, compared to patients in older age groups. Oral absorption of medicinal products may be less predictable in term infants compared to older paediatric patients.

1.1.4.3 Infants and Toddlers (28 Days to 23 Months)

During this period, significant advancements occur in CNS maturation, immune system development, and overall body growth. The hepatic and renal clearance pathways continue to undergo rapid maturation. The developmental pattern of maturation is highly reliant on the specific pathways of clearance. It is important to note that there is often substantial inter-individual variability in the maturation process.

Oral absorption of medicinal products becomes more relevant for children once weaned and when solids are introduced into their diet as this can enable dosage forms that need to be swallowed to be introduced.

1.1.4.4 Children Aged 2–11 Years

The majority of drug clearance pathways, including hepatic and renal pathways, are fully mature during this stage, often exceeding adult values. Changes in drug clearance can be influenced by the maturation of specific metabolic pathways. Children reach important milestones in psychomotor development, which can be adversely affected by CNS-active drugs. As children enter school and their cognitive and motor skills advance, their ability to participate in certain types of clinical efficacy studies may be impacted. Key factors for assessing the effects of medicinal products in children include skeletal growth, weight gain, school attendance, and school performance. Recruiting an adequate representation of patients across the age range within this category is crucial to ensure sufficient evaluation of younger patients. Age stratification within this category is often unnecessary, but it may be appropriate to stratify patients based on pharmacokinetic and/or efficacy endpoints. The onset of puberty varies widely, with girls experiencing puberty earlier, sometimes as early as 9 years of age. Puberty can affect the activity of drug-metabolising enzymes, leading to significant dose reductions for some medicinal products on a mg/kg basis (e.g., theophylline). In certain cases, it may be appropriate to specifically assess the impact of puberty on a medicinal product by studying pre- and postpubertal paediatric patients. The ability of children to swallow solid oral dosage forms will change extensively and be subject to much variability during this period and needs to be considered for formulation based trials.

1.1.4.5 Adolescents (12–16 (or 18 Years))

During adolescence, sexual maturation occurs, and medicinal products may potentially interfere with the actions of sex hormones and hinder development. In certain studies, it may be appropriate to conduct pregnancy tests and review sexual activity and contraceptive use. Medicinal products and illnesses that either delay or accelerate the onset of puberty can significantly impact the pubertal growth spurt and, by altering growth patterns, may affect final height. Evolving cognitive and emotional changes during adolescence can potentially influence the outcomes of clinical studies. Hormonal changes around puberty also have an influence on various diseases, such as increased insulin resistance in diabetes mellitus, seizure recurrence around menarche, and changes in the frequency and severity of migraine attacks and asthma exacerbations. Thus, hormonal changes can impact the results of clinical studies.

Within this age group, adolescents are gradually assuming responsibility for their own health and medication management. Noncompliance becomes a particular concern, especially when medicinal products (e.g., steroids) affect appearance.

Compliance checks on the products used are important in clinical studies. The upper age limit for adolescence varies among regions. In some cases, it may be possible to include older adolescents in adult studies, although issues of compliance can pose challenges. Due to the unique challenges of adolescence, it may be appropriate to consider conducting studies on adolescent patients in centres that possess expertise and specialised care for this population, regardless of whether they are included in adult studies or separate protocols.

1.2 Conclusions

There are many changes within a paediatric population, from birth through childhood and adolescence. The value of sub-stratification within a paediatric population for clinical research is recognised, yet a lack of consistency can make dosing instructions complex. There is still much to be discovered about biological and psychological development within a paediatric population and how these processes affect the effectiveness and efficacy of pharmacotherapy. Trial design that accounts for age differences and promotes consistency in reporting of age-related data is essential to ensure the validity and clinical usefulness of paediatric trial data.

References

1. UN General Assembly, The United Nations Convention on the Rights of the Child. 1990. https://www.unicef.org.uk/rights-respecting-schools/wp-content/uploads/sites/4/2017/01/UNCRC-in-full.pdf. Accessed June 2023.
2. The European Parliament and the Council of the European Union, Regulation (EC) No 1901/2006 of the European Parliament and of the Council of 12 December 2006 on medicinal products for paediatric use and amending Regulation (EEC) No 1768/92, Directive 2001/20/EC, Directive 2001/83/EC and Regulation (EC) No 726/2004. 2006. https://eur-lex.europa.eu/legal-content/en/ALL/?uri=CELEX:32006R1901. Accessed June 2023.
3. U.S. Department of Health and Human Services Food and Drug Administration, Pediatric Drug Development: Regulatory Considerations — Complying with the Pediatric Research Equity Act and Qualifying for Pediatric Exclusivity Under the Best Pharmaceuticals for Children Act Guidance for Industry Draft Guidance. 2023. https://www.fda.gov/regulatory-information/search-fda-guidance-documents/pediatric-drug-development-regulatory-considerations-complying-pediatric-research-equity-act-and. Accessed June 2023.
4. ICH, ICH Harmonised Guideline Addendum to ICH E11: clinical investigation of medicinal products in the pediatric population E11 (R1). 2017. https://database.ich.org/sites/default/files/E11_R1_Addendum.pdf. Accessed June 2023.
5. Children Act. https://www.legislation.gov.uk/ukpga/1989/41/section/105. 1989. Accessed June 2023.
6. OECD Family Database. https://www.oecd.org/social/family/database.htm. Accessed June 2023.
7. Gavin L, et al. Sexual and reproductive health of persons aged 10–24 years-United States, 2002–2007: Department of Health & Human Services. Atlanta: Centers for Disease Control and Prevention; 2009.

8. Casey BJ, Getz S, Galvan A. The adolescent brain. Dev Rev. 2008;28(1):62–77.
9. Olarinde ES, Bamidele I. Gillick v West Norfolk and Wisbech AHA: the right of adolescents to make medical decisions and the many shades of grey. JL Pol'y Global. 2016;54:13.
10. Williams K, et al. Standard 6: age groups for pediatric trials. Pediatrics. 2012;129(Suppl 3):S153–60.

Chapter 2
Paediatric Development and Pharmacotherapy

Hannah Batchelor

Abstract Growth and development are major aspects of paediatric populations that are not readily apparent in adults. The process of growth and development encompasses a series of biological processes which greatly influence the pharmacology of medications in children, specifically their pharmacokinetics, which refers to what the body does to a drug, and can be summarised using the acronym ADME: absorption, distribution, metabolism, and elimination. Such influence is linked to alterations in body composition, the maturation of organs and organ systems, and changes in their respective functions, for example, swallowing reflexes, excretion of digestive enzymes, intestinal motility, transit time and intestinal transporters and drug metabolising enzymes. For example, changes in drug efflux transporters result in a decrease or increase in expelling drugs back into the intestinal lumen and thereby in variation in oral bioavailability. Understanding how development can alter drug performance aids in guiding paediatric therapy.

Keywords Dose selection · Oral absorption · Gastric functionality · Small intestinal functionality · Drug elimination

2.1 Introduction

Development of age-appropriate medicines for children requires not only an understanding of their preferences for different formulations, flavours, and textures of products but also an understanding of the physical and biochemical differences between children and adults. The most obvious difference encountered in drug therapy is dose adjustment and the algorithms used to calculate dosages relevant to

H. Batchelor (✉)
Strathclyde Institute of Pharmacy and Biomedical Sciences, University of Strathclyde, Glasgow, Scotland
e-mail: Hannah.batchelor@strath.ac.uk

H. Batchelor, K. Rose (eds.), *Pediatric Formulations*, AAPS Advances in the Pharmaceutical Sciences Series 58, https://doi.org/10.1007/978-3-031-77240-5_2

sub-populations within the overall paediatric population. There is much emphasis on the idiom that "children are not just small adults" in terms of drug therapy. Indeed, growth and development are two major aspects of children not readily apparent in adults. The topic of human growth and development is extensive with many dedicated reference books (e.g., {Chamley, 2005 #1;McIntosh, 2008 #2; {Sinha, 2012 #136}}).

There is extensive data in the literature on anatomical and physiological parameters and growth of humans from all continents and many countries. However, the most comprehensive data are available from European populations or North America. Reference individuals, for example, mean values used in the calculation of radiological protection, are typically based on data from Western populations and may not be representative of all populations [1]. The variability in growth of children may be greater when considering populations in developing countries compared to Western populations as health and diet influence the rate of growth. There have been recent reviews that specifically consider oral absorption in children from low- and middle-income countries, including those with malnutrition [2, 3].

There are several comprehensive reviews on age-related factors that affect oral drug absorption [4–7], some specific reviews for neonates [8–10], and more general pharmacokinetics [11].

In terms of drug delivery, the growth from birth to adulthood is important as dose adjustments are typically in line with overall age, weight, or body surface area; therefore, there is a need to measure these values to ensure appropriate therapy. Furthermore, oral absorption processes also change, and as such, it is important to consider sub-divisions within the paediatric population to better understand the risks of extrapolation from adult anatomy/physiology. The concept of "extrapolation" in this context is only valid in cases when it can be assumed that the course of the disease and the expected response to a medicinal product would be sufficiently similar in the paediatric and reference (adult or other paediatric) population.

2.2 Dose Selection for Paediatric Populations

There are many alternative approaches that can be used for initial selection of a drug dose for paediatric patients. In 1940, Dawson [12] reported that doses increased less rapidly than predicted directly from weight based on findings that smaller species are generally more tolerant of drug treatment than larger species.

The simplest and oldest approach consists of normalising the dose to body weight (BW), since age and body weight are correlated.

$$\mathrm{Dose}_{\mathrm{child}} = \mathrm{Dose}_{\mathrm{adult}} \times \frac{\mathrm{BW}_{\mathrm{child}}}{70}$$

where $\mathrm{Dose}_{\mathrm{child}}$ is the dose given to the child of weight $\mathrm{BW}_{\mathrm{child}}$ (in kg) and $\mathrm{Dose}_{\mathrm{adult}}$ is the typical dose for an adult weighing 70 kg. This formula assumes a linear relationship between age and weight across infancy, childhood, and adolescence, which is not true as one can see from growth charts. Furthermore, it does not account for age-related changes in elimination pathways of drugs.

Body surface area (BSA) was subsequently proposed in 1950 to be a better algorithm for dose adjustments compared to body weight or age, particularly during infancy and childhood [13]. There are several methods for calculating body surface area from height and mass measurements (e.g., [14, 15]), and all follow a similar functional form that includes an exponential component. All examples contain complex mathematical operations and variables and are thus of limited practicability in daily routine, especially in case of emergency situations.

$$\text{BSA} = a_0 * \text{Height}^b \times \text{Weight}^c$$

where a, b, and c are constants.

The paediatric dose can be estimated as follows:

$$\text{Dose}_{\text{child}} = \text{Dose}_{\text{adult}} \times \frac{\text{BSA}_{\text{child}}}{1.73}$$

where $\text{Dose}_{\text{child}}$ is the dose given to a child of $\text{BSA}_{\text{child}}$ and $\text{Dose}_{\text{adult}}$ is the typical dose for an adult with a BSA of 1.73 m^2.

Humans have an extremely long growth period compared to other mammals, with height increasing up to approximately 16 years in females and 18 in males and weight increasing up to 25 years, although the rate of growth in weight is much slower after 18 years [1]. However, growth is known to occur in spurts and not as a single continuous linear phenomenon. The relationship between body mass, height, and surface area for Western populations is shown in Fig. 2.1.

The data in Fig. 2.1 show that there is wide variability in the ratio to adult value between weight, height, and body surface area in the young, and these values converge with age.

In terms of body surface area the relative contributions change with age. During growth the limbs lengthen; thus, especially in the early years of life, the infant is markedly elongating in stature. This is reflected in the relative contributions of the head and lower extremities to body surface area, which is presented in Fig. 2.2.

At birth the head is a quarter of the total body length, whereas in the adult it is one-seventh [16]. Also, the trunk is long with the upper limbs being longer than the lower limbs; from 6 months of age to puberty, the extremities grow more rapidly than the head. This is reflected in the postural changes of the infant, from a recumbent one to that of an upright position. This shows that the percentages normalise to close to adult values from 10 years of age.

It is important to understand the algorithm used in dose adjustment to ensure that the correct dosage adjustment is undertaken for each compound. However, it should be noted that all approaches assume that the child's weight, height, and body composition are age appropriate and normal and that the "reference" normal adult has a BW and BSA of 70 kg and 1.73 m^2, respectively [17]. There are several examples

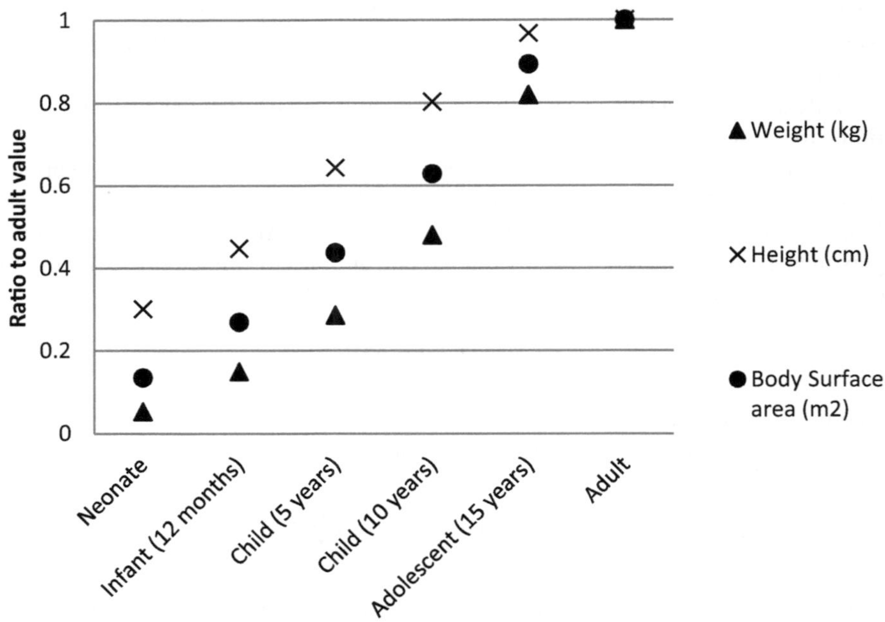

Fig. 2.1 Ratio of weight, height, and body surface area to adult values by age (ICRP data [1])

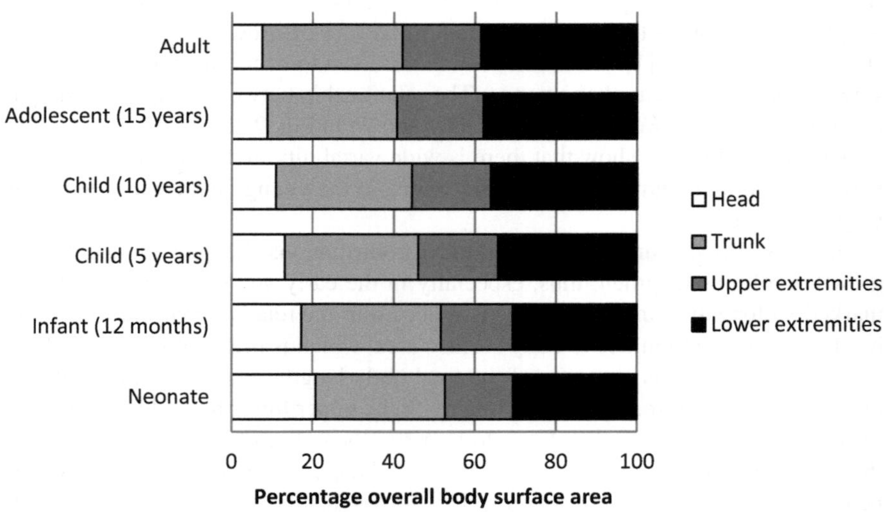

Fig. 2.2 Percentage distribution of body surface area as related to postnatal development

in the literature of instances where child growth is lower than expected with age in developing countries (e.g., [18–20]).

More recent models include a scaling factor for drug elimination, typically the allometric 3/4 power model [21] that has been found useful in normalising a large number of physiological and pharmacokinetic variables [22].

$$CL_{child} = CL_{adult} \times \left(\frac{BW_{child}}{70} \right)^{0.75}$$

Physiologically based pharmacokinetic modelling is now the preferred method to estimate a dose in paediatric populations as they can incorporate each organ independently into a system to better understand the impact on drug ADME.

Organs of interest during paediatric development that impact drug delivery include digestive, respiratory, and urinary organs. The pattern in growth of these organs is similar to the whole body with a rapid increase in mass in infancy and early childhood followed by a slower growth phase and a second phase of rapid growth at pre-puberty ending in a terminal phase of slow growth in adolescence. The ratio of the mass of these organs compared to adult values is shown in Fig. 2.3.

The plot in Fig. 2.3 shows that the organ growth is typically in line with overall body growth for these reference populations which suggests that a child can be considered to be a small adult. However, the relative size of an organ to the total body weight can provide an alternative view as in Fig. 2.4, where it is obvious that the liver and kidney are relatively much larger in younger children compared to adults.

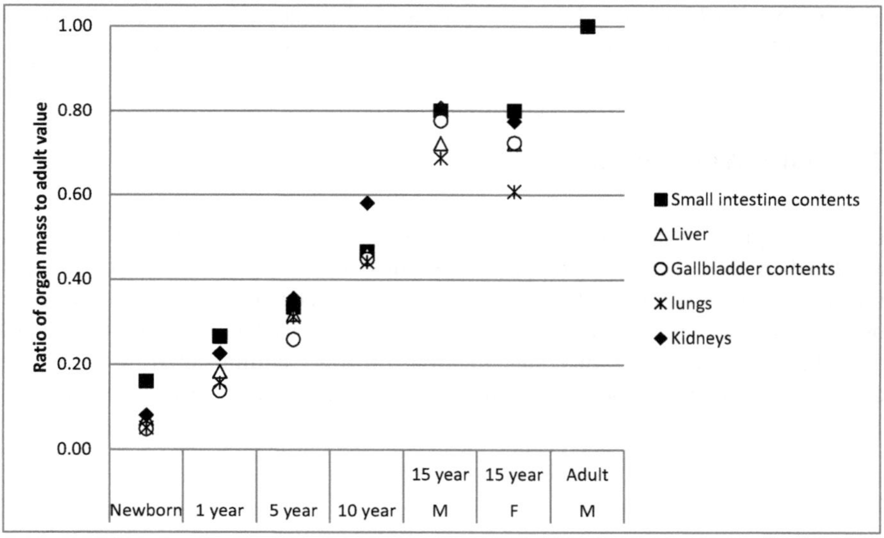

Fig. 2.3 Ratio of organ mass to adult values (by weight) with age [1]

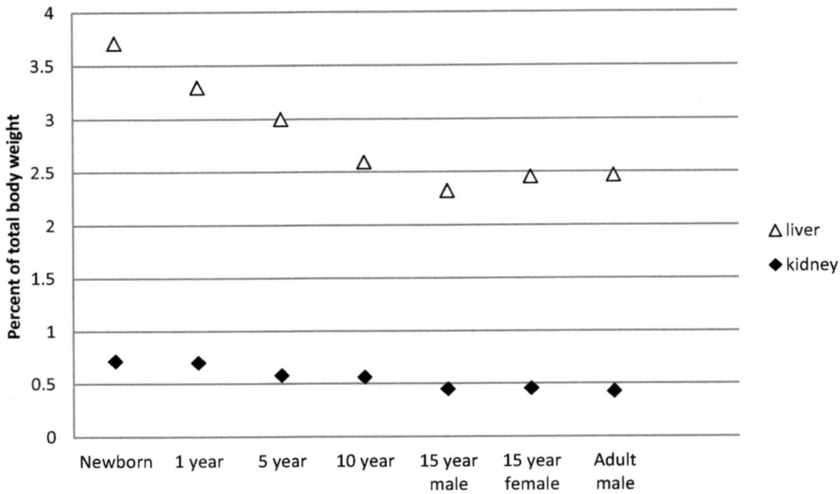

Fig. 2.4 Liver and kidney as per cent of total body weight

The growth of a child is complex where not only anatomic growth influences development but also the maturation of many associated physiological processes. As well as changes in the relative size of the liver and kidney with age, there is an associated change in functionality as elimination pathways mature. Drug elimination clearance, for example, may increase with weight, height, age, body surface area (BSA), and creatinine clearance [15]. All of these covariates may show a high degree of correlation, and they are not mutually exclusive; any one factor may or may not predict between subject differences in clearance. The data shows that at 15 years old, the relative weight of the liver and kidney is the same as that in adults.

2.3 Oral Absorption

Drug absorption in infants and children follows the same general principles as in adults where both the physiochemical properties of the drug and the patient-related factors will have an effect. Unique factors that influence drug absorption include blood flow at the site of administration, as determined by the physiologic status of the infant or child, gastrointestinal function for orally administered drugs, which changes rapidly during the first few days after birth, and skin permeability in topical drugs.

2.3.1 *The Swallowing Reflex*

The act of swallowing is the process whereby matter is conveyed from the mouth to the stomach. The oral phase of the swallow is voluntary and involves moving a bolus to the posterior wall of the pharynx. The second phase is involuntary when the bolus is forced into the pharynx by the tongue, at which point inhalation is automatically prevented. The third phase is also involuntary where the bolus transits the oesophagus.

Swallowing is observed in foetuses from 10 to 12 weeks' gestation with consistent swallowing by 22–24 weeks' gestation [23]. It has been estimated that the near-term human foetus swallows 500–1000 mL/day of amniotic fluid [24]. Prior to 4–5 months of age, infants are only able to swallow liquids due to the extrusion reflex, which prevents any non-liquids from entering the pharynx. With age the infant develops the anatomical and physiological capability to swallow solid foods. When drinking a liquid, an adult swallows 17 mL water with each swallow [25], whereas a child between 1.25 and 3.5 years swallows approximately 4.5 ml per swallow. The frequency of swallowing also varies during development with adults swallowing 585 times in a 24-hour period [26] compared to 43 times per hour in a near-term foetus [27]. About one-third of swallows accompany eating and drinking, and the remaining occur while breathing out. Relatively few swallows occur during sleep (<10%).

A detailed review on the changes in the oral physiology and how these changes are associated with mastication abilities of children is available [28]; although this is based on feeding and weaning habits, it offers insights for oral medicine administration. The aim of mastication is to form a cohesive bolus with saliva in order to facilitate swallowing; there are several physiological factors that determine the age at which mastication is feasible, including the number of teeth, occlusion contact area, bite force, and ability to control masticatory muscles. The ability to swallow purees and soft solids, reflecting mastication efficiency, was demonstrated at 6 months; whereas for harder foods, changes in efficiency occurred between 6 months and 2 years of age [28]. There was also an observation of infants allowing saliva to soften solid prior to swallowing which could have an effect on how medicines are handled in the mouth. In food research, chew and spit experiments can reveal the size of masticated material that a person would swallow; this may indicate the potential size of a dosage form that could be swallowed by a child. Typical sizes of 4–5 mm were identified in children from 3–8 years whereas adult studies reported a size of 2–3 mm [29]. This suggests that liquids or semi-solids are the most suitable dosage forms for neonates, infants, and young children unless solids can be manufactured at such small sizes.

2.3.2 Oesophageal Transit

The oesophagus is 25 cm long in adults and serves to move boluses of food and drink from the oral cavity into the stomach. Autopsy measurements on infants and children indicate an oesophageal length of about 8–10 cm at birth, 12 cm at age 1 year, 18 cm at 10 years, and 19 cm at 15 years [30]. The transit time for liquids is approximately 5 seconds in both adults and children as young as 3 months old with similar values reported in neonates [31, 32].

The normal transit time through the oesophagus is typically less than 10 s for solid materials. Oesophageal transit ranges from 3.4 ± 1 s for infants to 4.6 ± 1.9 s for patients 8–16 years of age, although gastro-oesophageal reflux or oesophagitis is associated with prolonged transit times [33].

It is well recognised that tablets or capsules taken by patients in the supine position may lodge in the oesophagus, causing damage and irritation [34]. If tablets are taken without water, the risk is greatly increased, and the units may remain lodged in the lower oesophagus until they disintegrate [35]; the FDA recommends 240 mL of water in adults. Sticking of solid dosage forms in the oesophagus is reported to be more common with gelatine capsules compared to enteric coated tablets in elderly patients [36]. A study that examined the effect of tablet size and shape concluded that, size and shape of tablets can affect oesophageal transit time after swallowing while in the standing position but not in the supine position; changing the shape of larger tablets from round to oval may aid oesophageal transit [35]. However, other studies have shown that body position does not affect the rate of oesophageal transit in adults [37]. The influence of body position on oesophageal transit in paediatric patients has not been reported, although the use of solid formulations in supine patients of any age should be minimised where possible.

2.3.3 Gastric Function and Motility

The structure of the stomach is largely developed by 14 weeks of gestation where the cells that will produce gastrin and pepsin are recognisable [38]. It is generally agreed that gastric pH is neutral at birth, although there is debate over the time taken for the pH to reduce following birth with reports of 24–48 hours to reach pH 3 [39] with a further rise to neutral after 72 hours; or 10 days at neutral followed by a decrease to acidic values comparable to adults at 2 years [40]. Investigations by Hyman et al. [1983] showed hypochlorhydria with gastric pH greater than 4 in 19% of neonates at 1 week of age, 16% at 2 weeks of age, and 8% at 3 weeks of age [41]. No infant demonstrated a basal gastric pH greater than 4 after 6 weeks of age. Thereafter, gastric pH falls to a value of 1.5–3.0, which is comparable to adult gastric pH (1.5–2.5). The pH values quoted by Hyman relate to basal acid output which was 12 ± 2 μmol/kg/hr. at 1 week of age and increased over the first 4 weeks to 30 ± 5 μmol/kg/hr. [41]. By the age of 3 years, the amount of gastric acid excreted per kilogram of body weight is similar to that excreted in adults. Pepsin secretion is 0.1 $mgkg^{-1}$ h^{-1} at birth to 3 months increasing to adult values of 1 $mgkg^{-1}$ h^{-1} at 18 months [42]. The activity of pepsin is reduced in young infants due to the increased pH in these populations.

Gastric juice, produced and secreted by the oxyntic and pyloric glands of the stomach, assists in the break-up of solid particles and solubilisation of agents. The daily secretion volume for adults is 1.5–2 L per day (equivalent to 0.9–1.2 mL/kg/h) [43], whereas these are about 1 mL/kg/h in the neonate and increase to 2–3 mL/kg/h by 1–2 years of age [42].

The composition of gastric fluid has been characterised in several studies [44, 45]. The pH of gastric fluids was reported to be 2.0–2.7 for neonates, 1.56 for children (aged 2–12 years), and 1.92 for adolescents [44]. Other data suggested a gastric pH range of 0.57–11.05 in children, although this variability may have been affected by sample processing and measurement techniques [45].

Extensive variability in the concentration of bile salts within the stomach has been reported for paediatric populations with values ranging from 0 to >5 mM with mean values approximately 0.2–0.3 mM [44, 45].

Low gastric fluid volumes have been reported in paediatric populations with values ranging from 1 to 45 mL in the fasted state [4, 46]. This volume can be important for orally administered dosage forms where the drug needs to dissolve in gastrointestinal fluids in order to be absorbed.

At birth the muscular layers of the stomach are thinner than in the adult, the pylorus is poorly developed, and gastric motility is very low [47]. Motor activity within the stomach is limited in the foetus prior to 30 weeks of gestation. Amniography has shown movement of markers out of the stomach and along the intestine from 30 weeks of gestation [48]. In the first few days of life peristaltic waves are shallow, widely spaced, and intermittent. In older infants these waves are prominent and pass from the body of the stomach to the pylorus in rapid succession [49]. In the fasted state the motility pattern in the stomach is regulated by the interdigestive migrating myoelectric complex (IMMC) that follows a three-phase cyclic pattern. In adults those three phases have been designated as phase I, 45–60 min of quiescence and essentially no movement of the gastric fluid; phase II, 30–45 min of irregular activity favouring dissolution in the stomach; followed by phase III, 2–10 min of intense contractile activity during which the stomach content is emptied into the small intestine ("housekeeper wave"). A study on term and preterm neonates demonstrated that by term, well-defined fasting motor activity was present with clearly discernible phases I, II, and III.

The majority of information regarding gastric motor activity in infants is gastric emptying of a variety of liquids. In the neonate, delayed gastric emptying has been reported with emptying times of 6–8 h [40] and adult values reached in 6–8 months. However, other studies show that feed composition can determine the rate of gastric emptying with gastric half-emptying times for meals of human milk and infant formula of 25.1 ± 11.5 and 51.9 ± 9.8 min, respectively, in preterm infants; the corresponding half-emptying times for term infants were 48 ± 15 and 78 ± 14 min [50].

Gastric half-emptying times were measured in healthy children using Chocolate Technecrispy cake. Children between 5 and 10 years of age gave mean time of 107.2 minutes [51], although this high value may be a result of the high calorific value of the meal ingested. Another study compared gastric emptying rates in children (mean age 9 years) and adults, using three different test meals: a low calorie pancake (150 kcal), a high calorie pancake (250 kcal), and 210 ml of milk (134 kcal) [52]. No significant difference was found between children and adults in the emptying rate of the pancake meals, although the high calorie meal emptied significantly slower in both groups. The milk test meal, however, was emptied at a faster rate in adults and at a slower rate in children compared with the low calorie solid test meal. Moreover, the emptying rate of milk in children was significantly slower than in adults; one theory for this reduced emptying rate is the presence of casein as whey-based milk formulae have shown shorter emptying times [53]. An alternative study reported that gastric emptying times for toddlers, young children, and adolescents are generally within the range of values determined for adults [54]. The half-life for gastric emptying in adults has been reported to be 20 minutes, although the exact age at which gastric emptying time approaches adult levels is unclear [39]. Gastric emptying time can dictate the onset of absorption for drugs; this is particularly important for drugs where gastric emptying is the rate-limiting step to absorption. Retention of medicines within the stomach for longer periods also increases the likelihood of acid degradation.

Particle size has been demonstrated to alter gastric emptying times in healthy adults with larger particles taking longer to empty [55]. A meal containing plastic particles of 1–2 mm in diameter showed a statistically similar gastric emptying time to a control meal in adults [56] which suggests that particles of this size are not retained within the stomach. 3D ultrasonography techniques supported this finding where it was demonstrated that material is ground to approximately 2 mm prior to exiting the stomach [57]. Previous reports state that solid particles must be 1 mm or less to pass through the pylorus for children [33]. However, there are also extensive reports of larger particles, including coins that are greater than 1 mm passing through the GI tract of children [58]. Limits in dimensions of 6 cm long and 2 cm wide have been suggested in the dimensions that can pass through the pyloric sphincter [59]. The disintegration of tablets into small particles can be a rate-limiting step in drug absorption; this disintegration phase may be slower in children due to the need for smaller particles to pass through the pyloric sphincter.

Enteric-coated tablets can show considerable variability in GI transit with multiple-unit enteric-coated pellets producing less intra-subject and inter-subject variation compared to single-unit tablets [60]. The gastric retention time for solid dosage forms, particularly erodible tablets, should be considered in children compared to adults due to the passage of particulate material into the small intestine.

2.3.4 Small Intestinal Function and Motility

The small intestine functions to absorb nutrients and is capable of this activity at birth; in terms of drug delivery, the small intestine is the major site of absorption of orally administered drugs. At the level of the intestinal mucosa, several surface factors, either non-immunologic (mucin and microvillus membrane) or immunologic (secretory IgA), determine whether and how luminal molecules are absorbed.

The absorptive surface is dependent upon the dimensions of the small intestine. There are contradictory reports as to the length of the intestine during development, with Neu (2007) [61] reporting a mean length of 275 cm at birth, whereas ICRP [30] reported that the small intestine grew from 45 cm at birth to 260–270 cm in adolescence. Weaver et al. (1991) determined the length of the small intestine from conception to adulthood using data taken from eight published reports of necropsy; they reported that mean length at term was 275 cm, at 1 year 380 cm, at 5 years 450 cm, at 10 years 500 cm, and at 20 years 575 cm [62]. One explanation for the differences in reported lengths is linked to the differences in anatomical and physiological lengths, with anatomical lengths usually recorded at autopsy, which are longer than the corresponding physiological lengths.

The diameter of the lumen of the small intestine changes with age and varies with location within the intestine; values in the range 1.2–2.6 cm have been estimated for the small intestine in newborns [30]. For the adult, estimates are in the range 3–6 cm for the first part of the small intestine and decrease to 1.5–2.5 cm for the last part [30].

Contradictory evidence exists relating to the morphology of the small intestine during development; Thomson et al. 1998 state that the mean villus surface area in adults and infants (up to 24 months) was similar in both groups yet mean crypt length was increased by 31% in infants (270 ± 56 μm) compared with adults (206 ± 29 μm); although, mean epithelial cell height was significantly less in infants (27.0 ± 3.0 μm) compared with that in adults (30.9 ± 4.6 μm) [63]. Changes in microvilli were not examined, but the lower enterocyte height in infants compared with that in adults in this study might indicate that enterocyte maturation will develop later in childhood and will include the development of microvilli, which are known to amplify the epithelial surface area another 15-fold [63]. However, Walker-Smith (1972) stated that at birth, the morphology of the small intestine is characterised by narrow villi and small crypts, and during development, villi become wider by 1 month of age, and crypts deepen at weaning [64]. At weaning, luminal mucosal growth is driven by crypt hyperplasia, which results in longer villi and a twofold increase of the villus surface area [63]. In children the villi of the small intestine tend to be broad leaf-shaped rather than finger-shaped projections (as in adults).

The conclusion from this contradictory literature is that the functional surface area of the small intestine is smaller in neonates. By adulthood, the intestinal surface area is approximately 200 m^2. The increase in surface area during development has significant implications in terms of nutrient absorptive capability [61]. In the absence of literature values of the surface area of the small intestine in paediatric populations, it is acceptable to assume that the ratio in length is proportional to the surface area in estimations for paediatric populations, although this is likely to underestimate the overall surface area due to the development pattern of villi.

Using magnetic resonance imaging it has been shown that the volume of intestinal fluid in fasted adults is highly variable ranging from 45 to 320 ml [65], yet the intestinal fluid is not homogeneously distributed along the gut but forms fluid pockets. Data from 32 fasted children (aged 0–16 years) and 23 fluid-fed children (aged 8–16 years) were evaluated, and the small intestinal volume was recorded to be 0–51 mL in the fasted and 6–91 mL in the fluid-fed state with an average number of 7.7 and 22.4 fluid pockets, respectively [46].

Intestinal fluids comprise pancreatic juices, bile Brunner's gland secretions, and other enterocyte secretions with a total volume of 4 L secreted daily in adults [30]. Pancreatic juice produced in the pancreas and released into the duodenum contains enzymes including α-amylase, chymotrypsin, and lipase; the amount of fluids secreted by the exocrine pancreas into the intestine increases with maturation. Bile produced by the liver, stored and concentrated in the gall bladder, and released into the duodenum reduces the surface tension of the GI fluids and can assist in solubilisation of drugs within the GI lumen; adults secrete 0.5–1 l of bile per day [43]. Bile secretion in the first 2–3 weeks of life is poor with luminal concentrations below the critical micelle concentration (2–4 $mmoll^{-1}$); the concentrations increase by four to seven times in the days following birth [66]. Very low birth weight infants have a lower duodenal concentration of bile acids due to lower synthesis and ileal reabsorption of bile. An indirect indication of the bile salt concentration is the surface tension, with bile salts reducing surface tension compared to water. Surface tension values of 33–46 mNm^{-1} in the fasted stomach and 28–33.6mNm^{-1} in the upper small intestine have been reported in adults [67, 68]. Although no values are reported in paediatric populations, the evidence suggests that the total bile concentration is likely to be lower in the very young.

The mean pH within the small intestine of adults and children from 8 to 14 years was measured using a pH-sensitive radiotelemetry capsule with results showing comparable pH values with pH 6.4 measured in the duodenum rising to 7.4 in the distal part of the small intestine [69, 70]. Measurements on aspirated fluids have also shown comparable pH values in adults and paediatrics ranging from neonates to adolescents [44, 45, 67, 71].

Intestinal transit can be prolonged in neonates due to reduced intestinal motility and peristalsis; however, in older infants, this can be reduced due to increased motility [40]. ICRP [30] reports paediatric small intestinal transit times of 4 hours in neonates and infants rising to 4–9.2 hours in children and 4 hours in adolescents compared to reports of 2.7–8.5 hours in adults [72].

There have been relatively few studies done on pre- and postnatal maturation of intestinal motility [73]. In infants, intestinal motor activity occurs less frequently than in adults, with a different pattern of rhythmic peristaltic activity [74]. In general, intestinal peristalsis in infants is irregular and partially dependent on food intake and feeding habits [73]. Reports on GI motility suggest that the segmental amplitude within the small intestine in adults oscillates between a mean minimum of 9.7 mm and a mean maximum value of 20.5 for the diameter of the small intestine [75]. It is likely that this oscillation is proportionally similar in older children.

Small intestinal motility has been compared between children and adults with Table 2.1 showing the key differences in migrating motor complexes (MMCs) [76]. MMCs were seen as bands of high amplitude rhythmic (11–13 cpm) pressure waves which propagated from the duodenum into the jejunum.

2.3.5 *Intestinal Permeability*

Intestinal permeability is reported to be high at birth and decreases progressively during the first week of life [77]. This may be related to the reduced surface area: volume ratio due to the villi being broader and providing a smaller overall surface area; this phenomenon is well documented in rats [78].

Sugar absorption tests are typically used to assess intestinal permeability in premature neonates; following enteral administration of a test solution containing lactulose and mannitol, the excretion of these sugars is measured in urine. The theory behind the sugar absorption test is that in a healthy intestine, monosaccharides (e.g., mannitol) are readily absorbed via the transcellular pathway, but larger disaccharides (e.g., lactulose) are only absorbed through the paracellular pathway. Therefore the ratio of lactulose:mannitol in urine is a measure of intestinal integrity. Using sugar absorption tests intestinal permeability is reported to be higher in preterm neonates than in healthy term neonates but only if measured within two days of birth suggesting that there is rapid postnatal adaptation of the small intestine [79]. The few bioavailability studies that have examined the absorption of drugs (e.g., phenobarbital, sulfonamides, and digoxin) and nutrient macromolecules (e.g., arabinose and xylose) suggest that the processes of both passive and active transport are fully mature in infants by approximately four months of age. Generally, the rate at which most drugs are absorbed is slower in neonates and young infants; however, the cause of this slower absorption is unknown.

The enteral absorption of drugs was studied in children; D(+)xylose which is absorbed by an active mechanism in the upper small intestine showed no difference in the amount absorbed with age [73]. However, the rate constant, Ka, for enteral absorption of D(+) xylose was non-linear with age where Ka was less for newborns and young babies (up to 150 days) compared to older children [73]. L(+) arabinose

Table 2.1 A comparison of the properties of MMCs within the small intestine of children (8 months–11 years) and adults [76]. Results show mean ± SEM

	Children ($n = 6$) (8 months to 11 years)	Adults ($n = 18$)
Duration (minutes)	5.5 ± 0.6	5.9 ± 0.4
Interval between MMC (minutes)	99.5 ± 19.4	112.5 ± 11.4
Propagation velocity (cm/min)	29.5 ± 22.9	11.30 ± 0.14

is absorbed by passive diffusion, and the pattern for absorption and absorption rate constant showed very similar results to that of D(+) xylose, although the intestinal absorption rate was significantly reduced in neonates. These results were again consistent for sulfonamides with phenobarbitol, digoxin and β-methyldigoxin all showing similar results [73]; these findings suggest that the rate of absorption is slower in neonates and children, yet the amount absorbed is similar (matched by mass). Prolonged gastric emptying time and reduced intestinal motility may be somewhat responsible for the similarity observed in total mass absorbed. A further study was conducted to measure the effects of intestinal motility on absorption using metoclopramide; the results showed an increase in Ka in both young and older infants, yet the ratio of Ka: age remained constant [73]. These results suggest that the reduced Ka observed was not solely due to longer transit times or reduced motility, but other factors are also involved.

The lack of high-quality pharmacokinetic studies undertaken in the paediatric population limits exhaustive knowledge regarding absorption mechanisms within this population. However, most studies conducted revealed that absorption in neonates and infants is slower than that of children and adults [80].

The membrane fluidity of intestinal cell membranes is age dependent in many species due to increasing cholesterol-to-phospholipids and protein-to-phospholipids ratios and changes in the spectrum of fatty acids [81]. As the changes of phospholipid environment alter the activity of various transporters, the developmental changes of membrane phospholipids become a potential regulator of intestinal transport. However, this hypothesis has not been confirmed conclusively.

In some instances, the lack of xenobiotic metabolising ability observed in infants is not due to absence of certain microflora but rather to immaturity of the bacterial enzyme systems in the gut lumen. For example, the extent of metabolism of several compounds such as digoxin, cholesterol, and methane increases with age, and this reflects the developmental processing in bacterial enzymes such as b-glucosidase and reductases [82].

Cytochromes P450 3A (CYP3A) and P-glycoprotein (P-gp) are mainly located in enterocytes and hepatocytes. The CYP3A/P-gp system contributes to the first-pass metabolism of many drugs, resulting in limited bioavailability. During the neonatal period, a shift between CYP3A7, the foetal form, and CYP3A4 occurs in the liver, but data on the expression of the CYP3A/P-gp complex in the intestine are very limited. A study to investigate localisation and expression of CYP3A and P-gp were studied in 59 normal duodenal biopsies from Caucasian children aged 1 month to 18 years; the result showed that CYP3A was expressed in all children 6 months and older and in half those up to 6 months; P-gp mRNA was found variably expressed from 0.08 to 1 with a significant increase between 6 and 12 months of age [83]. The clinical impact of these results is currently unknown. A study by Johnson et al. (2001) investigated expression of CYP 3A with age; an increase in CYP3A expression was observed that was mirrored by a corresponding change in CYP3A4 enzyme activity [84]. A recent study quantified small intestinal drug-metabolising enzymes (DMEs) and drug transporters (DTs) across the paediatric age range using surgical leftover intestinal tissues from paediatric and adult jejunum and ileum.

Samples from 58 children (48 ileums, 10 jejunums, age range: 8 weeks to 17 years) and 16 adults (8 ileums, 8 jejunums) were included in the study. The abundance of BCRP, MDR1, PEPT1, and UGT1A1 was significantly higher in the adult ileum as compared with the paediatric ileum. Jejunal BCRP, MRP2, UGT1A1, and CYP3A4 abundance was also higher in the adults compared with children 0–2 years of age [85].

2.3.6 *Impact of Food on Oral Absorption*

Food is a complicating factor in the absorption of drugs. In babies and infants, a wide range of drugs are often mixed with food by the caregivers prior to administration to ensure that the medication is acceptable to the patient [86]. Many medicines are co-administered with food or fruit juice to aid in their acceptability. However, manipulation of medicines with food can have an impact on efficacy and safety for a number of reasons, including effects on absorption, bioavailability [87] and metabolism, and through inaccurate dosing [40]. The impact on bioavailability of mixing drugs with soft food has been investigated in studies in healthy adults [88–93]; no differences in extent of exposure were observed. More recent work has explored dissolution and the integration of PBPK modelling to better understand the impact of co-administration of drugs with soft food vehicles [94, 95].

Food effects in paediatric populations have been investigated in a smaller number of studies (e.g., [96, 97] where typically there was a decrease in the peak plasma concentration or no significant effect. Other studies have been conducted that measure the degradation of drugs in soft food to predict any ex vivo impact of food manipulation on drug stability (e.g., [98–100]). Previous studies conducted in adults [101–103] have linked physicochemical properties of drugs to the likelihood of a food effect; however, there have been no similar predictions made for paediatric populations.

The lack of a truly fasted state in neonates due to the almost continuous presence of milk in the stomach may affect the absorption of drugs that are (i) lipid soluble due to the lack of lipases or (ii) bind to protein present in milk. It is clear that further studies are required to better understand and predict the effect of food on the absorption of drugs within a paediatric population.

2.4 Parenteral Absorption

Absorption following intramuscular or subcutaneous injection depends mainly, in neonates as in adults, on the rate of blood flow to the muscle or subcutaneous area injected. Physiologic conditions that might reduce blood flow to these areas are cardiovascular shock, vasoconstriction due to sympathomimetic agents, and heart failure. However, sick preterm infants requiring intramuscular injections may have very little muscle mass. This is further complicated by diminished peripheral perfusion to these areas. In such cases, absorption becomes irregular and difficult to predict, because the drug may remain in the muscle and be absorbed more slowly than expected. If perfusion suddenly improves, there can be a sudden and unpredictable increase in the amount of drug entering the circulation, resulting in high and potentially toxic concentrations of drug. Examples of drugs especially hazardous in such situations are cardiac glycosides, aminoglycoside antibiotics, and anticonvulsants.

The skin of the full-term neonate possesses an intact barrier function and is similar to that of an older child or adolescent. Thus, the neonate will be exposed to a relatively greater amount of drug topically than older infants, children, or adolescents. In contrast, data of human skin from preterm infants indicates an inverse correlation between permeability and gestational age. Permeability rates were 100- to 1000-fold greater before 30 weeks of gestation as compared with full-term neonates, with a three- to fourfold greater permeation rate seen beyond 32 weeks. In vivo studies suggest that this increased dermal permeability in preterm infants is a short-lived phenomenon with the permeability barrier of even the most premature neonates similar to that of full-term neonates by 2 weeks of postnatal life [104]. There are numerous reports in the literature underscoring the importance of skin absorption in neonates primarily showing toxicity after exposure to drugs or chemicals. Therefore, extreme caution needs to be exercised in using topical therapy in neonates and young infants. In contrast, the possibility of turning enhanced skin absorption of drugs to the infant's advantage was explored by using the percutaneous route to administer theophylline or caffeine for apnoea in preterm infants [105].

2.5 Drug Distribution

The volume of distribution of a drug can change with age due to changes in body composition. A neonate has a higher percentage of its body weight in the form of water (70–75%) compared to an adult (50–60%). Similarly, extracellular water is 40% of body weight in the neonate, compared with 20% in the adult. This difference in body water can affect the volume of distribution and hence concentration of certain drugs, for example, water-soluble drugs (such as aminoglycosides) [106].

Binding of drugs to plasma proteins is another major factor that affects the distribution of drugs. The concentrations of plasma proteins have been reported to be lower in neonates compared to adults with albumin levels at 75–80% of adult levels, while alpha 1-acid glycoprotein levels are at 50% of adult values [107]. Further research demonstrated that plasma levels of albumin and total proteins increased for the first 1–3 years of life [108]. These differences can impact the free fraction of drugs and are of particular significance for drugs that are highly protein-bound, for example, diazepam, phenytoin, ampicillin, and phenobarbital. Therefore, the concentration of free (unbound) drug in plasma is increased in neonates.

2.6 Drug Metabolism

Upon termination of the umbilical blood supply, the liver in the newborn takes on many biosynthetic and detoxification functions essential for adaptation to extrauterine life. These include aerobic metabolism, gluconeogenesis, synthesis of coagulation factors, and bile production and transport. Both liver size and volume relative to total body weight decrease during childhood (Fig. 2.4) [109]. About 80% of drugs in clinical use undergo metabolic reactions in the body.

There have been recent efforts using proteomic analysis to quantify the ontogeny of drug metabolising enzymes (DMEs) and drug transporter proteins in paediatric intestinal and liver samples [85, 110, 111]. The most significant DMEs (e.g., CYP1A2, 2A6, 2D6; 2E1, 3A4) have been identified to be low at birth, with external factors contributing to rapid maturation.

Longitudinal studies investigating certain genes, like CYP2D6, have revealed that the concordance between genotype and phenotype becomes apparent in infants and young children as early as two weeks after birth in full-term infants [112]. However, for other genes such as CYP2C19, this concordance is largely absent during infancy, as observed with pantoprazole, and phenotype prediction cannot be reliably determined through genotyping in such cases [113]. The colonisation of microbes in newborns begins at birth, and the composition of the microbiome is influenced by factors like the mode of delivery, breastfeeding versus formula feeding, hospitalisation, antibiotic usage, and diet [114]. Animal studies suggest that diet and other factors have the potential to influence the development of drug biotransformation pathways. Consequently, predicting drug clearance, both at the population level and for individual patients, is highly complex [113].

When administering medications to infants, it is crucial to take into account the process of maturation, particularly for drugs that are given over extended periods. Additionally, the neonate‘s exposure to drugs, such as phenobarbital, taken by the mother during pregnancy, can lead to early maturation of foetal hepatic enzymes. Consequently, the neonate may possess a greater ability to metabolise certain drugs than anticipated, resulting in reduced therapeutic effects and lower drug concentrations in the bloodstream when administering the usual neonatal dosage. In the age range from 12 to 36 months, the metabolic rate of many drugs surpasses that of adults, often necessitating higher doses per kilogram of body weight compared to later stages of life. The use of physiologically based pharmacokinetic models to predict the consequences of ontogeny has proved to be useful in this field [115, 116].

There are some excellent recent reviews on the ontogeny of drug metabolising enzymes, and the reader is directed to these for further insights [11, 117–119].

2.7 Drug Elimination

The mass of the kidney relative to age is severalfold greater in preschool-age children than in young adults [120]. Renal clearance is an important route of drug elimination. While during the neonatal period there is minimal glomerular filtration and active tubular secretion of drugs, there is a well-described rapid development in these processes in the post-neonatal period. The glomerular filtration rate is much lower in newborns than in older infants, children, or adults, and this limitation persists during the first few days of life. Calculated on the basis of body surface area, glomerular filtration in the neonate is only 30–40% of the adult value. The glomerular filtration rate is even lower in neonates born before 34 weeks of gestation. Function improves substantially during the first week of life. At the end of the first week, the glomerular filtration rate and renal plasma flow have increased 50% from the first day. By the end of the third week, glomerular filtration is 50–60% of the adult value; by 6–12 months, it reaches adult values (per unit surface area). Therefore, drugs that depend on renal function for elimination are cleared from the body very slowly in the first weeks of life. A less appreciated fact is that during toddlerhood, there is an "overshoot" of the glomerular filtration rate (GFR) well above the levels encountered in older children and adults, and there is an early achievement of adult levels in active drug secretion, which stays at a plateau throughout childhood and adulthood. Due to the high GFR in toddlers, dose requirements for renally excreted drugs in this age group are on a per-kilogram basis, much larger than in adults [120]. The need for higher doses of renally cleared drugs during early childhood reflects the enhanced excretory capacity of the kidney in this age group when normalised to body weight. In contrast to animal models, the developmental changes in the human kidney are not linear processes. In particular, GFR in prepubertal children is almost twofold higher compared to adult values, as is the expression and function of the P-glycoprotein transporter. In contrast, no similar "surpass" is seen with organic anionic or cationic transport. Hence, the current adult dosing cannot simply be extrapolated to children. Instead, developmental changes must be taken into account when designing appropriate dosage regimens of renally excreted drugs for infants and children. In particular, knowledge of the specific drug transporters involved in drug clearance is important for the therapeutic dose adjustment.

Subsequently, during toddlerhood, it exceeds adult values, often necessitating larger doses per kilogram than in adults, as described previously for drug-metabolic rate. Penicillins, for example, are cleared by preterm infants at 17% of the adult rate based on comparable surface area and 34% of the adult rate when adjusted for body weight. A decreased rate of renal elimination in the neonate has also been observed with aminoglycoside antibiotics (kanamycin, gentamicin, neomycin, and streptomycin). Since renal function in a sick infant may not improve at the predicted rate during the first weeks and months of life, appropriate adjustments in dosage and dosing schedules may be very difficult. In this situation, adjustments are best made on the basis of plasma drug concentrations determined at intervals throughout the course of therapy. Although great focus is naturally concentrated on the neonate, it is important to remember that toddlers may have shorter elimination half-lives of drugs than older children and adults, probably due to increased renal elimination and metabolism. For example, the dose per kilogram of digoxin is much higher in toddlers than in adults. The mechanisms for these developmental changes are still poorly understood.

2.8 Conclusions

Growth results from the interaction of genetics, health, and nutrition, and is often used as an indicator of wellbeing. Paediatric development is often measured in terms of physical growth with height and weight measurements being integral to many clinical examinations. As dose adjustments are made on the basis of height and weight, there is a need to carefully consider other parameters that may be mature even when height and weight are lower than anticipated for age; for example, liver and kidney function may be greater than anticipated for small for age children. An understanding of ADME is essential when predicting the pharmacokinetics of a drug in a paediatric population. There are still many gaps in ADME knowledge in paediatrics, particularly for neonates who are the most vulnerable within the paediatric population.

2.9 Further Reading

As stated at the outset there are several comprehensive reviews on age-related factors that affect oral drug absorption [4–7], some specific reviews for neonates [8–10], and more general pharmacokinetics [11].

References

1. ICRP. Basic anatomical and physiological data for use in radiological protection: reference values. A report of age- and gender-related differences in the anatomical and physiological characteristics of reference individuals. ICRP Publication 89. Ann ICRP. 2002;32(3–4):5–265.
2. Meesters K, et al. Pharmacokinetics of antimicrobials in children with emphasis on challenges faced by low and middle income countries, a clinical review. Antibiotics. 2023;12(1).
3. Freerks L, et al. A review of GI conditions critical to oral drug absorption in malnourished children. Eur J Pharm Biopharm. 2019;137:9–22.
4. Wollmer E, et al. Review of paediatric gastrointestinal physiology relevant to the absorption of orally administered medicines. Adv Drug Deliv Rev. 2022;181:114084.
5. Gleeson JP, Fein KC, Whitehead KA. Oral delivery of peptide therapeutics in infants: challenges and opportunities. Adv Drug Deliv Rev. 2021;173:112–24.
6. Lin W, et al. Pediatric physiologically based pharmacokinetic model development: current status and challenges. Curr Pharmacol Rep. 2018;4(6):491–501.

7. Batchelor HK, Fotaki N, Klein S. Paediatric oral biopharmaceutics: key considerations and current challenges. Adv Drug Deliv Rev. 2014;73:102–26.
8. Allegaert K, Mian P, van den Anker JN. Developmental pharmacokinetics in neonates: maturational changes and beyond. Curr Pharm Des. 2017;23(38):5769–78.
9. Bueters R, et al. Ontogeny and cross-species comparison of pathways involved in drug absorption, distribution, metabolism, and excretion in neonates (review): kidney. Drug Metab Dispos. 2020;48(5):353–67.
10. Neal-Kluever A, et al. Physiology of the neonatal gastrointestinal system relevant to the disposition of orally administered medications. Drug Metab Dispos. 2019;47(3):296–313.
11. Smits A, et al. Current knowledge, challenges and innovations in developmental pharmacology: a combined conect4children Expert Group and European Society for Developmental, Perinatal and Paediatric Pharmacology White aper. Br J Clin Pharmacol. 2022;88(12):4965–84.
12. Dawson WT. Relations between age and weight and dosage of drugs. Ann Intern Med. 1940;13(9):1594–615.
13. Crawford JD, Terry ME, Rourke GM. Simplification of drug dosage calculation by application of the surface area principle. Pediatrics. 1950;5(5):783–90.
14. Lack JA, StuartTaylor ME. Calculation of drug dosage and body surface area of children. Br J Anaesth. 1997;78(5):601–5.
15. Anderson BJ, Holford NHG. Mechanism-based concepts of size and maturity in pharmacokinetics. Annu Rev Pharmacol Toxicol. 2008;48(1):303–32.
16. Huelke DF. An overview of anatomical considerations of infants and children in the adult world of automobile safety design. Annu Proc Assoc Adv Automot Med. 1998;42:93–113.
17. Jones BL, Van Den Anker JN, Kearns GL. Chapter 25 - Pediatric clinical pharmacology and therapeutics. In: Principles of clinical pharmacology (3rd ed.). Academic Press; 2012. p. 417–36.
18. Avan BI, Kirkwood B. Role of neighbourhoods in child growth and development: does place matter? Soc Sci Med. 2010;71(1):102–9.
19. Jones LL, et al. A comparison of the socio-economic determinants of growth retardation in South African and Filipino infants. Public Health Nutr. 2008;11(12):1220–8.
20. Lee J, et al. Socioeconomic disparities and the familial coexistence of child stunting and maternal overweight in Guatemala. Econ Hum Biol. 2012;10(3):232–41.
21. Holford NHG. A size standard for pharmacokinetics. Clin Pharmacokinet. 1996;30(5):329–32.
22. Ritschel WA, et al. The allometric approach for interspecies scaling of pharmacokinetic parameters. Compar Biochem Physiol Part C. 1992;103(2):249–53.
23. Miller JL, Sonies BC, Macedonia C. Emergence of oropharyngeal, laryngeal and swallowing activity in the developing fetal upper aerodigestive tract: an ultrasound evaluation. Early Hum Dev. 2003;71(1):61–87.
24. Ross MG, Nijland MJ. Development of ingestive behavior. Am J Phys. 1998;274(4 Pt 2):R879–93.
25. Jones DV, Work CE. Volume of a swallow. Am J Dis Child. 1961;102:427.
26. Lear CS, Flanagan JB Jr, Moorrees CF. The frequency of deglutition in man. Arch Oral Biol. 1965;10:83–100.
27. El-Haddad MA, et al. In utero development of fetal thirst and appetite: potential for programming. J Soc Gynecol Investig. 2004;11(3):123–30.
28. Le Révérend BJD, Edelson LR, Loret C. Anatomical, functional, physiological and behavioural aspects of the development of mastication in early childhood. Br J Nutr. 2014;111(3):403–14.
29. Julien K, et al. Normal masticatory performance in young adults and children. Arch Oral Biol. 1996;41(1):69–75.
30. ICRP. Report of the task group on reference man. Ann ICRP. 1979;3(1–4):iii.
31. Guillet J, Wynchank S, Basse Cathalinat B. Pediatric esophageal scintigraphy. Results of 200 studies. Clin Nucl Med. 1983;8(9):427–33.

32. Cresi F, et al. Esophageal bolus transit in newborns with gastroesophageal reflux disease symptoms: a multichannel intraluminal impedance study. Pediatr Gastroenterol Hepatol Nutr. 2015;18(4):238–45.
33. Warrington JC, Charron M. Pediatric gastrointestinal nuclear medicine. Semin Nucl Med. 2007;37(4):269–85.
34. Channer KS, Virjee JP. The effect of formulation on oesophageal transit. J Pharm Pharmacol. 1985;37(2):126–9.
35. Channer KS, Virjee J. Effect of posture and drink volume on the swallowing of capsules. Br Med J (Clin Res Ed). 1982;285(6356):1702.
36. Perkins AC, et al. Impaired oesophageal transit of capsule versus tablet formulations in the elderly. Gut. 1994;35(10):1363–7.
37. Kasashi K, et al. The influence of size, specific gravity, and head position on the swallowing of solid preparations. Oral Sci Int. 2011;8(2):55–9.
38. Dubois PM, Paulin C, Chayvialle JA. Identification of gastrin-secreting cells and cholecystokinin-secreting cells in the gastrointestinal tract of the human fetus and adult man. Cell Tissue Res. 1976;175(3):351–6.
39. Strolin Benedetti M, Whomsley R, Baltes EL. Differences in absorption, distribution, metabolism and excretion of xenobiotics between the paediatric and adult populations. Expert Opin Drug Metab Toxicol. 2005;1(3):447–71.
40. Bartelink IH, et al. Guidelines on paediatric dosing on the basis of developmental physiology and pharmacokinetic considerations. Clin Pharmacokinet. 2006;45(11):1077–97.
41. Hyman PE, et al. Gastric acid secretory function in preterm infants. J Pediatr. 1985;106(3):467–71.
42. Agunod M, et al. Correlative study of hydrochloric acid, pepsin, and intrinsic factor secretion in newborns and infants. Am J Dig Dis. 1969;14(6):400–14.
43. DeSesso JM, Jacobson CF. Anatomical and physiological parameters affecting gastrointestinal absorption in humans and rats. Food Chem Toxicol. 2001;39(3):209–28.
44. Van Den Abeele J, et al. Gastric fluid composition in a paediatric population: age-dependent changes relevant for gastrointestinal drug disposition. Eur J Pharm Sci. 2018;123:301–11.
45. Pawar G, et al. Characterisation of fasted state gastric and intestinal fluids collected from children. Eur J Pharm Biopharm. 2021;158:156–65.
46. Papadatou-Soulou E, et al. Magnetic resonance imaging quantification of gastrointestinal liquid volumes and distribution in the gastrointestinal tract of children. Mol Pharm. 2019;16(9):3896–903.
47. Deren JS. Development of structure and function in the fetal and newborn stomach. Am J Clin Nutr. 1971;24(1):144–59.
48. Milla PJ. Intestinal motility during ontogeny and intestinal pseudo-obstruction in children. Pediatr Clin N Am. 1996;43(2):511–32.
49. MacLeod SM, Radde IC. Textbook of pediatric clinical pharmacology. Massachusetts: PSG Publishing Co; 1985.
50. Cavell B. Gastric emptying in infants fed human milk or infant formula. Acta Paediatr Scand. 1981;70(5):639–41.
51. Singh SJ, et al. Gastric emptying of solids in normal children--a preliminary report. J Pediatr Surg. 2006;41(2):413–7.
52. Maes BD, et al. Relation between gastric emptying rate and energy intake in children compared with adults. Gut. 1995;36(2):183–8.
53. Fried MD, et al. Decrease in gastric emptying time and episodes of regurgitation in children with spastic quadriplegia fed a whey-based formula. J Pediatr. 1992;120(4 Pt 1):569–72.
54. ICRP. Human alimentary tract model for radiological protection. ICRP Publication 100. A report of The International Commission on Radiological Protection. Ann ICRP. 2006;36(1–2): 25–327, iii.

55. Olausson EA, et al. Small particle size of a solid meal increases gastric emptying and late postprandial glycaemic response in diabetic subjects with gastroparesis. Diabetes Res Clin Pract. 2008;80(2):231–7.
56. McIntyre A, et al. Effect of bran, ispaghula, and inert plastic particles on gastric emptying and small bowel transit in humans: the role of physical factors. Gut. 1997;40(2):223–7.
57. Mundt MW. Assessment of gastric accommodation and gastric emptying using 3D- and Duplex ultrasonography. 2007.
58. Cheng W, Tam PK. Foreign-body ingestion in children: experience with 1,265 cases. J Pediatr Surg. 1999;34(10):1472–6.
59. Conners G. Pediatrics, foreign body ingestion. 2011. Available from: http://www.emedicine.com/emerg/topic379.htm.
60. Graffner C, Josefsson K, Stockman O. Intra- and intersubject variation of erythromycin absorption from single-unit and multiple-unit enteric-coated products. Biopharm Drug Dispos. 1986;7(2):163–71.
61. Neu J. Gastrointestinal maturation and implications for infant feeding. Early Hum Dev. 2007;83(12):767–75.
62. Weaver LT, Austin S, Cole TJ. Small intestinal length: a factor essential for gut adaptation. Gut. 1991;32(11):1321–3.
63. Thompson FM, et al. Epithelial growth of the small intestine in human infants. J Pediatr Gastroenterol Nutr. 1998;26(5):506–12.
64. Walker-Smith J. Variation of small intestinal morphology with age. Arch Dis Child. 1972;47(251):80–3.
65. Schiller C, et al. Intestinal fluid volumes and transit of dosage forms as assessed by magnetic resonance imaging. Aliment Pharmacol Ther. 2005;22(10):971–9.
66. Navarro, J. and J. Schmitz, Paediatric gastroenterology. 1992, Oxford University Press.
67. Perez de la Cruz Moreno M, et al. Characterization of fasted-state human intestinal fluids collected from duodenum and jejunum. J Pharm Pharmacol. 2006;58(8):1079–89.
68. Persson EM, et al. The effects of food on the dissolution of poorly soluble drugs in human and in model small intestinal fluids. Pharm Res. 2005;22(12):2141–51.
69. Evans DF, et al. Measurement of gastrointestinal pH profiles in normal ambulant human subjects. Gut. 1988;29(8):1035–41.
70. Fallingborg J, et al. Measurement of gastrointestinal pH and regional transit times in normal children. J Pediatr Gastroenterol Nutr. 1990;11(2):211–4.
71. Lindahl A, et al. Characterization of fluids from the stomach and proximal jejunum in men and women. Pharm Res. 1997;14(4):497–502.
72. Degen LP, Phillips SF. Variability of gastrointestinal transit in healthy women and men. Gut. 1996;39(2):299–305.
73. Heimann G. Enteral absorption and bioavailability in children in relation to age. Eur J Clin Pharmacol. 1980;18(1):43–50.
74. Raddle IC. Mechanisms of drug absorption and their development. In: McLeod, Raddle, editors. Textbook of pediatric clinical pharmacology. Littleton: PSG Publishing Co; 1985. p. 17–43.
75. Froehlich JM, et al. Small bowel motility assessment with magnetic resonance imaging. J Magn Reson Imaging. 2005;21(4):370–5.
76. Fenton TR, Harries JT, Milla PJ. Disordered small intestinal motility: a rational basis for toddlers' diarrhoea. Gut. 1983;24(10):897–903.
77. van Elburg RM, et al. Intestinal permeability in relation to birth weight and gestational and postnatal age. Arch Dis Child Fetal Neonatal Ed. 2003;88(1):F52–5.
78. Zakeri-Milani P, et al. Predicting human intestinal permeability using single-pass intestinal perfusion in rat. J Pharm Pharm Sci. 2007;10(3):368–79.
79. Corpeleijn WE, et al. Assessment of intestinal permeability in (premature) neonates by sugar absorption tests. Methods Mol Biol (Clifton, N.J.). 2011;763:95–104.

80. Edington AM, Fokati N. Oral drug absorption in pediatric populations. In: Dressman J, Reppas C, editors. Oral drug absorption: prediction and assessment. New York: Informa Healthcare; 2010. p. 108–26.
81. Schwarz SM, et al. Ontogeny of basolateral membrane lipid composition and fluidity in small intestine. Am J Phys. 1989;257(1 Pt 1):G138–44.
82. Ilett KF, et al. Metabolism of drugs and other xenobiotics in the gut lumen and wall. Pharmacol Ther. 1990;46(1):67–93.
83. Fakhoury M, et al. Localization and mRNA expression of CYP3A and P-glycoprotein in human duodenum as a function of age. Drug Metab Dispos. 2005;33(11):1603–7.
84. Johnson TN, et al. Enterocytic CYP3A4 in a paediatric population: developmental changes and the effect of coeliac disease and cystic fibrosis. Br J Clin Pharmacol. 2001;51(5):451–60.
85. Kiss M, et al. Ontogeny of small intestinal drug transporters and metabolizing enzymes based on targeted quantitative proteomics. Drug Metab Dispos. 2021;49(12):1038.
86. Akram G, Mullen AB. Paediatric nurses' knowledge and practice of mixing medication into foodstuff. Int J Pharm Pract. 2012;20(3):191–8.
87. Winstanley PA, Orme ML. The effects of food on drug bioavailability. Br J Clin Pharmacol. 1989;28(6):621–8.
88. Chun AHC, et al. Bioavailability of lansoprazole granules administered in juice or soft food compared with the intact capsule formulation. Clin Ther. 2002;24(8):1322–31.
89. Chun AHC, et al. Effect on bioavailability of admixing the contents of lansoprazole capsules with selected soft foods. Clin Ther. 2000;22(2):231–6.
90. Eliot L, et al. Pharmacokinetic evaluation of a sprinkle-dose regimen of a once-daily, extended-release morphine formulation. Clin Ther. 2002;24(2):260–8.
91. Fay MA, Sheth RD, Gidal BE. Oral absorption kinetics of levetiracetam: the effect of mixing with food or enteral nutrition formulas. Clin Ther. 2005;27(5):594–8.
92. Pentikis HS, et al. Methylphenidate bioavailability in adults when an extended-release multiparticulate formulation is administered sprinkled on food or as an intact capsule. J Am Acad Child Adolesc Psychiatry. 2002;41(4):443–9.
93. Thyssen A, Solanki B, Treem W. Randomized, open-label, single-dose, crossover, relative bioavailability study in healthy adults, comparing the pharmacokinetics of Rabeprazole granules administered using soft food or infant formula as dosing vehicle versus suspension. Clin Ther. 2012;34(7):1636–45.
94. Martir J. et al. Impact of food and drink administration vehicles on paediatric formulation performance part 2: dissolution of Montelukast sodium and Mesalazine formulations. AAPS PharmSciTech. 2020;21(7).
95. Martir J, et al. In vivo predictive dissolution testing of Montelukast sodium formulations administered with drinks and soft foods to infants. AAPS PharmSciTech. 2020;21(7):282.
96. Stevens RC, et al. Effect of food and pharmacokinetic variability on didanosine systemic exposure in HIV-infected children. Pediatric AIDS Clinical Trials Group Protocol 144 Study Team. AIDS Res Hum Retrovir. 2000;16(5):415–21.
97. Riccardi R, et al. Influence of food intake on bioavailability of oral 6-mercaptopurine in children with acute lymphoblastic leukemia. Pediatr Hematol Oncol. 1986;3(4):319–24.
98. Shah T, et al. Administration of melatonin mixed with soft food and liquids for children with neurodevelopmental difficulties. Dev Med Child Neurol. 2008;50(11):845–9.
99. Johnson DA, et al. Stability of esomeprazole capsule contents after in vitro suspension in common soft foods and beverages. Pharmacotherapy. 2003;23(6):731–4.
100. Carrier MN, Garinot O, Vitzling C. Stability and compatibility of tegaserod from crushed tablets mixed in beverages and foods. Am J Health Syst Pharm. 2004;61(11):1135–42.
101. Singh BN. A quantitative approach to probe the dependence and correlation of food-effect with aqueous solubility, dose/solubility ratio, and partition coefficient (Log P) for orally active drugs administered as immediate-release formulations. Drug Dev Res. 2005;65(2):55–75.
102. Gu CH, et al. Predicting effect of food on extent of drug absorption based on physicochemical properties. Pharm Res. 2007;24(6):1118–30.

103. Wu CY, Benet LZ. Predicting drug disposition via application of BCS: transport/absorption/elimination interplay and development of a biopharmaceutics drug disposition classification system. Pharm Res. 2005;22(1):11–23.
104. Ginsberg G, Hattis D, Sonawane B. Incorporating pharmacokinetic differences between children and adults in assessing children's risks to environmental toxicants. Toxicol Appl Pharmacol. 2004;198(2):164–83.
105. Barrett DA, Rutter N. Transdermal delivery and the premature neonate. Crit Rev Ther Drug Carrier Syst. 1994;11(1):1–30.
106. Pacifici GM. Clinical pharmacokinetics of aminoglycosides in the neonate: a review. Eur J Clin Pharmacol. 2009;65(4):419–27.
107. McNamara PJ, Alcorn J. Protein binding predictions in infants. AAPS PharmSci. 2002;4(1):E4.
108. Sethi PK, et al. Ontogeny of plasma proteins, albumin and binding of diazepam, cyclosporine, and deltamethrin. Pediatr Res. 2016;79(3):409–15.
109. Murry DJ, et al. Liver volume as a determinant of drug clearance in children and adolescents. Drug Metab Dispos. 1995;23(10):1110–6.
110. Ladumor MK, et al. Ontogeny of hepatic sulfotransferases and prediction of age-dependent fractional contribution of sulfation in acetaminophen metabolism. Drug Metab Dispos. 2019;47(8):818–31.
111. Bhatt DK, et al. Age-and genotype-dependent variability in the protein abundance and activity of six major uridine diphosphate-glucuronosyltransferases in human liver. Clin Pharmacol Ther. 2019;105(1):131–41.
112. Blake M, et al. Ontogeny of dextromethorphan O-and N-demethylation in the first year of life. Clin Pharmacol Ther. 2007;81(4):510–6.
113. Leeder JS, et al. Understanding the relative roles of pharmacogenetics and ontogeny in pediatric drug development and regulatory science. J Clin Pharmacol. 2010;50(12):1377–87.
114. Johnson CL, Versalovic J. The human microbiome and its potential importance to pediatrics. Pediatrics. 2012;129(5):950–60.
115. Naji-Talakar S, et al. Potential implications of DMET ontogeny on the disposition of commonly prescribed drugs in neonatal and pediatric intensive care units. Expert Opin Drug Metab Toxicol. 2021;17(3):273–89.
116. Leeder JS, et al. Ontogeny of scaling factors for pediatric physiology-based pharmacokinetic modeling and simulation: microsomal protein per gram of liver. Drug Metab Dispos. 2022;50(1):24.
117. Chapron BD, Chapron A, Leeder JS. Recent advances in the ontogeny of drug disposition. Br J Clin Pharmacol. 2022;88(10):4267–84.
118. Allegaert K, van den Anker J. Ontogeny of phase I metabolism of drugs. J Clin Pharmacol. 2019;59(S1):S33–41.
119. Verscheijden LFM, et al. Physiologically-based pharmacokinetic models for children: starting to reach maturation? Pharmacol Ther. 2020;211:107541.
120. Chen N, et al. Ontogeny of drug elimination by the human kidney. Pediatr Nephrol. 2006;21:160–8.

Chapter 3
General Considerations for Pediatric Formulation Development: Setting Target Product Profiles

Daniel E. Schaufelberger

Abstract Developing new drugs for children and the corresponding age-appropriate dosage forms and formulations are complex tasks. In this chapter, we describe the use of a simple but powerful methodology: developing a *Pediatric Quality Target Product Profile* (ped QTPP) to generate focus and clarity for the product to be developed. In this context, "Quality" is a synonym for "Chemistry, Manufacturing, and Controls" (CMC). A proposal for structure and content of the ped QTPP is provided, as well as suggestions on how this "living document" can be used by the formulation scientists and project teams to guide compound development strategies and plans.

Keywords Pediatrics · Target Product Profile · Quality Target Product Profile · Pediatric formulations

3.1 Introduction

Developing age-appropriate pharmaceutical products for children is a highly complex task, as illustrated by the challenges discussed in this book. Compared to developing products for the adult patient population, formulation experts and Chemistry, Manufacturing, and Control (CMC) as a whole play a much more *central* role when pediatric products are designed and developed. In this chapter we explain why we believe this is the case and share our thoughts on best practices around setting clear

D. E. Schaufelberger (✉)
Johns Hopkins University, School of Medicine, Neurology, All Children's Hospital, St. Petersburg, FL, USA

Head Pediatric Center of Excellence, NGT BioPharma Consultants LLC, East Lyme, CT, USA

President Schaufelberger Consulting LLC, Longboat Key, FL, USA
e-mail: dschauf1@jhmi.edu

H. Batchelor, K. Rose (eds.), *Pediatric Formulations*, AAPS Advances in the Pharmaceutical Sciences Series 58, https://doi.org/10.1007/978-3-031-77240-5_3

development targets. Furthermore, we describe in more detail a simple but powerful methodology, *the use of a Pediatric Quality Target Product Profile (ped QTPP).*

3.2 Establishing Development Targets, from TPP to the QTPP to a Pediatric QTPP

Pharmaceutical product development is supposed to start with a patient-centric mindset, thinking about what the patient (and caregivers) need, either to close gaps for unmet medical needs, to improve existing therapies (e.g., better safety profiles, more convenient administrations, and dosing regimens) or to introduce lower cost generic versions and thus enhance global access to essential drugs, e.g., the ones listed on WHO's Model List of Essential Medicines for Children [1]. This patient-centric approach is visualized in Fig. 3.1 with the patient at the center of the "target" and the (unmet) needs documented in a Target Product Profile (TPP), kind of a blueprint of what the new product should look like, not unlike the architectural

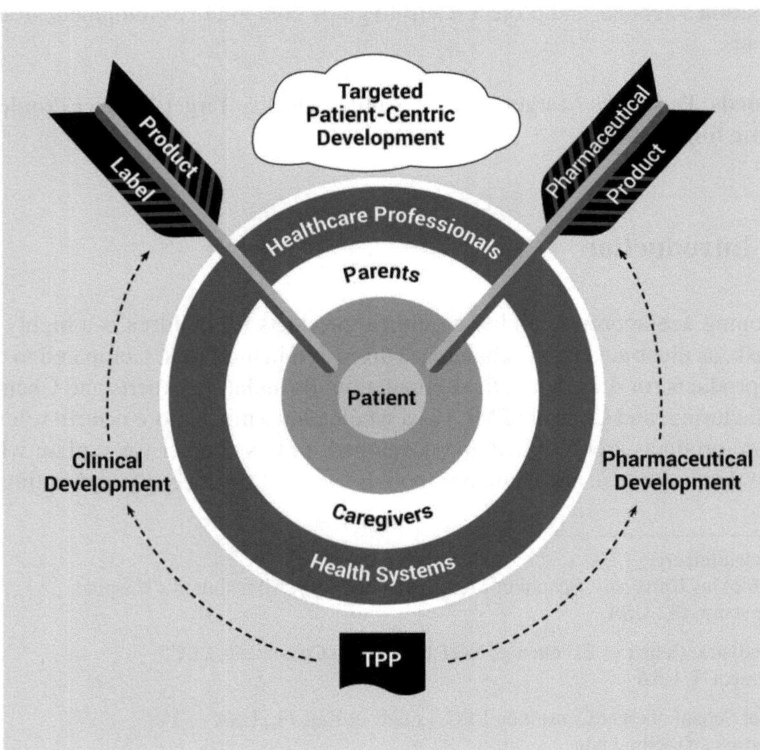

Fig. 3.1 Patient-centric drug development: Target Product Profile (TPP) drives clinical and pharmaceutical development

drawing when you are building a new home. The TPP then drives the design of the clinical studies and protocols that will eventually define the new product label with its indication(s) and corresponding safety and efficacy profiles. On the CMC side, the TPP drives the design of the new pharmaceutical product (dosage forms and formulations) that will deliver the active ingredient to the patient via the routes of administration defined in the original TPP.

The TPP is widely used in the pharmaceutical industry for the development of new adult products, and most of its content is centered around the clinical indications and regulatory pathways leading to a successful filing and approval, e.g., strategies around clinical trial designs. Although not formally structured in this way, the TPP in fact contains "open" and "closed" attributes. The "open" content of a TPP can be used for interactions with stakeholders, e.g., with health authorities (HAs), during pre-IND, end-of-Ph2, and pre-NDA meetings with the US FDA, and for the corresponding meetings with other HAs where a sponsor plans to apply for marketing authorizations. The "closed" portion of the TPP may contain company-sensitive commercial attributes, e.g., targets to achieve a competitive advantage against standard therapies or other drugs in development, pricing targets, as well as manufacturing-related targets around costs to produce the new drug, often referred to as "cost of goods sold" or COGS. It is also good practice to establish a TPP for a new pediatric development program; however, some of the attributes around competitiveness and manufacturing may be less relevant, considering the often low-commercial value and volume forecasts for pediatric products (see below). There are many different formats and approaches to a TPP; some also depend on the disease area and, like in the case of a new antiretroviral drug, may go further into details for the involvement of patient and advocacy groups. See Module 7 "Target Product Profiles" in Ref. [2].

While the "patient-centric" approach to drug development is key and typically the first step, for pediatric products, it is also worth considering a CMC-centric approach. This is justified by a couple of drivers that make the typical role (and influence) of CMC much bigger and puts CMC into the center of attention and decision-making. Some of those factors are as follows, but not limited to:

- Multiple dosage forms/formulations may be needed to cover different age-groups and need to be developed in parallel.
- Patient enrollment in pediatric studies can be very challenging (by nature of the disease, prevalence, etc.). Pediatric studies may take much longer than planned, requiring multiple manufacturing campaigns of clinical supplies, and, if feasible, the shelf-life extensions for clinical trial materials.
- In some cases, there is just a single efficacy study planned, maybe with adaptive design, thus dramatically shortening the time available for CMC development.
- Bioequivalence studies, often used in adult development programs to bridge Ph2 and Ph3 formulations, cannot be carried out in children.
- This all comes down for CMC to be ready with the final formulation for the first (!) pediatric study and leaving very limited room for later changes. Consequently,

the early CMC development work is often on the time-critical path, or very close to it.

- The specific starting and end dates of planned clinical studies are negotiated with health authorities (USA, EU) and become binding, thus adding more pressure to the pharmaceutical development departments to guarantee a timely start of clinical studies.

Because of the above, the CMC experts often find themselves in the "center of the storm" and need to play a very proactive role in the overall development process. Decisions on dosage forms, dose selection, age-ranges, and age-groups (or banding) become critical, e.g., for the timely selection of different strengths (solid dosage forms) or concentrations of liquids. In the author's experience, decision-making in large organizations and on project teams simultaneously focused on adult indications can be challenging at times, particularly for pediatric programs with an anticipated low commercial value. Therefore, we propose to take a CMC-centric approach, which could be like the one shown in Fig. 3.2, where the CMC expert/formulator is in close contact with the "surrounding" key functions in the inner circle, e.g., commercial and clinical. For each functional area, the CMC expert/formulator has numerous deliverables, but also depends on timely decisions (outer circles in Fig. 3.2) that are critical for advancing formulation development. The illustration in Fig. 3.2 could also serve as guidance and visual checklist for a person new to the CMC role on a cross-functional project team.

Regardless of the patient- or CMC-centric approach, we recommend that the following documents are formally established, as illustrated in Fig. 3.3. First, the TPP for the adult indication should be adapted for the corresponding pediatric program, resulting in a pediatric TPP. The ped TPP can then be "translated" by the CMC experts into the *pediatric Quality Target Product Profile (ped QTPP)*, similarly to the process of generating a QTPP based on an adult TPP. The pediatric QTPP will then guide the pharmaceutical development, like any other product following the ICH Q8 guidelines [3], and the preparation of several CMC documents, typically developed as part of an overarching quality (CMC) and control strategy, e.g., establishing critical quality attributes, critical process parameters, and the corresponding in-process and finished product control strategies (e.g., for analytical testing and specifications). Those, in turn, become the drivers of process validation and will be required for the ultimate regulatory submissions and pre-approval inspections.

3.3 Elements of the Pediatric Quality Target Product Profile (Ped QTPP)

The concept of a Pediatric Quality Target Profile has been recently proposed in more detail by the Formulation Expert Group of the CONECT4CHILDREN initiative [4]. The authors describe the ped QTTP and position this document as a contract between the CMC organization and the other functions normally represented

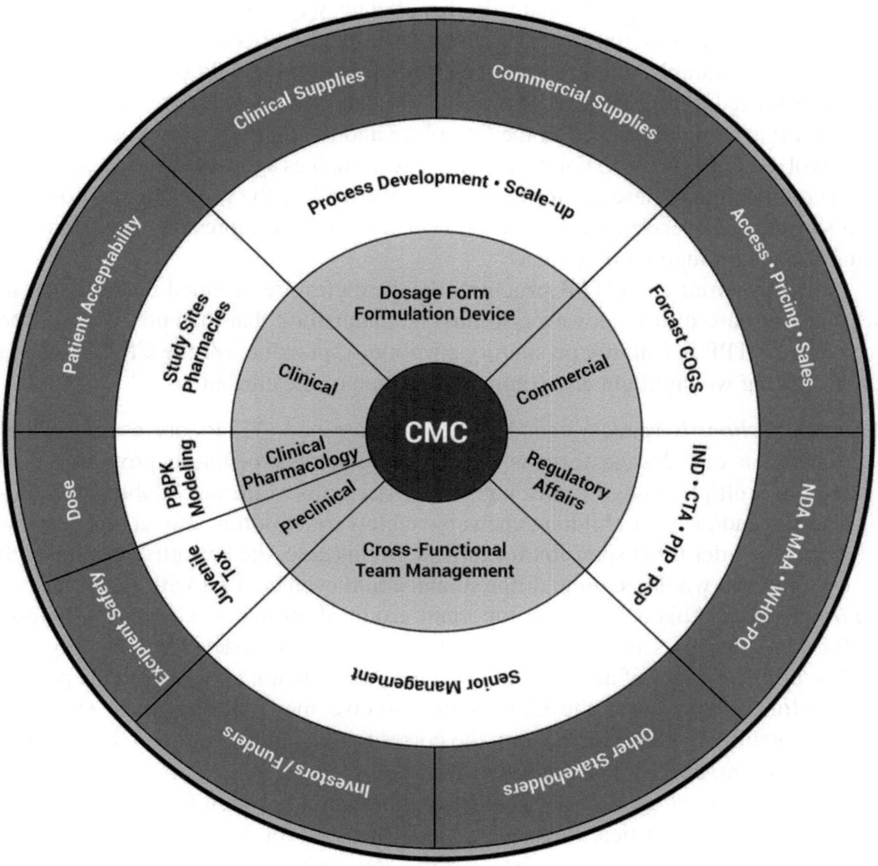

Fig. 3.2 CMC-centric approach to pediatric drug development

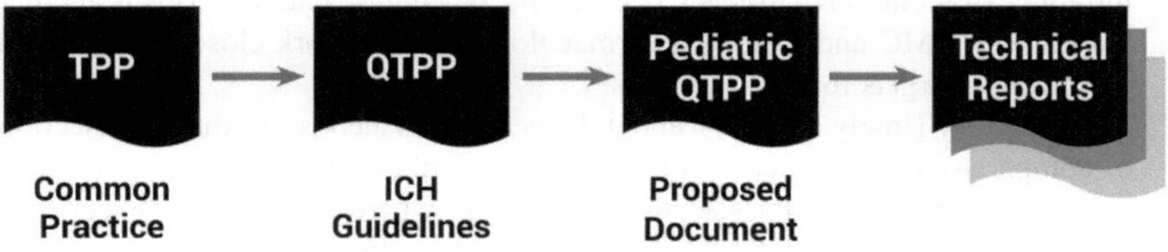

Fig. 3.3 Relationship between Target Product Profile (TPP), Quality Target Product Profile (QTPP), Pediatric Quality Target Product Profile (ped QTPP), and Technical Reports summarizing data required under ICH Q8 and other relevant guidelines

on a compound development team. In this chapter we have further expanded this ped QTPP and propose the structure shown in Table 3.1. It is important to note that this is just an example, and project-specific modifications should be considered and further adjusted during the lifecycle of the project. The key messages are (i) that a ped QTPP is being developed in the first place and (ii) that it is used by *all* the parties involved in the project. For critical attributes, such as age-range and age-groups, we recommend that these are formally acknowledged by the project team, possibly in a summarized version that would be signed off by the representatives from the clinical and commercial functions.

In the proposed ped QTPP structure, the attributes are grouped under different sections for ease of use. However, it must be understood that the authorship of the entire ped QTPP is with the person/organization responsible for the CMC work. In the following we highlight and comment on some of the attributes:

General Section: It is recommended that separate ped QTTPs are established if more than one dosage form is developed. Complex pediatric programs may require multiple dosage forms, e.g., lower strengths of the adult tablets for adolescents and/or older children, chewable tablets for children, and an oral suspension or powder for suspension for infants. In this case, the key attributes for each dosage form would be quite different and could even conflict with each other.

Patient Section: This is the most important group of attributes. Clarity and buy-in for those attributes are critical for the success of the entire development program. The ultimate proof of acceptability of a new formulation comes from the pediatric clinical study, and the CMC representative may have to make sure that "acceptability" is also included in the corresponding study protocol, like a clinical endpoint, but in a less formal way.

Clinical Pharmacology Section: A close collaboration between CMC and Clinical Pharmacology is critical for success [4] as the need for dosing flexibility can be different between clinical studies and the commercial product. In recent years, advances in PBPK modeling and other approaches have reduced the number of clinical studies and contributed to the acceleration of pediatric programs, e.g., through extrapolation of efficacy from adults to pediatric patients. This is another reason why CMC and Clinical Pharmacology need to work closely together in establishing targets for key attributes.

Clinical Section: Timely decisions about the need for placebos and the use/selection of comparator drugs are critical for CMC, particularly under accelerated development timelines. Also, the selection of countries where patients will be recruited has a critical impact on CMC. Due to the difficulties in achieving sufficient enrollment rates, clinical teams may increase the number of countries to conduct studies. This can have a huge impact on the clinical supply chain (logistics, procurement of comparator drugs) plus opens the door for multiple interactions with HAs. While the overall Pediatric Investigational Plan can be negotiated with the European Medicines Agency, it is worth noting here that the approval of the Clinical Trial Applications still happens at the national level. This can result in

Table 3.1 Pediatric Quality Target Product Profile (ped QTPP)—modified after [4] and further expanded

Focus	Attributes	Targets	Comments
General	Name of drug		
	Dosage form	Age-appropriate; early in development set direction for oral solid vs oral liquid (or both, when needed)	Dosage form must be suitable for use in proposed age group; define terminology for consistency
	Route of administration		Route depends on indication (disease state), drug properties, and age
Patient	Indication(s)		Indication can be different from adult indication
	Age-range	0–18 y	Adolescents can be included in original adult clinical studies and original submission, where justified
	Age-groups	Preterm and term newborn infants (0–27 d); neonates Infants (28 d–23 mo) Children (2 y–11 y) Adolescents 12 y–16 y (18 y)	Dosage forms may be different for different age groups; it is critically important to set target(s) to guide CMC development work. Consider two subcategories for "Children": preschool and school-aged children, where appropriate
	Dosing, dose flexibility	Pediatric dose range; dose increments Consider dose banding by weight or age	Higher dose flexibility may be needed during clinical studies; major driver for resulting number of SKUs
	Acceptability	Acceptable for the proposed patient population & care giver(s); appropriate for disease state	Acceptability is a relatively "soft target"; clinical evidence of "acceptability" needs to be generated. Consider healthy adult taste panels
Clin pharm	Desired release profile (PK)	Cmax, AUC, t ½	
	PKPD adult vs pediatric formulation	"Closeness" of pediatric PK with adult PK	Depends on development strategy, e.g., fully pediatric clinical program vs. modeling and extrapolation strategy. Close collaboration of CMC with Clinical Pharmacology and Clinical needed

(continued)

Table 3.1 (continued)

Focus	Attributes	Targets	Comments
Clinical	Comparator drug	Yes/no	For clinical studies; open-label vs blinding; need for over-encapsulation. Could become time-critical activity
	Placebo	Yes/no	For clinical studies
	Countries selected for clinical studies		Important to understand regulatory requirements for CMC
Preclin/tox	Safety of excipients	Tox qualified for intended age-group	Attention to novel excipients with potential need for additional tox studies
	Additional Tox studies	Juvenile tox studies	Gap assessment to identify need for additional studies, requirements for API/ formulation
CMC development	API characteristics	Morphology, particle size distribution, microbiology	Specify targets if API characteristics different from API used for adult product
	Dosage form	Patent protection/ freedom to operate	Assess potential for patent protection if innovative formulation concept
		Synergies with adult dosage form?	Existing knowledge (e.g., stability of adult formulation) vs starting fresh with new dosage form
	Intended strengths (solid dosage forms) or concentrations (liquids)	Meeting requirements for dosing flexibility	Typically, more strengths, more concentrations/ different volumes needed during clinical studies
	Administration, manipulations of dosage form	To enable instructions given in product label	There is also value in generating data to exclude certain manipulations

(continued)

Table 3.1 (continued)

Focus	Attributes	Targets	Comments
	Administration aids (devices)		E.g., indicate if application via nasogastric and gastric tubes required
	Dosing devices (drug/ device combination)		Clearly establish upfront and identify need for user requirements
	Sweeteners	Yes/no	If "yes", specify
	Flavors	Yes/no	
	Colorants	Yes/no	
	Food compatibility	When administration with specific food recommended	Good practice to identify food that should *not* be used for administration, e.g., for pH-sensitive formulations
	Impurity/degradation products	Anything requiring special attention?	As good or better than adult product?
	Dissolution rate/ profile (for solid oral dosage forms)	Tied to desired PK profile	Comparison with adult formulation
	Storage conditions	Commercially viable; minimize impact on supply chain	Cold storage may be acceptable for early/small clinical studies to accelerate timelines
	Stability (shelf life) in primary packaging material	Commercially viable; minimize impact on supply chain	Can be shorter or even different during clinical studies Consider all climatic zones, in particular for (neglected) tropical diseases
	Stability (in-use period) after manipulation, e.g., reconstitution of powders with liquids, sprinkling on food	Consider best practices, e.g., in hospital pharmacies; consider at-home use of final product	In-use stability: product to be administered within a specified period; consider practicality, e.g., time between preparation and administration
CMC manufacturing	Number of SKUs	Limited number of SKUs	
	Volume forecast	Based on commercial forecast	Decision factor for launch strategy: existing pilot plant vs investment in new manufacturing equipment/ facility
	Intended batch size	Calculated based on volume forecast	Optimize for supply chain; avoid obsolescence
	Cost of Goods Target	Driven by pricing	Pricing driven by payors; for neglected (tropical) diseases often driven by procurement agencies

(continued)

Table 3.1 (continued)

Focus	Attributes	Targets	Comments
Commercial	Intended markets		List countries/regions
	Commercial supply chain		Consider dual supply chain for life-saving products, e.g., oncology
	Supply chain integrity	Anticounterfeiting measures (overt/overt)	
	Access	Is voluntary licensing planned for LMICs?	Potential for sublicensing, tech transfers to low-cost country

multiple questions from local HAs and in different expectations, e.g., related to the use of excipients.

Preclinical Section: Unless the pediatric indication is the first indication to be developed, most of the preclinical studies with the new molecular entity have already been completed by the time a pediatric program starts. However, a critical gap assessment is still needed, e.g., for the requirement of juvenile tox studies. Selecting excipients that are toxicologically qualified for a given route of administration and for given age-groups is critically important. It should be defined upfront if commonly accepted excipients are to be selected, or if a novel dosage form may require the tox qualification of novel excipients. If the latter is the case, the impact on costs and timelines needs to be clearly understood by all parties.

CMC Development Section: This section is central to the ped QTPP and must include those attributes that are relevant for the non-CMC functions, e.g., specifics around administration and, if needed, the manipulation of the drug to be performed by patients and/or caregivers, storage conditions and shelf-life targets for supply chain and hold periods for reconstituted products (in-use stability). We also recommend listing targets for dissolution rate and profiles (for oral solid dosage forms) and other API and drug product-related attributes, if those are relevant for the project and require cross-functional alignment and buy-in. This includes the use of sweeteners, flavors, and colorants, having broader commercial and regulatory implications. Generally acceptable attributes and related specifications, e.g., for assay, content uniformity, microbial contamination, can be more appropriately captured in the analytical quality control strategy documents. We also recommend preparing mock-ups of the relevant sections of the *intended* product "label" as they would eventually appear in the prescribing information (USA) or the SmPC (EU). This approach creates focus and raises awareness of the product attributes that the CMC organization needs to get clarity around.

CMC Manufacturing Section: This section may have limited technical information; however, it is critical for the understanding of how many SKUs would be needed to achieve the desired dosing regimen(s). Cost targets are often not addressed early in the development process. Therefore, we recommend establishing those already in the first ped QTPP, even if only directional. To do this, estimates on commercial volume forecasts are required, even if hard to obtain at that point. This section is also the appropriate place to indicate whether local manufacturing (e.g., for a product to treat a neglected tropical disease) would be considered. For unique lifesaving drugs with vulnerable supply chains, it is recommended to decide on a second manufacturing site and the timing for its introduction.

Commercial Section: For new molecular entities (NCE), development activities are clinically focused, with time-to-market and meeting PIP and PSP requirements as the main drivers. From a CMC perspective, this can be best supported by selecting worldwide acceptable formulations. Consequently, not all wishes can be fulfilled, e.g., for an optimized formulation with respect to taste, flavor, and color. For products developed during the lifecycle or for generic products, the formulation as such may become a differentiating factor, and more attention and time can be given to CMC aspects. Identifying regions and countries where the product will be marketed will enable the CMC organization to identify regulatory requirements early on and to take those into account for the development of the product and of the appropriate testing/control strategies. Knowledge of any plans for an early voluntary licensing, e.g., via the Medicines Patent Pool [5], will help in planning for the necessary technology transfer activities in a timely manner.

3.4 Pediatric QTPP: Product Development and Life Cycle

While the primary focus of a ped QTPP is on the final commercial product, it can be looked at as a *living document* that evolves over time. Table 3.2 summarizes different phases in the development of a pediatric product and provides some examples of key attributes that may change over time, to illustrate the concept. The centerpiece of this remains the commercial product for which we recommend establishing "minimally acceptable" and "desirable" targets for key attributes, where feasible. For instance, cold storage of clinical supplies of a solid dosage form might be acceptable for small studies with a limited number of clinical sites. This would allow the program to proceed with limited stability data and without delays. On the other hand, cold storage would likely be rejected for a commercial, solid oral dosage form. Likewise, a shorter shelf life may be acceptable at launch if there are realistic plans in place for post-approval extensions. CMC activities are often on the time-critical path of pediatric development programs. It is therefore recommended to negotiate for certain enhancements to be considered as post-approval life cycle opportunities and document such agreements within the ped QTPP. The need for dosing flexibility may be significantly different between the clinical and commercial phases, respectively. It is likely that small dosing intervals are needed in a clinical study. The clinical data then may provide the justification to establish broader dosing bands, thus reducing the number of different strengths (for solid dosage forms) or volumes and concentrations for liquid dosage forms needed for the commercial product. Ultimately, this would limit the number of different SKUs and

Table 3.2 Pediatric Quality Target Product Profile over the development cycle

Attributes (selected examples)	Development phase up to PK study 1	Clinical safety & efficacy study(ies)	Commercial product		Life cycle management (post-approval)
			Minimally acceptable	Desired	
Dosage form	Multiple dosage forms may be considered	Selected dosage form, same as intended for commercial product	Age-appropriate Commercially viable	Age-appropriate Providing competitive advantage to brand	Additional dosage form(s) Additional strength(s) or concentrations(s)
Dosing Flexibility		To meet clinical dosing regimens as defined in clinical protocol	To meet ultimate dosing regimen as defined in the product label		To meet extended dosing regimens if different for new indications
Storage, e.g., for solid dosage form	Room temperature; cold storage acceptable	Room temperature (cold storage exception)	Room temperature	Same	
Shelf-life		Suitable for clinical supply management	2 years for launch	3+ years	Extend shelf-life
Number of Stock Keeping Units (SKU)	n/a	n/a		minimal	

1. Relative BA study in healthy adults, comparing the new pediatric formulation(s) with the commercial adult formulation.

make the supply chain of low-volume pediatric products more sustainable. In our experience, the CMC representative is uniquely positioned to understand and manage these changing and at times competing requirements. The pediatric QTPP is an ideal tool to facilitate this task and to document agreements reached over time across the various functions represented on a project team. The pediatric QTPP is also a useful tool to assess multiple products in a portfolio, as illustrated by the abbreviated pediatric QTPP recently proposed [6] to assess the age-appropriateness of currently used dosage forms listed on WHO's Essential Medicines List for Children (EMLc).

3.5 Conclusion

Pediatric product development is highly complex and characterized by multiple product attributes that need to be clearly defined upfront, but that also evolve over time and the entire project lifecycle. The pediatric QTPP presented in this chapter is a versatile planning tool that helps the formulation expert (i) as a check list of key attributes to consider, (ii) as a template for negotiations with other functions, e.g., clinical and commercial, (iii) to formally document agreements, and finally (iv) to provide direction to the dozen or so different CMC sub-disciplines typically involved in a pediatric product development effort.

The pediatric QTPP is best understood as a living document that will undergo multiple iterations. While a TPP may cover an entire pediatric program, it is recommended to develop separate pediatric QTPPs, should multiple dosage forms be required. It is also considered good practice to set different minimally acceptable and desirable targets for key attributes to help the CMC organization understand what is considered absolute "must-haves" vs "nice-to-haves." Like each child is different, it is expected that pediatric QTPPs will look differently for different projects. While the template described in this chapter covers multiple angles and aspects, the pediatric QTPP can be easily tailored toward the needs of a given project, the organizational settings and the working style of the cross-functional team tasked with the development of a new pediatric formulation.

Acknowledgments The author thanks Trupti Dixit, PhD, Seagen Inc.; Prof. Stephen Hoag, University of Maryland; Robert Ternik, PhD, formerly with Eli Lilly & Co.; and Jenny Walsh, PhD, of Walsh Consulting, for the very helpful discussions on a variety of pediatric formulation topics. The author also thanks Susan Lavender, Lavender Design, for the graphic design of the figures in this chapter.

References

1. World Health Organization model list of essential medicines for children – 8th List; 2021. https://www.who.int/publications/i/item/WHO-MHP-HPS-EML-2021.03.
2. Toolkit for research and development of paediatric antiretroviral drugs and formulations. Geneva: World Health Organization; 2018. Licence: CC BY-NC-SA 3.0 IGO. https://iris.who.int/bitstream/handle/10665/273151/9789241514361-eng.pdf.
3. ICH Q8 Pharmaceutical Development. https://www.fda.gov/media/71535/download.
4. Walsh J, Schaufelberger D, Iurian S, Klein S, Batchelor H, Turner R, Gizurarson S, Boltri L, Alessandrini E, Tuleu C. Path towards efficient paediatric formulation development based on partnering with clinical pharmacologists and clinicians. Brit J Clin Pharmacol. 2021:1–18. https://doi.org/10.1111/bcp.
5. https://medicinespatentpool.org
6. Walsh J, Masini T, Huttner BD, Moja L, Penazzato M, Cappello B. Assessing the appropriateness of formulations on the WHO model list of essential medicines for children: development of a paediatric quality target product profile tool. Pharmaceutics. 2022;14:473–89. https://doi.org/10.3390/pharmaceutics14030473.

Chapter 4
Acceptability of Formulations: Testing Strategies

Vallet Thibault, Bras Da Costa Sabrina, Wagner Leonie, and Ruiz Fabrice

Abstract In recent years, acceptability of formulations by pediatric patient populations has come to be considered as a key factor for patient adherence. Subsequently, due to the related impact of acceptability on efficacy and safety, the latest guidelines and recommendations from regulators and health agencies (EMA, FDA, WHO) highlight the importance of this key factor which must be considered in the design of medicines appropriate for pediatric populations. Acceptability is defined by the EMA as the overall ability and willingness of the patient to use and their caregiver to administer the medicine as intended. This multifaceted concept is likely to be driven by both the different product features and the patient's characteristics. From newborns to 17-year-olds, the pediatric patient population is particularly heterogeneous with regard to their perceptions, cognitive and physical capabilities. The assessment of patient acceptability, for these different sub-populations, requires a cross-functional effort to cover their different needs during the entire development process, from the early phases of formulation to the final marketing authorization. In this chapter, EMA recommendations are presented as a guide to address the various expectations. Ongoing efforts to investigate patient acceptability for oral dosage forms are summarized. Aspects such as shape and size, formulation type (e.g., tablets, granules, liquid preparations), palatability, and swallowability are covered. For various dosage forms, the method of administration, packaging configurations, and their overall usability will be discussed. Based on the literature, an overview of methods for assessing acceptability is provided and discussed. The chapter also identifies potential gaps in the field and elaborates on challenges faced by the industry to achieve adequate acceptability for a pediatric dosage form.

V. Thibault · R. Fabrice (✉)
ClinSearch, Malakoff, France
e-mail: fabrice.ruiz@clinsearch.net

B. D. C. Sabrina
Device and Packaging Development, F. Hoffmann-La Roche Ltd., Basel, Switzerland

W. Leonie
Formulation R&D Oral Dosage Forms, F. Hoffmann-La Roche Ltd., Basel, Switzerland

H. Batchelor, K. Rose (eds.), *Pediatric Formulations*, AAPS Advances in the Pharmaceutical Sciences Series 58, https://doi.org/10.1007/978-3-031-77240-5_4

Keywords Acceptability · Pediatric · Drug development · Evaluation

4.1 Introduction

Children often have a strong aversion to the bitter taste of medicine and may have difficulty swallowing oral ones; thus, it can be challenging to administer it as prescribed. Caregiver burden might also affect the correct course of treatment. These examples of challenges can result in noncompliance, leading to suboptimal treatment outcomes. Recognizing this, regulatory bodies such as the European Medicines Agency (EMA), the Food and Drug Administration (FDA), and the World Health Organization (WHO) have increasingly emphasized the importance of developing age-appropriate formulations and the necessity to consider users' needs throughout the drug development process [1–4]. Medicine acceptability has emerged as a key factor in ensuring patient compliance, which is crucial as it directly impacts treatment efficacy and safety. The EMA has defined acceptability as "the overall ability and willingness of the patient to use and its care giver to administer the medicine as intended" [1]. This multifaceted concept appears to be driven by the characteristics of the users (e.g., patient's age, health status, socio-cultural background), the medicines' features (e.g., palatability, swallowability, usability), as well as the context of use (e.g., method of administration, caregiver).

While the importance of medicine acceptability in pediatrics is widely acknowledged and despite the recent regulatory requirements, the knowledge and understanding of acceptability testing strategies are still fragmented. The choice of the method and the acceptance criteria are left to the applicant, resulting in a lack of harmonization. A range of methods and approaches are used, and this variability reflects the challenges associated with assessing this multifaceted concept in this heterogeneous population, given the diverse age groups with variations in perception, cognitive and physical capabilities. Furthermore, the methodologies for acceptability testing vary depending on the product development: from early formulation design to final approved formulation. It is essential to implement valid testing strategies to ensure medicine acceptability throughout pharmaceutical and clinical development.

In this chapter, we present an overview of methods for acceptability testing throughout the life cycle of the product. The methods are classified as follows: preclinical strategies, usability assessment, and observer or patient-reported outcomes (Fig. 4.1). Preclinical strategies involve laboratory-based studies that evaluate parameters of the active pharmaceutical ingredient (API) such as taste, appearance, or texture to guide formulation development. Usability assessments focus on the ease of use of medicinal products, including administration devices and packaging systems. Observer or patient-reported outcomes instruments involve reports coming directly from the patient or observer without interpretation of the response by anyone. Considering one dimension of acceptability or many aspects simultaneously, these methodologies will provide valuable information for drug development as well as insights into medicine acceptability in real-world settings.

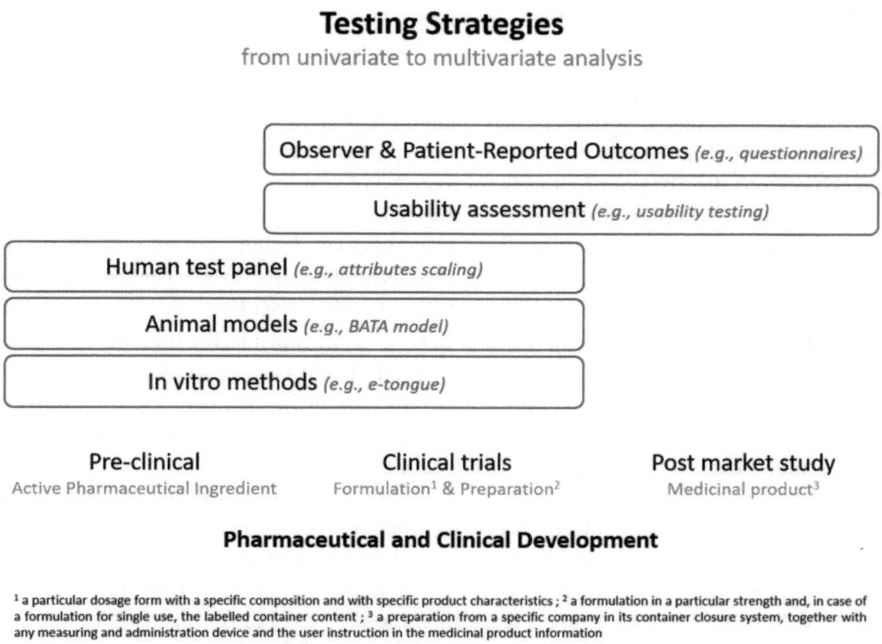

Fig. 4.1 Acceptability testing strategies for pharmaceutical development

4.2 Preclinical Strategies

At very early stages of the development of a medicine very little is known about both the active pharmaceutical ingredient and the finished product (e.g., tablet, or granules). It is hence very crucial to gather as much information as early as possible. There are multiple options to address questions about bitter taste early on in the lifecycle of an NME (New Molecular Entity) [5]. These options start from in silico modeling of taste and go over e-tongue measurements to in vivo tests with animals and hence human panel tests (refer to Chap. 5 for more details). Also, for the aspect of swallowability testing, there are options such as in vitro soft robotic tongue measurements [6] or dedicated placebo studies investigating swallowability of medicines [7]. Unfortunately, these tests mostly cover one aspect of acceptability which is taste, or in panels also mouthfeel, but not all aspects of the multifaceted concept of acceptability. To move in the right direction while developing a medicine tailored for the patient, the needs of the end user (patient/caregiver) must be known. One option to make sure that throughout the development phase of a molecule the needs of the end user are not missed is to use a QTPP (Quality Target Product Profile) [8], capturing all relevant aspects of the medicine, such as dosing regimen, patient population, route of administration, etc. Especially for the pediatric population, the pQTPP (pediatric Quality Target Product Profile) summarizes the key attributes of a medicine for use in the specific age group [9].

4.3 Usability Assessment

Usability Engineering (or Human Factors Engineering) is a discipline that helps understand user needs, anticipate use-related challenges, and assess the safety and effectiveness of the interaction between the pediatric user (and/or his/her parent/caregiver) and the drug product to be prepared and administered [10–12]. Although not identified as such, usability engineering activities enable and support the “ease of administration” characteristic of a drug product design (but not limited to), described as one of the most important characteristics determining patient acceptability as per EMA guidance [1, 13]. Usability may be considered as a key factor for acceptability, ensuring the drug product design is appropriate for the end-user use. The usability engineering process can be divided into two main steps: (1) understanding the patient/user needs and limitations supporting the drug product design requirements and (2) designing and evaluating the drug product design to ensure the drug product design meets the user needs and expectations.

The usability engineering process is usually recommended to be performed throughout the drug product development process. This allows identifying as early as possible the user needs and limitations and anticipating problems or difficulties that might impact the acceptability by the pediatric patient and their parent/caregiver of the combination product. It is recommended to plan for an iterative assessment and mitigate any risks or problems identified by design of the devices, packaging, and then instructions for use. Despite the very limited publications introducing and recognizing usability as a contributor to the acceptability of a drug product by pediatric, it seems that the need is not only on increasing awareness of the benefits of such a discipline but as well on identifying which specific usability methods could be identified as useful to support acceptability of a drug product [13, 14]. Usability testing is, for example, cited as one of the methods that one can refer to as a supplemental or even alternative method to a clinical study to get insights on the acceptance of a drug product by a patient/user [13]. There might be for example also possibilities to design standalone studies where different acceptability dimensions (e.g., swallowability, palatability, and usability) of a drug product design can be assessed jointly with representative pediatric patients and caregivers [15].

This chapter provides a general understanding of the steps as well as a list of recognized usability methods in order to support the design and assessment of acceptable drug product.

4.3.1 The Understanding of the Patient/ End-User Characteristics

Understanding of the patient/end-user characteristics is the key element to anticipate whether a pediatric and/or their parent/caregiver have the ability and willingness to use a drug product. And this ability or willingness can vary even more for the pediatric patient population as their use is often dependent on changing and evolving perceptual, cognitive, and physical factors [13]. It is well known that, for example, treatment adherence is one of the biggest issues in pediatrics when considering medicines acceptability. For oral drugs, for example, it is often related to an unpleasant taste of the drug (e.g., too bitter and/or irritating mouth and throat) which will not be swallowed by the child. An unpalatable drug can actually thwart the benefits of the most effective drug and lead to missing doses and even be life-threatening for the pediatric patient [16]. Nevertheless, there are also other characteristics to be considered that might have an effect on treatment adherence.

4.3.1.1 Disease Characteristics

The type of disease and its conditions can be a good indicator of how able and willing the pediatric patient will be to take the medicine and continue with the treatment. For example, children with disease from birth (e.g., diabetes or asthma) would probably be able and willing to self-administer drugs at an earlier age being used to it, whereas for some other periodic diseases pediatrics would be less adherent. Nevertheless, it is also important to consider that children might not perceive the risks of not taking a drug as an adult may do or may refuse to recognize their illness, especially for severe diseases. For example, an adolescent patient with HIV may be noncompliant with his medicines when he sleeps over at a friend's house, because of the stigma of taking medications or the embarrassment of admitting to his/her condition [17]. Thus, the complexity of the dose regimen, as well as timing and frequency of doses, may have an impact on the adolescent acceptability of the drug as she/he might not want to display it at school in front of other students, resulting in missed doses when social belonging is especially important for an adolescent. Adherence rates decrease as medication dose frequency increases [17]. Their willingness may often rely on whether or not the drug effects have an "immediate observable benefit" (e.g., treatment that helps reduce physical pain or allows physical movement flexibility vs treatment with no observed effect or even leading to cosmetic side effects such as acne or body weight). As an example, studies have demonstrated that the prevalence of noncompliance to treatment for pediatrics with cancer such as acute lymphoblastic leukemia or other malignancies (for which effects may not always be "directly observed" by children) is about 44% and that 47% of the relapses observed among these patients entering remission are attributable to nonadherence [18]. Additionally, forgetfulness is one of the most common self-reported reasons for missing doses in children with chronic disorders, and particularly those that do not result in immediate discomfort when scheduled doses are missed or delayed [19].

4.3.1.2 End-User Characteristics

The pediatric patient and their parent/caregiver capabilities and limitations should be considered for designing the product, especially when pediatric patients experience considerable developmental variability:

- Physical and physiological (e.g., anthropometrics, strength, flexibility, dexterity, and reach). For example, pediatrics might encounter difficulties in opening drug product packaging or depressing in a syringe plunger for self-administering a suspension when the hand strength and flexibility required is too high for their capabilities. Their skin thickness and body weight might have to be considered to determine which exposed needle length should an injector be equipped with as well as which part of the body it might be appropriate to inject.
- Perceptual (e.g., sight, hearing, taste, and touch). Children might perceive things in a "different light" than adults. For example, eating is potentially an intense experience for pediatrics in comparison to adults, as by the time they reach adulthood, only about a third of taste buds in the mouth remain on the tongue. Children may be reluctant and spit/vomit a "bad taste" drug preparation [18, 19].
- Cognitive (e.g., attention, recognition, language and communication, problem-solving, and memory). It can be expected that pediatrics interpret information for drug products use in ways that might make no sense to the adults who created the information. For example, at 4–5 years old, a child might not fully recognize a concept of time and lack attention span [20].
- Emotional and psychological (e.g., attitude toward the drug product, motivation, patience, expectations, stress/pain level, socio-economic and cultural influence). For example, as children get older, they tend to take on increasingly greater responsibility for their own safety but might also be sensitive and emotionally vulnerable [20, 21]. For example, when fear and/or pain toward a drug product (e.g., injectable) is experienced, they might move, cry/scream, and refuse to take their treatment. In addition, their acceptability of the medication is impacted by their relationship with the parent/caregiver, who is often responsible for giving the medication. In the case of older children or adolescents, parent/caregiver might be involved in medication administration to varying degrees, leading to partially assisted administration, which can be quite challenging (e.g., caregiver might intervene regularly and the child might feel pressure which can have higher chances for use errors) [19].

4.3.1.3 Context of Use Characteristics

An importance is given to the limitations of the use environment in which the pediatric patient lives (e.g., region of the world, culture and social context). Some aspects, such as accessibility to water or any other liquid and food necessary to swallow a pill or prepare a solution, might not be available at all times. In some countries, patients might have no possibility to store a drug product in a refrigerator.

Also, it is known that the concept of time is different from one culture/country to another; certain populations might not be using the concept of "time" the same way, and it is worth to consider this, especially when time for dosing, administering, and taking medicine is considered. Pediatric patient might not always be accompanied by a caregiver/parent responsible for taking their medicine. She/he would need to transport and take the medicine at school, which can be challenging, especially in older children who might want to hide their disease and, thus, medicines in front of others.

4.3.1.4 Drug Design Characteristics

The pediatric patient or their parent/caregiver's experience with the use of the drug product packaging, labeling, and devices can influence the acceptability of the drug product. The design of the instructions for use in the leaflet is important because that can play a role in engaging the children in taking their medicine. Pediatric patient and parent/caregiver might need to start or continue the therapy at home when a healthcare professional is not present to reassure or to explain how to prepare and administer the medicine adequately. It is known that training might not always be provided for home use. The pediatric and their parent/caregiver might not be comfortable with how to proceed, and this is when instructions for use need to be clear and easy to understand for both parents/caregivers and children. A systematic review focusing on adherence to medication stated that parents/caregivers often report misunderstanding of the instructions, leading to a refusal to administer the treatment to their child [19]. User-friendly training, educational materials, and instructions are considered as a way to provide information to the families, but also to engage them in the therapy as this is one major problem. Nevertheless, patient acceptability of drug products might not be sufficiently addressed by the user instructions in the package leaflet but primarily by the design of the packaging and devices [13]. It is important to consider which root of administration and "technology" is the most appropriate for the targeted pediatric population. For example, should the bottle of oral solution give visual feedback on the remaining dose to support accuracy but also recall how much of the solution remains to administer? Pediatric patient and its parent/caregiver experience with using other drug products can be supportive to target a more adequate and fitting drug product. A pediatric patient might refuse to take a medicine, if it is less convenient than another formulation or if the treatment is not easy to administer or appealing to him/her. A study conducted with European healthcare professionals showed that 40% of them recognized that pediatric patients and caregivers have difficulties in using oral administration/dosing devices and 70% were aware of difficulties encountered with the correct use of respiratory devices. It is shown, for example, that the second most common reason for admission of pediatric patients to the emergency room for asthma exacerbations is non-use or poor use of inhalers even before poor taste. Despite these challenges, only 30% of the healthcare professionals indicated they always spend time explaining how to use the device correctly, with 50% believing that the instructions provided are appropriate

[22, 23]. The design of devices, packaging and labeling, and most of all, the ease of preparation and administration of a drug do not only reduce risks but play an active part in the ability and willingness of a patient to treat himself.

All these usability characteristics help define the end-user needs and limitations to better design acceptable medicines. An adequate drug product usability may have a positive impact on the pediatric patient and/or parent/caregiver medication adherence and therapeutic outcomes which consequently leads to a safe and effective use of medicines [13].

4.3.2 Methods to Design and Assess the Drug Product

A list of methods presented below describes the most common methods used in Human Factors Engineering to ensure that the drug product design meets patient/end-user needs and expectations [12, 24, 25].

4.3.2.1 Contextual Inquiry (or Ethnographic Research)

An ethnographic field study that involves in-depth observation and interviews of a small sample of users to gain a robust understanding of their practices, interaction with items of interest, and behaviors. Typically, this method is used very early in the drug development process to identify and describe the patient/user needs and context of use and anticipate any use problems that might, for example, lead to nonacceptance of the drug.

4.3.2.2 Focus Groups

Similar to one-on-one interviews, focus groups engage multiple people at a time, usually patients/users providing the opportunity for them to discuss topics with each other during the group session. Typically led by a moderator and engaging four to eight participants. This technique is used for understanding the patient/user, their context of use, behaviors, perceptions, opinions, beliefs, and attitudes toward a user interface.

4.3.2.3 Expert Review and Heuristic Analysis

Reviews of the user interface by Human Factors experts (and other stakeholders such as clinical experts, for example), who evaluate to what extent recognized principles of user-centered design have been applied. This requires a prototype of the drug product device and/or packaging and some understanding of the user needs and context of use to be meaningful.

4.3.2.4 Usability Testing

Usability testing is the observation of the interaction of users with the user interface. This method can be used as a method to identify strengths and weaknesses of a drug product to ensure it meets the pediatric patient population needs and its safe and effective administration. Usability testing is a well-established method usually conducted to inform the development, by revealing potential shortcomings of the user interface and to validate its safe and effective use by its end-users under intended conditions. Whenever possible, it is valuable to conduct exploratory usability testing (generally with five to eight representative users) in early stages, for instance, prior to a clinical trial, in order to identify use errors with the drug product that might have a negative impact on the clinical study [26]. This type of testing is usually performed in representative conditions (e.g., at home or simulated environment) to better identify the final user's needs that might affect their acceptance of the treatment. A usability summative testing (also called Human Factors validation study) is conducted to demonstrate the safe and effective use of drug device combination product under usability regulation (with at least 15 representative users per user group) and, when possible, before phase 3 clinical trials, as this allows time for any changes to optimize the drug product prior to completing the clinical investigation. Training on the drug product administration might not always happen in real life or differently than expected, thus it is meaningful to evaluate the intuitiveness of a drug product without having trained the pediatric participants prior to this testing. The instructions for use and drug product labeling should also be part of it in order to evaluate the effectiveness of important warnings and critical information (e.g., to assess whether or not pediatric patients are able to locate, read, and understand the information) [11, 26, 27]. Table 4.1 gives an overview of the advantages and limitations of usability testing. Noting that in the case of pediatric user population, it is important to make clear the objective and the plan of the test. It is recommended to follow usability guidelines when testing with pediatric [28, 29]. For example, children older than 14 years of age will likely behave as adults in a testing situation and should be treated accordingly, whereas it cannot be expected that children aged 2 to 5 years old will be able to focus for a long time on one drug product. Instead, they may want to impress adults around to show how much they can do without being helped and not be able to really express their opinions.

These usability methods can mainly be categorized as inquiry techniques, inspection, and testing [24]. Table 4.1 gives some advantages and limitations of each of the methods presented above [12, 24, 28].

Table 4.1 Advantages and limitations of the most common usability methods

Category	Method	Advantages	Limitations
Inquiry techniques	Contextual inquiry	Meet patients/users and understand their needs	Multiple meetings to be set up as patients/users are met one by one and often in different locations in order to get a significant amount of data
		Real-life sense of what in the context of use can influence the therapeutic outcome or adherence/acceptability to the treatment	Requires training on how to run the observations and interviews and not introduce bias or influence the participant
	Focus group	Meet patients/users in group	Multiple focus groups might be needed to seek representative opinion
		Used to solicit spontaneous opinions and attitudes about an existing drug product or design concept, to explore attributes that are important, or to define user requirements that a design needs to meet to gain user or patient acceptance	Requires training to manage group dynamics effectively. Some individuals might have strong personalities, and unusual perspectives can dominate discussions
		Can be set up in different ways for different goals and using different techniques. Moderator can administer questionnaires, have participants provide ratings, compare competing drug products, or conduct hands-on testing with drug product design prototypes. Pediatric participants can, for example, be approached with different creative activities such as mood boards, drawings, etc. This helps pediatrics express their needs and opinions on a system [30]	Data collection and analysis might be demanding
Inspectiontechniques	Expert/ heuristic review	A checklist of validated usability criteria can be used for the heuristic analysis of the user interface Well-defined methodology and focused attention Identify weaknesses and strengths of drug product Can be performed quickly	Outcome varies according to the expert's expertise and the list of heuristics selected Requires a very good understanding of the users' needs/limitations and their context of use Requires a prototype of the drug product design Requires training

Testing	Usability testing	Discern patient/user needs and concerns while using drug product Confirm whether or not the product aligns with the actual expectations of the patient/user and their environment Gives more freedom than in a clinical setting where acceptability assessment can be limited Improves user experience. Engages and motivates patient/user to use drug product and provides recommendations to improve their experience Discovery of hidden issues that can compromise patient/user acceptance of the drug product design. Benefits for minimizing use-related risks for clinical trial outcome. This can play a role in the efficacy of drug and support foreseeing complexity in preparing and administering a drug Can be set up for different goals: Identify needs/limitations, testing a prototype/design, compare different designs (competitors), assess instructions for use, and training material to reinforce acceptability Can be set in different ways: Remotely (testing done in an environment familiar to the user) or in a simulated-use environment (room set up in such a way it represents the user's real environment)	Recruitment can be an issue, as in the case of pediatric, it is often about rare diseases, where representative pediatric patients and caregivers are difficult to find Testing with representative users. In the case of pediatric patients, they might be self-injecting or be partially assisted by their parent/caregiver in real life which can be a challenge for designing testing This method might not be beneficial if too early in the drug product development process, where: Formulation decision not taken At least a prototype of drug product and packaging is not available Dosing is not established Requires training and experience on how to run this test without introducing bias or influencing the participant. Usually performed by usability/human factors external suppliers Costs and time to collect, record, and analyze data depending on the objective of the testing

4.4 Observer and Patient Reported Outcomes

Assessing the patients' perspective is essential to thoroughly understand patients' experiences with their disease and treatment. For years, patient-reported outcome (PRO) instruments have been developed to help better understand the various aspects of patient conditions that are only known by the patients themselves, e.g., measure of symptoms, treatment benefit, health status, or quality of life. PRO is an umbrella term covering any health outcome that comes directly from the patient, without interpretation of the patient's response by an investigator or anyone else. The use of PRO instrument (i.e., a questionnaire plus the information and documentation that support its use) in pharmaceutical development has been the subject of particular interest from the research community, the industry, as well as the regulators. The US Food and Drug Administration (FDA) recognized the benefit of these measures in clinical trials and published recommendations for the development and validation of such questionnaires [31, 32]. Main steps of the iterative process could be briefly summarized as follows: defining the conceptual framework and developing a preliminary version of the questionnaire; using qualitative inputs from a small sample of patients to adjust the conceptual framework and ensure its relevance, completeness, and patients' understanding; testing the questionnaire on a large scale to assess its psychometric properties. It should be indeed valid—the extent to which the instrument measures the concept of interest—and reliable—the ability to yield consistent and reproducible measures.

Many concepts (i.e., the *thing* being measured) are measured using PRO, e.g., symptom relief, health-related quality of life (HRQL), adverse reaction, adherence, or satisfaction with treatment. Distinct PRO instruments have been elaborated to score the latter concept and better understand patients' perception of treatment. We can mention specific measures such as the Satisfaction With Antipsychotic Medication scale—SWAM® [33], and generic measures such as the Treatment Satisfaction Questionnaire for Medication—TSQM© [34–36] or the Treatment Satisfaction with Medicines Questionnaire—SATMED-Q© [37–39]. According to Shikiar and Rentz [40], symptom relief/efficacy, side effects, ease and convenience, impact on HRQL from a patient's perspective, and overall satisfaction are the different domains of patient satisfaction. Although acceptability can be defined as an acceptable degree of comfort—a minimum level of satisfaction—and usability—the physical and mental ability to do something as intended—satisfaction is likely to be a broader concept. As a continuation of this work a PRO was developed specifically to measure the acceptance of any long-term treatment: ACCEptance by the Patients of their Treatment questionnaire—ACCEPT© [41–43]. The tool was based on the following dimensions: the drug (preparation, route of administration, pharmaceutical form), duration of treatment, constraints (remembering to take, supply, transport, storage, amount taken, frequency of intake), side effects, effectiveness, and general opinion. These instruments investigating the patient experience were designed for adults and, consequently, cannot be used in pediatrics as specific considerations should be made with regard to this specific population.

Pediatric is a very heterogeneous population with great differences between individuals in terms of language comprehension, vocabulary, attention span, reasoning, decision making, or also duration of recall. Therefore, questionnaires need to be developed and validated specifically for pediatrics and, more especially, for subpopulations [44, 45]. Teenagers who are capable of adult evaluation techniques are indeed fundamentally different from young children who cannot read, write, and speak. Improvement of physical and cognitive skills with age impacts the assessment methods that can be used as well as patient needs and expectations regarding medication. Nevertheless, researchers should keep in mind that inter-individual variability, even within a fairly narrow age group, could remain important. Although the segmentation depends on several factors beyond age (e.g., concept studied, characteristics of the studied population, format and mode of administration) the International Society For Pharmacoeconomics and Outcomes Research—ISPOR [44] proposed 4 age groups as a working base: children under 5 years of age, for whom there is no evidence of the reliability and validity of self-reports; children aged 5–7, for whom very simple concepts could be asked even if reliability and validity of self-assessments remain an issue; children aged 8–11, for whom self-reports can be used if rigorous qualitative research demonstrates satisfactory reliability, validity, and understanding of the questionnaire; children aged 12–18, for whom self-reports are preferred.

From a single-item PRO instrument to a more sophisticated conceptual framework, different tools have been developed to assess acceptability in pediatrics. The TASTY scale was, for example, developed for pediatric self-report during palatability testing [46]. This child-friendly 7-point faces scale appeared to be preferred by both children and parents when compared with other hedonic scales for taste assessment. However, as we might expect, the scale is less reliable in children under the age of 5 years for whom self-assessment remains an issue. Focusing only on palatability, a crucial but not unique dimension of acceptability, this tool could not fully grasp this multifaceted concept. The Pediatric Oral Medicines Acceptability Questionnaire (P-OMAQ) is another PRO aiming to achieve such a goal [47]. Based on a literature review, panel discussions with experts, and interviews with patients and caregivers, the following six key attributes emerged: taste, aftertaste, swallowability, mouth feel, size or amount, and smell. The questionnaire takes into account these key attributes and additional ones, i.e., dosing frequency, overall happiness with/acceptability of the medicine, how much the medicine helps, the importance of taking the medicine, and willingness to keep taking the medicine, and compliance. The P-OMAQ-P is a 12-item questionnaire designed for young people for whom self-reports can be used, those from 8 to 17 years of age. For younger children who cannot make reliable and valid assessments, Observer-Reported Outcomes (ObsRO) are recommended to be used instead of or in addition to PRO [31, 44]. Assessing directly observed behavior should be preferred to proxy measures which require interpretation about the subjective experience of patients [31, 44]. Observers are unable to provide reliable and valid reports on a patient's pain intensity but can report patient behaviors caused by pain, for example. Thus, there is another version of the questionnaire: the P-OMAQ-C, which is a 19-item questionnaire designed for adult caregivers of young people aged 6 months to 17 years. This version combines caregiver-reported outcomes—what the caregiver has directly observed or what the child has said—and additional items to assess the acceptability of the caregiver himself. For younger children measuring using a five-point numerical rating scale, without adult interpretation, how happy/unhappy the child is with the number of times taken daily, size or amount, or feeling in the mouth after taking the medicine may be challenging. The European Medicines Agency in the Guideline on pharmaceutical development of medicines for pediatric use [1] listed palatability, the required dose, or the required dosing frequency but also many other product aspects which might determine acceptability. While identified during the literature reviews and discussions with experts, some of those product aspects (e.g., appearance, packaging) were not kept in the questionnaire following the qualitative patient and caregiver interviews. This interesting instrument remains at an early stage of development. Validation was based on limited data, and the psychometric performance should be assessed at a larger scale [47]. Furthermore, validating a scoring approach is essential for implementing the tool in clinical development [47].

Combining swallowability and palatability proven scores based on observer reports, Wargenau et al. [48] proposed a composite acceptability assessment tool for oral drug formulations in the pediatric population. Acceptability is defined as "high," "good," "low," or "no" depending on swallowability observation score distinguishing between "completely swallowed," "partially swallowed," "spat out," "swallowed the wrong way," or "refused to take" and combined rater palatability score (two blinded raters of video observation) defined as "pleasant," "neutral," "unpleasant," or "contradictory." Observing the result of the intake and the patient reaction were common ways to assess acceptability of oral medicine in pediatrics. Researcher observation of predefined facial expressions (e.g., pursed lips, nose wrinkle) and behaviors (e.g., child spits out medicine, child cries) could be used [49, 50]. Indicator of success for the result of the intake could also be the administered volume [51, 52], the ability to swallow the medicine without prior modification [53], with a clean mouth after ingestion [54], or within a predefined period of time [55]. Other researchers used an anchored three-point scale filled out by parents [56, 57] or a similar five-point scale filled out by trained investigators [58–62]. For collecting patient reaction there were also various scales used: Visual Analogue Scale (VAS) [56, 57, 63, 64], four- [65] or five-point facial hedonic scale [51, 66–68], as well as scale with two [69], three [51, 52, 69–72], four [73, 74], five [75], or six items [76]. The selection of an appropriate response scale appeared to be critical. Using larger scales for improving the power of discrimination among products is of interest when studying concepts such as patient preference, but this is not really the case for acceptability. To classify a medicine as accepted, it was sometimes necessary to define a two-group score, e.g., for a five-point facial hedonic scale grouping the response options from neutral to positive against the negative ones [50, 77]. Aiming to summarize acceptability dimensions into an intelligible single score that can be compared across age categories and drug formulations, this composite acceptability assessment tool is of real interest. Indeed, pooling data from different questionnaires, or different age-specific versions of a questionnaire, to support a claim in pediatrics may be challenging [44, 45]. Palatability and swallowability are crucial in pediatrics, but acceptability is driven by many other product aspects. If the required dose of a medication were to be fully taken without any negative reaction, that might be considered as accepted. However, is this truly the case if the caregiver needed, for example, to crush the tablet, mix the drug with food to mask its aversive taste, used a specific administration device, and consequently, required a long time to prepare and administer the mixture?

Simultaneous consideration of different measures appeared to be essential for assessing acceptability. Furthermore, in the lack of a gold standard, this multifaceted concept should not be quantified on its own [14], and consequently, assessment would need to be relative. Considering various patients and medicines from the wide spectrum of products on the global market would thus be an integral part of the acceptability evaluation. On this basis, a data-driven approach has been developed in recent years: the ClinSearch Acceptability Score Test® (CAST) [78–87]. The idea was to mine large batches of real-life observer-reported outcomes using multivariate exploratory data analysis, in order to design a reference framework.

In addition to observations of the result of the intake and patient's reaction commonly used, additional objective information was collected to describe the many aspects of acceptability. The time taken for medicine intake [51, 52, 73] or the time needed for medicine preparation describing the burden for the caregiver [88] reflects the ease of medicine use. Activities undertaken to ease/achieve administration, including modifications made to medications prior to administration, such as crushing tablets, are also reported [57, 88], or the use of food and drink to overcome palatability or swallowability issues [89–92]. In addition, the use of a delivery device not supplied with the medication promising a reward to the patient, or imposing administration upon the patient, were reports also collected [51, 52, 57, 93]. The questionnaire provides, for each evaluation of a medicine intake, a particular combination of observed measures for the nine observed behaviors/events reflecting the ability and willingness of patients and caregivers to use (preparing and administering) medicines as intended. Data mining techniques—mapping and clustering processes—summarized and visualized the key information contained in the large set of data from various countries (e.g., France, the United Kingdom, Germany, Norway, Poland, Morocco, India, Japan, Peru) into a three-dimensional map (3D-map) juxtaposing two acceptability profiles, "Positively accepted" and "Negatively accepted." All the combinations of observed measures are plotted on the 3D-map according to their similarity, between the extreme combinations, i.e., the combination reflecting a medicine use without any problem and a combination with many negatively connotated observations. The set of evaluations was then gathered into clusters according to their position on the 3D-map defining an effective cut-off between positively and negatively accepted medicines.

This acceptability reference framework permits standardized acceptability evaluation. Oral medicine could be assessed using the standardized questionnaire. The oral product under investigation is then positioned on the 3D-map at the barycenter of its evaluations. Confidence ellipses surrounding the barycenter for all dimension pairs will define an area containing its true position in the population of interest, with 90% probability. If the barycenter, along with the entire confidence ellipsis surrounding it, belongs to the green area of the map, materializing "positively accepted" profile, the medicine is classified as positively accepted. Application of the tool should be relative, and consequently, the acceptability score of the medicine should be compared within the reference framework. The clinical trial comparator drug could be used. In case of a single-arm trial, or if this is not relevant, the recommendation is to collect additional evaluations on acceptability of a carefully selected comparator product (or reference set) in an ad hoc acceptability study. Scoring comparison between similar products from different companies or between drug products from the same company, but with a different type of dosage form, route of administration, or variations in the pharmaceutical design such as composition, taste, appearance, or packaging could be relevant. In the absence of such evaluations from the trial or from an ancillary study, or if no relevant comparator is available on the market, knowledge from the large dataset that gave rise to the acceptability reference framework can be used. Acceptability score of the oral product under investigation could be compared with the average acceptability of a specific type of oral medicines in the sub-population of interest. In addition, the evaluations of the medicine intake collected in the study could be partitioned into subgroups according to any explanatory variables of interest—e.g., sex or age of the patients—and acceptability scores should be obtained for each subgroup in order to highlight any influence on acceptability. The study sample size should be defined depending on the factors of interest, as a minimum of 30 evaluations are required to obtain a reliable score with a satisfactory precision using the CAST methodology.

The distinctive feature of this data-driven approach is that user or product characteristics are considered as explanatory variables. Based on objective information, the reference framework allows to study acceptability variations due to changes among those characteristics and, consequently, improve knowledge on medicine acceptability. For example, the taste/aftertaste is not directly rated by observer, but the impact of aroma can be explored using the reference framework. Similarly, the impact of administration device could be investigated while there is no direct item on this product's feature in the questionnaire. To investigate what drives acceptability difference, the different components of the scores—the nine observed measures—should be studied. A study, investigating acceptability drivers of oral antibiotics in pediatrics illustrated such examples [84]: in toddlers and preschoolers different flavors used to mask the taste of a drug could significantly impact acceptability, even leading to contrasting effect on girls versus boys, for certain aromas such as lemon; place of administration and administration devices may also impact acceptability, as antibiotics appeared to be better accepted in hospital compared to at home and those provided with oral syringes were better accepted than those provided with measuring spoons.

To our knowledge, the first research article on acceptability evaluation in pediatrics was published in 1965 [94]. Since then, numerous tools have been used, and knowledge on acceptability testing has improved. A state-of-the-art study shows that providing a comprehensive and validated score to thoroughly grasp this multifaceted concept was definitely not low hanging fruit. Covering all attributes listed in the EMA's guideline [1] in a single score remains challenging. The difficulty of such multidimensional evaluations is compounded in pediatrics, a very heterogeneous population that requires specific testing techniques. Although none is perfect, different interesting tools are being developed, varying in reliability, validity, comprehensiveness, or usability in trials. On May 2023, The EMA published a letter of support for CAST in relative acceptability testing for oral medicines in children under 12 years of age [95]. Although some outstanding issues should be addressed, the EMA highly appreciated the attempt to concretize and quantify the "catch-all" concept of acceptability and further experience of the CAST score in clinical development plans will be welcomed. Finally, regardless of the study and the assessment tool, any interpretation of an acceptability study would be made in the context of the medicine's benefit-to-risk evaluations, including the commercial availability of alternative medicines and depending on the therapeutic indication and disease severity.

4.5 Conclusions and Perspectives

From the earliest stages of pharmaceutical development, it's important to bring together all the players who will come into contact with the product, throughout its development, and even after it has been marketed. At every stage, important information could be gathered on the acceptability profile of the future product. From the first formulations, through clinical trials, to commercial packaging, each player can, if they have been made aware of the issue, collect important information that will enable acceptability to be assessed. To the best of our knowledge, there is currently no single method that covers the entire development process. Stakeholders will therefore have to choose the method best suited to the stage of development concerned, in order to best assess the acceptability profile and, if necessary, improve it as far upstream as possible.

References

1. European Medicine Agency. Guideline on pharmaceutical development of medicines for paediatric use. EMA/CHMP/QWP/805880/2012 Rev. 2. London: EMA; 2013.
2. World Health Organization. Toolkit for research and development of paediatric antiretroviral drugs and formulations. Module 5: Acceptability. 2018.
3. U.S. Department of Health and Human Services. Food and Drug Administration. Size, shape, and other physical attributes of generic tablets and capsules guidance for industry. Revision 1. Fed Regist. 2022.
4. Administration USDoHaHSFaD. Pediatric study plans: content of and process for submitting initial pediatric study plans and amended initial pediatric study plans. Guidance for industry. Fed Regist. 2020.

5. Clapham D, Bennett J, Cram A, Discihnger A, Inghelbrecht S, Pensé-Lhéritier A-M, Ruiz F, Salunke S, Schiele J, Soto J. Proposed tool to compare and assess the applicability of taste assessment techniques for pharmaceuticals. J Pharm Sci. 2022;111:1219–23.
6. Lavoisier A, Avila-Sierra A, Timpe C, Kuehl P, Wagner L, Tournier C, Ramaioli M. A novel soft robotic pediatric in vitro swallowing device to gain insights into the swallowability of mini-tablets. Int J Pharm. 2022;629:122369.
7. Thabet Y, Klingmann V, Breitkreutz J. Drug formulations: standards and novel strategies for drug administration in pediatrics. J Clin Pharmacol. 2018;58:S26–35.
8. International Conference on Harmonisation of Technical Requirements for Registration of Pharmaceuticals for Human Use considerations (ICH). Guideline Q8 (R2) on pharmaceutical development. 2009.
9. Walsh J, Schaufelberger D, Iurian S, Klein S, Batchelor H, Turner R, Gizurarson S, Boltri L, Alessandrini E, Tuleu C. Path towards efficient paediatric formulation development based on partnering with clinical pharmacologists and clinicians, a conect4children expert group white paper. Br J Clin Pharmacol. 2022;88:5034–51.
10. International Organisation for Standardization. ISO 9241-210:2019. Ergonomics of human-system interaction - Part 210: Human-centred design for interactive systems. Geneva: ISO; 2019.
11. FDA US. Human factors studies and related clinical study considerations in combination product design and development. US FDA; 2016.
12. International Organisation for Standardization. IEC/TR 62366-2:2016. Medical devices - Part 2: Guidance on the application of usability engineering to medical devices. ISO: Geneva; 2016.
13. Ranmal SR, O'Brien F, Lopez F, Ruiz F, Orlu M, Tuleu C, Walsh J, Liu F. Methodologies for assessing the acceptability of oral formulations among children and older adults: a systematic review. Drug Discov Today. 2018;22:30474–7.
14. Drumond N, van Riet-Nales DA, Karapinar-Carkit F, Stegemann S. Patients' appropriateness, acceptability, usability and preferences for pharmaceutical preparations: results from a literature review on clinical evidence. Int J Pharm. 2017; https://doi.org/10.1016/j.ijpharm.2017.02.029.
15. Katarzyna Hofmanová J, Bennett J, Coupe A, A. Bartlett J, Monahan A, Batchelor HK. A novel oral syringe for dosing and administration of multiparticulate formulations: acceptability study in preschool and school children. Pharmaceutics. 2020;12:806.
16. Mennella JA, Beauchamp GK. Optimizing oral medications for children. Clin Ther. 2008;30:2120–32.
17. El-Rachidi S, Larochelle JM, Morgan JA. Pharmacists and pediatric medication adherence: bridging the gap. Hosp Pharm. 2017;52:124–31.
18. Gupta S, Bhatia S. Optimizing medication adherence in children with cancer. Curr Opin Pediatr. 2017;29:41.
19. Matsui D. Current issues in pediatric medication adherence. Paediatr Drugs. 2007;9:283–8.
20. Lueder R, Rice VJB. Ergonomics for children: designing products and places for toddler to teens. CRC Press; 2007.
21. Laughery KR, Lowoll DR, McQuilkin ML. Allocation of responsibility for child safety. In: Proceedings of Proceedings of the Human Factors and Ergonomics Society Annual Meeting. p. 810–3.
22. Alessandrini E, Walsh J, Salunke S. Usability of administration devices for oral and respiratory medicines: views from a UK primary school. Eur J Pharm Biopharm. 2022;178:150–8.
23. Walsh J, Math M-C, Breitkreutz J, Zerback T, Wachtel H, Initiative EPF. Devices for oral and respiratory paediatric medicines: what do healthcare professionals think? Int J Pharm. 2015;492:304–15.
24. Gulati A, Dubey SK. Critical analysis on usability evaluation techniques. Int J Eng Sci Technol. 2012;4:990–7.
25. International Organisation for Standardization. ISO/TR 16982:2002. Ergonomics of human-system interaction - Usability methods supporting human-centred design. Geneva: ISO; 2002.

26. Notenboom K, Leufkens HG, Vromans H, Bouvy ML. Learning from patients: identifying design features of medicines that cause medication use problems. Int J Pharm. 2017;517:128–34.
27. Eichler H-G, Abadie E, Breckenridge A, Flamion B, Gustafsson LL, Leufkens H, Rowland M, Schneider CK, Bloechl-Daum B. Bridging the efficacy-effectiveness gap: a regulator's perspective on addressing variability of drug response. Nat Rev Drug Discov. 2011;10:495.
28. Khanum MA, Trivedi MC. Take care: a study on usability evaluation methods for children. arXiv preprint arXiv:1212.0647. 2012.
29. Barendregt W, Bekker M. Guidelines for user testing with children. Eindhoven: Tech. Rep; 2003.
30. Bruseberg A, McDonagh D, Wormald P. The use of images to elicit user needs for the design of playground equipment. In: Proceedings of design and emotion. p. 114–8.
31. U.S. Department of Health and Human Services. Food and Drug Administration. Guidance for industry: patient-reported outcome measures: use in medical product development to support labeling claims. Fed Regist. 2009;74:65132–3.
32. U.S. Department of Health and Human Services. Food and Drug Administration. Guidance for industry: patient-reported outcome measures: use in medical product development to support labeling claims: draft guidance. Health Qual Life Outcomes. 2006;4:79.
33. Rofail D, Gray R, Gournay K. The development and internal consistency of the satisfaction with Antipsychotic Medication Scale. Psychol Med. 2005;35:1063–72.
34. Atkinson MJ, Sinha A, Hass SL, Colman SS, Kumar RN, Brod M, Rowland CR. Validation of a general measure of treatment satisfaction, the Treatment Satisfaction Questionnaire for Medication (TSQM), using a national panel study of chronic disease. Health Qual Life Outcomes. 2004;2:12.
35. Atkinson MJ, Kumar R, Cappelleri JC, Hass SL. Hierarchical construct validity of the treatment satisfaction questionnaire for medication (TSQM version II) among outpatient pharmacy consumers. Value Health. 2005;8:S9–S24.
36. Bharmal M, Payne K, Atkinson MJ, Desrosiers MP, Morisky DE, Gemmen E. Validation of an abbreviated Treatment Satisfaction Questionnaire for Medication (TSQM-9) among patients on antihypertensive medications. Health Qual Life Outcomes. 2009;7:1477–7525.
37. Ruiz MA, Pardo A, Rejas J, Soto J, Villasante F, Aranguren JL. Development and validation of the "Treatment Satisfaction with Medicines Questionnaire" (SATMED-Q). Value Health. 2008;11:913–26.
38. Rejas J, Ruiz MA, Pardo A, Soto J. Minimally important difference of the Treatment Satisfaction with Medicines Questionnaire (SATMED-Q). BMC Med Res Methodol. 2011;11:1471–2288.
39. Rejas J, Ruiz M, Pardo A, Soto J. Detecting changes in patient treatment satisfaction with medicines: the SATMED-Q. Value Health. 2013;16:88–96.
40. Shikiar R, Rentz AM. Satisfaction with medication: an overview of conceptual, methodologic, and regulatory issues. Value Health. 2004;7:204–15.
41. Marant C, Longin J, Spizak C, Patrick D, Saussier C, Arnould B, Van Ganse E. What does acceptability mean for patients and how should it be measured? Qualitative steps for the development of a new measurement instrument for pharmacies: the "ACCEPT©" questionnaire. In: Proceedings of ISPOR 11th Annual European Congress. Athens, Greece.
42. Marant C, Longin J, Gauchoux R, Arnould B, Spizak C, Marrel A, Patrick DL, Van Ganse E. Long-term treatment acceptance: what is it, and how can it be assessed? Patient. 2012;5:239–49.
43. Arnould B, Gilet H, Patrick DL, Acquadro C. Item reduction, scoring, and first validation of the ACCEPTance by the patients of their treatment (ACCEPT(c)) questionnaire. Patient. 2016;25:25.
44. Matza LS, Patrick DL, Riley AW, Alexander JJ, Rajmil L, Pleil AM, Bullinger M. Pediatric patient-reported outcome instruments for research to support medical product labeling: report of the ISPOR PRO good research practices for the assessment of children and adolescents task force. Value Health. 2013;16:461–79.

45. Arbuckle R, Abetz-Webb L. "Not just little adults": qualitative methods to support the development of pediatric patient-reported outcomes. Patient. 2013;6:143–59.
46. Wagner JA, Pabon G, Terrill D, Abdel-Rahman SM. Examining a new scale for evaluating taste in children (TASTY). J Pediatr Pharmacol Ther. 2020;25:131–8.
47. Turner-Bowker DM, An Haack K, Krohe M, Yaworsky A, Vivas N, Kelly M, Chatterjee G, Chaston E, Mann E, Reaney M. Development and content validation of the Pediatric Oral Medicines Acceptability Questionnaires (P-OMAQ): patient-reported and caregiver-reported outcome measures. J Patient-Report Outcomes. 2020;4:1–18.
48. Wargenau M, Reidemeister S, Klingmann I, Klingmann V. A composite endpoint for acceptability evaluation of oral drug formulations in the pediatric population. Ther Innov Regul Sci. 2022;56:1–7.
49. Bracken L, McDonough E, Ashleigh S, Wilson F, Shakeshaft J, Ohia U, Mistry P, Jones H, Kanji N, Liu F. Can children swallow tablets? Outcome data from a feasibility study to assess the acceptability of different-sized placebo tablets in children (creating acceptable tablets (CAT)). BMJ Open. 2020;10:e036508.
50. Mistry P, Stirling H, Callens C, Hodson J, Batchelor H. Evaluation of patient-reported outcome measurements as a reliable tool to measure acceptability of the taste of paediatric medicines in an inpatient paediatric population. BMJ Open. 2018;8:e021961.
51. Kraus DM, Stohlmeyer LA, Hannon PR. Infant acceptance and effectiveness of a new oral liquid medication delivery system. Am J Health Syst Pharm. 1999;56:1094–101.
52. Kraus DM, Stohlmeyer LA, Hannon PR, Freels SA. Effectiveness and infant acceptance of the Rx medibottle versus the oral syringe. Pharmacotherapy. 2001;21:416–23.
53. Mekmullica J, Pancharoen C. Acceptability of oral typhoid vaccine in Thai children. Southeast Asian J Trop Med Public Health. 2003;34:334–6.
54. Babbitt RL, Parrish JM, Brierley PE, Kohr MA. Teaching developmentally disabled children with chronic illness to swallow prescribed capsules. J Dev Behav Pediatr. 1991;12:229–35.
55. Beck MH, Cataldo M, Slifer KJ, Pulbrook V, Guhman JK. Teaching children with attention deficit hyperactivity disorder (ADHD) and autistic disorder (AD) how to swallow pills. Clin Pediatr. 2005;44:515–26.
56. Van Riet-Nales DA, de Neef BJ, Schobben AF, Ferreira JA, Egberts TC, Rademaker CM. Acceptability of different oral formulations in infants and preschool children. Arch Dis Child. 2013;98:725–31.
57. Van Riet-Nales DA, Ferreira JA, Schobben AF, de Neef BJ, Egberts TC, Rademaker CM. Methods of administering oral formulations and child acceptability. Int J Pharm. 2015;491:261–7.
58. Spomer N, Klingmann V, Stoltenberg I, Lerch C, Meissner T, Breitkreutz J. Acceptance of uncoated mini-tablets in young children: results from a prospective exploratory cross-over study. Arch Dis Child. 2012;97:283–6.
59. Klingmann V, Spomer N, Lerch C, Stoltenberg I, Fromke C, Bosse HM, Breitkreutz J, Meissner T. Favorable acceptance of mini-tablets compared with syrup: a randomized controlled trial in infants and preschool children. J Pediatr. 2013;163:1728–32.
60. Klingmann V, Seitz A, Meissner T, Breitkreutz J, Moeltner A, Bosse HM. Acceptability of uncoated mini-tablets in neonates-a randomized controlled trial. J Pediatr. 2015;7:00732–5.
61. Klingmann V, Pohly CE, Meissner T, Mayatepek E, Möltner A, Flunkert K, Breitkreutz J, Bosse HM. Acceptability of an orodispersible film compared to syrup in neonates and infants: a randomized controlled trial. Eur J Pharm Biopharm. 2020;151:239.
62. Kluk A, Sznitowska M, Brandt A, Sznurkowska K, Plata-Nazar K, Mysliwiec M, Kaminska B, Kotlowska H. Can preschool-aged children swallow several minitablets at a time? Results from a clinical pilot study. Int J Pharm. 2015;28:00196–9.
63. Lucas-Bouwman M, Roorda R, Jansman F, Brand P. Crushed prednisolone tablets or oral solution for acute asthma? Arch Dis Child. 2001;84:347–8.
64. Herd D, Salehi B. Palatability of two forms of paracetamol (acetaminophen) suspension: a randomised trial. Paediatr Perinat Drug Ther. 2006;7:189–93.

65. Martinez JM, Bartoli F, Recaldini E, Lavanchy L, Bianchetti MG. A taste comparison of two different liquid colecalciferol (vitamin [D. sub. 3]) preparations in healthy newborns and infants. Clin Drug Investig. 2006;26:663–6.
66. Sáez-Llorens X, Yogev R, Arguedas A, Rodriguez A, Spigarelli M, Castrejón TDL, Bomgaars L, Roberts M, Abrams B, Zhou W. Pharmacokinetics and safety of famciclovir in children with herpes simplex or varicella-zoster virus infection. Antimicrob Agents Chemother. 2009;53:1912–20.
67. Wollner A, Lecuyer A, De La Rocque F, Sedletzki G, Derkx V, Boucherat M, Elbez A, Gelbert-Baudino N, Levy C, Corrard F, et al. Acceptability, compliance and schedule of administration of oral antibiotics in outpatient children. Arch Pediatr. 2011;18:611–6.
68. El-Chaar GM, Mardy G, Wehlou K, Rubin LG. Randomized, double blind comparison of brand and generic antibiotic suspensions: II. A study of taste and compliance in children. Pediatr Infect Dis J. 1996;15:18–22.
69. Almenrader N, Passariello M, Coccetti B, Haiberger R, Pietropaoli P. Steal-induction after clonidine premedication: a comparison of the oral and nasal route. Paediatr Anaesth. 2007;17:230–4.
70. Jahnsen T, Thorn P. An acceptability study of two pivampicillin mixtures in children in general practice. Scand J Prim Health Care. 1987;5:241–3.
71. Gooch W, Philips A, Rhoades R, Rosenberg R, Schaten R, Starobin S. Comparison of the efficacy, safety and acceptability of cefixime and amoxicillin/clavulanate in acute otitis media. Pediatr Infect Dis J. 1997;16:S21–4.
72. Verrotti A, Nanni G, Agostinelli S, Alleva E, Aloisi P, Franzoni E, Spalice A, Chiarelli F, Coppola G. Effects of the abrupt switch from solution to modified-release granule formulation of valproate. Acta Neurol Scand. 2012;125:e14.
73. Uhari M, Eskelinen L, Jokisalo J. Acceptance of antibiotic mixtures by infants and children. Eur J Clin Pharmacol. 1986;30:503–4.
74. Van de Vijver E, Desager K, Mulberg AE, Staelens S, Verkade HJ, Bodewes FA, Malfroot A, Hauser B, Sinaasappel M, Van Biervliet S. Treatment of infants and toddlers with cystic fibrosis–related pancreatic insufficiency and fat malabsorption with pancrelipase MT. J Pediatr Gastroenterol Nutr. 2011;53:61–4.
75. Dagan R, Shvartzman P, Liss Z. Variation in acceptance of common oral antibiotic suspensions. Pediatr Infect Dis J. 1994;13:686–90.
76. Moniot-Ville N, Chelly M, Consten L, Rosenbaum M. The acceptability, efficacy and safety of a new paediatric oral suspension of roxithromycin in respiratory tract infections. J Int Med Res. 1998;26:144–51.
77. Cohen R, de La Rocque F, Lecuyer A, Wollner C, Bodin MJ, Wollner A. Study of the acceptability of antibiotic syrups, suspensions, and oral solutions prescribed to pediatric outpatients. Eur J Pediatr. 2009;168:851–7.
78. Ruiz F, Vallet T, Pense-Lheritier AM, Aoussat A. Standardized method to assess medicines' acceptability: focus on paediatric population. J Pharm Pharmacol. 2017;69:406–16. https://doi.org/10.1111/jphp.12547.
79. Vallet T, Ruiz F, Lavarde M, Pense-Lheritier AM, Aoussat A. Standardized evaluation of medicine acceptability in paediatric population: reliability of a model. J Pharm Pharmacol. 2018;70:42–50. https://doi.org/10.1111/jphp.12829.
80. Ruiz F, Vallet T, Dufay Wojcicki A, Belissa É, Fontan J-E, de Pontual L, Nathanson S, Chevallier A, Laribe-Caget S, Boudy V. Dosage form suitability in vulnerable populations: a focus on paracetamol acceptability from infants to centenarians. PLoS One. 2019;14:e0221261.
81. Vallet T, Elhamdaoui O, Berraho A, Cherkaoui LO, Kriouile Y, Mahraoui C, Mouane N, Pense-Lheritier A-M, Ruiz F, Bensouda Y. Medicines acceptability in hospitalized children: An ongoing need for age-appropriate formulations. Pharmaceutics. 2020;12:766.
82. Emeryk A, Vallet T, Wawryk-Gawda E, Jędrzejewski A, Durmont F, Ruiz F. Acceptability of a sublingual drug formulation for respiratory tract infections in children aged 3 to 5 years. Pharmaceutics. 2021;13:294.

83. Saito J, Miyamoto S, Yamada M, Yamatani A, Ruiz F, Vallet T. Adherence and acceptability of an oral antibiotic used for the prevention of pediatric urinary tract infection in Japan. Pharmaceutics. 2021;13:345.
84. Vallet T, Bensouda Y, Saito J, Mathiesen L, Pokharkar V, Klingmann V, Peak M, Elhamdaoui O, Yamatani A, Ivanovic I, et al. Exploring acceptability drivers of oral antibiotics in children: findings from an international observational study. Pharmaceutics. 2021;13:1721.
85. Pokharkar V, Sajith M, Vallet T, Akshantal S, Shah R, Ruiz F, Salunke S. Acceptability of different oral dosage forms in paediatric patients in hospital setting. Arch Dis Child. 2021:archdischild-2021-322604. https://doi.org/10.1136/archdischild-2021-322604.
86. Perez F, Vallet T, Bravo Z, Callahan K, Ruiz F. Acceptability of Mebendazole chewable tablet in children aged 2 to 4 years in Peru. Pharmaceutics. 2022;14:27.
87. Klingmann V, Vallet T, Münch J, Stegemann R, Wolters L, Bosse H-M, Ruiz F. Dosage forms suitability in pediatrics: acceptability of analgesics and antipyretics in a German hospital. Pharmaceutics. 2022;14:337.
88. Skwierczynski C, Conroy S. How long does it take to administer oral medicines to children? Paediatr Perinat Drug Ther. 2008;8:145–9.
89. Shah T, Tse A, Gill H, Wong I, Sutcliffe A, Gringras P, Appleton R, Tuleu C. Administration of melatonin mixed with soft food and liquids for children with neurodevelopmental difficulties. Dev Med Child Neurol. 2008;50:845–9.
90. Akram G, Mullen AB. Paediatric nurses' knowledge and practice of mixing medication into foodstuff. Int J Pharm Pract. 2012;20:191–8.
91. Adams LV, Craig SR, Mmbaga EJ, Naburi H, Lahey T, Nutt CT, Kisenge R, Noel GJ, Spielberg SP. Children's medicines in Tanzania: a national survey of administration practices and preferences. PLoS One. 2013;8:e58303.
92. Martir J, Flanagan T, Mann J, Fotaki N. Recommended strategies for the oral administration of paediatric medicines with food and drinks in the context of their biopharmaceutical properties: a review. J Pharm Pharmacol. 2016;69:384–97.
93. Bryson SP. Patient-centred, administration friendly medicines for children - an evaluation of children's preferences and how they impact medication adherence. Int J Pharm. 2014;469:257–9.
94. Jones DJ, Kronemer N. Parents' response to a drug study: a clinical study of the acceptability and efficacy of pediatric forms of erythromycin. Clin Pediatr. 1965;4:137–40.
95. European Medicines Agency. Letter of Support for an Acceptability Score Test in relative acceptability testing for oral medicines in children under 12 years of age. Available online https://www.ema.europa.eu/en/documents/other/letter-support-acceptability-score-test-relative-acceptability-testing-oral-medicines-children-under_en.pdf. Accessed 03 Feb 2023.

Chapter 5
Palatability of Oral Dosage Form: Considerations for Pediatric Patients

Anne-Marie Pensé-Lhéritier and Leonie Wagner-Hattler

Abstract The sense of taste is defined as a part of the sensory system in which taste receptors are able to detect a variety of chemical molecules and provide valuable information allowing the discrimination of five basic tastes: sweet, sour, salty, umami, and bitter. Humans have about 25 different receptors for bitter taste that belong to the G-protein-coupled receptor superfamily. They can detect at least 1000 chemicals and herbal compounds. Among these, many active pharmaceutical ingredients (API) are characterized by an unpleasant taste that hinders patients' willingness to take their medicinal products. Based on the FDA draft guidelines, palatability has been defined as "the quality of a pharmaceutical product that makes it agreeable or acceptable in taste, aftertaste, odor and texture." It is therefore essential during the development of a pediatric drug to evaluate whether an adverse taste is expected, as well as the mouthfeel, in order to ensure formulation palatability.

This chapter focuses on the mechanisms of interaction between the API, the galenic forms, and all the orosensory receptors and provides a comprehensive overview of the state-of-the-art research on palatability of oral dosage forms. Moreover, an overview is given about available in vivo test methods in order to present the meaning of evaluating sensory perception.

These are the Human Taste Panel, Sip and Spit Studies, and the BATA model (Brief-access taste aversion). For the in vitro tools, e-tongue model and cell-based model are shown. Less common methods such as amoeba-based assessment or methods using zebrafishes are also discussed. An additional part will discuss the options of in silico modeling of taste.

Keywords Taste · Smell · Mouthfeel · Human panel · In vitro/In vivo testing

A.-M. Pensé-Lhéritier (✉)
Independent Consultant, FRM GaleSens, Cergy, France
e-mail: annemarielheritier@frmgalesens.fr

L. Wagner-Hattler (✉)
Pharma Technical Development, F. Hoffmann-La Roche Ltd., Basel, Switzerland
e-mail: leonie.wagner@roche.com

H. Batchelor, K. Rose (eds.), *Pediatric Formulations*, AAPS Advances in the Pharmaceutical Sciences Series 58, https://doi.org/10.1007/978-3-031-77240-5_5

5.1 Introduction

Human senses are divided into five categories, and hence, sensory receptors are widespread throughout the body. With these clever systems it is possible to apprehend outside stimuli and convert them into signals accessible to our brain. Especially for children such stimuli are of high importance. As their sensory system is still developing, they might process these stimuli differently from adults.

Table 5.1 summarizes the different sensory organs and examples of receptors and stimuli [1].

The first two senses of vision and audition allow distant interaction. Indeed, activated by sound or light waves, these senses allow an appreciation of the stimuli without a physical contact with it. The third sense, olfaction is an intermediate sense because the smell can be perceived by being at a certain distance from the olfactory source. The last two senses lead to direct interaction; in fact, the stimulus must be in contact with the body or mouth.

Stimulation of sensory receptors results in the opening or closing of ion channels and generation of a potential by the motion of ions. Information is transmitted via afferent pathways to the central nervous system, which integrates such information and transmits signals back to effector systems [2]. The senses are the real gateways to the information essential to mental activity. With time, this cascade of stimuli becomes memories and learning. Subjected to these continuous external stimuli, the brain builds or makes internal experiences at the same time. A child constructs their reality of the outside world by the perceptions of all their senses.

Moreover, all senses are capable of generating basic emotions: anger, disgust, fear, sadness, and enjoyment [3]. These emotions are organized by brain systems that adaptively respond to two basic types of stimulation, appetitive and aversive, and support an integrated model of emotion [4]. But in any case, before any emotion of disgust, it is imperative that the sensory organ is activated by a sufficient number of stimuli.

Pediatric medicines are made up of a mixture of active pharmaceutical ingredients (API) and excipients. All these ingredients and the dosage form represent many stimuli for the sensory receptors. However, the most important senses activated during peroral drug administration are the ones activated by stimulation in the mouth: the so-called orosensory senses [5]. When a patient puts a drug in the mouth, a range

Table 5.1 A summary of the sensory organ with examples of receptors and stimuli

Sense	Organs	Examples of receptors	Examples of stimuli
Vision	Eye	Rod and cone	Light
Audition	Ear	Hair cells	Sound
Smell	Nose	Olfactive receptors	Molecules with low pressure tension
Touch	Skin and mucosa	Merkel cells/ thermoTRPs	Pressure/temperature
Taste	Mouth	Taste receptors	Molecules soluble in saliva such as salt, acid, etc.

of sensory inputs almost simultaneously reach the central nervous system via the peripheral gustatory, olfactory, and somatosensory pathways. The final decision to ingest or reject the drug will take place following an analysis of the combination of each input but also on the actual combination of segregated intakes from the oral cavity.

First, some definitions are provided that will be used throughout this chapter. Then the orosensory experiences in the child will be discussed with a greater focus on taste and bitterness. The methods that make it possible to qualify and quantify sensory experiences will then be summarized before ending with a reflection on the challenges of palatability.

5.2 Definitions

In order to reduce/avoid any misunderstanding and ambiguity the definition of important terms, such as palatability, and the different sensory features influencing it are provided.

5.2.1 Palatability

In the Guideline on pharmaceutical development of medicines for pediatric use "Palatability is defined as the overall appreciation of an (often oral) medicinal product in relation to its smell, taste, aftertaste and texture (i.e. feeling in the mouth)" [6]. The FDA noted another definition in their draft guideline about the use of liquids and/or soft food that states the following: "Palatability is defined as the quality of a drug product that makes it pleasant or acceptable in terms of taste, after-taste, smell, and texture" [7]. Both regulators agree on palatability's criticality in importance for the successful treatment with oral dosage forms, especially in the pediatric field. It is one of the factors affecting patient adherence [8, 9].

5.2.2 Taste

Taste is an important aspect influencing palatability in particular in the pediatric population. The perception can be divided into five tastes: sweetness, sourness, saltiness, bitterness, and umami. The most crucial one in drug development is the bitterness (see Chap. 20). Interestingly, there are some physiological differences between adult and pediatric taste perception. The size of children's papillae is smaller and higher in density compared to adults [10, 11]. Therefore, it is suspected that children have a higher sensitivity to sweetness during childhood.

5.2.3 Aftertaste

Aftertaste is the stimulus intensity perceived in the moments immediately following removal of the product. In other areas of perception, such as visual and auditory, this type of information is not meaningful because the sensation of the stimulus after disappearance declines very rapidly. In taste perception, however, there is a clear aftereffect that can vary strongly over time. Aftertaste also revealed significant temporal effects concerning bitterant substances. Reasons for these effects are not known but may suggest differential expression of bitter taste receptors or differences in bitter agonist-receptor binding affinity across tongue regions [12].

5.2.4 Smell

Smell is defined as the ability to perceive odor or scent through the nose, having olfactory nerves process the chemicals. The sense of smell impacts many fields of life. The ability to smell/detect harmful odors from the environment can help to protect against dangerous situations [13]. Also, when we enjoy a meal or smell the atmosphere of fresh rain, positive emotions in the brain are created. Research shows that olfaction also plays a role in the formation of memories [14]. Hence, the ability to smell contributes to health and quality of life. The function of the sense of smell can be categorized as a range of normal sensations (normosmic), diminished sensations (hyposmic), or absent ability (anosmic) to detect or label scents [13]. For a substance to be detected by the olfactory nerve, two pathways exist: one directly through the nose and the second one where the odorants reach the olfactory nerve via nasopharynx. The second pathway plays an important role in detecting smells from food or drinks; hence, it is closely interconnected with taste sensation [15]. That is the reason why olfaction also plays a role in palatability.

5.2.5 Texture

Texture has been defined by the International Standards Organization (ISO) in their standard vocabulary for sensory analysis as "All the rheological and structure (geometrical and surface) attributes of a food product perceptible by means of mechanical, tactile, and where appropriate, visual and auditory receptors" [16].

In pediatrics, multiparticulates have attracted a lot of attention because of the improved ease of swallowing and flexible dose titration. However, oral grittiness (i.e., rough mouthfeel) may arise from the presence of particles in the mouth, limiting palatability and indicating that taste is not the unique system determining the acceptability of a medicine.

5.3 The Pediatric Population and Sensory Experiences

The sensory experience of children in contact with an oral medicine has multiple aspects because, on the one hand, the oral cavity is covered with many receptors characterizing the taste and texture of the preparation, and on the other hand, the nasal cavity including olfactory receptors detects odorous molecules [17]. The sensory system develops in the fetal stage of life. Taste receptors in the taste buds start developing from the 13th week in fetuses and are mature before birth. In the human embryo, the taste buds can be spotted very early, and their number increases until birth. The fetus discovers the taste qualities of its amniotic fluid through its ingestion and regurgitation, which is a physiological process during pregnancy. The sense of touch develops around the 14th week of gestation. A large number of mechanical stimulation receptors are present on a large part of the skin surface of the fetus, particularly on the face and the perioral region. Electrophysiological stimuli in preterm infants show that the afferent somesthetic pathways are functional up to the primary cortex at 25 weeks of gestation. We can therefore assume that this is also true for receptors in the mouth. During the third trimester of gestation, all the chemosensory structures of the nose (olfactory and trigeminal perception) appear ready to function in our species, but there is no direct demonstration of fetal olfactory reactivity [18]. Only recently, the first direct evidence of human fetal responsiveness to flavors transferred via maternal consumption was published in 2022 [19]. On the other hand, behavioral responses to chemosensory stimulations are observed in preterm infants, tested shortly after birth, from 6 months of gestational age. A hedonic responsiveness of at least 31 weeks post-conception newborns to odors has been evidenced: smelling vanilla induces respiratory acceleration and more appetitive reactions (e.g., licking, sucking), whereas smelling butyric acid induces a decrease of respiration rates and facial expressions typical of disgust (e.g., wrinkling nose, raising upper lip, turning head away) [20].

The taste system allows the detection of five sensory qualities: salty, sweet, bitter, sour, and umami. The taste of chemical is detected by taste cells that transmit information to the gustatory nerves and subsequently to the central nervous system. The principal organs of chemosensoric recognition are the taste buds. It is estimated that a single human's taste papillae contain 2000 to 5000 taste buds. Children have substantially smaller papillae than adults but significantly higher papilla densities and consequently significantly more taste buds [10]. This difference in density may explain a greater sensitivity to taste for children. The existence of several different taste qualities implies that each taste quality has a specific coding mechanism mediated by specialized taste receptors. Reception of taste qualities that humans describe as sweet, umami, and bitter involves proteins from the T1R and T2R families. The interaction between bitter substances and their receptor is described in more detail in the next paragraph. Sour taste is detected by a small subset of cells that are distributed across all taste buds called Type III taste receptor cells. H+ ions (protons) that are abundant in sour substances can directly enter the Type III taste cells through a proton channel [21]. In contrast, cells and signaling pathways for salty taste are still incompletely understood. Latest research leads to the hypothesis that in the oral cavity Na + ions enter taste receptor cells through epithelial sodium channel, leading to membrane depolarization and the generation of action potentials [22].

The somatosensory system [23] encompasses mechanoreceptors and thermoreceptors in the oral cavity that give information about the shape, size, texture, and temperature of material during oral exploration. The qualities of mechanoreceptors are similar to fingertips, and they are evenly organized in sensory plexuses with a very high density on the tongue and the palate. Four mechanoreceptors are impacted in pressure sensing and spatial localization of mechanical inputs: Meissner's corpuscles, Pacini's corpuscles, Merkel's discs, and Ruffini's corpuscles [24]. Their actions enable characterization of the grittiness, the hardness, and the viscosity of the product in the mouth, features influencing the swallowability.

Whereas mechanosensitive neurons have receptor endings specialized to detect specific mechanical input, thermoreceptors are believed to have free nerve endings that respond to non-noxious heat and cold variation but also to various types of noxious stimuli, including those of chemical origin. Thus, the primary function of oral thermoreceptors may be exteroceptive—to represent the temperature of material present in the mouth. A secondary function appears to involve the chemesthesis system, defined as the sensitivity of mucosal surfaces to environmental chemicals. Chemosensation refers to a biochemical and sensory process of recognition of chemical molecules, by specialized transduction pathways in addition to those of smell and taste. Chemicals that elicit chemesthesis activate the transient receptor potential (TRP) family of cation channels expressed in Table 5.2.

The evaluation of the taste of drugs involves a last sensory system that is not located in the mouth but in the nose: the olfactory system. Olfactory perception originates in the interaction between volatile chemical compounds transported by inhaled air, directly or retro nasally, and odor receptors (ORs). The family of transmembrane proteins believed to be olfactory receptors (ORs), along with some of their genes [26], were elucidated in 1991 by Linda Buck and Richard Axel. The presence of a repertoire of more than 1000 ORs enables the identification of a wide range of odorants, further amplified by the distinct nature of the OR–odorant interaction. As a matter of fact, each OR can recognize several odorants, and, in turn, each odorant can bind and activate several ORs [27]. Several studies were carried

Table 5.2 Thermo- and chemosensitive transient receptor potentials' (TRPs) Temperature sensitivity and Chemical sensitivity [25]

Receptors	Temperature sensitivity	Chemical sensitivity
TRPV1 (VR1)	> 43 °C	Capsaicin, acid, ethanol, eugenol, camphor
TRP V2 (VRL-1)	> 52 °C	
TRPV3	> 30 °C	Camphor, carvacol, eugenol, thymol
TRPV4	> 24 °C	Acid
TRPM8	< 25 °C	Menthol
TRPA 1	< 18 °C	Gingerol, allicin

out to try to see if the detection threshold of olfactory molecules was different between children and adults. It has been shown that the children score in a similar manner than the adults, and sometimes even young children (from 4 years old) can be less discriminating than adults. Finally, children are also mature concerning the evaluation of olfactive irritant and potentially toxic chemicals, thanks to the chemosensory trigeminal function [28]. The taste system, acting in concert with the olfactory and trigeminal systems, indicates whether a product should be ingested.

5.4 Interaction API and Bitterness Receptor

The perception of bitterness is a vital mechanism for humans from birth onwards to avoid the ingestion of dangerous substances found in the environment. This perception of disgust was shown in 1979 by Steiner, who video recorded young babies prior to their first feeding experience. They tasted sweet, sour, and bitter solutions. Infants in contact with the bitter solution expressed behaviors like tight closing of the eyes, gaping mouth, and sudden turn of the head [29] (Fig. 5.1).

After this experiment it took a few decades until the bitterness receptors were identified. Only in 1988 was it possible in connection with the decoding of the human genome. Publications on this topic began in the 2000s, and so far about 25 receptors have been shown to detect bitter substances. They belong to the human TAS2 receptor (TAS2R) gene family [30], a specific family within the G-protein-coupled receptor superfamily. At the time of discovery, it was assumed that they are only located in the taste buds of the tongue [31]. Since then, research has indicated

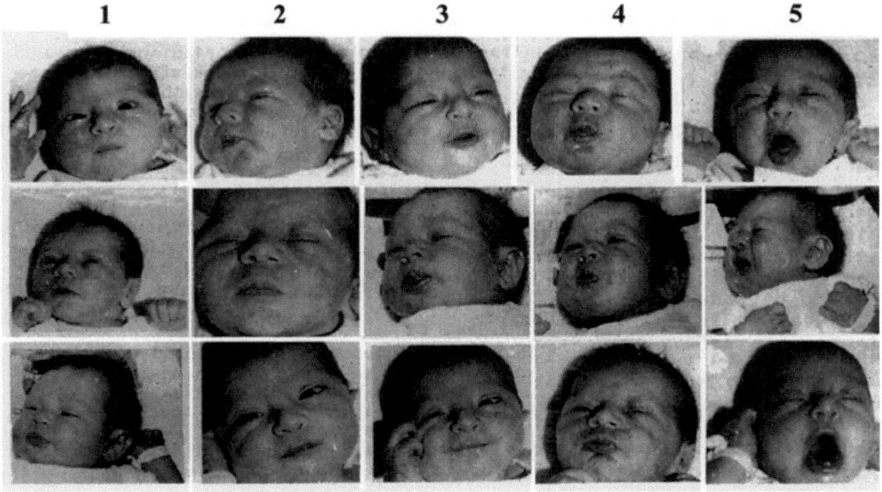

Fig. 5.1 Neonates displaying variable facial responses to basic tastes (Reproduced from Steiner. Ask permission). Key: 1 rest, 2 water, 3 sweet, 4 sour, 5 bitter

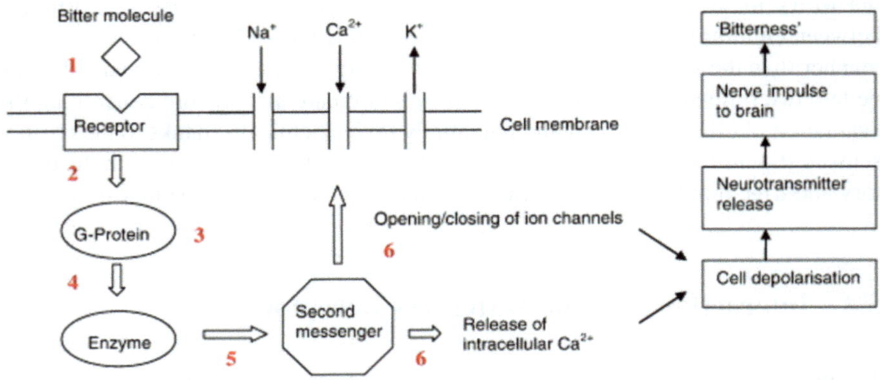

Fig. 5.2 Taste transduction. (Derived from McGregor, 2007 [15, 34])

that these receptors are also present in other parts of the body such as the respiratory system or the gastrointestinal tract [32].

The activation of TAS2R is summarized [33] (Fig. 5.2):

- The activation of TAS2Rs by bitter agonists leads to activation of a G-protein. Upon dissociation of activated G-protein, βγ subunits activate phospholipase Cβ2 (PLCβ2), resulting in the production of the second messenger inositol 1,4,5-trisphosphate (IP3).
- As a next step IP3 binds to its receptor residing in the membrane of the endoplasmic reticulum, which, when opened, allows the flow of calcium ions stored in the lumen of the reticulum into the cytoplasm [6].
- The increased cytosolic calcium level triggers the opening of the channel in the plasma membrane, resulting in an influx of extracellular sodium ions causing depolarization of the cell, which is followed by voltage-gated action potentials.

Overall, it is estimated that several substances, including fatty acids, peptides, amino acids, amines, amides, alkaloids, glycosides, and flavonoids, are capable of activating the transduction mechanism of bitterness receptors [35]. So how can it be explained that only 25 bitterness receptors have been identified so far? This is because each single bitter compound is capable of activating multiple TAS2Rs, and each TAS2R can be activated by multiple bitter compounds.

Among those receptors, TAS2R10 and TAS2R46 stand out because each of them responds in vitro to a large set of tested bitter compounds. Although TAS2R10 exhibits only 34% amino acid sequence identity with TAS2R46, it also responds to these compounds [35].

A Bitter database (http://bitterdb.agri.huji.ac.il) was introduced in 2012 as a central resource for information on bitter-tasting molecules and their receptors. BitterDB holds over 1000 bitter molecules, including the information of the bitter receptor associated with the substance (if available). For 270 molecules, at least one associated bitter taste receptor is reported. Among those, a few active

Table 5.3 List of TAS2Rs receptors known for some API (http://bitterdb.agri.huji.ac.il)

API	Receptors
Haloperidol	TAS2r 10, TAS2r 14
Chloroquine	TAS2r 3, TAS2r 7, TAS2r 10, TAS2r 14, TAS2r 39
Quinine	TAS2r 4, TAS2r 7, TAS2r 10, TAS2r 14, TAS2r 39, TAS2r 40, TAS2r 43, TAS2r 44, TAS2r 46,
Erythromycine	TAS2r 10
Caffeine	TAS2r 7, TAS2r 10, TAS2r 14, TAS2r 43, TAS2r 46

pharmaceutical ingredients (APIs) are presented with their receptors (see Table 5.3), but none of them have been qualified with respect to their sensory qualities.

Due to the lack of data concerning the bitterness of pharmaceutical active ingredients, it is important to be able to obtain this information in order to improve the palatability of drugs. Different methods exist and are presented in the following paragraphs of this chapter.

5.5 Evaluation Methodologies for Palatability

he necessity to measure the palatability of pediatric medicinal products has led to the emergence of a broad range of evaluation methods. The following paragraph shows a summary of methods.

5.5.1 Human Panel

The historical method to evaluate the taste, odor, or texture of medicine is sensory evaluation which is defined as "Examination of organoleptic attributes of a product by the sense organs" [16]. Sensory evaluation requires a group (panel) of human volunteers (panelists) who take part in a sensory test. The panelists can be "naive assessors" who do not have any training in sensory test experience. Another approach is to have "expert panelist" assessing the substance or formulation of interest. They are trained, discriminant, and reproducible in their evaluation. The tests can be carried out but adult or children panels. Some example sensory panels are listed in Table 5.4.

This method is recognized as the most effective way to guide development and validate sensory aspects [39]. However, there are some limitations: panelists participating in sensory research for medicinal products need to be recruited, not only on their sensory acuity but also on additional strict medical inclusion and exclusion criteria [40] (health status, use of other medication, consumption of alcohol/drugs, pregnancy, smoking, allergies). In addition, for obvious ethical reasons, the regulatory authorities suggest using adult panels. The differences in sensation between

Table 5.4 : examples of sensory studies for pediatric medicine

Sensation	API and dosage form	Sensory method
Bitterness	Mefloquine hydrochloride microparticles [36]	Analytical method: Quantitative analysis of bitterness—20 adults: Rating scale in 8 points from tasteless to strongly bitter
Grittiness, mouthfeel, taste, and sample volume	Placebo multiparticulate formulation	Affective method: Hedonic test—71 children, 62 adults. 5-point hedonic scale [37]
Bitterness, sweetness, roughness	Loperamide hydrochloride preparation marketed (granule, dry syrup, powder)	Analytical method: quantitative analysis of sensory attributes—41 adults. Structured rating scale from 0 to 10 [38]

adults and children are still poorly understood, but the literature certainly indicates a greater sensitivity in young children [41]. Finally, for safety reasons of certain APIs, it is not possible to have access to the sensory qualities of an active ingredient at the beginning of its development with a human panel. In order to overcome these limitations, alternative methods have been developed.

5.6 In Vitro Testing

5.6.1 E-tongue

Electronic tongue measurements are described in literature as one of the main in vitro methods for palatability evaluation for medicines [42]. They are used to assess taste-masking options in formulation development but also present a valid method to assess blinding, when the API has unique taste characteristics that the placebo should possess as well.

Two types of commercially available e-tongue systems are reported in literature, the Alphamos and the Insent technology, a comparison is shown in Table 5.5 [39, 43–45], whereas elsewhere up to six different technologies are mentioned [43]. Although both systems are based on potentiometric measurements, the technology used for the sensors differs [46]. In all cases, the sample needs to be in a liquid state and particle free. That means, for the investigation of solid formulations, the test formulation needs to be dissolved prior assessment. Literature shows that there is a good correlation between the in vitro electronic tongue measurements and the human taste assessment [39, 46]. For investigating a novel substance, usually a set of different concentrations is measured in order to see whether there is a linear relationship between sensor response and testing solution or whether the sensor is saturated by the substance [46]. In order to put the results obtained with the e-tongue in perspective a well-known reference substance such as Quinine HCl is used as a reference, where also human taste data is available in literature. For investigational molecules it is not always possible to deduce the human response solely from the

Table 5.5 Comparison of the Insent and the Alphamos setup e-tongues

Sensors	Insent	Alphamos
Handling	Demounting; special storage	No demounting; storage in air
Lifetime[a]	Up to one year	One to three months
Limitations in terms of sensor damage	Only pH 2–8, no alcoholic solutions > 40%	Not observed
Performance qualification	Passed according to the ICH guideline Q2 [47]	Not passed according to the ICH guideline Q2
Measurement		
Principle	One sensor indicating one taste sensation, cross-selective	Cross-selective
Preparation	Sensor check (30 min)	"Conditioning," "calibration," "diagnostic" (5–24 h)
Sample volume	2× 40 ml	25 ml
Duration of each analysis	Exchange of sensors → duration ↑	Several short cycles
	2× 7 h for all sensors	6 h

Adapted from [46]

e-tongue results; prior knowledge about taste/palatability is necessary to properly interpret the obtained data. Furthermore, the sensors used in the commercially available systems are not sensitive to all APIs and/or sweeteners [39].

5.6.1.1 Insent Technology

This system can be equipped with different bitter receptors (up to three [46]), for cationic, neutral, and anionic bitterness, as well as a sensor for saltiness, umami, sweetness, and astringent substances. These sensors are complemented by an Ag/AgCl- reference electrode [43]. Even though the sensors are assigned to the different qualities of taste, every sensor can detect multiple taste qualities but is most sensitive to the quality of its designation [46]. With this technology the initial taste and also the aftertaste can be measured. The sensors of this system are temperature sensitive as the pores of the membranes loosen under higher ambient conditions [39].

5.6.1.2 Alphamos Technology

The sensors of this system are not assigned to specific taste characteristics. Each sensor therefore detects substances from the sample but at different intensities. The electrodes of the Astree2 electronic tongue use the chemically modified field-effect transistor technology (ChemFET). This technology is related to the seion selective FET technology with the exception that the electrodes are coated with specific materials [46]. On this system (Astree2) there are seven sensors and one reference sensor installed.

5.7 In Vivo Testing (Nonhuman)

Different in vivo models have been developed for evaluating the bitterness of several compounds from vertebrates (BATA, Zebrafish) to invertebrates (Amoeba) [39, 48].

5.7.1 BATA: Brief-Access Taste Aversion

The brief-access taste aversion (BATA) model is a test setup that allows to assess the taste of APIs as well as formulations. It is used in different fields [49–52]. The concept of assessing aversiveness is based on giving water-deprived rodents (mice or rats) brief access to the test solution. This access is done in a lickometer that allows to measure the number of licks that the rodent makes during the short period of time it has access to the solution. The period the rodent has access to the test solution in different sipper tubes is between 5 and 10 s. The number of licks the animal made of the test solution is then compared to the number of licks it makes with drinking water. Low numbers of licks represent aversiveness. Usually, different concentrations are tested, resulting in an aversion-concentration curve. This is done in very few animals [53]. Research has shown that the results of BATA tests in rates are well correlated with human taste assessments [54]; hence, it can be seen as a good predictive model for human taste perception. With this method it is possible to assess both pure API and also formulations [55, 56]. The suitability of using other models to investigate taste-related responses, particularly for identifying compounds with bitter taste liability, should be considered. The limitation of the method can be seen when a weakly basic compound is tested in a buffer system at low pH as the rat might perceive the acidic buffer medium already as aversive, independent of the compound.

5.7.2 Zebrafish

The zebrafish is a useful vertebrate model for the development of new drug molecules, due to their short generation span and ease of manipulation. The T2R receptors for bitter tastants have also been characterized in zebrafish as well as their rejection of certain bitter molecules such as denatonium. The model was therefore considered to screen bitter molecules by following the consumption of food containing different molecules [57].

5.7.3 Amoeba

Non-vertebrates such as the amoeba Dictyostelium are suitable to characterize new bitter tastants and drugs. To perform the evaluation, cell behavior is monitored by taking images prior to and after compound addition. The response of Dictyostelium to bitter tastants is to lose the typical amoeboid shape and round up, and in so doing, block membrane protrusion formation. The method with Dictyostelium allows the obtention of IC50 of bitterness. These values are in a similar range of compounds in the rat in vivo BATA test, showing a significant positive correlation between the two models [58].

5.8 In Silico Prediction

As aforementioned, bitter taste is one of the main challenges in drug development. Therefore, in silico prediction of aversive taste of active pharmaceutical ingredients as early as possible is key to the success of drug product development. Recent research has shown that it is possible to predict bitterness of compounds based on their structural elements [59].

5.9 Palatability: Challenges in the Pediatric Population

The challenges to formulate palatable drug products for children are versatile. Looking at the option to mask bitter taste in a formulation poses the question of what aroma can be used that is proven to not interact with the active ingredient and is also globally accepted from a regulatory standpoint. The assessment of regulatory acceptance can become tedious as aroma/flavors consist of many different components that need to be assessed individually. Of course there are also other options to be used to fight bitter sensations of APIs, and these can be barrier coatings or complex formers like cyclodextrins to be used with the API. Additionally, bitter blockers can be used, which block the receptor responsible for the sensation [60]. However, the availability of such materials that can be used in pediatric products is limited. Additionally, the pricing is a challenge as these ingredients are niche products.

Another challenge that researchers face is that the existing test methods are mostly unidimensional (e.g., taste assessment without any other information). Only human taste panels have the potential to deliver more information about the drug product (e.g., mouthfeel, swallowability, stickiness, etc.). A sensory panel is used at late timepoints during the drug product development; hence, if there are issues detected, this might lead to longer times to market and higher costs as changes in the formulation might trigger additional clinical studies.

5.10 Conclusions

The tools that can be used to assess palatability are limited, and not all of them work as well for both pure APIs and formulations. This increases the challenges in interpreting the results and also cross-referencing results obtained from different methods.

The availability of palatability data for APIs and/or formulations is very limited. Most of the time the palatability is assessed in adult panels, and literature has shown that adult and pediatric palatability can differ from each other.

References

1. Marzvanyan A, Alhawaj AF. Physiology, sensory receptors [Updated 2021 Aug 27]. StatPearls [Internet]. Treasure Island: StatPearls Publishing; 2022.
2. Mulroney SEP, Myers AKP. Sensory physiology. In: Netter's Essential Physiology; 2016. p. 57–73.
3. Mikels JA, et al. Emotional category data on images from the International Affective Picture System. Behav Res Methods. 2005;37(4):626–30.
4. Lang PJ, Bradley MM, Cuthbert BN. Emotion, motivation, and anxiety: brain mechanisms and psychophysiology. Biol Psychiatry. 1998;44(12):1248–63.
5. van Eck A, Stieger M. Oral processing behavior, sensory perception and intake of composite foods. Trends Food Sci Technol. 2020;106:219–31.
6. E.E.M.A.E.C.Q.R. Pharmaceutical development of medicines for paediatric use. 2012.
7. FDA. Use of liquids and/or soft foods as vehicles for Drug Administration: General considerations for selection and in vitro methods for product quality assessments. 2018.
8. Baguley D, L.E., Bevan A, et al. Prescribing for children—taste and palatability affect adherence to antibiotics: a review. Arch Dis Child. 2012;97:293–7.
9. Tucker CM, et al. Associations with medication adherence among ethnically different pediatric patients with renal transplants. Pediatr Nephrol. 2002;17(4):251–6.
10. Segovia C, et al. A quantitative study of fungiform papillae and taste pore density in adults and children. Dev Brain Res. 2002;138(2):135–46.
11. Stein N, Laing DG, Hutchinson I. Topographical differences in sweetness sensitivity in the peripheral gustatory system of adults and children. Dev Brain Res. 1994;82(1):286–92.
12. Higgins MJ, Hayes JE. Regional variation of bitter taste and aftertaste in humans. Chem Senses. 2019;44(9):721–32.
13. Boesveldt S, et al. Anosmia—a clinical review. Chem Senses. 2017;42(7):513–23.
14. Herz RS. The emotional, cognitive and biological basics of olfaction. In: Sensory marketing: Research on the sensuality of products; 2010. p. 87–107.
15. Patel RM, Pinto JM. Olfaction: anatomy, physiology, and disease. Clin Anat. 2014;27(1):54–60.
16. ISO. I.O.f.S., Sensory analysis - Vocabulary ISO 5492. 2008: p. 148.
17. de Araujo IE, Geha P, Small DM. Orosensory and homeostatic functions of the insular taste cortex. Chemosens Percept. 2012;5(1):64–79.
18. Schaal B, Marlier L, Soussignan R. Olfactory function in the human fetus: evidence from selective neonatal responsiveness to the odor of amniotic fluid. Behav Neurosci. 1998;112(6):1438–49.
19. Ustun B, et al. Flavor sensing in utero and emerging discriminative behaviors in the human fetus. Psychol Sci. 2022;33(10):1651–63.

20. Schaal B, Hummel T, Soussignan R. Olfaction in the fetal and premature infant: functional status and clinical implications. Clin Perinatol. 2004;31(2):261–85.
21. Decoursey TE. Voltage-gated proton channels and other proton transfer pathways. Physiol Rev. 2003;83(2):475–579.
22. Liman ER. Salty taste: from transduction to transmitter release, hold the calcium. Neuron. 2020;106(5):709–11.
23. Haggard P, de Boer L. Oral somatosensory awareness. Neurosci Biobehav Rev. 2014;47:469–84.
24. Park B, Kim SJ. Cooling the skin: understanding a specific cutaneous thermosensation. J Lifestyle Med. 2013;3(2):91–7.
25. Simons CT, Carstens E. 3.20 - Oral chemesthesis and taste, in the senses: a comprehensive reference 2nd ed., Fritzsch B, edtor. Oxford: Elsevier; 2020. p. 398–422.
26. Miller G. Axel, Buck share award for deciphering how the nose knows. Science. 2004;306(5694):207.
27. Nara K, et al. A large-scale analysis of odor coding in the olfactory epithelium. J Neurosci. 2011;31(25):9179–91.
28. Hummel T, et al. Intranasal trigeminal function in children. Dev Med Child Neurol. 2007;49(11):849–53.
29. Steiner JE. Human facial expressions in response to taste and smell stimulation. Adv Child Dev Behav. 1979;13:257–95.
30. Meyerhof W. Elucidation of mammalian bitter taste. Rev Physiol Biochem Pharmacol. 2005;154:37–72.
31. Meyerhof W, et al. The molecular receptive ranges of human TAS2R bitter taste receptors. Chem Senses. 2009;35(2):157–70.
32. Rozengurt E. Taste receptors in the gastrointestinal tract. I. Bitter taste receptors and alpha-gustducin in the mammalian gut. Am J Physiol Gastrointest Liver Physiol. 2006;291(2):G171–7.
33. Behrens M, Ziegler F. Structure-function analyses of human bitter taste receptors—where do we stand? Molecules. 2020;25(19):4423.
34. Jennifer W, et al. Playing hide and seek with poorly tasting paediatric medicines: do not forget the excipients. Adv Drug Deliv Rev. 2014;73:14–33.
35. Meyerhof W, et al. The molecular receptive ranges of human TAS2R bitter taste receptors. Chem Senses. 2010;35(2):157–70.
36. Shah PP, et al. Design and optimization of mefloquine hydrochloride microparticles for bitter taste masking. AAPS PharmSciTech. 2008;9(2):377–89.
37. Lopez FL, et al. Acceptability of placebo multiparticulate formulations in children and adults. Sci Rep. 2018;8(1):1–10.
38. Inoue Y, et al. Human sensory testing of loperamide hydrochloride preparations for children to improve their palatability. World J Pharm Pharm Sci. 2015:570–9.
39. Clapham D, et al. Proposed tool to compare and assess the applicability of taste assessment techniques for pharmaceuticals. J Pharm Sci. 2022;111(5):1219–23.
40. Seince N, Sheehan L, Shine M. Evaluating the sensory properties of medicinal products. In: Nonfood sensory practices. Elsevier; 2022. p. 331–45.
41. Mennella JA, Bobowski NK. The sweetness and bitterness of childhood: insights from basic research on taste preferences. Physiol Behav. 2015;152:502–7.
42. Guedes MDV, et al. The use of electronic tongue and sensory panel on taste evaluation of pediatric medicines: a systematic review. Pharm Dev Technol. 2021;26(2):119–37.
43. Pein M, et al. Independent comparison study of six different electronic tongues applied for pharmaceutical analysis. J Pharm Biomed Anal. 2015;114:321–9.
44. Zheng JY, Keeney MP. Taste masking analysis in pharmaceutical formulation development using an electronic tongue. Int J Pharm. 2006;310(1):118–24.
45. Immohr LI, et al. Suitability of E-tongue sensors to assess taste-masking of pediatric liquids by different beverages considering their physico-chemical properties. AAPS PharmSciTech. 2017;18(2):330–40.

46. Woertz K, et al. A comparative study on two electronic tongues for pharmaceutical formulation development. J Pharm Biomed Anal. 2011;55(2):272–81.
47. Woertz K, et al. Performance qualification of an electronic tongue based on ICH guideline Q2. J Pharm Biomed Anal. 2010;51(3):497–506.
48. Mohamed-Ahmed AHA, et al. Non-human tools for the evaluation of bitter taste in the design and development of medicines: a systematic review. Drug Discov Today. 2016;21(7):1170–80.
49. Devantier HR, et al. Quantitative assessment of TRPM5-dependent oral aversiveness of pharmaceuticals using a mouse brief-access taste aversion assay. Behav Pharmacol. 2008;19(7):673–82.
50. Clapham D, et al. Assessing taste without using humans: rat brief access aversion model and electronic tongue. Int J Pharm. 2012;435(2):137–9.
51. Glendinning JI, et al. Contribution of α-gustducin to taste-guided licking responses of mice. Chem Senses. 2005;30(4):299–316.
52. Contreras RJ, Carson CA, Pierce CE. A novel psychophysical procedure for bitter taste assessment in rats. Chem Senses. 1995;20(3):305–12.
53. Soto J, et al. Development of a model for robust and exploratory analysis of the rodent brief-access taste aversion data. Eur J Pharm Biopharm. 2015;91:47–51.
54. Soto J, et al. Assessing the bitter taste of medicines: a comparison between rat taste panels (via the brief-access taste aversion (BATA) model) and human taste panels. Int J Pharm. 2016;511(2):1127–8.
55. Münster M, et al. Comparative in vitro and in vivo taste assessment of liquid praziquantel formulations. Int J Pharm. 2017;529(1):310–8.
56. Soto J. Assessing the feasibility of using an animal model for in vivo taste assessment of pharmaceutical compounds and formulations. UCL (University College London); 2016.
57. Oike H, et al. Characterization of ligands for fish taste receptors. J Neurosci. 2007;27(21):5584–92.
58. Cocorocchio M, et al. Bitter tastant responses in the amoeba Dictyostelium correlate with rat and human taste assays. ALTEX. 2016;33(3):225–36.
59. Bahia MS, Nissim I, Niv MY. Bitterness prediction in-silico: a step towards better drugs. Int J Pharm. 2018;536(2):526–9.
60. Andrews D, et al. Bitter-blockers as a taste masking strategy: a systematic review towards their utility in pharmaceuticals. Eur J Pharm Biopharm. 2021;158:35–51.

Chapter 6
Taste Masking Technologies

Kavil Patel, Rand Turki, Samar Issa, and Fang Liu

Abstract The taste and flavour perception of active pharmaceutical ingredients (APIs), particularly bitterness, is a contributing factor to acceptability of pharmaceutical oral dosage forms. This is especially significant for paediatric patients who have an increased sensitivity to aversive tastes compared to adults. Various strategies can be employed to mask the aversive taste of oral formulations, as such improving acceptability of paediatric medicines. These strategies can be categorised as either obscuring taste sensors or taste concealing. For obscuring the taste sensors, the taste pathway can be exploited using blocking or competing agents, such as bitter blockers, or sweeteners and flavouring systems which are the most commonly applied taste masking approaches in approved oral paediatric products, the majority being liquids. Taste concealing can be achieved by physical separation of the API and taste receptors via entrapment or encapsulation technologies, for example, film coating, use of lipids, ionic resin complexation, and spray congealing. It is critical to evaluate the effectiveness of the taste masking approach during product development by employing in vivo (e.g. human panels and animal models) and/or in vitro methods (e.g. dissolution and sensor-based approaches). Standardisation of the methods requires further exploration for reproducible and reliable assessments in paediatric medicine development.

Keywords Palatability · Taste masking · Coating · Paediatrics

K. Patel
Fluid Pharma Ltd, Maclaurin Building, Bishop Square, Hatfield, UK

R. Turki · F. Liu (✉)
University of Hertfordshire, Department of Clinical, Pharmaceutical and Biological Sciences, College Lane, Hatfield, UK
e-mail: f.liu3@herts.ac.uk

S. Issa
Ecole de Biologie Industrielle - EBI, UPR EBInnov®, Cergy-Pontoise, France

H. Batchelor, K. Rose (eds.), *Pediatric Formulations*, AAPS Advances in the Pharmaceutical Sciences Series 58, https://doi.org/10.1007/978-3-031-77240-5_6

6.1 Introduction

Most active pharmaceutical ingredients (APIs) have unpleasant or obnoxious tastes, including bitter, salty, or metallic tastes. This can influence patient acceptance of the medicine, and therefore, taste masking is an important aspect to consider when developing oral dosage forms, especially for children who have an increased sensitivity to aversive tastes compared to adults [1].

The human taste pathway involves the sensory buds, considered as chemoreceptors, where the ingested food or drugs are chemicals dissolved in saliva. Shaikh et al. describe the different factors that make taste a multidimensional experience involving cultural, genetic, and memorised gustatory experiences [2]. The taste physiology pathway is described in Fig. 6.1:

Taste masking is a known strategy to modify or prevent the interactions of the API with the taste receptors and thus intercept the taste pathway. Prior to taste masking, various factors must be understood, including the API solubility, dose, detectable/aversive concentration in the oral cavity, particle size/morphology, and the desired release profile. Understanding these factors brings insight to select the appropriate taste masking strategy, ensures effective taste masking without compromising bioavailability, and influences the choice of the final product presentation.

An acceptable taste may be achieved by modifying the physicochemical properties of the API, for example, by changing the salt form to reduce solubility. This

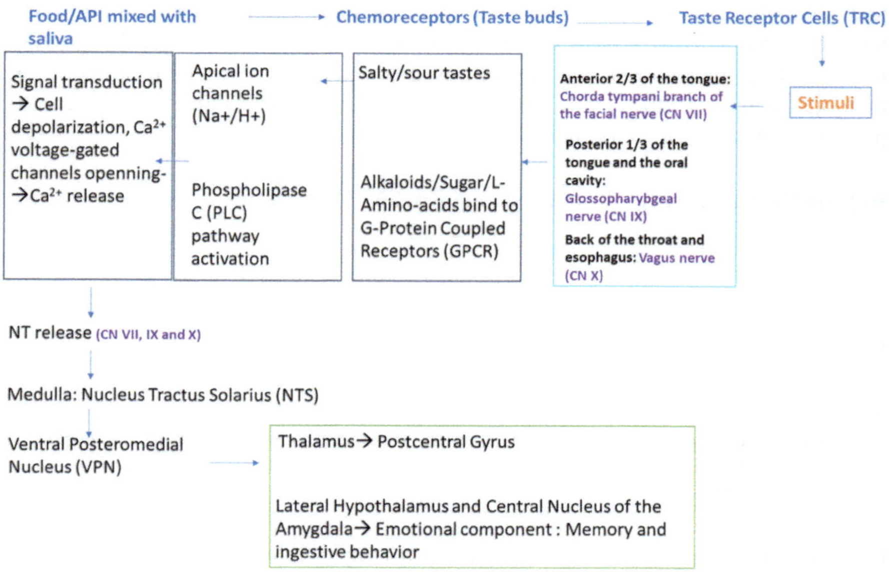

Fig. 6.1 Human physiological taste pathway. (Adapted from the text described by Shaik et al. 2023) (CN Cranial Nerve) [2]

taste masking approach by chemical modification of the API is not included in this chapter. Formulation related to taste masking techniques include:

1. Obscuring the taste sensors through
   - Bitter taste blockers to compete with the API for binding with taste receptors
   - Strong pleasant sensations to compensate the aversive taste of the API (e.g. using sweeteners, flavouring systems, and food)
2. Taste concealing via creating a barrier between the API and the taste receptors including
   - Molecular barriers such as ionic exchange resin and cyclodextrin complexations
   - Physical barrier technologies such as utilising lipids, film coating, coacervation, spray congealing, and spray drying

6.2 Obscuring the Taste Sensors

6.2.1 Bitter Blockers and Taste Modifiers

Type 2 Taste Receptors (T2Rs) are responsible for the bitter taste and are involved in different physiological responses, including defence response as well as the coordination of the metabolic regulation and behaviour [3].

The bitter taste is an alert stimulus to prevent threats related to consuming poisonous food or toxic ingested substances [4]. T2Rs are G-Coupled Protein Receptors (GCPR), not only identified on the tongue but in other tissues, especially expressed in the nose (solitar chemosensory cells), enteroendocrine cells (EECs) of the gastrointestinal tract, thyroid gland, ciliated epithelial cells in the bronchi, and urethra. The signal transduction (Fig. 6.1) of the T2Rs may be triggered, for some, by different bitter molecules, while some T2Rs are specifically activated by single ligands [4].

Bitter blockers are compounds which can directly antagonise T2Rs or the taste transduction cascade to reduce the bitter taste perception. Specifically, the activation of gustducin (taste-cell-specific G protein) is prevented, therefore leading to the inhibition of the taste perception. These compounds are usually close structural analogues of bitter ligands which allow them to still bind to these receptors. Examples of bitter blockers include salt types such as sodium acetate and endocannabinoids which have been shown to improve sweetness.

6.2.2 Sweeteners

Taste masking technologies using sweeteners are the foremost and simplest approaches, especially for paediatric formulations. There are several natural and synthetic sweeteners available; however, their effectiveness is limited for highly bitter APIs.

Different sweetener types are available such as:

- Sucrose: Most used sweetener, which is a disaccharide that can be readily hydrolysed within the intestine to the absorbable monosaccharides: fructose and glucose. Sucrose can cause a decrease in dental plaque pH and promote tooth enamel damage; therefore, formulations made with large quantities of sucrose should be avoided. For paediatric formulations, it should ideally be replaced by sugar-free sweeteners.
- Aspartame: It is an aspartic acid dipeptide and a phenylalanine methyl ester which is up to 200 times sweeter than sucrose.
- Sorbitol and xylitol: They are monosaccharides which are not readily absorbed from the gut and could be regarded as safe for patients with diabetes. They should be avoided in fructose-intolerant patients as they could lead to liver damage.
- Fructose: It causes a steady increase in blood glucose levels and should be avoided for patients with diabetes.

Sweeteners are not always sufficient as a lone approach and are typically used in combination with other taste masking technologies such as flavouring systems and film coatings. Using sweeteners with paediatric formulations has some limitations in terms of safety and regulatory aspects, especially for neonates and infants, due to differences in pharmacokinetics compared to adults. The regulatory bodies' guidelines should be followed when selecting sweeteners within a formulation intended for the paediatric population.

6.2.3 Flavouring Systems

Flavouring systems can be classified as natural flavours (such as raspberry juices, vanilla, peppermint, lemon, and orange flavours) and synthetic flavours (alcoholic solutions, powders, and aqueous solutions).

Syrups of sweet-tasting agents such as cinnamon, orange, cherry, and raspberry can be utilised in effectively masking salty and bitter tastes within drug products. Some flavours such as menthol and mint have a cooling effect which aids in reducing after-taste perception.

Any synthetic flavours selected for use within drug products must be deemed as Generally Recognised as Safe (GRAS) [5]. The acceptable daily intake (ADI) of sweeteners or flavours should also be considered. Exceeding the limits could lead to carcinogenic effects and higher blood sugar, amongst other adverse effects which should be avoided in paediatric patients.

Examples of paediatric products which utilise flavouring systems to improve palatability are shown below (Table 6.1).

6.2.4 *Others: Food*

Other techniques include the administration of medicines mixed with small amounts of food; according to Smith et al., some recommendations could be provided to caregivers and parents [6].

In general, recommendations would be provided to numb the taste buds with cold treats (ice cream, ice blocks), coat the taste buds with fatty food (peanut butter or chocolate), or prepare the children favourite drink to be given following the medicine administration.

Crushed tablets or capsule content (if approved by the local pharmacist or doctor) could be added on foods (custard or yoghurt) or small amounts of juice. Oral liquids can be mixed with small amounts of juice, under the condition that the entire dose is taken.

It is important to note that whether a medicine is suitable to be administered with food should be considered and investigated during product development, such as drug-food interactions and food effect on pharmacokinetic profiles. Supporting evidence is then used to form the basis of product labelling. In clinical practice, the administration method (such as tablet crushing, or sprinkling capsule contents on food) should follow the product labelling, e.g., the Summary of Product Characteristics (SmPC). Any unauthorised dosage form alteration could result in unlicensed use of medicines, leading to legal implications to the prescriber and/or administrator of the medicine.

6.3 Taste Concealing

6.3.1 *Film Coating*

Film coating technologies provide an opportunity to separate the bitter-tasting API from taste receptors using established manufacturing techniques. Additionally, film coating can improve acceptability by refining formulation appearance, odour, and mouthfeel.

To achieve effective taste masking, the film coating process should produce a complete and robust barrier between the API and taste receptors as well as resisting penetration of saliva which can initiate API dissolution. Ideally, this barrier should remain intact during formulation residence in the oral cavity but should not inhibit

Table 6.1 Examples of commercial paediatric products utilising flavouring systems (information taken from Summary of Product Characteristics)

Trade name	Active pharmaceutical ingredients	Sweetening and flavouring agents	Formulation	Intended patient age group	Year of authorisation
Vyvanse®	Lisdexamfetamine dimesylate	Mannitol, sucralose, artificial strawberry flavour	Chewable tablet	6–12 years	2017
Prexxartan™	Valsartan	Grape flavour, sodium citrate dihydrate, sucralose	Oral solution	>6 years	2017
Xatmep™	Methotrexate sodium	Sodium citrate, citric acid, sucralose	Oral solution	<18 years	2017
Xyzal® Allergy 24 h	Levocetirizine dihydrochloride	Saccharin sodium, grape or tutti frutti flavours	Oral solution	Perennial allergic rhinitis: 6 months–2 years Chronic idiopathic urticaria >6 years	2017
Macrilen™	Macimorelin acetate	Saccharin sodium	Oral suspension	2–18 years	2017
Carospir®	Spironolactone	Anhydrous citric acid, sodium citrate dihydrate, saccharin sodium, banana flavour	Oral suspension	0–17 years	2017
Emflaza®	Deflazacort	Sorbitol	Oral suspension	> 2 years	2017
Qmiiz™	Meloxicam	Anhydrous citric acid, aspartame, orange flavour	Orally disintegrating tablet	> 2 years	2018
Epidiolex®	Cannabidiol	Strawberry flavour, sucralose	Oral solution	> 2 years	2018
Diacomit®	Stiripentol	Aspartame, sorbitol	Oral suspension	3–18 years	2018
Evekeo™	Amphetamine sulfate	Anhydrous citric acid, malic acid, sucralose	Orally disintegrating tablet	> 3 years	2019
Ozobax®	Baclofen	Anhydrous citric acid, natural grape flavour, sodium citrate dihydrate, sucralose	Oral solution	> 12 years	2019
Katerzia™	Amlodipine benzoate	Citric acid, sodium citrate, sucralose	Oral suspension	> 6 years	2019
Xeljanz®	Tofacitinib citrate	Natural grape flavour, sucralose, xylitol	Oral solution	2–18 years	2020
Qdolo®	Tramadol hydrochloride	Citric acid, grape flavour, sucralose	Oral solution	> 12 years	2020

Evrysdi®	Risdiplam	Isomalt, mannitol, strawberry flavour, sucralose	Oral solution	2–10 years	2020
Fintepla®	Fenfluramine hydrochloride	Cherry flavour, citric acid, sucralose	Oral solution	> 2 years	2020
Elyxyb™	Celecoxib	Acesulfame potassium, banana flavour, bubble gum flavour, peppermint flavour, sucralose	Oral solution	6–12 years	2020
Dificid®	Fidaxomoxin	Citric acid, mixed berry flavour, sodium citrate, sucralose	Oral suspension	> 6 months	2020
Xarelto®	Rivaroxaban	Anhydrous citric acid, mannitol, sucralose, sweet and creamy flavour	Oral suspension	> 6 months	2021
Noxafil® Powdermix Kit	Posaconazole	Artificial cherry flavour, citric acid, sodium citrate dihydrate	Oral suspension	> 2 years	2021

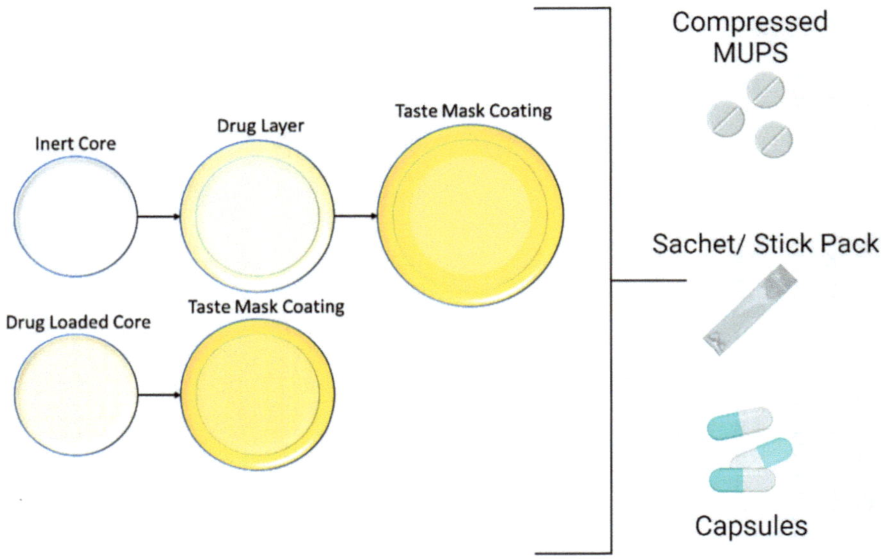

Fig. 6.2 Schematics of taste masked multiparticulate formulations (MUPS Multi-unit Pellet System). (Created with Biorender.com)

immediate and complete API release from the dosage form in the gastrointestinal (GI) tract. From a manufacturing point of view, taste masking should ideally be achieved using low levels of safe excipients and relatively simple techniques which can be industrially scaled up.

Taste masking coating can be applied on singular/monolithic dosage forms, such as tablets; however, multiparticulate systems are more appropriate for paediatric applications, because of low GI irritation, increased dosing flexibility, and improved swallowability/acceptability. A film-coated multiparticulate formulation involves the inclusion of a core substrate which can either contain the API in a matrix structure or use inert starting cores for the API to be spray-layered onto (Fig. 6.2). If the core is to be drug-loaded in a matrix, techniques such as extrusion spheronisation, hot melt extrusion, or direct pelletisation, via wet or dry granulation in a fluid bed system, can be utilised.

With regard to cores for drug layering, various materials are available which can be relatively inert, where no/little chemical interactions with the API and no/little impact on the drug release process are observed (Table 6.2). Alternatively, cores, made from tartaric acid, for example, can be functional where the material can influence drug solubility and bioavailability (Table 6.2). Soluble cores, such as sucrose, lactose, and mannitol, are chemically inert, but they can impact drug release by osmotic pressure generation. For film coating processes, microcrystalline cellulose cores (MCCs) are commonly applied due to their high sphericity and narrow size distribution which improves the likelihood of obtaining a uniform coating layer. MCC cores are known to swell during drug release, and the cellulosic fibres can

Table 6.2 Summary of substrates available for drug layering (information taken from manufacturers' websites)

Material	Water solubility	Brand(s)	Example suppliers	Smallest particle size range (μm)	Largest particle size range (μm)
Sucrose	Soluble	Suglets®	Colorcon	250–355	1400–1700
		Pharm-a-spheres™	Pharm-a-spheres	90–150	1700–2000
		Surinerts®	IPC Process-Center	355–425	1180–1700
		Vivapharm®	JRS Pharma	250–425	1400–1700
		Sprayspheres™-SS	Umang Pharmatech	180–250	2000–5000
		Sugar spheres SANAQ®	IPC Process-Center	212–300	1000–1180
Tartaric acid		TAP®		355–425	710–1000
		Tartaric acid pellets	Umang Pharmatech	180–250	600–850
		Espheres TA	Ideal Cures	425–600	600–850
		InstaSpheres TA (seal coated)		600–850	600–850
Isomalt		galenIQ™	Beneo	340–1290	340–1290
Xylitol		Xylinert	IPC Process-Center	710–1700	710–1700
Mannitol		Mannitol spheres SANAQ®	Pharmatrans SANAQ	75–150	1000–1180
		Sprayspheres™-MAS	Umang Pharmatech	150–180	850–1000
Lactose		Sprayspheres™-LS		150–180	850–1000
Microcrystalline cellulose (MCC)	Insoluble	Vivapur® MCC	JRS Pharma	100–200	1000–1400
		Cellets®	IPC Process-Center	73–115	1000–1400
		Celphere™	Asahi Kasei	75–212	500–710
		Sprayspheres™-MS	Umang Pharmatech	180–250	1180–1680
Anhydrous dibasic calcium phosphate		PharSQ® Spheres CM	Chemische Fabrik Budenheim	500–710	500–710
Silicon dioxide		Sprayspheres™-AS	Umang Pharmatech	75–212	850–1000
Calcium carbonate		Calcium carbonate SANAQ®	Pharmatrans SANAQ	75–212	710–850
		Sprayspheres™-CS	Umang Pharmatech	75–212	850–1000
Starch		Sprayspheres™-STS		150–180	1400–1680

Table 6.3 Examples of commercially available water-soluble polymers for film coating (information taken from manufacturers' website)

Polymer name	Coating system	Brand name	T_g (°C)[a]	Manufacturer/ supplier
Polyvinyl pyrrolidone (PVP)	Aqueous and organic	Kollidon®	108	BASF
Methyl cellulose (MC)	Aqueous	Metolose® SM	100	Shin-etsu
Hypromellose (HPMC)		Pharmacoat®	165	
		Methocel™	165	Colorcon
		Spectracel™	165	Sensient
Hydroxyl ethyl cellulose (HEC)		Natrosol™	127	Ashland
Sodium carboxymethyl cellulose (CMC sodium)		Walocel™ CRT	121	DuPont

[a]Glass transition temperature (T_g)

adsorb drugs, hindering release. Soluble sugar (sucrose) spheres can be used to overcome these challenges; however, the consumption of sugars should be restricted in paediatric patients where possible. Polyols, such as isomalt, provide similar properties to sugar spheres with favourable properties as they have less glycaemic effect. Water-soluble cores are not always applicable as aqueous drug layering can cause changes in the core shape, size, and surface morphology [7].

A variety of film coating polymers are available that can be applied for taste masking purposes and can be divided into three groups: water-soluble polymers, polymers with pH-dependent solubility, and insoluble polymers (Tables 6.3, 6.4 and 6.5). For water-soluble polymers, the challenge for taste masking is to manufacture a film coating layer which will not prematurely dissolve or facilitate API leakage before exiting the oral cavity. Each polymer within this category is available in varying grades differing in viscosities (Table 6.3). Higher viscosity will provide longer-lasting barriers which are beneficial for taste masking compared to less viscous polymers. However, applying viscous polymers is challenging for processing due to a high tendency of substrate aggregation during spraying.

Polymers with pH-dependent solubility can be categorised as enteric or reverse enteric polymers (Table 6.4). Enteric polymers dissolve above a specific pH and are conventionally applied for delayed release or targeted drug release to a segment of the intestine such as the small intestine or the colon. In contrast, reverse enteric polymers only dissolve in acidic conditions (e.g. $<$ pH 5.0) and can be advantageous for taste masking applications, considering the pH of the oral cavity is near neutral (e.g. $\approx$7.5 in paediatrics) [8], whereby polymer dissolution is prevented in the mouth and occurs in the stomach. When formulating for paediatric patients, the specific nuances with regard to the conditions of the oral cavity and gastrointestinal (GI) tract should be considered [9]. The significance of these findings may eliminate some polymers which are appropriate for taste masking in adults.

Water-insoluble but permeable polymers are commonplace for sustained release applications which can also be used to achieve taste masking at low application

Table 6.4 Examples of commercially available polymers with pH-dependent solubility for film coating (information taken from manufacturers' website)

Category	Polymer name	Dissolving pH	Coating system	Brand name	T_g (°C)[a]	Manufacturer/ supplier
Reverse enteric	Methacrylate amino ester copolymer	<5.0	Organic/ Aqueous	Eudragit® E 100/12.5/PO	45	Evonik
	Amino diethyl-methacrylate copolymer	<5.5	Aqueous	Kollicoat® Smartseal 30 D	70	BASF
Enteric	Cellulose acetate phthalate (CAP)	>5.5	Organic	CAP-NF	160–170	Eastman
	Cellulose acetate phthalate (CAP)	>5.5	Aqueous	Aquacoat® CPD	160–170	Colorcon
	Shellac	>7.0		SSB® 55 Pharma	40–50	SSB Pharma
	Polyvinyl acetate phthalate (PVAP)	>5.0		Sureteric®	42.5	Colorcon
	Hypromellose phthalate (HPMCP)	>5.0		HP-50	143–147	Shin-etsu
		>5.5		HP-55	143–147	
	Hypromellose acetate succinate (HPMCAS)	>5.5		AQOAT® AS-LF	120–135	
		>6.0		AQOAT® AS-MF	120–135	
		>6.5		AQOAT® AS-HF	120–135	
	Methacrylic acid methyl acrylate and methyl methacrylate copolymer	>7.0		Eudragit® FS 30 D	48	Evonik
	Methacrylic acid ethyl acrylate copolymer	>5.5		Eudragit® L 30 D-55	96	
		>5.5		Kollicoat® MAE 30 DP	113	BASF
		>5.5	Organic	Eudragit® L 100 D-55	96	Evonik
	Methacrylic acid methyl methacrylate copolymer	>5.5		Eudragit® L 100/12.5	>130	
		>7.0		Eudragit® S 100/12.5	>130	

[a]Glass transition temperature (T_g)

levels (coating thicknesses) relative to water-soluble polymers, due to the permanency of the film coating during drug release. The insoluble polymers available for film coating can be differentiated by their permeability (Table 6.5). With highly permeable coatings, there would be a higher risk of API leakage in the oral cavity,

Table 6.5 Examples of commercially available water-insoluble, semipermeable polymers for film coating (information taken from manufacturers' website)

Polymer name	Permeability	Coating system	Brand name	T_g (°C)	Manufacturer/ supplier
Ethyl cellulose (EC)	Low	Aqueous	Surelease®	45–55	Colorcon
		Organic	Ethocel™	130–133	
		Aqueous	Aquacoat® ECD	90	
Polyvinyl acetate (PVA)			Kollicoat® SR 30 D	30	BASF
Poly(ethyl acrylate, methyl methacrylate, trimethyl ammonio ethyl methacrylate chloride) 1:2:0.1			Eudragit® RS 30 D	58	Evonik
		Organic	Eudragit® RS 100/12.5/PO	58	
Cellulose acetate (CA)	Moderate	Aqueous	Eastman CA	130	Eastman
Poly(ethyl acrylate-co-methyl methacrylate) 2:1			Eudragit® NE 30 D	12	Evonik
			Eudragit® NM 30 D	9	
Poly(ethyl acrylate, methyl methacrylate trimethyl ammonio ethyl methacrylate chloride) 1:2:0.2	High		Eudragit® RL 30 D	63	
		Organic	Eudragit® RL 100/12.5/ PO	63	

[a]Glass transition temperature (T_g)

whereas coatings with low permeability increase the likelihood of incomplete release in the GI tract even at low coating levels.

Various techniques can be utilised for film coating. The most common ones are pan coating for tablets and fluid bed coating processes for multiparticulates. Aqueous solutions are typically used for water-soluble polymers, while water-insoluble polymers are either dispersed in an aqueous vehicle or dissolved in organic solvent. Aqueous dispersions are preferred as they do not require organic solvents and are less viscous compared to polymer solutions; they can therefore be sprayed at higher spray rates. Films prepared using aqueous dispersions typically require additional curing to promote polymer chain coalescence and complete film formation, subsequently increasing long-term stability. To achieve effective curing, the product must be held at a temperature exceeding the polymer's glass transition temperature (T_g) for an extended period ranging from 2 to 24 h. This is often carried out using drying ovens but can also be achieved "in-process", for example, during fluid bed drying.

For both aqueous and organic coatings, the addition of plasticisers may be required to lower the polymer T_g and allow film formation under conditions which require low levels of energy input. Plasticisers are hydrophilic or hydrophobic, the latter of which can reduce drug leakage when the product is held in the oral cavity (Table 6.6). In addition to plasticisers, anti-tacking agents are sometimes required to

Table 6.6 Example plasticisers used in film coating

Plasticiser	Hydrophilic/ hydrophobic	Usual concentration used (% w/w based on dry polymer)
Polyethylene glycol (PEG)	Hydrophilic	10–20
Triethyl citrate (TEC)		5–20
Triacetin		10–35
Dibutyl sebacate (DBS)	Hydrophobic	10–30
Diethyl phthalate (DEP)		
Dibutyl phthalate (DBP)		5–30
Acetyl triethyl citrate		5–20
Acetyl tributyl citrate		

Table 6.7 Example anti-tacking agents used in film coating

Anti-tacking agent	Usual concentrations (% w/w based on dry polymer)	Notes
Talc	50–100	Prone to sedimentation
Magnesium stearate	30–50	Suitable for organic coating
Glycerol monostearate (GMS)	5–20	Emulsification steps required
Silicon dioxide	5–20	Effect on taste masking is grade dependent

reduce aggregation of substrates. This is especially true when using polymers which are inherently tacky, highly viscous coating liquids and/or substrates which are small or have rough surfaces and prone to mechanical interlocking (Table 6.7). The inclusion of anti-tacking agents could affect drug release. For example, silicon dioxide or silica (SiO_2) is available as hydrophilic and/or mesoporous grades which can increase drug release compared to hydrophobic and/or colloidal grades, and the consequential effect on taste masking should be considered [10].

For coating multiparticulates, fluid bed techniques, such as Würster coating, are preferred due to the high heating, mixing, and drying efficiency which can minimise coating run times and provide uniform coatings critical for taste masking. Other variations of fluid bed coating include top spray and tangential spray; however, there are drawbacks to be considered for both when used for taste masking. Films prepared using top spray tend to be varied in terms of thickness which is not ideal for the consistency of product performance. For tangential spray, the forces exerted on the particulates are high, and therefore, damage to their surfaces is more likely which can lead to taste masking failure.

There are various examples in literature and as approved products which demonstrate how film coating of multiparticulate-based formulations can be developed for paediatric patients (Tables 6.8 and 6.9).

Table 6.8 Literature examples of taste mask-coated multiparticulates intended for paediatric patients

Active pharmaceutical ingredients	Coating polymer(s)	Core type	Particle size (μm)	Final formulation	Reference
Famotidine	CA and/or CAB:PVP (80:20)	API, PVP, and lactose granules via rotor granulation	150–400	Chewable tablets	[11]
Acetaminophen (APAP)	Ethyl cellulose/ HPMC	API crystals	Not mentioned	ODT	[12]
Cimetidine	Eudragit® E	API crystals	Not mentioned	ODT	[12]
APAP/Codeine	Eudragit® E and Eudragit® NE 30 D	APAP crystals	Not mentioned	ODT	[12]
Ibuprofen	Ethyl cellulose	API crystals	Not mentioned	ODT	[12]
Doxycycline	Eudragit® E	API layered sugar cores	Not mentioned	ODT	[12]
APAP, Ibuprofen, Famotidine, Loperamide	Eudragit® E with CA/CAB	API granules	<500	Chewable tablets	[13]
Nabutamone	Eudragit® RL 30 D	API granules	180–420	Chewable tablets	[14]
Clarithromycin	Eudragit® L 30 D-55	Granule with HPMC via wet granulation	180–425	Suspension	[15]
Etorcoxib	HPMC, polymethacrylate	API containing microspheres	Not mentioned	Suspension	[16]
Ranitidine, famotidine, cetirizine; fexofenadine, sumatriptan, electriptan, zolmitriptan, ondansetron, granisetron, tiagabine, tizanidine, zolpidem, zaleplon, zafirlukast, montelukast, sildenafil, tadalafil	Ethyl cellulose and calcium carbonate	Sugar core	400	ODT	[17]
APAP, ibuprofen, acetylsalicylic acid	Kollicoat® SR	API granules via rotor granulation	<500–1000	Chewable tablets	[18]

(continued)

Table 6.8 (continued)

Active pharmaceutical ingredients	Coating polymer(s)	Core type	Particle size (μm)	Final formulation	Reference
Fexofenadine	Eudragit® E	API layered inert core	90–150	ODT	[19]
Quinine sulfate	Eudragit® E	API pellets	300–800	Pellets	[20]
Clarithromycin	Eudragit® L 30 D-55	API granules via granulation with poloxamer	200–500	Oral suspension	[21]
Ibuprofen/riboflavin	Ethyl cellulose	API granules via wet granulation	<850	Granules	[22]
Naproxen	Eudragit® E	API granules	130–250	Granules	[23]
Ornidazole	Kollicoat® Smartseal 30 D	API granules	150	Oral suspension	[24]
Diclofenac	Eudragit® E	API granules via wet granulation	Not mentioned	ODT	[25]
APAP	HPMCAS	API layered MCC spheres	170	ODT	[26, 27]
APAP	Surelease®/ Opadry® (85:15)	APAP granules	196	Granules	[26, 27]
Clindamycin	Eudragit® E	API layered MCC spheres	250	ODT	[28]
APAP/caffeine	Eudragit® E	API crystals	125–355	ODT	[29]
Lafutidine	Ethyl cellulose/ HPMC (67/33)	API layered lactose spheres	210 (d_{90})	ODT	[30]
Losartan	HPMC	API granules	2000	Minitabs	[31]

Acronyms (*CA* Cellulose acetate, *CAB* Cellulose acetate butyrate, *PVP* Polyvinyl pyrrolidone, *ODT* Orally disintegrating tablet, *HPMC* Hydroxypropyl methyl cellulose, *APAP* Acetaminophen)

6.3.2 Lipids

Lipids comprise a heterogenous group of substances which share the common characteristics of having little to no solubility in water. Unlike polypeptides and polysaccharides, lipids do not form macromolecular structure and have relatively low molecular mass. Lipids are applied in pharmaceuticals utilising their unique composition, crystalline structure, melting properties, and ability to associate with water and other non-lipid molecules [32, 33].

According to their structure, lipids can be classified as simple, complex, or derived which are further divided based on the number of products by hydrolysis (Fig. 6.3) [34, 35]. Simple lipids can be divided further into acylglycerols which are predominantly made up of oils and fats comprised of three alcohol groups bound to

Table 6.9 Examples of approved taste masked products for paediatrics comprising film-coated multiparticulates (information taken from Summary of Product Characteristics)

Active pharmaceutical ingredients	Indication	Intended patient age group	Dosage form	Coating polymer(s)	Brand name	Market authorisation holder	Market authorisation year
Tenofavir	HIV-Hep B	>2 years	Granules	Ethyl cellulose	Viread®	Gilead Sciences	2002
Terbinafine	Tinea capitis	>4 years	Granules	Basic butylated methacrylate copolymer	Lamisil®	Novartis	2007
Raltegavir	HIV-1	>2 years	Chewable tablets	Ethyl cellulose and hypromellose	Isentress	Merck Sharp and Dohme Corp	2007
Pyronaridine-artesunate	Malaria	>2 months	Granules	Ethyl cellulose and hypromellose	Pyramax®	Shin Poong Pharmaceutical Co., Ltd	2012
Sodium phenylbutyrate	Urea cycle disorders	From birth	Pellets	Ethyl cellulose and hypromellose	Pheburane®	Eurocept International BV	2013
Hydrocortisone	Adrenal insufficiency	From birth	Granules (in sprinkle capsules)	Ethyl cellulose	Alkindi®	Diurnal	2018
Raltegavir	HIV-1	>4 weeks	Granules for oral suspension	Ethyl cellulose and hypromellose	Isentress®	Merck Sharp and Dohme Corp	2021

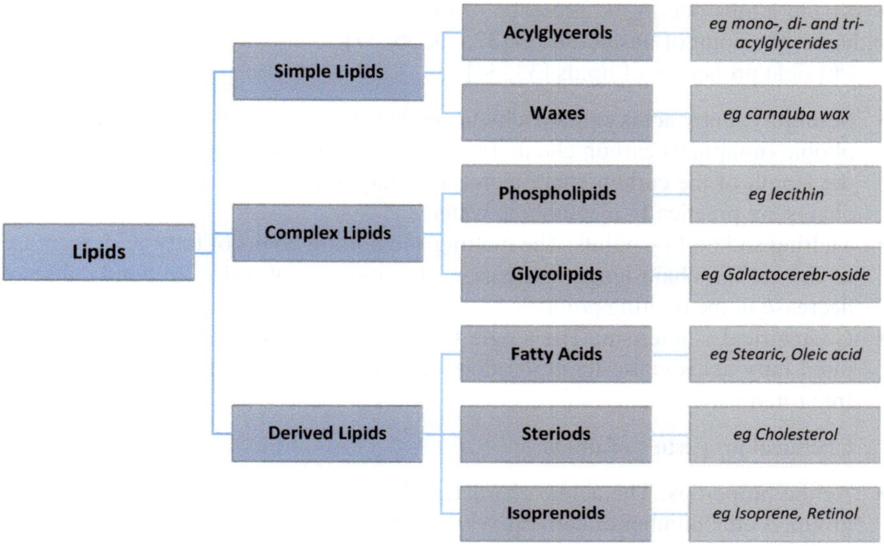

Fig. 6.3 Classification of lipids with brief examples

their molecules as well as waxes which are esters of long-chain monohydric alcohols with high molecular weights. Complex lipids can be divided into phospholipids that have a phosphoric acid and glycolipids with carbohydrates instead of phosphates in their molecules. Finally, derived lipids are divided into fatty acids where the formula includes a linear carbon chain with a monocarboxylic acid group, steroids formed by four aromatic rings derived from sterane, and finally isoprenoids derived from a five-carbon compound, isoprene.

Lipid molecules consist mainly of repeating units of fatty acids (monocarboxylic organic acids) which are internal building blocks. The physiochemical state of a lipid is vital to determine its behaviour, such as the solid fat content, which can influence the spreadability, firmness, stability, and mouthfeel of the lipid. The solid fat content can be determined by measuring the density of a lipid over a range of temperatures, since the density of a solid fat is much higher when compared to the density of a liquid fat. Properties of a lipid relevant to drug delivery include the melting point, the clear point (temperature at which a fat completely melts and becomes transparent), the slip point (temperature at which the fat starts to move downward due to its weight), and the cloud point (temperature at which crystallisation starts in a liquid fat). Other properties include thermal stability which can affect the composition of the lipids (as temperature decreases, the lipid is expected to become more unsaturated) and their water sorption. Lipids also differ in their degree of saturation or unsaturation due to the number of double bonds present within their structure. A saturated lipid has a fatty acid chain that is held by carbon atoms with hydrogen atoms attached to each carbon, while an unsaturated lipid has missing hydrogen atoms due to the presence of one or more carbon double bonds (which indicates the point of unsaturation) [32, 33, 36, 37].

Physical properties of lipids [33, 34]:

- Solubility: Fatty acids contain a hydrophilic carboxyl group (polar) and a hydrophobic (nonpolar) carbon chain. The water solubility of fatty acids decreases as the length of the carbon chain increases. Fatty acids with more than six carbon chains are practically insoluble in water.
- Melting and boiling points: The melting and boiling points of fatty acids increase as the carbon chain length increases. The presence of a double bond causes a decrease in the melting point.
- Geometrical isomerism: The single bonds between the carbon atoms on saturated fatty acids cause them to adopt various spatial arrangements and lead to free rotations.

Chemical properties of lipids:

- Acidic properties: The acidic character of a fatty acid depends on the carboxyl group. As the number of carbons within the chain increases, the acidic character of the fatty acid decreases.
- Oxidation: Fatty acids with a carbon-carbon double bond (unsaturated) are easily oxidised.

Tables 6.10 and 6.11 list examples of lipids used for taste masking and their applications reported in the literature. Lipids have many advantages when used for masking the bitter taste of a drug, for example [48, 49]:

Table 6.10 Lipid examples that could be used for taste masking and their properties

Lipid [38]	Melting point (°C)	Mol weight	Structure	Solubility	Reference
Glycerol monostearate	74.0	358.6	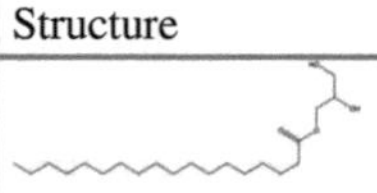	Insoluble in water; soluble in hot oils and organic solvents	[39]
Lecithin	236.1	758.1	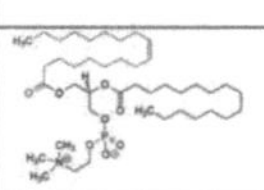	Partly soluble in water and alcohol	[40]
Carnauba wax	82–86	351.7	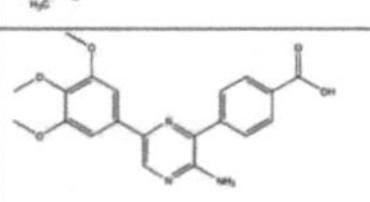	Practically insoluble in water and practically insoluble in alcohol	[41]
Glyceryl tristearate	72–75	891.5	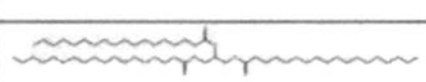	Insoluble in water, soluble in chloroform/carbon disulfide	[42]

Table 6.11 Example lipids used in the literature for taste masking

Lipid(s)	Active pharmaceutical ingredients	Dosage form	Method of preparation	Reference
Lecithin	Acetaminophen	Powders	Spray-drying using aqueous dispersions	[43]
Carnauba wax and stearic acid	Mesalazine	Microcapsules	Drug containing lipids prepared by spray-congealing followed by spray drying to dry coat the microparticles	[44]
Glyceryl distearate and trimyristate	Metformin HCl	Orally disintegrating tablets	Direct compression tablets based on solid lipid pellets	[45]
Palmitic acid and stearic acid	Benzoic acid	Microspheres	Microspheres prepared by a spray chilling process	[46]
Vegetable fat	Proanthocyanidin-rich cinnamon extract	Solid lipid microparticles	Solid lipid microparticles via spray chilling technique using vegetable fat as carrier	[47]
Carnauba wax	Quercetin	Microcapsules	Drug microencapsulated by hot melt extrusion	[41]

- Lipids have acceptable taste and good flavour.
- Due to their hydrophobic nature, some lipids are digested and absorbed within the body.
- Lipids can alter the dissolution behaviour of a drug and therefore can be used for taste masking without the need of additional polymers/excipients.
- Lipids are considered safe as excipients within pharmaceutical products.

There are various methods for taste masking using lipids, including the following key approaches:

Spray drying/coating [50]—a process where the drug, lipids, and other excipients are dissolved in organic solvents and introduced into a coating chamber via spraying using an atomiser. The organic solvent is then evaporated due to the hot air/gas within the chamber. Subsequently, the lipid layer will form a physical barrier around the dosage form and hence preventing the release of the drug.

Spray congealing [51]—a method consisting of spraying molten material such as lipids (API either dissolved or dispersed in the lipid) through a nozzle into a coating chamber containing cool air to produce uniform shaped spherical particles. Some of the vital formulation/processing parameters include the melting point of the lipid, formulation viscosity and the temperature of the cool air.

Hot Melt Extrusion [52, 53]—a process of converting a powder mixture into a shaped compact mixture through a hot barrel at high pressure. The mixture of drugs, lipids, thermal lubricants, and other excipients is melted and forced through an extruder at high temperatures. The materials are then cooled to produce extrudates with desirable properties.

Hot Melt Coating [54, 55]—a solvent-free method where a lipid or mixture of lipids acts as the coating agent for taste masking at temperatures above melting point of the lipid(s). The lipids must be maintained in liquid state to be properly atomised and sprayed. Specialised equipment is usually used to ensure that the temperature is controlled throughout the process, including a specific nozzle with insulation which is placed as close as possible to the powder bed to reduce the distance between the lipid droplets and the substrate. The transporting tubes used usually have a double layer with heated compressed air going through, to ensure the lipid stays in molten form.

Melt granulation [56]—a process within a fluidised bed where molten binder (lipids) is added to a powder mixture to form agglomerates (granules) with desired properties. The powder mixture consists of the drug and other excipients such as plasticisers and lubricants.

In terms of safety, lipids are generally considered safe and are compatible with most excipients and APIs [5]. However, the long carbon chain on lipids can increase the time of residence in the GI tract which may cause safety and toxicity issues for drugs with a narrow therapeutic index [57]. Some limitations for lipids may arise during the development of a drug product with a high melting point, where the lipids may reduce the drug absorption which could lead to a reduction in bioavailability and therefore compromise the therapeutic efficacy [48, 58].

To expand on using lipids for taste masking, some factors need to be considered such as [48]:

- The melting point of lipids is a useful factor in selecting the appropriate processing method (lipids with low melting points are suitable for heat-free coating process).
- The stability of lipids is another important factor, and saturated lipids are known to be more stable since the degree of unsaturation of a lipid acyl chain is more likely to be degraded by oxidation. However, saturated lipids have formulation issues due to high melting points.
- Lipids with an optimum melting point and lower degree of unsaturation are preferred for taste masking.

6.3.3 Spray Drying and Spray Congealing

Spray drying and spray congealing are related processes that offer the potential for taste concealing bitter material via the creation of matrix particles. Spray drying is a method where a drug solution/suspension is sprayed into a hot air stream, producing dry and spherical particles. Several liquids such as solution, emulsions, dispersions, and slurries can be directly transformed into solid particles with preferred properties such as size, distribution, shape, porosity, and density by this method [51, 59].

Generally, the spray drying process consists of the following steps:

1. Feed preparation (emulsion, suspension, or solution)
2. Atomisation/droplet formation
3. Particle shape formation and drying
4. Separation of dried product

Each phase composition and processing condition can affect the spray drying process efficacy and the characteristics of the final dried product. One benefit of using spray drying for taste masking is that the technique is capable of producing amorphous solid dispersion of the API, hence improving the dissolution rates and bioavailability of poorly soluble drugs [60]. On the other hand, the rapid release of the API may compromise taste masking in the oral cavity. Table 6.12 lists some literature examples where spray drying is used for taste masking.

Spray-congealing involves the melting, dispersing, or dissolving of an API in waxes, gums, fatty acids, and other solids with low melting points. The dispersion is then sprayed into a stream of air and other gases with a temperature below the melting point of formulation components to produce spherical congealed pellets [51, 65]. Zinnat™ suspension is a marketed product where stearic acid is used to mask the unpleasant taste of cefuroxime axetil via the spray congealing process [66].

Table 6.12 Examples of spray drying for taste masking in the literature

Active pharmaceutical ingredients	Polymer(s)	Dosage form	Purpose	Reference
Enalapril maleate	Eudragit® E	Microparticles	Developing drug-loaded microparticles using spray-drying and assessing their taste masking potential	[61]
Ondansetron hydrochloride	Chitosan, Methocel™ E15 LV, and Eudragit® E100	Microspheres	Developing microspheres of drug via spray-drying using three polymers to assess the effect on taste masking and release properties	[62]
Roxithromycin	Eudragit® L 30D-55	Microcapsules	Developing drug-loaded microcapsules via spray drying to evaluate taste masking using the electronic tongue	[63]
Famotidine	Eudragit® E	Orally disintegrating tablet	Developing orodispersible famotidine tablets via co-spray drying (drug/mannitol solution was initially spray dried before a second spray drying run with the polymer)	[64]

Table 6.13 Advantages and disadvantages of spray drying and spray congealing processes [67–69]

	Spray drying	Spray congealing
Advantages	Simple and rapid process	
	Reproducible and scalable	
	Producing free-flowing and spherical particles	
	Particles can be compressed into tablets	
	Particle size, shape, porosity, and density control	Uses minimal time and energy
	Enhanced dissolution rate of drugs	Can be solvent-free
Disadvantages	High consumption of energy due to high temperatures	High maintenance cost
	Not applicable for temperature-sensitive drugs	Small particle size range

The process involves the following steps:

1. Preparation of the fluid consisting of the molten carrier and the active ingredient (excipients with 35–90 °C melting temperatures are suitable).
2. The molten fluid stream is broken into small droplets via an atomiser.
3. The molten droplets solidify upon cooling in the chamber to produce the final solid spherical particles.

Generally, spray drying equipment can be used for spray congealing processes with modifications. Both processes yield matrix particles with the drug dissolved or dispersed within the matrix. Although both processes have many advantages, there are several critical concerns with regard to high energy consumption and particle size limitations (Table 6.13) [51].

6.3.4 Coacervation

Coacervation is an encapsulation method based on colloid chemistry in which aqueous colloidal solutions are divided into two liquid phases, one rich in colloid (coacervate) and the other poor in colloid.

The method involves the formation of three immiscible chemical phases (Fig. 6.4): Initially the core material (such as API particles) is dispersed into the liquid phase of the coating material (A), the polymeric solution formed is separated (B) which is then coated as a uniform layer around the dispersed core particles (C). The coating is then solidified (D) [68].

There are two types of coacervation: simple and complex coacervation (depending on the involved phase separation mechanisms). Simple coacervation refers to cases where only one polymer is involved which is salted out by the action of electrolytes or via de-solvation by adding a water-immiscible solvent (such as ethanol). Complex coacervation entails a reaction that takes place due to the attraction forces of oppositely charged polymers (protein and polysaccharide). This is also called the

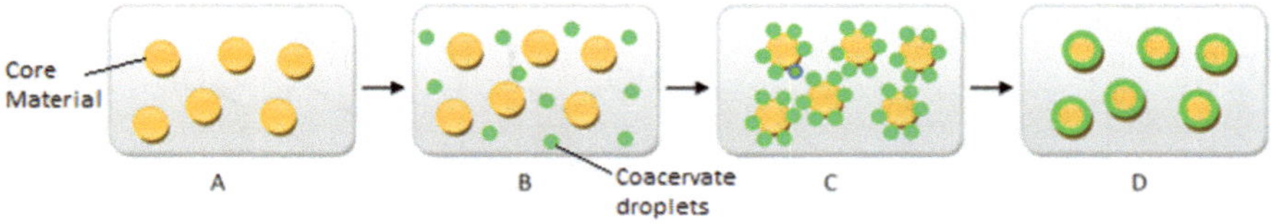

Fig. 6.4 Schematic representation of the coacervation method. (**a**) Core material dispersed in polymer (coating) solution; (**b**) separation of polymer coacervate from solution; (**c**) microdroplets of coacervate coat core material; and (**d**) coalescence of coacervate to form solidified shell around core material

polymer-polymer interaction method. The hydrocolloids that are used to produce complex coacervates include numerous proteins and polysaccharides such as gelatin, carrageenan, whey proteins, alginate, pea protein, pectin, and gum arabic [69, 70].

The method offers a potential to taste conceal and sustain the delivery of soluble APIs dosed for sustained delivery. One example shows masking the bitter taste of paracetamol (acetaminophen, APAP) by preparing solid dispersions using coacervation technique based on dissolving the drug and polymeric carrier in pH-adjusted water. Gradually changing the pH level led to the precipitation of the amorphous solid dispersion granules [71]. Coacervation depends on many factors such as pH, concentration, ionic strength, polymer type, and the ratio of biopolymers [72]. The use of coacervation for taste concealing is limited by drug solubility and coacervate chemistry due to challenges in finding the appropriate polymer systems [72].

6.3.5 Cyclodextrins

Cyclodextrins (CDs) are a family of cyclic oligosaccharides (sugar molecules bound together in rings) produced from starch that have been widely used in pharmaceutical applications. CDs possess the ability to include guest molecules inside their cavities, forming inclusion complexes which can mask the bitter taste of a drug by decreasing its oral solubility on ingestion or decreasing the amount of drug molecules exposed to taste buds, thereby reducing the perception of bitter taste. CDs are comprised of at least six D-(+) glucopyranose units connected by α-(1,4) glucosidic bonds that differ in their ring size and solubility (Fig. 6.5). There are three main naturally occurring subtypes: α-CD, β-CD, and γ-CD with six, seven, or eight dextrose units, respectively (Table 6.14). These subtypes form a hollow truncated cone structure which is composed of a hydrophilic outer surface and a hydrophobic inner surface. This unique structure enables CDs to be water soluble and the ability to form complexes with entrapped hydrophobic molecules (CD complexes are typically formed in a liquid environment). CD complexes are usually in a 1:1 pairing of a guest molecule within a cyclodextrin ring (complexes of one guest with two or more cyclodextrin rings can also occur). β-cyclodextrin is the commonly used

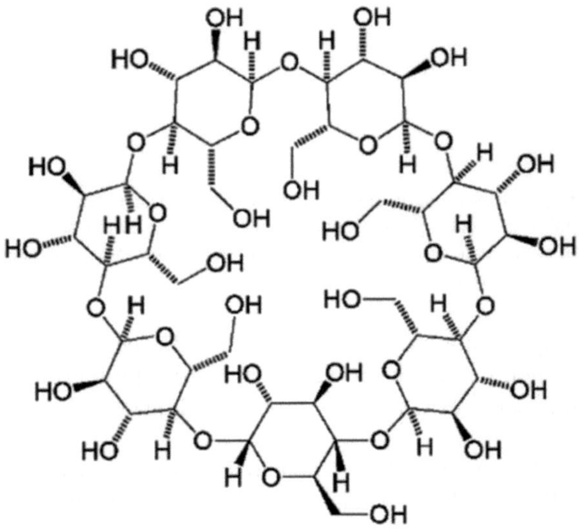

Fig. 6.5 Structure of cyclodextrin

Table 6.14 Properties of natural cyclodextrins and selected cyclodextrin derivatives [73, 74]

Material	Glucose units	Molecular weight	Water solubility (mg/mL)
α-Cyclodextrin	6	973	145
β-Cyclodextrin	7	1135	18.5
γ-Cyclodextrins	8	1297	232
2-Hydroxypropyl-γ-cyclodextrins	8	Dependent on extent of substitution	800
2-Hydroxypropyl-β-cyclodextrins	7	Dependent on extent of substitution	>1200

complexing agent for taste masking due to its sweet taste and non-toxic nature [73, 75].

Limitations of naturally occurring CDs include low water solubility and a number of toxicological issues which led to the development of derivatives of the native CDs with improved physicochemical properties. CD derivatives that are of interest in the pharmaceutical industry include 2-hydroxypropyl and sulfobutylether CDs which serve as solubilising excipients in several US Food and Drug Administration (FDA)-approved medications [74, 76]. An example is Ozalin® (0.2% midazolam oral solution which is utilised for moderate sedation prior to a procedure with the paediatric population), whereby midazolam formed inclusion complexes with γ-cyclodextrins to enhance solubility/stability and to mask its bitter taste [77]. Table 6.15 lists some examples where CDs were used for taste masking within the literature. Limitations of cyclodextrin drug inclusions include the possibility of inducing drug polymorphism or co-crystal formation depending on the drug properties and the other formulation components [73].

Table 6.15 Example taste masking applications using cyclodextrins

Active pharmaceutical ingredients	Cyclodextrin	Purpose	Reference
Ranitidine hydrochloride	2-Hydroxypropyl-β-cyclodextrin	Assess the role of inclusion complexation in taste masking and the complex formation process	[78]
Famotidine	β-Cyclodextrin	Evaluate the potential taste masking properties of ternary complexation consisting of drug, cyclodextrin, and a polymer	[79]
Diltiazem hydrochloride	β-Cyclodextrin	Evaluating the taste masking of a drug via complexation method to form an orodispersible dosage form. The complexes of the drug with the cyclodextrin (1:1 ratio) were prepared via kneading, co-evaporation, co-grounding, freeze-drying, and melting techniques	[80]
Donepezil	2-Hydroxypropyl-β-cyclodextrin	Investigating the effect of cyclodextrin on taste masking of donepezil orodispersible film via complexation of drug with cyclodextrin. The complexes were then integrated into the dosage form via a solvent casting technique	[81]

6.3.6 *Ionic Resin Complexation*

The use of ion exchange resin complexation, where an ionisable drug and suitable resin form a drug-resinate complex, is another potential approach for taste masking. The ion exchange resins are water-insoluble cross-linked polymers containing a self-forming group at repeating positions on the polymer (with cationic and anionic functional groups). The drug is attached to the oppositely charged resin substrate, to form insoluble resonates via weak ionic bonding [82]. For taste concealing, the ionic bonding prevents the dissociation of the drug-resin complex under salivary pH conditions. The drug is then released gradually, following oral administration through the displacement of ions of the same charge present in the GI fluids. Ion exchange resins can be classified into four major groups: Strong acid cation-, weak acid cation-, strong base anion-, and weak base anion-exchange resins [83, 84]. Various ion exchange resins exist on the market which can be applied for taste masking (Tables 6.16 and 6.17).

Strong acid cation/base anion exchange resins can operate throughout the entire pH range and can be used to mask the bitter taste of basic/acidic drugs. Weak acid cation-exchange resins function at pH levels above 6.0, while weak base anion-exchange resins function well below pH of 7.0 [90]. There are several drugs on the market that have successfully applied ion exchange resin complexations such as paroxetine [91], tinidazole [92], and omeprazole [93]. Ion exchange resins do have some limitations, including the extended steps and processing time which could be costly.

Table 6.16 Types of ion exchange resins with commercial examples [82, 84–86]

Type	Functional group	Effect of pH	Resin salts	Commercial resin
Strong cation	$-HSO_3^-$	Mostly independent	Stable	AmberLite™ IR 120/IRP69
Weak cation	-COO-	Negligible in acid solution	Hydrolyse on washing	AmberLite™ IRC 50/IRP88/IRP64, Tulsion® 335
Strong anion	$-N^+R_3$	Mostly independent	Stable	AmberLite™ IR 400, Dowex® 1, Duolite™ AP143
Weak anion	$-HN^+R_2$	Negligible in alkaline solutions	Hydrolyse on washing	AmberLite™ IR 4B, Dowex® 2

Table 6.17 Example of taste masking applications using ion exchange resins

Active pharmaceutical ingredients	Ion exchange resin(s)	Dosage form	Reference
Clarithromycin	Indion 204/234, Tulsion® 335	Chewable tablets	[87]
Nizatidine	AmberLite™ IRP-69, Dowex®-50	–	[88]
Ciprofloxacin	Copolymerisation of acrylic acid with cross linking agents	–	[89]

6.4 In Vitro and In Vivo Evaluation of Taste Masking Technologies

To assess the taste masking effectiveness, various methods are available which either involve in vitro testing, similar to dissolution, or in vivo evaluation, using human or animal models. In vivo taste testing is used only when necessary or as the last step of the development due to the ethical quandaries involved but is valuable as the intricacies and complexities of in vivo processes are difficult to mimic in vitro.

In vivo taste evaluations utilise human panels and/or animal models. For human panels, the subject groups vary based on various demographics such as age and level of training (naïve, trained, and expert panels). Age affects sensitivity to bitterness; for example, children aged 4–8 years old were more sensitive to the aversive taste of quinine at low concentrations, a compound which has historically been used as a standard for bitterness detection, while older children (9–11 years old) were found to discriminate bitterness similarly to untrained adults at higher quinine concentrations [1]. This suggests employing naïve young adults for taste evaluation might provide more translatable taste evaluation for the target paediatric patient group than older subjects, without the practical, ethical, and regulatory challenges involved in testing in children [1, 94, 95]. Other challenges of using human panels involve the subjective nature of the results and the lack of discrimination between "measurements" [96]. Trained panellists who can discriminate between small differences in taste could be useful in defining taste thresholds for drug solutions but may not correlate with the taste experiences of young children for pharmaceutical formulations.

One example of animal models used for taste evaluation is the Rodent Brief-Access Taste Aversion (BATA) model. This involves the production of a set of standard solutions which are left for rodents to voluntarily lick from a liquid feeder. The number of licks within a timeframe can be used to determine aversiveness thresholds [97–99]. A good correlation between the BATA model and human taste assessments was demonstrated in literature [94, 95, 100]; however, it is useful to couple animal testing with human panels for validation of taste masked formulations. Additionally, animal models are still ethically contentious due to concerns regarding animal welfare, and in vitro methods are recommended for formulation screening purposes.

Table 6.18 provides literature examples of in vivo taste assessments. To improve reproducibility and reliability of results, considerations should be made in the methodology design of human or animal models for taste masking assessments. These include:

- Test subject inclusion/exclusion criteria
- Pre-test acclimatisation or preparation steps for baseline standardisation
- Test protocol including sample volume, exposure times, and cleansing methods
- Range of outcomes for taste description—bitterness scores/adjective selection

Table 6.18 Literature examples of in vivo taste assessments

	Pre-test preparation	Applied method(s)	Method notes	Result type	Reference
Humans ($n = 59$)	No food or drink 1 h pre-test	Single blind triadic forced-choice ascending method	Two methods of liquid application were tested: liquid drops directly onto the tongue; saturated filter paper	Taste strength (0–8) and taste quality (up to 23 adjectives)	[101]
Humans ($n = 15$)	None	Rinse and spit study	Quinine used as model compound for "calibration"	Bitterness score given	[102]
Gerbils ($n = 48$)	Subjects anaesthetised for nerve exposure	Chorda tympani nerve exposed	Stimuli delivered in 2 mL samples by gravity flow system with 1 min interstimulus rinsing with deionised water	Integrated chorda tympani response	[101]
Rats ($n = 40$)	2 weeks acclimatisation with water and standardised diet	Test material mixed with food	Various food types tested, and for each type, four groups were used. Food alone, food with placebo resin, food with active resin, food with drug	Food consumption weighed	[103]

- Number of repeated tests to improve result reliability

In vitro methods are less subjective and can have a higher degree of discrimination and standardisation than in vivo methods; however, it is challenging to closely represent the oral cavity with regard to salivary volume, salivary flow, humidity, and the stresses applied to the formulation to achieve in vitro in vivo correlation (IVIVC). In vitro methods can be divided into methods to obtain samples for drug concentration measurements (in vitro assay models) or systems designed to mimic the tongue (biomimetic taste sensing systems) which are also known as lipid-based taste sensors (LBTSs) or "electronic tongues".

There are various assay methods demonstrated in literature which can be used to determine taste masking effectiveness. To mimic the oral cavity, low volumes of test media and appropriate levels of agitation should be applied to the formulation. The use of simulated saliva as a test media is extensively studied, and using a consistent formula for simulated saliva preparation is essential. Examples of simulated saliva compositions are shown in Table 6.19 which were developed to have a close correlation to human saliva with regard to pH, buffer capacity, and viscosity [106, 107]. Media to mimic paediatric saliva has not yet been reported. When exploring the use of in vitro methods, the following considerations should be made:

- Testing fluid type and volume
- Interaction between the formulation and the test liquid
- Sample collection and preparation for testing
- Methods of analysis, i.e., HPLC, UV-Vis, electronic tongue
- Predetermined taste threshold
- Correlation with in vivo taste testing

LBTSs such as the Astree e-tongue (Alpha MOS, France) and TS-5000Z system (INSENT, Japan) are made of lipid–polymer membranes which are fitted into multichannel detection electrodes comprising silver wires plated with Ag/AgCl. The electrodes are connected to a scanner which transmits the information obtained to a detector. The systems use ChemFET (Chemical modified Field Effect Transistor) technology which is based on non-covalent interactions with dissolved molecules. The Astree e-tongue system is a fully automated taste-sensing instrument equipped with a seven-sensor probe assembly for both qualitative and quantitative analysis. The probes are made of silicon transistors with proprietary organic coatings which determine parameters for sensitivity and selectivity. A potentiometric difference is measured between each of the seven sensors and the Ag/AgCl reference electrode. Each sensor has a specific organic membrane which can interact with ionic or neutral compounds present in the liquid sample which subsequently generates different taste profiles. The TS-5000Z system is a newer version of the lipid-based electronic tongue equipped with seven lipid membrane sensors indicating different taste descriptors: bitterness, umami sensitivity, saltiness, sourness, astringency, and sweetness.

When implementing LBTSs for taste masking assessment, the following steps are required:

Table 6.19 Simulated saliva examples outlined in literature [104, 105]

Simulated saliva 1 [104]		Simulated saliva 2 [104]		Simulated saliva 3 [106]		Simulated saliva 4 [106]	
Ingredient	Quantity	Ingredient	Quantity	Ingredient	Quantity	Ingredient	Quantity
Potassium dihydrogen phosphate	1.63 g	Potassium dihydrogen phosphate	0.19 g	pH 7.0 buffer	6.211 mL	pH 7.4 buffer	6.892 mL
Sodium chloride	2.34 g	Sodium chloride	8.0 g	Tween® 20	5.6 µL	Tween® 20	5.6 µL
Calcium chloride	0.17 g	Sodium dihydrogen phosphate	2.38 g	Xanthan gum	0.05 w/v%	Xanthan gum	0.05 w/v%
Sodium hydroxide	To pH 6.2–6.75–7.4	Sodium hydroxide	To pH 6.75	Porcine gastric mucin	10 mg/mL	Porcine gastric mucin	10 mg/mL
Demineralised water	To 1 L	Demineralised water	To 1 L	Porcine pancreatic alpha-amylase	1.0 mg/mL	Porcine pancreatic alpha-amylase	1.0 mg/mL

1. Sampling—Only dissolved solids can be analysed, so liquid dosage forms can be directly analysed, whereas solid dosage forms require dissolving before liquid samples are taken.
2. Set up—Samples are set in sequence.
3. Analysis—The selected sensors and probes are automatically inserted into each sample, and measurements are generated.

Various examples of in vitro evaluation for taste masking are shown in Table 6.20.

Table 6.20 Literature examples of in vitro taste assessments

Formulation	Device and brief method	Media composition	Media volume (mL)	Test duration	Analysis Method	Reference
Granules	Beaker and stirrer	Phosphate buffer pH 6.8	25	5 min	UV-Vis	[50]
Granules	Beaker and stirrer	Phosphate buffer pH 6.8	25	5 min	UV-Vis	[62]
ODTs	25 mL Erlenmeyer flask and stirring at 50 rpm	Simulated saliva pH 6.75	10	2 min	UV-Vis	[29]
Microparticles	Glass tube and stirrer	Deionised water	10	30 s	HPLC	[108]
Microspheres	Touch Mixer at speed 7	Phosphate buffer pH 6.8 and deionised water (1:1)	7	10 s	HPLC-UV	[22]
API-Resinate complex	25 mL bottles, no stirring	Phosphate buffer pH 6.8	5	2 min	HPLC-UV	[97]
Microspheres	Glass beaker, shaken	Simulated saliva pH 6.2	10	1 min	HPLC-UV	[98]
Pellets	1 mL column fitted with peristaltic pump	Simulated salivary fluid	–	5 min	HPLC-UV	[109]
Pellets/tablets	125 mL Erlenmeyer flask, shaken at 50 rpm	Phosphate buffer pH 6.8	50	2 min	HPLC-UV	[28]
Powder	Unpacked Omega column tube heated to 37 °C	Phosphate buffer saline pH 7.4	1 mL/min by a syringe pump	Up to 30 min	HPLC and Astree e-tongue	[99]

(continued)

Table 6.20 (continued)

Formulation	Device and brief method	Media composition	Media volume (mL)	Test duration	Analysis Method	Reference
Pellets	Glass beaker	Deionised water	100		Astree e-tongue with Bitterness Prediction Module	[20]
API Solution	–	Deionised water	–	N/A	SA402B Insent Inc system with TecLabS sensors/ electrodes	[110]
ODTs	Glass beaker	Deionised water	450	30 s	SA402B Insent Inc system with TecLabS sensors/ electrodes	[111]
API solutions	Glass beaker	Maltodextrin-deionised water	–	N/A	TS-5000Z, Astree e-tongue and UV-Vis	[112]
Extrudates	Glass beaker	Deionised water	25	60 s	Astree e-tongue with Alpha M.O.S sensor set #2	[104]
ODTs	Glass beaker and stirrer	Deionised water	100	3 min	TS-5000Z electronic tongue	[25]
Microspheres	Glass beaker	Simulated saliva pH 6.2	100	3 min	Astree II with 7 liquid cross-selective sensors	[105]
Microparticles	Glass beaker	Deionised water	100	7 min	16 potentiometric standard ion-selective electrodes	[113]

Acronyms (*API* Active pharmaceutical ingredient, *ODT* Orally disintegrating tablet)

6.5 Conclusions

Taste masking of oral dosage forms for paediatrics poses a unique challenge by the heightened taste sensitivities of children. Many factors affect the selection of the appropriate approach for taste masking, including the API properties, formulation choices, excipient safety, and manufacturability. The simple way of adding sweeteners and flavours remains the dominating approach in approved paediatric products, but its effectiveness is limited for APIs with high water solubility and/or strong aversive tastes. More complex formulation techniques are advancing both in literature research and in marketed products, showing promise in taste masking effectiveness and applicability to a range of drugs. The manufacturing complexity, scalability, and associated cost are aspects that impact their industrial adoption. In parallel to technological advancement, progresses are made in in vitro and in vivo methods to assess the taste masking effectiveness of formulations, and a consensus on their standardisation is required to improve the reliability of results and ensure product performance.

References

1. Ranmal SR, Nhouchi Z, Keeley A, Adler L, Lavarde M, Pensé-Lhéritier AM, et al. Taste assessment for paediatric drug development: a comparison of bitterness taste aversion in children versus naïve and expert young adult assessors. Int J Pharm. 2023;647:123494.
2. Shaikh FH, Shumway KR, Soni A. Taste physiology. In: StatPearls. Treasure Island: StatPearls Publishing; 2023.
3. Harmon CP, Deng D, Breslin PA. Bitter taste receptors (T2Rs) are sentinels that coordinate metabolic and immunological defense responses. Curr Opin Physio. 2021;20:70–6.
4. Jaggupilli A, Howard R, Upadhyaya JD, Bhullar RP, Chelikani P. Bitter taste receptors: novel insights into the biochemistry and pharmacology. Int J Biochem Cell Biol. 2016;77:184–96.
5. Food and Drug Administration (FDA). Generally Recognized as Safe (GRAS) [Internet]. 2023 [cited 2023 Dec 1]. Available from: https://www.fda.gov/food/food-ingredients-packaging/generally-recognized-safe-gras.
6. Smith L, Leggett C, Borg C. Administration of medicines to children: a practical guide. Aust Prescr. 2022;45:188–92.
7. Craig D, Kállai-Szabó N, Lengyel M, Farkas D, Tibor Barna Á, Fleck C, et al. Review on starter pellets: inert and functional cores. Pharmaceutics. 2022;14:1299.
8. Alamoudi N, Farsi N, Faris J, Masoud I, Merdad K, Meisha D. Salivary characteristics of children and its relation to oral microorganism and lip mucosa dryness. J Clin Pediatr Dent. 2004;28:239–48.
9. Wollmer E, Ungell AL, Nicolas JM, Klein S. Review of paediatric gastrointestinal physiology relevant to the absorption of orally administered medicines. Adv Drug Deliv Rev. 2022;181:114084.
10. Waters LJ, Hanrahan JP, Tobin JM, Finch CV, Parkes GMB, Ahmad SA, et al. Enhancing the dissolution of phenylbutazone using Syloid® based mesoporous silicas for oral equine applications. J Pharm Anal. 2018;8:181–6.
11. Roche EJ, Papile SM, Freeman EM. Rotogranulations and taste masking coatings for preparation of chewable pharmaceutical tablets - US5215755A. US. 1993.
12. Cousin G, Bruna E, Gendrot E. Rapidly disintegratable multiparticular tablet - US5464632A. US. 1995.
13. Hoy M, Roche E. Taste mask coatings for preparation of chewable pharmaceutical tablets - US5489436A. US. 1993.
14. Hayward M. Palatable pharmaceutical compositions - US5837277A. US. 1998.
15. Nandi I, Guo M, Gassert CM. Taste masking antibiotic composition - US20050084540A1. US. 2005.
16. Dumont H, Thibert R. Flavored taste-masked pharmaceutical formulation made using a one-step coating process - US20060228410A1. 2006.
17. Lai J, Venkatesh GM, Qian KK. Taste-masked pharmaceutical compositions with gastrosoluble pore-formers - EP1802285B1. 2006.

18. Kolter K, Scheiffele S, Einig H, Bodmeier R. Use of film coating as taste-masking coating of oral dosage forms - US20010007680A1. 2006.
19. Chenevier P, Marechal D. Taste-masking coated particles, process for the preparation thereof and orodispersible tablets containing said coated particles - EP1587496B1. 2006.
20. Kayumba PC, Huyghebaert N, Cordella C, Ntawukuliryayo JD, Vervaet C, Remon JP. Quinine sulphate pellets for flexible pediatric drug dosing: formulation development and evaluation of taste-masking efficiency using the electronic tongue. Eur J Pharm Biopharm. 2007;66:460–5.
21. Schwarz FX, Kosilek I. Taste-masking pharmaceutical compositions - WO2006040112A2. 2008.
22. Hamashita T, Matsuzaki M, Ono T, Ono M, Tsunenari Y, Aketo T, et al. Granulation of core particles suitable for film coating by agitation fluidized bed II. A proposal of a rapid dissolution test for evaluation of bitter taste of ibuprofen. Chem Pharm Bull (Tokyo). 2008;56:883–7.
23. Stange U, Führling C, Gieseler H. Taste masking of naproxen sodium granules by fluid-bed coating. Pharm Dev Technol. 2013;19:137–47.
24. Chivate A, Sargar V, Nalawade P, Tawde V. Formulation and development of oral dry suspension using taste masked Ornidazole particles prepared using Kollicoat® Smartseal 30 D. Drug Dev Ind Pharm. 2013;39:1091–7.
25. Guhmann M, Preis M, Gerber F, Pöllinger N, Breitkreutz J, Weitschies W. Development of oral taste masked diclofenac formulations using a taste sensing system. Int J Pharm. 2012;438:81–90.
26. Pein M, Preis M, Eckert C, Kiene FE. Taste-masking assessment of solid oral dosage forms–A critical review. Int J Pharm. 2014;465:239–54.
27. Kaushik D, Dureja H. Recent patents and patented technology platforms for pharmaceutical taste masking. Recent Pat Drug Deliv Formul. 2004;8:37–45.
28. Cantor SL, Khan MA, Gupta A. Development and optimization of taste-masked orally disintegrating tablets (ODTs) of clindamycin hydrochloride. Drug Dev Ind Pharm. 2015;41:1156–64.
29. Drašković M, Medarević D, Aleksić I, Parojčić J. In vitro and in vivo investigation of taste-masking effectiveness of Eudragit E PO as drug particle coating agent in orally disintegrating tablets. Drug Dev Ind Pharm. 2016;43:723–31.
30. Nishiyama T, Ogata T, Ozeki T. Preparation of bitter taste-masking granules of lafutidine for orally disintegrating tablets using water-insoluble/soluble polymer combinations. J Drug Deliv Sci Technol. 2016;32:38–42.
31. Elezaj V, Lura A, Canha L, Breitkreutz J. Pharmaceutical development of film-coated mini-tablets with losartan potassium for epidermolysis bullosa. Pharmaceutics. 2022;14:570.
32. Blanco A, Blanco G. Lipids. In: Medical biochemistry. Academic Press; 2017. p. 99–119.
33. Domínguez R, Pateiro M, Purriños L, Munekata PES, Echegaray N, Lorenzo JM. Introduction and classification of lipids. In: Food lipids: sources, health implications, and future trends; 2022. p. 1–16.
34. Zurao P, Landge S, Jaiswal SB, Chandewar AV. Lipids - The instrumental excipient pharmaceutical dosages form. Int J Res Pharm Chem. 2011;1:992–7.
35. Jambrak AR, Škevin D. Lipids nutraceutical and functional food components; 2017. p. 127–51.
36. Aldred EM. Pharmacology: a handbook for complementary healthcare professionals. Philadelphia: Elsevier Ltd.; 2009.
37. Khatri R, Varia U, Katariya H, Detholia K. Oral lipid based drug delivery system: a review. Int J Pharm Sci Res. 2020;11:5386–03
38. Food and Drug Administration (FDA). Food Additive Status List [Internet]. 2023 [cited 2023 Dec 1]. Available from: https://www.fda.gov/food/food-additives-petitions/food-additive-status-list.
39. National Center for Biotechnology Information. Glyceryl monostearate [Internet]. 2023 [cited 2023 Dec 1]. Available from: https://pubchem.ncbi.nlm.nih.gov/compound/Glyceryl-monostearate.

40. National Center for Biotechnology Information. Lecithin [Internet]. 2023 [cited 2023 Dec 1]. Available from: https://pubchem.ncbi.nlm.nih.gov/compound/5287971.
41. Khor CM, Ng WK, Kanaujia P, Chan KP, Dong Y. Hot-melt extrusion microencapsulation of quercetin for taste-masking. J Microencapsul. 2017;34:29–37.
42. National Center for Biotechnology Information. 1-Palmitoyl-2-linoleoylphosphatidylcholine [Internet]. 2023 [cited 2023 Dec 1]. Available from: https://pubchem.ncbi.nlm.nih.gov/compound/1-Palmitoyl-2-linoleoylphosphatidylcholine.
43. Hoang Thi TH, Morel S, Ayouni F, Flament MP. Development and evaluation of taste-masked drug for paediatric medicines—application to acetaminophen. Int J Pharm. 2012;434:235–42.
44. Balducci AG, Colombo G, Corace G, Cavallari C, Rodriguez L, Buttini F, et al. Layered lipid microcapsules for mesalazine delayed-release in children. Int J Pharm. 2011;421:293–300.
45. Petrovick GF, Kleinebudde P, Breitkreutz J. Orodispersible tablets containing taste-masked solid lipid pellets with metformin hydrochloride: influence of process parameters on tablet properties. Eur J Pharm Biopharm. 2018;122:137–45.
46. Qi S, Deutsch D, Craig DQM. An investigation into the mechanisms of drug release from taste-masking fatty acid microspheres. J Pharm Sci. 2008;97:3842–54.
47. Tulini FL, Souza VB, Echalar-Barrientos MA, Thomazini M, Pallone EMJA, Favaro-Trindade CS. Development of solid lipid microparticles loaded with a proanthocyanidin-rich cinnamon extract (Cinnamomum zeylanicum): potential for increasing antioxidant content in functional foods for diabetic population. Food Res Int. 2016;85:10–8.
48. Banerjee S, Joshi U, Singh A, Saharan VA. Lipids for taste masking and taste assessment in pharmaceutical formulations. Chem Phys Lipids. 2021;235:105031.
49. Silva LAD, Teixeira FV, Serpa RC, Esteves NL, Dos Santos RR, Lima EM, et al. Evaluation of carvedilol compatibility with lipid excipients for the development of lipid-based drug delivery systems. J Therm Anal Calorim. 2016;123:2337–44.
50. Bora D, Borude P, Bhise K. Taste masking by spray-drying technique. AAPS PharmSciTech. 2008;9:1159–64.
51. Bertoni S, Dolci LS, Albertini B, Passerini N. Spray congealing: a versatile technology for advanced drug-delivery systems. Ther Deliv. 2018;9:833–45.
52. Maniruzzaman M, Boateng JS, Snowden MJ, Douroumis D. A review of hot-melt extrusion: process technology to pharmaceutical products. ISRN Pharm. 2012;2012:1–9.
53. Lang B, McGinity JW, Williams RO. Hot-melt extrusion—basic principles and pharmaceutical applications. Drug Dev Ind Pharm. 2014;40:1133–55.
54. Jannin V, Cuppok Y. Hot-melt coating with lipid excipients. Int J Pharm. 2013;457:480–7.
55. Lopes DG, Salar-Behzadi S, Zimmer A. Designing optimal formulations for hot-melt coating. Int J Pharm. 2017;533:357–63.
56. Passerini N, Calogerà G, Albertini B, Rodriguez L. Melt granulation of pharmaceutical powders: a comparison of high-shear mixer and fluidised bed processes. Int J Pharm. 2010;391:177–86.
57. Wagner D, Spahn-Langguth H, Hanafy A, Koggel A, Langguth P. Intestinal drug efflux: formulation and food effects. Adv Drug Deliv Rev. 2001;50:S13–31.
58. Nanjwade BK, Patel DJ, Udhani RA, Manvi FV. Functions of lipids for enhancement of oral bioavailability of poorly water-soluble drugs. Sci Pharm. 2011;79:705–28.
59. Jacobs IC. Atomization and spray-drying processes. In: Microencapsulation in the food industry: a practical implementation guide. Academic Press; 2014. p. 47–56.
60. Ferreira MPA, Martins JP, Hirvonen J, Santos HA. Spray-drying for the formulation of oral drug delivery systems. In: Nanotechnology for oral drug delivery: from concept to applications. Academic Press; 2020. p. 253–84.
61. Georgieva Y, Kassarova M, Kokova V, Apostolova E, Pilicheva B. Taste masking of enalapril maleate by microencapsulation in Eudragit EPO® microparticles. Pharmazie. 2020;75:61–9.
62. Kharb V, Saharan VA, Kharb V, Jadhav H, Purohit S. Formulation and evaluation of lipid based taste masked granules of ondansetron HCl. Eur J Pharm Sci. 2014;62:180–8.

63. Jańczyk M, Kutyła A, Sollohub K, Wosicka H, Cal K, Ciosek P. Electronic tongue for the detection of taste-masking microencapsulation of active pharmaceutical substances. Bioelectrochemistry. 2010;80:94–8.
64. Stavras FM, Partheniadis I, Nikolakakis I. Formulation of taste-masked orodispersible famotidine tablets by sequential spray drying and direct compression—bitterness evaluation. J Drug Deliv Sci Technol. 2023;81:104290.
65. Karami A, gholami A, abroumand, Babaloo H, Farhadian N. Encapsulation: spray chilling and cooling. In: Principles of biomaterials encapsulation, vol. 1. Woodhead Publishing; 2023. p. 109–30.
66. Favaro-Trindade CS, De Matos Junior FE, Okuro PK, Dias-Ferreira J, Cano A, Severino P, et al. Encapsulation of active pharmaceutical ingredients in lipid micro/nanoparticles for oral administration by spray-cooling. Pharmaceutics. 2021;13:1186.
67. Sosnik A, Seremeta KP. Advantages and challenges of the spray-drying technology for the production of pure drug particles and drug-loaded polymeric carriers. Adv Colloid Interf Sci. 2015;223:40–54.
68. Napiórkowska A, Kurek M. Coacervation as a novel method of microencapsulation of essential oils—A review. Molecules. 2022;27:5142.
69. Thies C. Microcapsules. In: Encyclopedia of food sciences and nutrition. Academic Press; 2003. p. 3892–903.
70. Manjanna KM, Shivakumar B, Pramod Kumar TM. Microencapsulation: an acclaimed novel drug-delivery system for NSAIDs in arthritis. Crit Rev Trade Ther Drug Carrier Syst. 2010;27:509–45.
71. Al-Kasmi B, Alsirawan MHDB, Paradkar A, Nattouf AH, El-Zein H. Aqueous and pH dependent coacervation method for taste masking of paracetamol via amorphous solid dispersion formation. Sci Rep. 2021;11:1–11.
72. Bayraktar O, Erdogan I, Köse MD, Kalmaz G. Nanocarriers for plant-derived natural compounds. In: Nanostructures for antimicrobial therapy: nanostructures in therapeutic medicine series. Elsevier; 2017. p. 395–412.
73. Jose A, Sreeja V, Sadanandan P, Athira KV. Cyclodextrin in drug delivery. In: Natural biopolymers in drug delivery and tissue Engineering; 2023. p. 179–222.
74. Puskás I, Malanga M. Benefits and limitations of using cyclodextrins in drug formulations. International Symposium of Drug Delivery Systems. Woodhead Publishing; 2017
75. Shinde B, Khot P, Patil D, Doshi P, Gautam M, Gairola S. Cyclodextrins (CDs) derived from natural source as an essential component in biopharmaceutics. In: New horizons in natural compound research. Academic Press; 2023. p. 361–71.
76. Frey CR, Scott MJ. Liquid formulations. AAPS Adv Pharm Sci Ser. 2014;11:141–51.
77. Lyseng-Williamson KA. Midazolam oral solution (Ozalin®): a profile of its use for procedural sedation or premedication before anaesthesia in children. Drugs Ther Perspect. 2019;35:255–62.
78. Chay SK, Keating AV, James C, Aliev AE, Haider S, Craig DQM. Evaluation of the taste-masking effects of (2-hydroxypropyl)-β-cyclodextrin on ranitidine hydrochloride; a combined biosensor, spectroscopic and molecular modelling assessment. RSC Adv. 2018;8:3564–73.
79. Patel AR, Vavia PR. Preparation and evaluation of taste masked famotidine formulation using drug/β-cyclodextrin/polymer ternary complexation approach. AAPS PharmSciTech. 2008;9:544–50.
80. Jagdale SC, Gawali VU, Kuchekar BS, Chabukswar AR. Formulation and in vitro evaluation of taste-masked oro-dispersible dosage form of diltiazem hydrochloride. Braz J Pharm Sci. 2011;47:907–16.
81. Liu T, Wan X, Luo Z, Liu C, Quan P, Cun D, et al. A donepezil/cyclodextrin complexation orodispersible film: effect of cyclodextrin on taste-masking based on dynamic process and in vivo drug absorption. Asian J Pharm Sci. 2019;14:183–92.
82. Golden LS. Ion exchange resins: characterization of. In: Encyclopedia of separation science. Academic Press; 2000. p. 3172–9.

83. Tan DCT, Ong JJ, Gokhale R, Heng PWS. Hot melt extrusion of ion-exchange resin for taste masking. Int J Pharm. 2018;547:385–94.
84. SenGupta AK. Commercial ion exchangers. In: Ion exchange in environmental processes. Wiley; 2017. p. 445–56.
85. Arrad O, Sasson Y. Commercial ion exchange resins as catalysts in solid-solid-liquid reactions. J Org Chem. 1989;54:4993–8.
86. Sharma S, Lewis S. Taste masking technologies: a review. Int J Pharm Pharm Sci. 2010; https://doi.org/10.22270/jddt.v9i4-A.3590.
87. Dasankoppa FS, Komal S, Sholapur HN, Nanjundaswamy NG, Sajjanar VM. Design, optimization and evaluation of chewable tablets of clarithromycin using ion exchange resins. Indian J Pharm Sci. 2017;78:818–26.
88. Panraksa P, Boonsermsukcharoen K, Hwang KM, Park ES, Jantrawut P. Taste masking of Nizatidine using ion-exchange resins. Processes. 2019;7:779.
89. Michael Rajesh A, Bhatt SA, Brahmbhatt H, Anand PS, Popat KM. Taste masking of ciprofloxacin by ion-exchange resin and sustain release at gastric-intestinal through interpenetrating polymer network. Asian J Pharm Sci. 2015;10:331–40.
90. Singh I, Rehni A, Kalra R, Joshi G, Kumar M, Aboul-Enein H. Ion exchange resins: drug delivery and therapeutic applications. FABAD J Pharm Sci. 2007;32:91–100
91. Leonard GS, Cooper D. Oral liquid compositions containing paroxetine resinate - US5811436A. 1995.
92. Mehta K, Tu Y-H. Modified release formulations containing drug-ion exchange resin complexes - US8747902B2. 2007.
93. Jewell R. Omeprazole. In: xPharm: the comprehensive pharmacology reference; 2007. p. 1–5.
94. Soto J, Winzenburg G, Turner R, Desset-Brèthes S, Sheng Y, Orlu-Gul M, et al. Assessing the bitter taste of medicines: a comparison between rat taste panels (via the brief-access taste aversion (BATA) model) and human taste panels. Int J Pharm. 2016;511:1127–8.
95. Yoo O, von Ungern-Sternberg BS, Lim LY. Paediatric medicinal formulation development: utilising human taste panels and incorporating their data into machine learning training. Pharmaceutics. 2023;15:2112.
96. Guedes M, Marques M. The use of electronic tongue and sensory panel on taste evaluation of pediatric medicines: a systematic review. Pharm Dev Technol. 2021;26(2):119–37. Taylor & Francis
97. Shukla D, Chakraborty S, Singh S, Mishra B. Fabrication and evaluation of taste masked resinate of risperidone and its orally disintegrating tablets. Chem Pharm Bull (Tokyo). 2009;57:337–45.
98. Yan YD, Woo JS, Kang JH, Yong CS, Choi HG. Preparation and evaluation of taste-masked donepezil hydrochloride orally disintegrating tablets. Biol Pharm Bull. 2010;33:1364–70.
99. Thi THH, Morel S, Ayouni F, Flament M-P. Development and evaluation of taste-masked drug for paediatric medicines—application to acetaminophen. Int J Pharm. 2012;434:235–42.
100. Clapham D, Belissa E, Inghelbrecht S, Pensé-Lhéritier AM, Ruiz F, Sheehan L, et al. A guide to best practice in sensory analysis of pharmaceutical formulations †. Pharmaceutics. 2023;15:2319.
101. Schiffman SS, Zervakis J, Suggs MS, Budd KC, Iuga L. Effect of tricyclic antidepressants on taste responses in humans and gerbils. Pharmacol Biochem Behav. 2000;65:599–609.
102. Legin A, Rudnitskaya A, Clapham D, Seleznev B, Lord K, Vlasov Y. Electronic tongue for pharmaceutical analytics: quantification of tastes and masking effects. Anal Bioanal Chem. 2004;380:36–45.
103. Cotterill JV, Massei G, Cowan DP. Masking the taste of the conditioned taste aversion agent levamisole using an ion-exchange resin, for practical application in wildlife management. Pest Manag Sci. 2006;62:120–5.
104. Maniruzzaman M, Boateng JS, Bonnefille M, Aranyos A, Mitchell JC, Douroumis D. Taste masking of paracetamol by hot-melt extrusion: an in vitro and in vivo evaluation. Eur J Pharm Biopharm. 2012;80:433–42.

105. Yi EJ, Kim JY, Rhee YS, Kim SH, Lee HJ, Park CW, et al. Preparation of sildenafil citrate microcapsules and in vitro/in vivo evaluation of taste masking efficiency. Int J Pharm. 2014;466:286–95.
106. Ali J, Bong Lee J, Gittings S, Iachelini A, Bennett J, Cram A, et al. Development and optimisation of simulated salivary fluid for biorelevant oral cavity dissolution. Eur J Pharm Biopharm. 2021;160:125–33.
107. Ali J, Zgair A, Hameed GS, Garnett MC, Roberts CJ, Burley JC, et al. Application of biorelevant saliva-based dissolution for optimisation of orally disintegrating formulations of felodipine. Int J Pharm. 2019;555:228–36.
108. Tung NT, Tran CS, Nguyen TL, Hoang T, Trinh TD, Nguyen TN. Formulation and biopharmaceutical evaluation of bitter taste masking microparticles containing azithromycin loaded in dispersible tablets. Eur J Pharm Biopharm. 2018;126:187–200.
109. Keeley A, Teo M, Ali Z, Frost J, Ghimire M, Rajabi-Siahboomi A, et al. In vitro dissolution model can predict the in vivo taste masking performance of coated multiparticulates. Mol Pharm. 2019;16:2095–105.
110. Woertz K, Tissen C, Kleinebudde P, Breitkreutz J. Performance qualification of an electronic tongue based on ICH guideline Q2. J Pharm Biomed Anal. 2010;51:497–506.
111. Harada T, Uchida T, Yoshida M, Kobayashi Y, Narazaki R, Ohwaki T. A new method for evaluating the bitterness of medicines in development using a taste sensor and a disintegration testing apparatus. Chem Pharm Bull (Tokyo). 2010;58:1009–14.
112. Woertz K, Tissen C, Kleinebudde P, Breitkreutz J. A comparative study on two electronic tongues for pharmaceutical formulation development. J Pharm Biomed Anal. 2011;55:272–81.
113. Amelian A, Szekalska M, Ciosek P, Basa A, Winnicka K. Characterization and taste masking evaluation of microparticles with cetirizine dihydrochloride and methacrylate-based copolymer obtained by spray drying. Acta Pharma. 2017;67:113–24.

Chapter 7
Oral Liquid Formulations

David Harris and Peter Kühl

Abstract Oral liquids—principally solutions and suspensions—are among the most widely used pediatric formulations. They enable flexible dosing and are swallowable by children of all ages, and can also be administered through feeding tubes if necessary. The principal challenges in oral liquid formulation selection include the solubility of the drug, the physical and chemical stability of the product, the need for microbial presentation of multidose presentations, the taste of the formulation, and the usability of the product by the parent or caregiver. This chapter reviews the principles of formulating oral solutions and suspensions, as well as their packaging. Also discussed are considerations of biopharmaceutics and dosing devices, as well as the particular formulation challenges associated with poorly soluble drugs and modified-release products. This chapter highlights the complexity of pediatric formulation selection, the trade-offs inherent in this decision, and the complexity of the formulation-dosing device interface.

Keywords Formulation · Oral liquid · Solution · Suspension · Syrup

7.1 Introduction to Oral Liquid Formulations

Oral liquids, including solutions and suspensions, are probably the most widely used pediatric formulations [1, 2]. They are the dosage form of choice for newborns and infants who are unable to take solids, and are well-suited for toddlers and school-age children and also for older patients who have difficulty swallowing solid

D. Harris (✉)
Merck & Co., Inc., Rahway, NJ, USA
e-mail: david.harris@merck.com

P. Kühl (✉)
F. Hoffmann-La Roche AG, Basel, Switzerland
e-mail: peter.kuehl@roche.com

H. Batchelor, K. Rose (eds.), *Pediatric Formulations*, AAPS Advances in the Pharmaceutical Sciences Series 58, https://doi.org/10.1007/978-3-031-77240-5_7

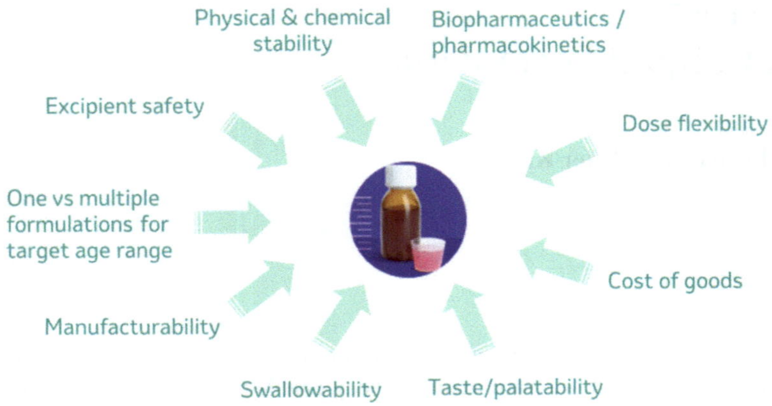

Fig. 7.1 Illustration of the factors that constrain design and development of oral pediatric formulations

dosage forms. Liquids are suitable for adolescents, although solid dosage forms are more likely to be preferred by this age group [3, 4]. The chief limitations of oral liquids are the need for antimicrobial preservation of multi-dose presentations, their technical complexity, especially for poorly water-soluble drugs, and their higher transportation costs relative to solid dosage forms. There is a strong preference, from some stakeholders, that the use of antimicrobial preservatives in pediatric formulations should be avoided, favoring the development of solid dosage forms and unit-dose formulations for pediatric medications that can nonetheless support flexible dosing [5, 6]. However, oral liquids offer unparalleled dose flexibility and generally require minimal manipulation by the parent or caregiver at the point of use and thus remain a mainstay of pediatric formulation development [1–3, 5].

The selection of an appropriate formulation platform for a pediatric product is challenging; Fig. 7.1 illustrates the multiple factors that should be taken into account in making that decision. It must also be emphasized that many of these design considerations are in conflict with one another, and so the ultimate formulation platform decision is usually a compromise; there is clearly no "best" formulation choice in most situations. While oral liquids have many advantages, these may not be the most suitable pediatric formulation in all cases.

Oral liquid dosage forms can principally be divided into solutions and suspensions. Solutions may be aqueous or (less commonly now) solvent or co-solvent based; syrups are solutions that contain a high sugar content. Both solutions and suspensions may be provided as "ready-to-use" products or as powders or granules for constitution with liquid—usually water—at the point of use. The principal types of oral liquid products and the principles guiding their development are described in detail in the following sections; some less common oral liquid products are also described in summary form.

7.2 Oral Liquid Formulation Selection

The advantages and disadvantages of oral liquids relative to other formulation platforms are discussed at length elsewhere. In principle, though, the two key benefits of oral liquids over other oral formulations are:

1. Dose flexibility—the ability for the caregiver to make adjustments to the dose
2. Swallowability—the ability of the patient to swallow the dose, or if necessary, for it to be administered through a feeding tube

These two characteristics clearly set oral liquids apart from oral solid dosage forms, for which dose flexibility is constrained and for which swallowability is always a consideration, particularly for infants and children otherwise unable to swallow tablets or capsules.

When formulating an oral liquid, the first and most fundamental decision to be made is whether to formulate as a solution or a suspension. A solution (or syrup) will generally be the preferred dosage form so long as all of the following three conditions can be met:

1. The drug has adequate solubility, such that the intended dose can be dissolved in a suitable volume of an aqueous vehicle or can be solubilized by means of pH adjustment or use of cosolvents or solubilizers at acceptable levels.
2. The drug has acceptable chemical stability in the vehicle.
3. The taste of the dissolved drug is acceptable or can be adequately masked.

Drugs that cannot be adequately solubilized will, by necessity, be formulated as suspensions. Drugs with inadequate chemical stability may be formulated as suspensions, or may be formulated as powders or granules for constitution at or closer to the time of use; these are termed powder-for-suspension or, less commonly, powder-for-solution products. The European Medicines Agency recommends dose volumes of not more than 5 mL for children under 5 years and not more than 10 mL for children 5 years and older [5]. Dose volumes are further limited for infants, and even more so for neonates, in whom stomach volumes are extremely limited.

The functionality and selection of excipients for use in oral liquid formulations, and the particular considerations for use of these in pediatric formulations, are discussed in [3, 7, 8]. Excipient selections are scrutinized by regulatory agencies much more closely today than has been the case in past decades; concerns have been raised regarding the safety of many excipients and excipient classes, including cosolvents, sugars, artificial sweeteners, antimicrobial preservatives, and colorants. A European Medicines Agency guideline [9] details the cautionary information that is required to be included in product labeling for certain excipients and the use thresholds above which this labeling is required. The details of excipient selection are addressed later in this chapter, where specific dosage forms are discussed, and in the chapter on excipients elsewhere in this volume.

7.2.1 Solutions and Syrups

Solutions and syrups are suitable dosage forms for pediatric patients of all ages, provided the active substance can be adequately solubilized in the vehicle, has acceptable chemical stability, and has an acceptable taste profile. At its simplest, a solution formulation may consist of only the active substance, a solvent, an antimicrobial preservative, and perhaps a buffer. In practice, most solution formulations are more complex than this and may also include flavors and/or sweeteners, thickeners, solubilizers, antioxidants, and dyes. Syrups are viscous solutions that contain high concentrations of sugar, traditionally sucrose, although other sugars may be used instead. The high sugar concentrations and correspondingly low water activities in syrups can render them self-preserving, thus avoiding the need for an antimicrobial preservative. The sweetness of the sugar can also serve to mask the taste of some active substances and thus improve the palatability of the formulation, as discussed later in this chapter.

The unique elements of oral solution formulation development are summarized in the following sections.

7.2.1.1 Drug Solubility and Vehicle Selection

Entirely aqueous vehicles are obviously preferred for oral solution formulations, although this can be difficult to achieve for all but the most soluble drug substances, considering the target dose and the limited liquid volumes that can be dosed to children. Mixed organic solvent systems or aqueous/organic solvent systems can also be used to solubilize the drug; a substantial number of oral liquid products contain organic cosolvents, in particular ethanol, glycerin, polyethylene glycol (PEG), and propylene glycol [1, 2]. However, there are acute and chronic safety concerns associated with the use of these materials in children, particularly in young children. For example, the metabolism of propylene glycol is limited in children under 4 years of age, which manifests as a longer half-life and thus a greater degree of accumulation with repeat dosing [3]. In the current regulatory environment and with the heightened focus on excipient safety, it is questionable whether all of the currently marketed products that contain organic cosolvents would be approvable today. It should be noted that these limitations make it particularly difficult to formulate oral solution formulations for infants, which is the age group most in need of formulations that support flexible dosing and are readily swallowed.

A distinct category of oral solution formulations is syrups; these are viscous, predominantly aqueous solutions containing high concentrations of sugar, traditionally sucrose, although other sugars and oligo/polysaccharides (e.g., sorbitol, sorbitol solution, xylitol, liquid glucose) may also be used. Syrup NF contains 85% w/v sucrose and is self-preserving (i.e., resistant to microbial growth) by virtue of its low water activity. The sweetness imparted by sucrose, and to varying extents by other sugars, provides some degree of taste masking. One of the main drawbacks of sucrose is that it supports tooth decay and thus it is not favored for therapies intended for chronic use.

Sorbitol and xylitol—6-carbon and 5-carbon monosaccharides, respectively—are also used in a number of syrup and oral solution formulations. Both are sweet-tasting and yield clear viscous solutions; these are not fermented by microbes in the oral cavity and thus do not support tooth decay. Sorbitol solution is an aqueous solution that is composed primarily of sorbitol but also contains lower levels of oligosaccharides and polysaccharides; it is formed by the hydrolysis of starch. Liquid glucose is a similar material, composed mostly of glucose along with maltose and dextrins; it is formed by the hydrolysis of corn starch. Like other syrups, these materials are clear, viscous, and sweet-tasting; both are used in a number of oral liquid formulations.

It should also be noted that non-absorbed or poorly absorbed sugar alcohols such as mannitol and sorbitol can exert an osmotic effect and accelerate small intestinal transit, thereby reducing the time available for the drug to be absorbed in the small intestine. Mannitol has been shown to have an effect on small intestinal transit time [10] at levels that are relevant to formulations and clearly below the levels needed to induce a laxative effect. This presents an obvious risk for drugs that are primarily or exclusively absorbed from the small intestine; studies have, in fact, shown a reduction in the oral bioavailability of some drugs when formulated with higher amounts of sugar alcohols [11, 12]. The uses of such sugar alcohols in pediatric formulations are thus limited to around 150–300 mg/kg/day [5]. PEG 400, which is also poorly absorbed and frequently used as a cosolvent, has also been shown to speed small intestinal transit [13], which is perhaps unsurprising given that higher molecular weights of PEG are used clinically as a laxative.

The aqueous solubility of ionizable drugs can be increased by adjusting the pH of the vehicle with the aid of acidulants or buffer systems, although it should be recognized that vehicle pH can also affect the chemical stability of the drug and the effectiveness of the antimicrobial preservative system, as discussed later. Drug solubility can also be increased by means of surfactants and cyclodextrins [14].

7.2.1.2 pH and Buffer Capacity

In addition to its role in drug solubilization, the pH of a liquid formulation is also an important determinant of drug stability, since many degradation reactions, particularly hydrolytic or oxidative reactions, are pH-dependent. In addition, the effectiveness of many antimicrobial preservatives is pH-dependent (see Section 7.3.2, Selection of Antimicrobial Preservatives). It is therefore essential that the pH of the vehicle be appropriately selected and that the pH be maintained over the shelf life of the product. A buffer system is generally necessary to ensure the pH does not drift over time. The most commonly used buffer systems are based on citrate or phosphate salts or combinations of these two; buffer salt concentrations are usually within the range of 0.1 to 1.0 Molar.

7.2.2 Suspensions

A suspension is a system in which particles of the drug substance are uniformly dispersed and suspended in a liquid—most commonly aqueous—vehicle. As outlined above, drugs are usually formulated as suspensions because they do not have adequate solubility in acceptable vehicles or—less commonly—to mitigate chemical instability or the taste of the drug substance. The drug substance in suspension formulations is generally crystalline, with a median particle size of 0.5 μm to 1 μm or greater; particle sizes significantly smaller than 1 μm may be considered nanosuspensions or colloidal suspensions and are a special case, not often used as an oral liquid formulation for pediatric administration. The principles of suspension formulation, including excipient selection, have been detailed by a number of authors [8, 15].

Many of the elements of suspension formulation development—consideration of the chemical stability of the drug substance, antimicrobial preservation, palatability—are similar to those of solution formulation development, as described elsewhere. There are some additional and critically important elements in suspension formulation related to the physical stability of the suspension and the dispersed nature of the product.

7.2.2.1 Suspension Stability

Most pharmaceutical suspensions are inherently physically unstable; i.e., in the absence of agitation, the dispersed phase will tend to settle out over time to form a sediment at the bottom of the container. The rate of sedimentation is determined by Stokes' Law, which defines the rate of sedimentation of a particle to be related to the square of its diameter, the difference in density between the particle and the medium, and the viscosity of the medium. Thus the smaller the particle, the smaller the density difference and the higher the viscosity of the vehicle, the slower will be the rate of sedimentation. Carlin [15] clearly defined the different approaches that can be taken to formulate a pharmaceutically acceptable suspension and the characteristics of the resulting products, which are briefly summarized in the following sections. It should be apparent that proper understanding and control of the settling and flocculation behavior of the drug particles in suspension are essential to ensure dose uniformity, both in the short term (to permit a dose to be accurately withdrawn after dispersing) and in the long term (to ensure suspension stability or redispersibility over shelf life).

7.2.2.2 Structured Vehicles

Structured vehicles result in permanent suspensions, in which the drug substance does not sediment over the shelf-life of the product. These are achieved through the use of suspending agents such as clays and combinations of microcrystalline cellulose and sodium carboxymethylcellulose (Avicel® RC-591 and Avicel® CL-611). These vehicles have a gel-like structure at rest with a defined yield stress, which prevents the drug substance from settling. These materials also exhibit pseudo-plastic and/or thixotropic behavior, meaning they show a reduction in viscosity when exposed to shear; this facilitates pouring or otherwise dosing the product. It should be noted that the structured vehicle approach is not widely used in oral suspension development.

7.2.2.3 Settling Suspension

The more widely used approach to suspension formulation is the use of polymeric and other materials to increase the viscosity of the vehicle, which slows the rate of drug sedimentation, but it is essential to recognize that drug sedimentation will still occur in these systems, and the vehicle viscosity determines the rate of sedimentation, but does not prevent it. Once settled, these suspensions must be redispersed by shaking prior to dosing; too high a viscosity will make it more difficult to resuspend the settled drug particles by shaking, and so the viscosity of the system should be selected with care. The most commonly used suspending agents for this purpose are cellulosic polymers (methyl cellulose, hydroxypropyl cellulose, hydroxypropyl methylcellulose, or hypromellose) and gums (xanthan gum, carrageenan). In addition to retarding the settling of the drug particles, these materials also provide rheological properties suitable for administration of the product; these agents are also generally pseudo-plastic and/or thixotropic.

7.2.2.4 Flocculated Systems

Flocculation refers to the association of particles in suspension to form larger aggregates or flocs, driven by interparticulate forces. Flocculation leads to more rapid settling of suspensions, due to the larger size of the particle aggregates, but—crucially—the resulting sediment is less dense and thus more easily redispersed by agitation (shaking); non-flocculated drug particles that settle slowly to form a dense sediment may not be readily redispersed by shaking. Flocculation in pharmaceutical suspensions is achieved by modulating the surface charge on the particle surfaces by means of adding electrolytes, surfactants, or polymers to the formulation and/or by adjusting the pH of the formulation. The flocculation approach, too, is not widely used in the development of oral liquid suspension formulations.

7.2.2.5 Ostwald Ripening

Ostwald ripening is a phenomenon that occurs in dispersed systems where small particles grow at the expense of larger particles over time. This process is driven by the difference in solubility between the smaller and larger particles. Drug particles in suspension are generally not monodisperse; in practice, there is generally a distribution of particles present. The free energy at the surface of smaller particles is higher than that at the surface of larger particles. There is thus a thermodynamic driving force for the drug to dissolve from the smaller particles and crystallize on the surface of the larger particles. This is exacerbated by temperature fluctuations, which cause the solubility of the drug in the vehicle to also fluctuate over time. This leads over time to an increase in the size of the larger particles in the system and a loss of the smaller particles. The risk of Ostwald ripening can be minimized by minimizing the solubility of the drug in the vehicle (e.g., by selecting the pH appropriately), by ensuring the particle size range is as narrow as possible, and possibly also by use of surfactants or polymers, which can adsorb on the particle surface and slow crystal growth.

7.2.2.6 Wettability

Poorly soluble drugs tend to have high interfacial tensions and be poorly wettable with aqueous media; this can cause practical difficulties in incorporating the drug substance during manufacturing and can lead to aggregation of drug particles with one another in the product and adsorption of drug particles to surfaces. This can be mitigated by the addition of a surfactant to the formulation to reduce the interfacial tension at the particle surface; the surfactants used in pediatric oral formulations have been reviewed elsewhere [1, 2].

7.2.3 Powders/Granules for Solution/Suspension

In contrast to the syrup/solution and suspension products described above, which are all supplied as ready-to-use liquids, powders/granules for solution/suspension are supplied as solids and constituted with liquid at or close to the time of use. These consist of a somewhat diverse range of products and presentations [1, 2]. These may be considered simple powder blends or granulations, depending on the manufacturing process by which they are prepared; also, tablets for constitution are increasingly being developed for pediatric patients. Upon constitution with liquid, these may form solutions or suspensions; although in practice, these are most commonly suspensions. One of the main reasons for using this formulation approach is that it enables the delivery of drugs that are not adequately stable in ready-to-use solutions or suspensions. The liquid used for constitution is most often water, for which potable water is usually suitable; there are a few examples of products that require purified water or a "custom" vehicle for constitution. When developing a product for resource-challenged settings, the availability of potable water for constitution should also be considered. These products may be formulated and provided as multidose bottle presentations or as unit-dose presentations; this decision has significant implications for the development, manufacturing, and use of the product.

7.2.3.1 Multidose Presentations

These are generally presented in a bottle format, intended to be constituted by the pharmacist or other healthcare professional at the time of dispensing or, less commonly, by the parent or caregiver. This platform is widely used for antibiotics for pediatric use. Once constituted, as multidose products, these should meet essentially all of the design criteria described above for ready-to-use liquid products. The exact amount of diluent (usually water) to be added to the bottle is either measured separately in an appropriate measuring device or assessed by filling the bottle to a mark molded on the bottle or printed on the label. This constitution step is a critical step, since any error in the volume of diluent added will affect the drug concentration and thus lead to dosing errors. Due to the impact of this risk, it is usually preferred to instruct the diluent volume to be measured separately; for the same reason, it is also preferred that such constitution steps are performed by a pharmacist or other healthcare professional and not by parents/caregivers.

Some of the unique aspects of multidose formulations for constitution are as follows:

- These formulations must be preserved so as to prevent microbial growth, as described elsewhere in this chapter.
- Formulation of the powder blend or granulation must assure that the blend is sufficiently free-flowing to enable accurate filling into bottles, and that the blend is not overly sensitive to segregation during processing such that the filled bottles are uniform with respect to both drug and antimicrobial preservative(s).
- The blend must readily wet and disperse upon addition of water, and achieve a pharmaceutically acceptable suspension with a shake of reasonable intensity and duration.
- It is necessary that the resulting product can be dosed accurately; if a suspension, it must remain homogeneous for long enough after shaking for a parent or caregiver to extract an accurate dose, and if settling occurs between doses, it must be readily resuspended with a simple shake.

7.2.3.2 Unit-Dose Presentation

These are generally packaged as sachets or "stick packs" such that the contents of one (or more) unit are combined with vehicle by the parent or caregiver, some or all of the resulting product is dosed to the patient, and any remainder is discarded. Some of the unique aspects of this formulation approach are as follows:

- No antimicrobial preservation is needed, since the product is dosed immediately after preparation.
- The contact time between the powder or granulation and the vehicle is short, enabling the dosing of formulations that are highly unstable in aqueous media. An important example is amorphous solid dispersions of poorly water-soluble drugs, which are designed to yield supersaturated (and thermodynamically unstable) solutions of the drug upon contact with aqueous media.
- The liquid formulation is both prepared and dosed by the parent or caregiver; it is thus essential that the preparation steps are clearly explained and readily executed by lay users so as to minimize the risk of misdosing. It is important to consider that if only a portion of the constituted solution or suspension is to be dosed, the preparation must be sufficiently uniform to assure that the withdrawn volume will contain the correct dose.
- This approach can offer a cost-effective means of making low-commercial-volume products available to pediatric patients in an age-appropriate format. At their simplest, these powders or granulations may be derived from those used for an adult solid dosage form, with little if any modification.

7.2.4 Modified-Release Oral Liquid Formulations

Oral solution and suspension formulations are inherently immediate-release. Only one technology has been successfully developed and commercialized to provide shelf-stable modified-release oral liquid dosage forms, and that utilizes ion-exchange resin complexes coated with release-controlling polymer membranes. In these formulations, the ionized drug molecule is complexed with an oppositely charged ion exchange resin. Typical cation-binding resins for basic drugs are sodium polystyrene sulfonate and polacrilin potassium; anion-binding cholestyramine containing a quaternary ammonium group can be used for acidic drugs. Once complexed onto the resin, the drug remains bound until exposed to electrolytes, at which point the drug will be displaced from the resin and pass into solution. The rate of drug release can be modulated by application of a polymer coat onto the particles, most often by means of a bottom-spray Wurster process. These drug-loaded and coated resin particles can be formulated into shelf-stable suspension products, which only begin to release the drug when exposed to the electrolyte-rich environment of the gastrointestinal tract. An example of this technology is the Delsym® extended-release oral liquid formulation of dextromethorphan, currently marketed in the United States [1]. The formulation of these highly specialized products is detailed elsewhere [16–18] and will not be discussed further here.

7.3 Antimicrobial Preservation

Antimicrobial preservation is obligatory for all oral liquid formulations that are designed as multidose products for repeated dose withdrawal with an in-use period of more than a few hours, to prevent microbial growth resulting from contamination of the product during dosing [19]. Furthermore, since oral liquid formulations are not manufactured to be sterile and have an intrinsic bioburden, ready-to-use liquids also require preservation during shelf life. The use of antimicrobial preservatives requires justification, including a risk-benefit assessment, and the actual level selected must be the lowest feasible level as demonstrated with appropriate experimental data.

7.3.1 Antimicrobial Effectiveness Testing

The definitive experiments to demonstrate antimicrobial efficacy of the formulation containing a certain concentration of preservative are the compendial preservative effectiveness tests, performed according to Ph.Eur. 5.1.3., JP <19>, and USP <51>. The principle of these tests is to show the necessary reduction in microbial counts under standardized conditions after inoculation with high counts of microbes, while under actual use conditions, product contamination is more likely to occur with relatively lower counts. In these tests, microorganism counts of 10^5 to 10^6 per mL are inoculated into the test formulation, and the reduction (or growth) of each organism is tested after 14 and 28 days. The criteria for acceptable antimicrobial efficacy are specified as log 10 reductions in the counts of each organism relative to the initial counts. The test microorganisms specified by the compendia and the major classes of contaminants that they represent are listed in Table 7.1.

The compendial test monographs are not fully harmonized; the Ph.Eur. Chapter 5.1.3. sets more stringent requirements for the reduction of organisms after 14 days compared to the USP test. Once the minimally effective concentration of

Table 7.1 Compendial challenge microorganisms for antimicrobial effectiveness testing

Organism	Class	Compendial reference
Pseudomonas aeruginosa	Gram-negative bacillus	Ph.Eur./USP/JP
Staphylococcus aureus	Gram-positive coccus	Ph.Eur./USP/JP
Candida albicans	Yeast	Ph.Eur./USP/JP
Aspergillus brasiliensis	Mold	Ph.Eur./USP/JP
Escherichia coli	Gram-negative bacillus, facultative anaerobe	USP/JP

antimicrobial preservative has been determined in this way, a quantitative analytical procedure must be established to assure that the preservative content in each unit of drug product is within appropriate limits at the time of release, and that the preservative level remains above the minimally effective concentration for the duration of the shelf life and in-use period of the product.

The microorganisms listed in Table 7.1 are generally considered to be representative of possible microbial contamination under real-use conditions but may be supplemented by additional resistant organisms, such as *Burkholderia cepacia*, *Bacillus subtilis*, or *Zygosaccharomyces rouxii*. Furthermore, the *Burkholderia cepacia* complex (Bcc) represents a group of objectionable microorganisms that require exclusion from aqueous nonsterile products; these are opportunistic pathogens that may develop resistance against preservatives and have been noted for causing serious respiratory infections in immunosuppressed patient populations [20]. In reaction to these reports and product recalls, the FDA has introduced the monograph USP <60> for microbial testing for Bcc. Since oral liquids and the associated dosing devices could potentially present a source for contamination with Bcc, development scientists should consider applying microbial challenge tests for Bcc during product development.

7.3.2 Selection of Antimicrobial Preservatives

There are only a few preservatives established and commonly used in oral liquid medicines, principally methyl-, ethyl-, and propyl-parabens, potassium sorbate, propionic acid, benzoic acid and sodium benzoate, and benzyl alcohol. The many considerations that go into selection of a preservative have been described in detail by Elder and Crowley [21]. Many of these preservatives have optimal efficacy at particular pH ranges, as noted earlier in this chapter; microbial organisms themselves have varying tolerance to different vehicle pH. Therefore, the pH of an oral liquid formulation is a very important factor in selecting the antimicrobial preservative and optimizing the preservative level. It should be noted that other excipients can also have significant effects on the efficacy of the preservative, such as cyclodextrins due to their complexing nature, surfactants, chelating agents, and viscosity enhancers. As noted earlier, some stakeholders, including the EMA, have expressed the view that the use of antimicrobial preservatives in pediatric formulations should be avoided [5]. There are particular safety risks with some preservatives in children and infants, as described in more detail below. The selection of a suitable antimicrobial preservative for a pediatric liquid formulation is thus very challenging.

Sodium benzoate, along with its corresponding free form, benzoic acid, is highly effective as a preservative, provided the pH is not higher than about pH 5. Concentrations of 0.1–0.2% w/v are usually effective. However, benzoic acid and sodium benzoate (and also benzyl alcohol) can present a safety risk for neonates under 8 weeks of age; the immaturity of their metabolic enzyme systems can lead to an accumulation of benzoic acid, which can in turn displace bilirubin from albumin, resulting in hyperbilirubinemia. This can lead to acute bilirubin encephalopathy and may evolve to kernicterus (bilirubin-induced brain dysfunction) if left untreated. Accordingly, sodium benzoate should, if possible, be avoided in products intended for neonates [22–24].

Potassium sorbate/sorbic acid systems are also used in some oral liquid formulations, although less commonly than the benzoate and paraben systems. Concentrations of 0.05–0.2% w/v are usually effective. Like the sodium benzoate/benzoic acid system, potassium sorbate/sorbic acid systems are only effective at a pH of about 5 and below. Potassium sorbate/sorbic acid systems are considered safer than many of the other preservatives and are considered suitable for oral use in children; however, their relatively weak efficacy against bacteria and chemical incompatibilities and instability limit their broader use [21].

Combinations of methylparaben and propylparaben comprise the other preservative system frequently used in oral liquid formulations. Parabens (parahydroxybenzoate esters and their sodium salts) are effective over a wide pH range; concentrations totaling around 0.1–0.2% w/v are usually effective. The antimicrobial activity of these agents increases with their alkyl chain length, but their aqueous solubilities correspondingly decrease, hence their use in combination with one another. Estrogenic effects have been observed in preclinical species with propylparaben which have limited the use of this agent; allergenic effects have also been observed. A more recent assessment, which considered the findings of some more recent safety studies, has concluded that propylparaben may be safe to use, although it should be used at the lowest feasible level [25]. A PDE (Permitted Daily Exposure) value of 2 mg/kg/day has been calculated for propylparaben in pediatric as well as in adult patients; a similar exposure level is also considered acceptable for methylparaben [25].

Finally, formulation vehicles with low water activity (e.g., syrups) and vehicles containing sufficient quantities of cosolvents such as ethanol or propylene glycol may be inherently self-preserving. Cosolvents alone are not generally used for this purpose due to the high levels at which they need to be used and the safety concerns associated with these materials. However, cosolvents as well as low molecular weight polyols, sugar alcohols, and salts that decrease in the water activity can have auxiliary effects that increase the efficacy of other chemical preservatives.

7.4 Taste-Masking

While acceptability risks for younger children due to swallowability issues can be avoided with liquid forms, the palatability risks related to aversive taste are more prominent. Liquid formulations spread easily in the mouth and throat and rapidly expose unpleasant-tasting drug or excipients directly to the taste buds. Dissolved drug, which is present to some extent even in suspensions, can interact directly with taste receptors and give rise to an unpleasant product taste or aftertaste. With suspension formulations, residual drug particles may adhere to the mucosa or lodge elsewhere in the oral cavity, resulting in prolonged taste impact. It is almost inevitable that some measures are necessary to mask the taste of oral liquid formulations. Taste-masking is discussed at length elsewhere in this book, and has been comprehensively reviewed by Walsh et al. [26]; that content will not be repeated here. This section highlights the elements of taste-masking that are particularly relevant or unique to oral liquid formulations.

The most common taste-masking approaches that are used to some extent in almost all oral liquid formulations are the use of sweeteners and/or flavors to compete with and to cover the taste of the drug, or sometimes of adverse-tasting excipients. Bulk sweeteners or sugars (e.g., sucrose, sorbitol, xylitol, fructose) contribute sweetness and also mouth-feel to the product; these are generally used at relatively high levels in the formulation. High-intensity sweeteners, which may be synthetic (e.g., saccharin, aspartame, acesulfame potassium, sucralose) or naturally derived (e.g., ammonium glycyrrhizate, steviol glycosides), tend to be much sweeter than sugars on a weight-for-weight basis and are thus used at much lower levels. These sweeteners vary considerably with respect to taste quality, duration of sweetness, chemical stability, safety, and regulatory acceptability. Oral solutions sweetened with high-intensity sweeteners may be formulated with a polymer (e.g., a cellulosic) to increase the viscosity of the vehicle and thus provide a syrup-like mouthfeel. Flavors are most commonly fruit-derived or otherwise plant-derived and are often natural in origin, although some may contain synthetic agents (e.g., vanillin, cinnamaldehyde). In principle, at least, the flavor selected should be based on consideration of the taste characteristics of the drug or the vehicle. Taste preference varies with age, with children understood to prefer sweeter flavors than adults; taste preference also varies by country or region. The chemical stability of the flavor system must also be considered. Natural flavors, particularly fruit extracts, can contain tens or even hundreds of constituents, and loss or degradation of some of these can result in changes to the flavor profile, discoloration of the formulation or other product stability issues.

The second broad category of taste-masking approaches is to use formulation approaches to coat or otherwise sequester the drug to prevent access of the drug to the taste buds. Coating approaches are technically challenging for liquid formulations for two reasons. First, the particle size of coated particles should be kept sufficiently small—under 200 μm or so—so as not to be perceived as gritty in the mouth; taste-masking coatings can be applied to particles in this size range, but drug loading and processability present significant limitations. Second, the ability of the coating to prevent drug leakage during storage must be considered; practically, these coated particulate systems are often better suited to unit-dose presentations of powders or granules for suspension, where the contact time between powder and liquid or vehicle will be short (i.e., on the order of minutes or at most hours). Other sequestration approaches include the use of cyclodextrins and ion-exchange resins. All of these technological approaches to taste-masking have been reviewed elsewhere in this book and by other workers [18, 26 27].

7.5 Biopharmaceutical Aspects

Even though oral liquid formulations are not as dependent on dosage form disintegration or drug dissolution as is the case for solid dosage forms, there are multiple ways in which the biopharmaceutics of the drug and, thus, clinical effectiveness can be affected by formulation and excipient choices.

Oral liquids disperse and distribute in the GI tract more rapidly than solid dosage forms since there is no disintegration step necessary. Liquids empty from the stomach more quickly than solid materials in both the fasted and fed states as long as they are not highly viscous. In the fed state, low-viscosity liquids are able to exit the stomach readily even while solid materials are retained; this is sometimes termed the Magenstrasse or "stomach road" [28]. Due to this difference in the transit of liquid and solid materials, oral liquid formulations often show more rapid drug absorption relative to solid dosage forms, provided the dissolution of drug particles in suspension is not rate-limiting.

Formulation of poorly soluble (BCS Class 2/4) drugs as oral liquids can be challenging. Salunke and coworkers [14] reviewed the available approaches, including salt formation, cyclodextrin solubilization, and lipid solubilization. BCS Class 2/4 drugs may also be solubilized by means of cosolvents, such as ethanol, glycerin, PEG, and propylene glycol, or by pH adjustment, as described earlier. Fine particle size suspensions or even nanosuspensions can be formulated, leveraging the high specific surface area of finely divided materials to maximize the dissolution rate. The oral bioavailability that can be attained with such formulations is dictated by the solubility and the degree of supersaturation that can be achieved in the small intestine. Another well-established approach for the formulation of poorly soluble drugs is the formation of amorphous solid dispersions of drug in polymer, most commonly by means of spray-drying or hot-melt extrusion techniques; these metastable systems dissolve in aqueous media, delivering supersaturated drug solutions which can greatly boost the absorption of such drugs relative to administration of crystalline drug. These metastable solid dispersions are not amenable to the formulation of ready-to-use liquid formulations but can be provided as unit-doses of powders or granules for constitution, as reviewed earlier. The resulting liquids usually need to be dosed within a short time of preparation (hours or even minutes) to avoid the risk of drug precipitation and the consequent loss of bioavailability. Finally, lipophilic drug compounds can be formulated as lipid-based solutions, based either on a single lipid as solvent or, more robustly, as a self-emulsifying or even self-micro-emulsifying system containing surfactant and sometimes also cosolvents. Such lipid-based formulations are most often used as fills in soft gelatin capsules, but there are a few examples of commercial pediatric liquid products based on this approach [14], despite the expected adverse taste of these systems.

Formulation of weakly basic drugs can be similarly challenging, due to their differing solubilities at gastric (acidic) and intestinal (approximately neutral) pHs. Basic drugs formulated in solution at low pH may precipitate upon exposure to intestinal pH. Basic drugs formulated as suspensions may dissolve under gastric conditions, although the extent of dissolution is likely to depend, among other things, on the gastric residence time and the pH of the gastric contents, both of which may vary with patient age, disease state, presence of food, etc. The phenomenon of precipitation of such drugs on entry to the small intestine is widely recognized as a risk in adults [29, 30], although the true extent of the risk for any given drug is less well understood even in adults. This uncertainty is magnified in pediatric formulation development, given the relative paucity of data on the composition of gastrointestinal media (compositions, volumes, and pH) in children and the likely differences in gastrointestinal media and motility patterns as functions of age. Differences in bioavailability have been observed in pediatric populations due to changes in formulation, release profile, or excipient, and even relatively minor composition changes can result in large differences in drug absorption [31].

Since the dissolution step is not relevant for drugs formulated as oral solutions, the requirements for bioequivalence or bioavailability studies to bridge formulation changes may be waived. However, these waiver provisions [32, 33] stipulate that the drug concentration should be the same, and that there be no difference in the content of excipients that may affect the gastrointestinal transit time or absorption, such as sugars that can affect small intestinal transit, or surfactants that may affect membrane permeability. For suspensions, a biorelevant dissolution assay should ideally be developed to establish a physiologically based PK model that enables mechanistic prediction of drug absorption as a function of patient age. Changes to suspension formulations during development, however, are more likely to require bioequivalence or bioavailability studies to bridge.

7.6 Primary Package

Definition of the primary package is an integral part of the product development process for an oral liquid. The choice of drug product presentation—whether a multidose bottle or a unit-dose presentation, and whether a ready-to-use product or a powder/granule for constitution—is interrelated with the formulation and is an essential element of the Target Product Profile. Related to this decision is the selection of the devices and accessory items needed to support dosing and administration. These selections define the requirements for the manufacturing, secondary packaging, transportation, and shipping/storage conditions of the product and have implications for product stability for assurance of accurate dosing of the product by parents and caregivers. Considerations of product stability need to be balanced with considerations of usability and safety risks; thus, an early understanding of potential stability challenges is important to identify the appropriate container-closure system in the context of the desired Target Product Profile.

7.6.1 Multidose Bottles

Multidose formulations are typically packaged in glass or plastic bottles. Often, the need for a high degree of dosing flexibility or the desire to cover a wide patient age range with a single presentation dictates a multidose presentation, but this decision is associated with some drawbacks. Bottles are bulkier and heavier than unit-dose presentations, with implications for storage and transportation within the supply chain and also in the hands of the parent/caregiver. In particular, ready-to-use liquids are heavier and may require more strict temperature control during transportation and storage (e.g., protection from freezing); glass bottles also carry a higher risk of breakage.

Glass bottles are often selected since they provide the best possibility to optimize the chemical and physical stability. Glass is an inert, impermeable material that, in contrast to plastic, does not include any processing additives. There are no extractables or leachables to be assessed during product development, with the exception of the contact with the closure liner, and any ancillary devices such as press-in bottle adapters. The appropriate minimized level of antimicrobial preservatives can be more readily determined since there is no risk of absorption of the preservative by the bottle material over shelf life. The permeation of moisture and oxygen is dictated by the composition of the cap and its sealing performance. There are three types of glass used in medicinal products, which are differentiated by their hydrolytic resistance. The most resistant is borosilicate glass, which is typically not used for oral liquid products because their manufacturing process does not typically involve elevated temperatures that require high hydrolytic resistance. Type glass II is soda-lime silicate glass that offers improved surface resistance. Typically, however, Type III glass without surface treatment is used for oral liquids; this represents also the least expensive bottle material. Glass bottles for oral liquids are often amber colored, achieved by the inclusion of metal oxide as additives, for photoprotection. The compendial monographs USP <660> and Ph.Eur. 3.2.1 specify a maximum light transmission of 10% for any wavelength between 290 nm and 450 nm for colored glass. For highly photosensitive drugs, amber glass, even in combination with storage of the product in the secondary carton, may not provide sufficient photoprotection; in this case, opacified bottle materials are necessary.

Bottles made of plastic are lighter and less prone to breakage. Diverse types of plastics, such as HDPE, PP, PC, and PET, can be used; these have different mechanical attributes, transparencies, and oxygen and water vapor permeabilities. A commonly used alternative to glass is PET, which provides comparable transparency, which is an important feature for oral liquids, particularly for powder-for-constitution and suspension products. All plastic materials exhibit a certain degree of moisture vapor permeability. Very oxygen- or moisture-sensitive products may have inferior stability in plastic bottles compared to glass bottles. Since no single plastic material provides optimal protection against oxygen and moisture uptake, laminated bottles combining the protective properties of different plastics layers can be used to optimize product stability. However, these bottles are not visually transparent. HDPE and PP provide low moisture permeation rates, which make bottles made of these materials suitable for ready-to-use liquids, but protection against oxygen is poor. Plastic materials include processing additives that make it necessary to test for extractables and leachables and for compatibility with the liquid formulation. Excipients such as preservatives or flavor components may be adsorbed by the bottle wall; potential losses must be assessed and compensated for, particularly for antioxidants and antimicrobial preservatives, to assure the chemical and microbial stability of the product during shelf life and in use.

It is preferred that bottles for certain oral liquid formulations have some degree of transparency. For powder-for-constitution formulations, transparency of the bottle enables the product to be observed after shaking to confirm dispersion of the powder and that a homogeneous suspension or a clear solution has been attained. Similarly, suspensions that require shaking to resuspend prior to each dosing, and liquids that are susceptible to air bubble entrainment or foaming, can be visually inspected prior to dosing so as to reduce the risk of dosing errors.

An advanced bottle system, which may overcome the challenges of conventional powder-for-constitution configurations, is the dual-chamber bottle. A powder or liquid reservoir is filled into a sealed cavity of the cap, which is mounted on the bottle containing the liquid for constitution or mixing. The content of this reservoir is emptied into the liquid by means of a plunger integrated in the cap after a tamper-evident ring is removed. This mechanism allows the medication to be constituted without opening the bottle, thus eliminating the risk of airborne dust exposure during constitution. For highly potent drugs, this exposure risk might otherwise require the constitution operation to be performed in a controlled setting, thus limiting the distribution of such medicines to specialized pharmacies. This system can also enable oral liquid formulations of drugs with stability challenges for which customized vehicles are required for constitution [34]. The drawback of such a dual-chamber system is the cost of designing, manufacturing, and filling these components, and the need for specialized filling lines, for what may be relatively low commercial volumes.

7.6.2 Closures

The bottle closure system typically consists of a child-resistant screw cap with a push-and-twist mechanism that is equipped with a tamper-evident band either as part of the cap or as shrink foil applied over the cap. The screw cap has a disk liner which represents the product-contact surface, often of PE; this liner is usually a multilayered laminate, including a foamed layer to provide the necessary sealing force at the zone of contact with the rim of the bottle. The tightness of the bottle closure is critical for prevention of product leakage and to assure the stability of the product over shelf life. A consistent application torque has to be used for cap closure during the packaging process. Container-closure integrity testing such as vacuum decay tests or blue dye test may be used as in-process controls to ensure the required closure integrity. The cap removal torque range should be adequate to ensure that the product meets child-resistant test criteria, but should also enable the cap to be opened by the intended patient or caregiver population, including seniors. Where a press-in bottle adapter (PiBA) is to be used to enable the product to be withdrawn from the bottle using an oral dosing syringe, the design of the screw cap must also meet the closure integrity requirements with the PiBA in place. The design of the PiBA may also impact the functionality of the cap.

7.6.3 *Unit-Dose Packages*

Unit-dose presentations of powders or granules for constitution are usually packaged as sachets or stickpacks. The packaging material is usually a laminate consisting of at least an inner seal layer, a barrier layer, which is commonly aluminum, and an outer layer such as PET, providing the necessary mechanical protection and the basis for printing. Sachets are square or rectangular-shaped units heat-sealed along all four sides. Stickpacks are more elongated, heat-sealed at both ends, and with third longitudinal or "fin" seal forming a tube-like configuration; stickpacks have somewhat higher capacities than sachets for their size. A wide range of powder and granule fills can be achieved using these platforms. The heat-sealing step is critical to assuring the integrity of the sealed units and thus product stability; similarly, maintaining the integrity of the aluminum barrier layer during processing and handling is also important. Seal integrity tests such as vacuum decay tests are essential as in-process controls to assure the quality of the product. Sachets and stickpacks may be fabricated with tear lines or printed with cut lines for product dispensing. Laser technology can enable perforation of the outer layer without compromising the barrier function of the aluminum foil layer. Child resistance of these units must also be considered; for units which are intended to be opened by tearing, this can be achieved by requiring two independent motions to open the pack. The development of pediatric products in sachets and stickpacks provides an option for increased convenience for parents/caregivers where technical constraints dictate a unit-dose presentation, but the worldwide availability of such products is still limited since the manufacturing technology is more specialized and is more complex than bottle filling.

7.7 Devices

Devices relevant to oral liquids encompass those used to measure the dose volume (e.g., syringes, dosing spoons, droppers) and also, in the case of unit-dose powders/granules for constitution, to prepare the liquid formulation for administration (e.g., mixing cups). These components are regulated as medical devices in both the EU and the US markets; under both EU and US regulations, they are categorized as Class I devices, which represent the lowest risk.

In the EU, Class I devices are self-certified by the supplier or manufacturer with a "CE" label to denote conformance with the safety and performance requirements of the Medical Device Regulation (EC/2017/745 Annex 1). A "notified body" has to assess the conformance before the device enters the market, and sign a declaration of conformity. The device components, such as oral syringes or spoons, may be co-packed with the medicinal product, in which case these components are considered part of a combination product.

In the United States, the FDA distinguishes the level of control based on classification into three different risk categories. Measuring devices for oral liquids typically belong to class 1, which are exempted from premarket notification based on their similarity to already-approved devices.

Products that comprise a device component for dosing and administration, either as a single physically integrated drug/device unit or as co-packed, separate components, may be regulated as "combination products." As the regulatory frameworks and the exact definition and classification of combination products differ in the regional and national regulations, the development and global registration of such products is complex. For instance, in the United States, an FDA office assigns the review center that will have lead responsibility for the review based on the "primary mode of action" of the product. Oral liquid products are typically regulated as drug products. In the EU, the medical device regulations lay down a framework for medical devices that are "intended to administer medicinal product" with two cases: a single integral product and a co-packaged combination of medicinal product and medical device. In the latter case, the device component is governed by the medical device framework and requires CE certification.

Although the development of a combination product increases the complexity of development and registration and makes the drug product applicant also responsible for the quality and conformance of co-packed device items, there is in many cases a strong benefit in supplying the items for dosing and administration together with the drug product. For instance, the preparation and administration of oral liquids prepared from granules or powders can result in safer and more reliable product dosing by the use of company-supplied components. Cups equipped with a mark for the required water for constitution and optionally also with lid for mixing by shaking may be co-packed with the product, since households are typically lacking appropriate measuring devices and small containers. Even if unit-dose presentations do not require further dose measurement, the use of an oral dispenser supports more reliable dose administration in small children.

Historically, oral liquid medicines for children were most frequently dosed with spoons or measuring cups. With the evolving awareness of the risks of dosing errors and the higher expectations of Health Authorities for safe and accurate dosing, oral liquid products are increasingly dosed with syringes, and those are often co-packaged with the product. This is particularly true for oral liquids with dose volumes of less than 5 mL, for which measuring cups are considered insufficiently accurate [35]. Even at a volume of 5 mL, significantly higher rates of dosing errors were found with measuring cups compared to syringes. Household teaspoons vary in volume, but even medical dosing spoons are prone to dosing errors due to errors in appropriately aligning the liquid level with the markings on the device; the ease and accuracy of determining the liquid level are further dependent on properties of the formulation such as viscosity and surface tension. Good measuring accuracy is obtained when cylindrical spoons with a graduated cavity are used [35], although dose administration from a spoon has a higher risk of spillage and dose loss relative to a syringe, especially for younger children. Pipettes with calibrated graduations for volumes up to 5 mL, sometimes also termed droppers, may also be suitable, in terms of both measurement accuracy and ease of administration for younger children.

Droppers are typically presented as a pipette integrated into the screw cap or as a fixed insert in the bottle neck. Small dose volumes of less than 1 mL can be dosed by counting a prescribed number of drops. The drop radius, and thus the drop volume and also the dose, are dictated by the diameter of the dropper tip, and also influenced by properties of the formulation such as surface tension. Use of a different dropper may result in a different drop size and thus a different dose. Although designed for application of small volumes, which are particularly needed for newborns and infants, the dosing accuracy for droppers is not reliable, since droppers need to be held vertically for an accurate drop size; inclination of the dropper leads to a reduced drop volume [36]. For these reasons, oral syringes may present a safer alternative for dosing very small volumes [37].

Oral syringes, when used appropriately, enable accurate measurement of dose volumes and safe dose administration to small children. A wide range of doses can be realized for an oral liquid formulation of one strength, particularly if different syringe sizes are provided. Oral syringes are commonly coupled to a press-in bottle adaptor inserted in the bottle neck to permit the dose to be withdrawn with the bottle inverted. These steps add to the complexity of use of the product, and care must be taken to manage the reduced pressure in the bottle as larger volumes are withdrawn, but with the bottle held at eye level, good visual control of the plunger position in the syringe can be achieved. For dosing, the tip of the syringe should be placed in the cheek and the plunger depressed slowly to minimize the risk of choking. It is essential that the graduation markings on the syringes correspond to the dose volumes specified in the dosing instructions; the use of syringes with different volume markings or an unsuitable graduation scale or of an inappropriate size has been shown to increase the risk of dosing errors. A best practice for products that are co-packaged with more than one syringe size is to use color coding of the syringes to minimize the risk of dosing errors. Accuracy requirements for syringe graduations, overall length of the scale, and the scale intervals are defined in the ISO standard 7886–1:1997 set into force specifically for hypodermic syringes but also established as a standard for oral syringes. The minimal accurate volume for each syringe size can be derived from the maximum permissible tolerances according to ISO 7886-1. For a 1 mL syringe, the minimum volume with not more than a 10% deviation from the nominal volume has been shown to be 0.1 mL [38]. Despite the high dosing accuracy that can be achieved with oral syringes, studies have shown that the erroneous use of these devices contributes significantly to the overall actual inaccuracy of dosing and suggest that training of caregivers is necessary [39].

The syringes traditionally used for oral dosing, and the corresponding PIBA, use a conical slip-fit design on the syringe tip. These are compatible with the Luer slip-tip design used for parenteral administration, and historically, parenteral syringes have been used for both oral and parenteral dosing. In response to cases of inadvertent administration of oral liquids by the parenteral route, a new international standard (ISO 80369-3:2016) introduced an enteral standard connector called ENFit™ for syringes and other components (e.g., feeding tubes) intended for oral use, which are incompatible with components intended for parenteral use. Compliance with the ENFit™ standard thus requires oral dosing syringes as well as the associated PIBA and any feeding tubes to be used to be compatible with this tip design. The different (wider) tip design of ENFit™-compatible syringes can have an impact on the volume of liquid delivered, and thus a low-dose tip design has been introduced to increase accuracy for small dose volumes [40]. It has been reported, however, that these low-dose-tip syringes can introduce overdosing errors at low (1 mL and below) doses, depending on the dosing technique due to the presence of a void space or "moat" around the tip of the syringe [41].

Compatibility of oral liquid products with feeding tubes should also be evaluated if this is a likely clinical dosing scenario. Liquid products that exhibit high viscosity may require high pressure to pass the product through a small diameter feeding tube, such as may be used for infants and newborns in particular. Oral suspension products should be evaluated to determine if aggregates or flocculates of the suspended drug or suspending agent can block the tube. Recovery tests should be performed to assess adsorption or absorption of drug by the tube material, and to determine the volume and type of liquid used to flush the tube. The risk of physical incompatibility between the flushing liquid and the product (e.g., due to pH differences), leading to precipitation of drug, should also be considered. Flushing with breast milk or infant formula may be necessary for neonates and infants may require flushing with milk to limit their intake of non-nutritive liquid. The US FDA has issued a guidance [42] reflecting in vitro testing needed to support the administration of oral drug products via enteral feeding tubes.

7.8 Summary

In conclusion, several challenges should be apparent from the above discussion.

1. First, there is no "one-size-fits-all" formulation approach for pediatric products; every formulation decision represents a compromise, taking into consideration the particular needs of the disease, the patient population, and the molecule in question. It is thus necessary for the development scientist to understand all of the technical factors influencing product design and the needs of the target patient population, and to appropriately balance those considerations in the final

formulation selection. Oral liquids are among the most widely used dosage forms for children, although these may not be the most suitable formulations in all cases.

2. Second, when formulating liquid products for infants and small children—the age group most in need of formulations that support flexible dosing and can be readily swallowed—excipient safety considerations can severely limit the options for formulation design. For example, co-solvents and solubilizers may be desirable to achieve a solution or syrup formulation, and antimicrobial preservatives may be desirable to achieve a multi-dose liquid presentation, but there are safety concerns associated with the use of many of these materials in pediatric formulations, especially those intended for infants. It is important that all stakeholders recognize this trade-off and recognize that an unconditional focus on one aspect of product development—in this case, excipient safety—can result in products that are sub-optimal in other aspects, such as dose flexibility or usability.
3. The interface between the drug product and the device components needed to dose the product, and the parallel development of these, can be complex. The development scientist must consider (i) technical aspects related to product compatibility and suitability of the components; (ii) usability considerations, taking into account the steps that need to be performed by the parent or caregiver, and the supporting instructions-for-use; and (iii) regulatory expectations that may be different in different geographic regions, and in some cases may be continuing to evolve. A further complication is that the appropriate dosing for pediatric patients may not be known until very late in the product development program. It is critical, therefore, that all of the functions engaged in pediatric product development collaborate closely throughout the development process.

References

1. Strickley RG, Iwata Q, Wu S, Dahl TC. Pediatric drugs - a review of commercially available oral formulations. J Pharm Sci. 2008;97:1731–74.
2. Strickley RG. Pediatric oral formulations: an updated review of commercially available pediatric oral formulations since 2007. J Pharm Sci. 2019;108:1335–65.
3. European Medicines Agency. Reflection paper: formulations of choice for the paediatric population. 2006. EMEA/CHMP/PEG/194810/2005.
4. Van der Veken M, Brouwers J, Budts V, Lauwerys L, Pathak SM, Batchelor H, Augustijns P. Practical and operational considerations related to paediatric oral drug formulation: an industry survey. Int J Pharm. 2022;618:121670.
5. European Medicines Agency. Guideline on pharmaceutical development of medicines for paediatric use. 2013. EMA/CHMP/QWP/805880/2012 Rev. 2.
6. World Health Organization. Development of paediatric medicines: points to consider. 2012. Technical Report Series 970 Annex 5.
7. Orlu M, Salunke S, Tuleu C. Paediatric pharmaceutics - the science of formulating medicines for children. In: Tovey GD, editor. Pharmaceutical formulation: the science and technology of dosage forms. Royal Society of Chemistry; 2018.

8. Kulshreshtha AK, Singh ON, Wall GM, editors. Pharmaceutical suspensions: from formulation development to manufacturing. Springer; 2010.
9. European Medicines Agency. Annex to the European Commission guideline on 'Excipients in the labelling and package leaflet of medicinal products for human use' (SANTE-2017-11668). 2019. EMA/CHMP/302620/2017 Rev. 1.
10. Adkin DA, Davis SS, Sparrow RA, Huckle PD, Phillips AJ, Wilding IR. The effect of different concentrations of mannitol in solution on small intestinal transit: implications for drug absorption. Pharm Res. 1995;12:393–6.
11. Matsuia K, Nakagawab T, Okumurac T, Yamanea M, Tokunagad Y, Yokotad S. Potential pharmacokinetic interaction between orally administered drug and osmotically active excipients in pediatric polypharmacy. Eur J Pharm Sci. 2021;165:105934.
12. Chen M-L, Straughn AB, Sadrieh N, Meyer M, Faustino PJ, Ciavarella AB, Meibohm B, Yates CR, Hussain AS. A modern view of excipient effects on bioequivalence: case study of sorbitol. Pharm Res. 2007;24:73–80.
13. Schulze JDR, Waddington WA, Ell PJ, Parsons GE, Coffin MD, Basit AW. Concentration-dependent effects of polyethylene glycol 400 on gastrointestinal transit and drug absorption. Pharma Res. 2003;20:1984–8.
14. Salunke S, O'Brien F, Tan DCT, Harris D, Math MC, Ariën T, Klein S, Timpe C. Oral drug delivery strategies for development of poorly water soluble drugs in paediatric patient population. Adv Drug Deliv Rev. 2022;190:114507.
15. Carlin BA. Suspension quality by design. In: Tovey GD, editor. Pharmaceutical formulation: the science and technology of dosage forms. Royal Society of Chemistry; 2018.
16. Raghunathan Y, Amsel L, Hinsvark O, Bryant W. Sustained-release drug delivery system I: coated ion-exchange resin system for phenylpropanolamine and other drugs. J Pharm Sci. 1981;70:379–84.
17. Frey CR, Madsen JS. Liquid formulations. In: Bar-Shalom KRD, editor. Pediatric formulations: a roadmap. AAPS Press; 2014.
18. Guo X, Chang R-K, Hussain MA. Ion-exchange resins as drug delivery carriers. J Pharma Sci. 2009;98:3886–901.
19. European Medicines Agency. Excipients in the dossier for application for marketing authorisation of a medical product. 2006. EMEA/CHMP/QWP/396951/2006.
20. Tavares M, Kozak M, Balola A, Sá-Correia I. Burkholderia cepacia complex bacteria: a feared contamination risk in water-based pharmaceutical products. Clin Microbiol Rev. 2020;33:1–25.
21. Elder DP, Crowley PJ. Antimicrobial preservatives part two: choosing a preservative. Am Pharm Rev. 2017;20
22. European Medicines Agency. Benzyl alcohol and benzoic acid group used as excipients. 2017. EMA/CHMP/272866/2013.
23. European Medicines Agency. EMA/CHMP/302620/2017, 2019, Annex to the European Commission guideline on 'Excipients in the labelling and package leaflet of medicinal products for human use' (SANTE-2017-11668). 2019. EMA/CHMP/302620/2017.
24. European Medicines Agency. Questions and answers on benzoic acid and benzoates used as excipients in medicinal products for human use. 2017. EMA/CHMP/508189/2013.
25. European Medicines Agency. Reflection paper on the use of methyl- and propylparaben as excipients in human medicinal products for oral use. 2015. EMA/CHMP/SWP/272921/2012.
26. Walsh J, Cramb A, Woertz K, Breitkreutz J, Winzenburg G, Turner R, Tuleu C. 2014, Playing hide and seek with poorly tasting paediatric medicines: do not forget the excipients. Adv Drug Deliv Rev, 73, pp. 14–33.
27. Szejtli J, Szente L. Elimination of bitter, disgusting tastes of drugs and foods by cyclodextrins. Eur J Pharm Biopharm. 2005;61:115–25.
28. Grimm M, Scholz E, Koziolek M, Kühn J-P, Weitschies W. Gastric water emptying under fed state clinical trial conditions is as fast as under fasted conditions. Mol Pharm. 2017;14:4262–71.

29. Kostewicz ES, Wunderlich M, Brauns U, Becker R, Bock T, Dressman JB. Predicting the precipitation of poorly soluble weak bases upon entry in the small intestine. J Pharm Pharmacol. 2004;56:43–51.
30. Carlert S, Pålsson A, Hanisch G, von Corswant C, Nilsson C, Lindfors L, Lennernäs H, Abrahamsson B. Predicting intestinal precipitation - a case example for a basic BCS class II drug. Pharm Res. 2010;27:2119–30.
31. Pawar G, Wu F, Zhao L, Fang L, Burckart GJ, Feng K, Mousa YM, Naumann F, Batchelor HK. Development of a pediatric relative bioavailability/bioequivalence database and identification of putative risk factors associated with evaluation of pediatric oral products. AAPS J. 2021;23:57.
32. U.S. Code of Federal Regulations. Criteria for waiver of evidence of in vivo bioavailability or bioequivalence. § 320.22.
33. European Medicines Agency. Guideline on the investigation of bioequivalence. 2010. CPMP/EWP/QWP/1401/98 Rev. 1.
34. Malori A. Oral drug reconstitution: making it easy and accurate via packaging innovation. ONdrugDelivery. 2018;88:18–21.
35. Breitkreutz J, Wesse T, Boos J. Dosage forms for peroral drug administration to children. Paediatr Perinatal Drug Ther. 1999;3:25–33.
36. Brown D, Ford JL, Nunn AJ. An assessment of dose-uniformity of samples delivered from paediatric oral droppers. J Clin Pharm Ther. 2004;29:521–9.
37. Walsh J, Bickmann D, Breitkreutz J, Chariot-Goulet M. Delivery devices for the administration of paediatric formulations: overview of current practice, challenges and recent developments. Int J Pharm. 2011;415:221–31.
38. Arenas-López S, Gurung K, Tibby SM, Ángel M, Hernández C, Tuleu C. Accuracy of enteral syringes with commonly prescribed paediatric liquid medicines. Arch Dis Child. 2017;102:655–9.
39. Sobhani P, Christopherson J, Ambrose PJ, Corelli RL. Accuracy of oral liquid measuring devices: comparison of dosing cup and oral dosing syringe. Ann Pharmacother. 2008;42:46–52.
40. Walsh J, van Riet-Nales D, Hermans E, de Vries R, Hilton G, Blowers P, Salunke S. European paediatric formulation initiative workshop report: improving the administration of oral liquid medicines in paediatrics using dosing syringes and enteral accessories. Eur J Pharm Biopharm. 2020;151:91–7.
41. Guha S, Niswander W, Herman A, Antonino MJ, Venkataraman Rao P, White K, Haugen S, Vesnovsky O. Assessing dosing errors in legacy and low dose tip ENFit™ syringes. J Clin Pharm Ther. 2022;47:218–27.
42. FDA. Draft guidance for industry: oral drug products administered via enteral feeding tube: in vitro testing and labeling recommendations. 2021.

Chapter 8
Paediatric Solid Formulations

Sejal R. Ranmal, Susan A. Barker, and Catherine Tuleu

Abstract Oral drug delivery is the predominant and favoured route of administration across all age populations, given its convenience for patients and cost-effective manufacturing methods. Solid oral formulations, with their diverse range of dosage forms and delivery system designs, represent the most common modality for drug delivery. Conventional adult strength tablets and capsules rarely meet the dosing and administration needs of the heterogeneous paediatric population, ranging from neonates to adolescents. Nevertheless, (smaller) monolithic dosage forms tailored to age-appropriate strengths and sizes might be preferable for (younger) children and adolescents, and patients with chronic conditions who require more frequent medication intake. More contemporary oral dosage forms, including mini-tablets, multiparticulates, (oro-)dispersible, and chewable preparations, offer several advantages related to patient acceptability, safety, and access. These formulations can facilitate easier swallowing and provide dosing flexibility, two key attributes which have traditionally favoured liquid formulations over solids for younger children. With heightened efforts to develop better medicines for children, evidence and experience of how different solid oral dosage forms can be used in this population are steadily growing. This chapter describes the rationale for different solid oral dosage forms, current regulatory recommendations related to their pharmaceutical development, and emerging evidence of their acceptability and use across the paediatric population.

S. R. Ranmal
UCL School of Pharmacy, London, UK

Sciense Ltd, London, UK
e-mail: sejal.ranmal@ucl.ac.uk

S. A. Barker
University of Greenwich, Kent, UK
e-mail: s.barker@greenwich.ac.uk

C. Tuleu (✉)
UCL School of Pharmacy, London, UK
e-mail: c.tuleu@ucl.ac.uk

H. Batchelor, K. Rose (eds.), *Pediatric Formulations*, AAPS Advances in the Pharmaceutical Sciences Series 58, https://doi.org/10.1007/978-3-031-77240-5_8

Keywords Dosage form · Solid formulation · Solid dosage form · Pharmaceutical development · Paediatric formulation

8.1 Introduction

Oral drug delivery remains the most widely accepted and preferred route of administration in paediatric and adult populations alike, both from a manufacturing and patient perspective [1, 2]. Advantages include convenience of drug administration, patient preference, cost-effectiveness, and ease of large-scale manufacturing. For too long, adult strength tablets and capsules have needed to be physically altered by patients, their caregivers, and healthcare professionals to aid administration, obtain the appropriate paediatric dose, or both. This can be achieved through pharmaceutical compounding, namely, creating a custom-made medication tailored for individual patients, or pragmatically adapting dosage forms at the point of administration, for example crushing tablets or opening capsules (commonly referred to as a manipulation) [3–5]. The related dangers of these modifications, from both a safety and efficacy perspective, are well recognized [6, 7]. The lack of appropriate paediatric formulations has been a well-acknowledged problem in the field and recognized as part of the global regulatory legislations introduced initially in the United States (US) and in Europe (EU). In line with these initiatives, research and development into formulations for children is gaining momentum, and important advances continue to be made.

The development of paediatric formulations poses several unique challenges compared to adult drug formulations; a topic that has been thoroughly reviewed [8–11]. An age-appropriate paediatric medicine is one whose pharmaceutical design makes it suitable for use in the target age group(s) [12]. This chapter reviews the significance of solid oral formulations as a crucial drug delivery modality for children, examining both the evidence and current regulatory recommendations to inform their development.

8.2 Regulatory Guidance on Age-Appropriate Formulations

At the time worldwide paediatric regulations were first introduced, evidence to support the pharmaceutical design of paediatric formulations was scarce and fragmented. Acknowledging this, the European Medicines Agency (EMA) first published a reflection paper in 2006 on "Formulations of choice for the paediatric population" which was intended to communicate available information and current thinking at the time to support all stakeholders [13]. The "Guideline on pharmaceutical development of medicines for paediatric use" published in 2013 describes aspects that need to be considered and justified as part of the pharmaceutical drug product design for children's medicines [12]. This guidance should apply across all

European drug licensing procedures and for any new application, whereas the principles may also be considered for paediatric investigation plans (PIPs) [12, 14]. Relevant details from the EMA guideline to support the development of solid formulations are highlighted throughout the chapter.

Aside from the physico-chemical characteristics of the drug substance and other pre-formulation factors, formulators also need to give due consideration to the unique needs of the paediatric patient during the pharmaceutical design of medicines. As defined by the EMA, the pharmaceutical design of a medicine encompasses "*the composition, dosage form, route of administration, dosing frequency, packaging, measuring or administration device and the user instruction of a medicine*" while a paediatric formulation describes "*the composition of a particular dosage form of a medicine for paediatric use*" [12].

This heterogeneous population, spanning neonates to adolescents, undergoes rapid growth and development and presents unique physiological characteristics and varying medication requirements. The magnitude of doses required through childhood can vary 100-fold, and the ability to handle different dosage forms can also vary considerably [13]. Given the extensive physiological and developmental differences across the population, it is unlikely that one single formulation could fulfil all requisites and be deemed appropriate for all intended ages. As such, initial pharmaceutical development should focus on a minimum number of acceptable dosage forms which are capable of meeting the needs of the majority of the children in the target age group(s) [12]. Rather than a "one size fits all" approach, investigators should aim to tailor each formulation to cover the widest age range, ensuring pharmaceutical development strategies are effective and economical whilst providing authorized products for each target subset that allow simple, accurate, and safe dosing.

8.3 The Rationale for Solid Formulations

Given the variety of dosage form types and delivery system designs, solid oral formulations emerge as one of the predominant modalities for drug delivery. Figure 8.1 shows a range of solid oral products for children [15]. The selection of paediatric formulations and their pharmaceutical design is dependent on multiple factors related to patient acceptability, safety, and access, as reviewed by Walsh et al. [16]. Desirable characteristics of oral medicines related to these three facets are highlighted in Table 8.1 and reflect the needs of a global community.

Historically, oral liquids, including solutions, syrups, suspensions, and emulsions, were considered the "gold standard" in paediatric formulations. Ease of ingestion and dose flexibility are key advantages of liquid medicines which have traditionally led to their precedence over solid formulations for children. In comparison to solid dosage forms, however, liquid formulations are notoriously more challenging and expensive to formulate, and have limitations related to taste-masking, excipient safety, and shelf-life. Several thorough review papers have

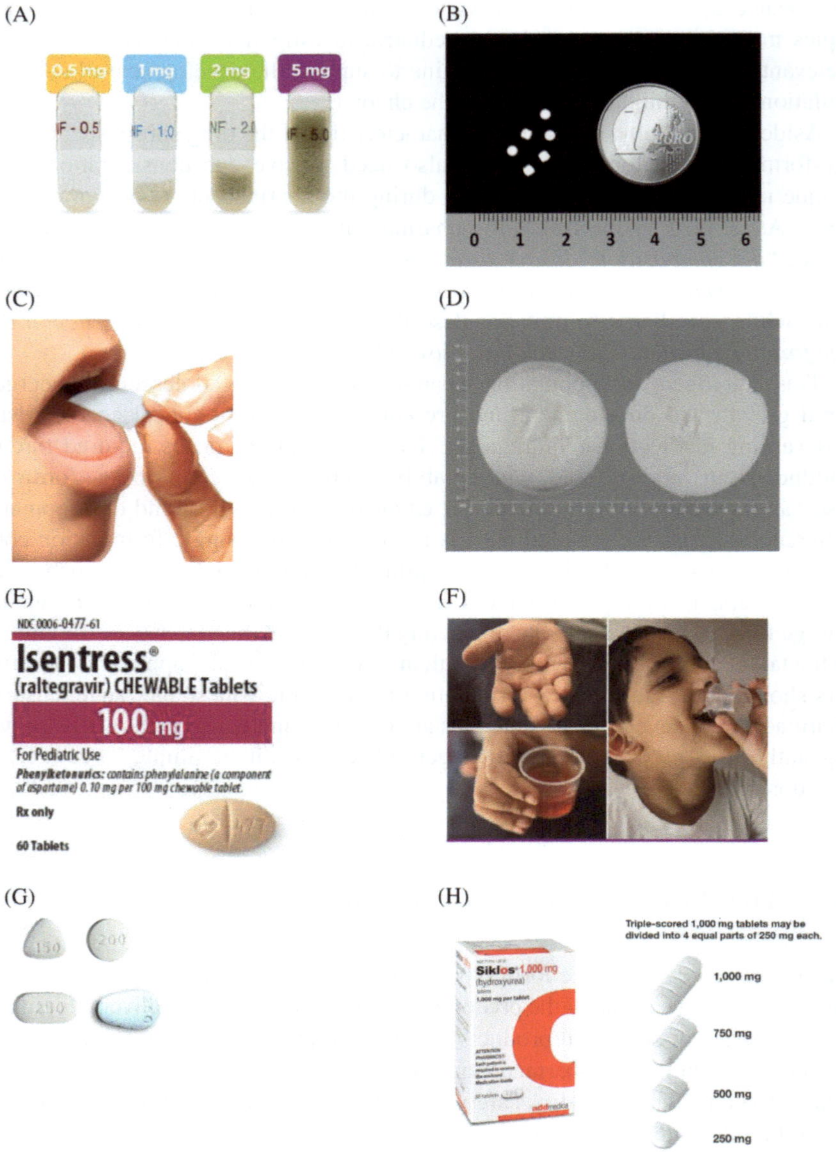

Fig. 8.1 Pictures of solid oral dosage forms for children from Litalien et al. (2022) [15]. (**a**) Alkindi oral granules (hydrocortisone) supplied in 0.5-, 1-, 2- and 5-mg capsules; (**b**) six uncoated mini-tablets in relation to a 1 Euro coin; (**c**) orodispersible films; (**d**) orodispersible tablet, Zofran ODT (ondansetron); (**e**) chewable tablets for paediatric use, Isentress (raltegravir); (**f**) dispersible tablet containing fixed dose combination of isoniazid, rifapentine, and rifampin for the treatment of tuberculosis in children dissolved in water; (**g**) conventional tablets, Viread (tenofovir disoproxil fumarate); (**h**) scored tablets for paediatric use, Siklos (hyroxyurea) 1000 mg triple-scored tablet

Table 8.1 Desirable attributes of oral paediatric formulations related to patient acceptability, safety, and access

Patient acceptability	Easy to ingest (suits patient capabilities including ability to swallow)
	Palatable (preferably with a neutral taste where possible)
	Easy and convenient to administer (preferably ready-to-use, otherwise requiring minimal handling)
	Minimal impact on lifestyle, including minimum dose frequency and transportable
	Acceptable among end-users (patients, caregivers, and health care professionals) across socio-cultural settings
Safety	Enables safe, accurate, and reliable dose administration (minimal risk of mis-dosing through incorrect handling, measurement, and/or administration)
	Dose uniformity and dose appropriate/adjustable to intended age
	Achieves adequate bioavailability
	Minimal, non-toxic excipients
	Stable (while in storage, in use, and across variable global climates)
Access	Readily supplied with a suitable administration device (where needed) and clear information for use
	Consistently available and accessible to patients
	Affordable (including cost to manufacture and procure/supply)
	Commercially viable

assessed the key considerations, as well as comparative advantages and limitations of liquid versus solid formulations for paediatric use [9, 14, 15, 17, 18]. Aside from traditional tablets and capsules, the emergence of more innovative drug delivery technologies has led to a paradigm shift towards research and development into solid oral dosage forms (SODFs) for use across the heterogeneous patient population. Further discussion is provided on certain key advantages associated with the desirable attributes outlined in Table 8.1.

8.3.1 Patient Acceptability

The EMA defines patient acceptability as "*the overall ability and willingness of the patient to use and its care giver to administer the medicine as intended*" [12]. Further details on this crucial aspect of paediatric drug development can be found in Chap. 4. The acceptability of a medicine can be influenced by both the pharmaceutical characteristics of the medicinal product (including swallowability and palatability) and the user (such as age, ability, disease type, and state) [19]. Users in the case of the paediatric population can include caregivers who play an integral part in medicine administration, as well as patients themselves. Formulation factors which can impact the acceptability of oral medicines have been well reviewed by multidisciplinary academic and industry members of the European Paediatric Formulation Initiatives (EuPFI) [20, 21]. Nevertheless, the EMA acknowledges that some children may not be able or willing to take solid oral formulations as intended, even when the dosage form is generally considered as age-appropriate [12]. For this reason, the agency advocates bringing different preparations to the market or exploring alternative strategies to aid their administration.

The ability to swallow conventional SODFs can vary among individuals across all ages, even within the adult population. Nevertheless, infants, toddlers, and many young children are simply unable to swallow conventional large single-unit SODFs such as tablets and capsules. Verified modifications (approved by the regulatory authority) may be one strategy to facilitate administration of these SODFs; however, supporting data should demonstrate that modifying the original preparation will not adversely change its pharmaceutical characteristics in a way that compromises the safety and/or efficacy of the medicine [12]. Alternatively, more novel dosage forms, including mini-tablets, multiparticulates, (oro-)dispersible, and chewable preparations, should be explored. These negate the need to swallow large units intact and can potentially provide a flexible and individualized approach to drug delivery, with prime application in paediatrics.

Palatability is one of the main elements that impacts patient acceptability of oral formulations. It is defined as "*the overall appreciation of an (often oral) medicinal product in relation to its smell, taste, aftertaste and texture (*i.e. *feeling in the mouth)*" and is determined by the characteristics of the active substance, excipients, and formulation [12]. Taste-masking approaches and their importance as discussed in Chaps. 5 and 6. Many drugs exhibit an unpleasant taste, and many studies have reported poor palatability as a barrier to acceptability and medication adherence [22–24]. Taste-masking of liquid formulations can pose significant challenges [25, 26]. Compounds with high solubility often cannot be easily formulated as suspensions, and even suspensions of poorly soluble compounds can exhibit poor palatability if small amounts of the drug are perceptible or if mouthfeel is compromised [26]. SODFs generally offer superior taste-masking strategies, including encapsulation or application of polymer coatings. The development of chewable and (oro-) dispersible formulations may encompass the same taste-masking challenges that are encountered with liquids, especially for high solubility drugs which may dissolve rapidly in the mouth.

Dosing flexibility is critical as weight-based dosing is the most common approach in paediatric pharmacotherapy and weights can vary as much as 140-fold (from a 500 g premature neonate to a 70 kg adolescent) [15]. SODFs may be divided into those which provide flexible dosing, such as mini-tablets and multiparticulates, and those which provide unit doses, such as tablets and capsules. Another key advantage is the opportunity for the development of functionalized formulations, such as modified, prolonged, and delayed-release systems, which is technically more challenging with liquids. This not only gives the potential for targeted drug delivery but can also benefit patients by reducing dose frequency and minimising burden on lifestyle (e.g. the need to administer medicines at nursery or school). Attributes of SODFs, such as size, shape, and colour, can be adapted to provide aesthetic final products that can be more easily differentiated to reduce these risks. Though care must be taken to ensure these are not too attractive or bear a strong resemblance to confectionery [12, 27].

8.3.2 Safety

Important safety considerations relate to incorrect dosing of the active, risk of medication errors, and safety of excipients [28]. Limitations of liquid formulations can include chemical, physical, and/or microbial instability, sedimentation, and inhomogeneous distribution of the drug substance, and emulsion splitting which can compromise the quality and safety of the medicinal product [14]. The addition of excipients such as preservatives, stabilizers, suspending agents, and solubility enhancers (both in quantity and type) is complicated by the fact that there is limited evidence of their safety and toxicity in children of different ages, as discussed further in Chap. 18. Excipients such as propylene glycol, benzyl alcohol, surfactants, and ethanol can lead to adverse effects due to inherent toxicities and incomplete maturation of metabolic function, especially in neonates [29].

Liquid medicines also require a dosing device which presents risks of errors when measuring the dose. In the case of both age extremities of the population, small volumes can be difficult and inaccurate to measure, while larger dose volumes for older children would be inappropriate. Target dose volumes of $\leq$5 mL for children under 5 years and $\leq$10 mL for those 5 years and over have previously been proposed, and in most cases, this would necessitate the provision of multiple dosage form strengths [13]. There is also limited control to ensure complete dose intake, particularly for very young children who may spit the medicine out. Liquid medicines also require delivery devices (e.g. spoons, cups, or oral syringes) which have previously been linked to medication administration errors [30–32]. Dose accuracy can also be compromised by use of inappropriate measuring devices (e.g. teaspoons) and limited health literacy among caregivers.

When administered as intended and licensed, SODFs provide an easy, safe, and convenient means of dose delivery. Complete and accurate dose delivery is facilitated by individual, uniform dose units or packages (e.g. sachets or capsules for multiparticulates). Dose adaptation with single unit dosage forms for ingestion intact is limited, and numerous strengths may need to be developed when flexibility is required. However, pharmaceutical development and product design strategies require a careful balance of risks and benefits associated with this. An appropriate number of formulations of different strengths should be available to cover the needs of the target age groups; however, a disproportionate number may be a risk factor for medication errors in practice.

8.3.3 Access

Development of appropriate formulations is a global health challenge, and as such, the distinct requirements in resource-limited settings must also be considered [33–35]. The World Health Organization (WHO) in 2008 recommended prioritizing the development of "flexible" SODFs [1]. These are formulations which can be administered in more than one manner (e.g. dispersed or taken orally as a whole) and suitable for use in low- and middle-income countries [36]. One such example is tablets that are orodispersible or alternatively could be used to prepare oral liquids suitable for younger children, e.g. dispersible and soluble tablets. An important basis for this rationale is cost; oral liquid medicines are generally more expensive than solid formulations [37–39], whereas it is feasible to manufacture (oro-)dispersible tablets in settings with conventional tableting facilities [40]. Nevertheless, proprietary technologies, special attention towards packaging to overcome moisture uptake, and the need for specific excipients may all bear some cost limitations.

In relation to resource-limited settings, SODFs have the added advantage of superior stability and have low bulk and weight, thus being easy to transport and store. In addition to aiding the logistics of procurement and distribution, the added advantage of being less conspicuous minimizes problems with confidentiality and social stigma. These practical advantages in transportation and conspicuousness have reportedly led to caregivers preferring tablets over syrups in such settings, and the effect of social factors on overall formulation acceptance should not be overlooked [41]. Other studies have shown that better acceptability of tablets over syrups has facilitated both adherence and clinical outcomes for malaria [42] and HIV [43] treatments among children in Sub-Saharan Africa. Low uptake of novel paediatric formulations has been observed in some regions [44], although modest use here may be influenced by a variety of factors, including lack of specific healthcare initiatives, historical use, and familiarity among all stakeholders, including prescribers.

Formulations intended for reconstitution to a liquid form (e.g. reconstituted antibiotics) may be inappropriate considering availability and access, or lack thereof, to clean drinking water and refrigerators in many parts of the world. Nevertheless, some evidence suggests that parents and caregivers invest in clean water (e.g. by purchasing bottled water) when administering medicines to children [45]; the most economical use of this would be for use when swallowing medicines intact or dispersing medicines in a small volume (e.g. dispersible or soluble tablets).

8.4 Solid Oral Dosage Forms for Children

The availability of SODFs is increasing. A 2008 review by Strickley et al. identified 17 different commercial oral paediatric formulations in the USA including ready-to-use formulations (solutions/syrups/suspensions, tablets, scored tablets, chewable tablets, orally disintegrating tablets, and thin strips) and those requiring some form of preparation, referred to as manipulation by the authors, (powder or drops for constitution to a suspension or solution, tablets for constitution to a suspension, concentrated solution for dilution and effervescent tablet) [46]. The most common formulations at the time were liquid solutions and powder for suspension. An updated review by the same author reviewing formulations marketed in the USA, EU, and Japan from 2007 to 2018 identified 51 total new paediatric oral formulations, of which 21 were ready-to-use and 30 requiring manipulation [47]. The most common formulations were multiparticulates ($n = 12$ in total, including 6 sprinkle capsules, 2 oral powders, 3 oral granules, and 1 manipulated mini-tablet known as oral granules). In the author's opinion, mini-tablets were deemed the future of paediatric oral formulations [47]. Herein, current evidence related to the design and use of these different SODFs is reviewed.

8.4.1 Tablets and Capsules

Tablets and capsules, designed to be ingested intact, have been a robust and widespread means of delivery since the nineteenth century. They remain the most commonly prescribed dosage forms considering their relatively non-complex development process, and cost-effective, ease of manufacture (using routine manufacturing and packaging processes with standard equipment) [16]. Their use may be somewhat limited in the paediatric population, due to inability of very young children to swallow monolithic dosage forms intact. A long-standing question in the field has been the age from which tablets and capsules can safely be prescribed for paediatric patients. Moreover, there is very limited evidence to demonstrate how geometric attributes such as size and shape need to be addressed, to render their design appropriate for children.

The initial EMA reflection paper proposed an average age of 6 years old from which children could swallow SODFs, whilst recognising the inherent variability amongst children and the influence of individual patient and disease-related factors [13]. This recommendation was based on anecdotal feedback and supported by a few studies reporting behavioural training outcomes [48]. Studies evaluating prescribing patterns have shown that children typically move from liquid to SODFs from 5 to 7 years of age [49, 50].

An earlier draft of the EMA guideline provided some guidance on acceptable tablet sizes as a function of age, as shown in Table 8.2. Undefined "small" capsules were similarly deemed appropriate for children from the age of 6 years. Following public consultation, this guidance was removed from the adopted guideline due to the lack of strong evidence-based rationale. Adapting geometric dosage form attributes could enable monolithic dosage forms to be tailored to the requirements and capabilities of specific age subsets; however, evidence to guide this dosage form design is lacking.

In a small study of 55 participants, the swallowability of three 6–0 mm placebo tablets of standard round biconvex shape was evaluated in children aged 4–12 years old [51]. Nearly all older children aged 9–12 years successfully swallowed all three sized tablets, while most of the younger children aged 4–8 years who attempted to swallow the 6, 8, and 10 mm tablets successfully did so (67%, 91% and 95%, respectively). Some younger participants did not attempt the larger sizes, and many participants who were unsuccessful had never taken tablets before.

Table 8.2 Proposal for acceptable tablet dimensions (width or length whichever is longest) in the initial draft EMA guideline on pharmaceutical development of medicines for paediatric use. This guidance was removed from the adopted guideline

Age subset	Acceptable tablet dimensions
6 months to <2 years	None; multiparticulates acceptable
2–5 years	≤3–5 mm (small tablets)
6–11 years	≤5–10 mm (medium tablets)
12–18 years	≤10–15 mm (large tablets)

In addition to age, the acceptability and preferences for different SODFs can be influenced by health status and background, experience with medicines use, and culture. In a large survey of children's preferences conducted across seven European countries, monolithic SODFs were mostly chosen by adolescents and children with a chronic disease taking medicines frequently [52]. Healthy children, with less frequent medicines use were more inclined to choose liquids or effervescent tablets over SODFs, while those with chronic conditions taking one or more medicines daily preferred tablets or capsules over liquids. These findings were similar to an earlier study where adolescents and their caregivers showed more positive attitudes towards tablets versus other SODFs, including mini-tablets, multiparticulates, and orodispersibles [53].

The EMA highlights that for chronic diseases, the acceptability of tablets with a particular size and shape in children may be improved by adequate training [12]. Interventions such as behavioural training or use of swallowing aids may enhance acceptability of these dosage forms further [48, 54–56]. Interestingly, in the UK, various initiatives introduced in children's hospitals involve training staff and embedding a system of interventions (e.g. a children's "pill school") to support transition from liquids to SODFs [57, 58]. In a small study in children as young as 3 years old, 87% (26/30) of children successfully learnt how to swallow pills after one training session (mean duration 14.5 min), and 92% were discharged on pills [57].

In the absence of alternative dosage forms, other strategies to aid administration can be considered (e.g. dispersing or crushing tablets, opening of capsules, mixing with food or drinks). These "verified modifications" need to be approved by the regulatory agency by providing any type of adequate evidence that there will be no negative impact on safety and/or efficacy (e.g. using (bio)analytical data or evidence from existing practices) [12]. Where appropriate, functional scoring (or breaklines) may add some level of flexibility by enabling the dose to be adapted, but this depends on the criticality of the dose. These should be appropriately labelled to be distinct from scorelines intended to aid administration only. Apart from the size and shape, the suitability of these SODFs should be further justified in relation to different health conditions or disease development profiles and the risks associated with under- or overdosing, choking, aspiration, and chewing [12].

8.4.2 Dispersible, Soluble, and Effervescent Preparations

Dispersible, soluble, and effervescent preparations are presented as solid formulations that are intended to be dispersed or dissolved in water prior to administration. Dispersible and soluble tablets should disintegrate within 3 min in a small amount of water to yield a homogeneous dispersion or solution. These forms would be further advantageous for APIs and excipients compatible and suitably palatable when dispersed in breast milk, as they could be suitable for use with children younger than 6 months old [1]. Effervescent powders, granules, or tablets are also added to water to produce a draught, but large volumes are usually required. For this dosage form, it is important to consider the overall solubility of the API and how this may affect biopharmaceutical performance [9].

Dispersible and effervescent tablets benefit from a non-complex development process with standard tabletting equipment but may need modified tooling and low humidity conditions. Generally, good stability is observed, although sensitivity to moisture necessitates protective primary packaging, while solutions/dispersions exhibit limited stability [16]. Dosing accuracy can be compromised, especially with large volumes, if children are unable or unwilling to drink the full volume of the solution/dispersion and/or residue. According to the EMA guideline, it is generally not acceptable to take a portion of the liquid to adjust the dose, requiring additional verification on the ease of preparing the liquid, its homogeneity, and the accuracy of measuring the correct volume [12]. Moreover, as these preparations exist as liquids prior to administration, ensuring acceptable palatability is also essential and flavourings and/or sweeteners may be required. The sodium, potassium, and bicarbonate content should also be considered [16].

While these formulations require minimal preparation prior to use, the health literacy of caregivers may be an important factor to consider in their use. Clear instructions should detail appropriate diluents and volumes to dissolve or disperse in, and caregivers should be instructed not to administer the solution before effervescence has subsided, to minimize the ingestion of hydrogen carbonate. Further in relation to their global use, access to clean water and stability issues in humid climates during manufacturing and storage need to be considered. Although the application of these preparations as flexible SODFs is advocated, in cases where these formulations are not suitable for swallowing intact, the potential risks associated with administration prior to dispersion or dissolution need to be assessed.

Nevertheless, use of these dosage forms in paediatrics is highlighted by the success of Coartem® Dispersible tablets (Novartis Pharma AG, Basel, Switzerland), an artemisinin-based combination therapy (ACT) which has been a life-saving medicine for millions of infants and children worldwide. Early development studies assessed palatability as well as bioavailability [59], and clear instructions for use on packaging, as illustrated in Fig. 8.2, have facilitated its use [60]. The EMA guideline encourages companies to consider the use of novel packaging like this, to support children's acceptance and adherence, as well as being convenient for caregivers and reducing the risk of errors [12].

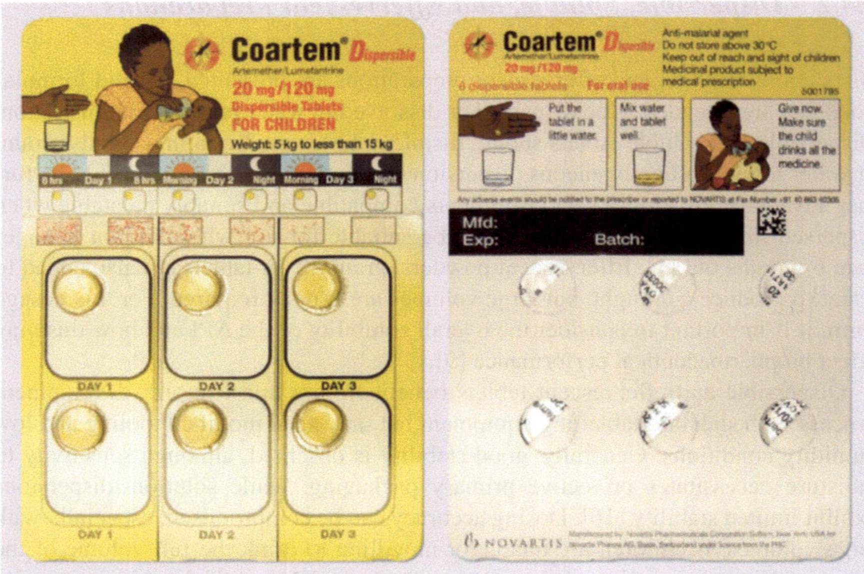

Fig. 8.2 The Coartem® Dispersible packaging, illustrating how to make up the formulation and administer it to infants to provide a 3-day treatment course

In a randomized open-label trial in Kenya, Coartem® tablets were compared to film-coated tablets of another ACT therapy where formulations were dispersed in a small volume of water or milk [61]. Adherence to the treatment regimen was higher in the Coartem® arm than the tablet arm (93.6% vs. 85.6%), with more caregivers reporting it simple to use (82% vs. 67%) and that the taste was liked (72% vs. 56%) [61]. Caregivers preferred the dispersible tablet formulation compared to a syrup formulation of the same therapy (76.8% vs. 16.8%) [61]. Other studies have demonstrated that zinc dispersible tablets are less costly to produce, easier to distribute and store, have a longer shelf-life, and are easier for caregivers to handle and track administration days [62].

8.4.3 *Orodispersible Formulations*

Orodispersible formulations can include tablets (ODTs) as well as oral lyophilisates and thin oral dispersible films (ODFs). They are intended to be placed directly in the mouth, where they rapidly disintegrate in saliva, usually within seconds, thus surpassing the need to swallow and the need for water. These formulations are well-suited for APIs with high aqueous solubility; however, their application may be restricted by limited drug loading. ODTs can function as flexible forms if, in addition to disintegration in the mouth, they could also be administered in a small amount of liquid prior to administration, or they could be swallowed intact. Enabling flexibility with regard to dose is limited, although some ODTs have functional scorelines, and in the case of ODFs, this could potentially be achieved if films could be cut or in other ways divided (however, no ODFs with this capability are currently available).

During pharmaceutical development, particular attention must be paid to ensure organoleptic properties and palatability are acceptable. For dosage forms to be dispersed (solubilized) in the mouth, primary critical attributes to be included in acceptability testing include taste and aftertaste, mouth feel, smell, and time needed to dissolve/disperse (duration of administration) [19]. Taste-masking may be further challenging due to the limited quantity of flavouring agents and sweeteners which can be incorporated into the dosage forms, particularly ODFs [20]. Formulations may be moisture sensitive, while the low compression force needed to produce ODTs may lead them to be friable and thus difficult to handle and package. Complexity of manufacture depends on the technology used and can be a routine manufacturing process with standard equipment (compressed ODTs) or specialist process and equipment (lyophilisates) [16].

Rapid disintegration or adherence to mucosa may facilitate their use in younger children, since this makes them unlikely to be spat out. However, any risks associated with their unintended use must also be considered (e.g. if they are chewed or swallowed whole), particularly for very young children who may lack the cognitive capacity to understand the instructions for their correct use and comply [12]. The rapid absorption, improved bioavailability, and rapid action achieved using these dosage forms can be advantageous, but the risks of higher drug absorption leading to poisoning in children have also been reported [37].

Orodispersible preparations may be applicable for use across the paediatric population, including infants and young children, and there is evolving evidence of their use and acceptability in children. Placebo ODTs, 5 mm in diameter, were found to be acceptable in a study of 40 children aged 2–10 years old [63]. Acceptable administration was observed in 37/40 (93%) of children, while two children showed some reluctance to take the ODT and one spat it out [63]. Acceptability has also been demonstrated with ODTs containing desmopressin and ondansetron in children 5 years and older [64, 65].

The acceptability of placebo ODFs (3 cm × 2 cm) has been investigated in infants as young as 6 months old up to 5-year-old children [66]. After the ODFs were administered, 79% of infant caregivers and 86% caregivers of preschool children positively rated their child's acceptance of the ODF [66]. However, partial loss was observed in 15% of children where the ODF became stuck to teeth, lips, or palate, resulting in it being partially spat out or removed by the child using their finger. In a randomized cross-over study with 150 neonates and infants aged 2 days to 12 months, ODFs (2 × 3 cm) administered with a drink of choice (most commonly milk) were compared to glucose syrup (0.5–3 mL) administered with an oral syringe [67]. The overall swallowability of the ODF in all age groups was 70.0% vs 48.7% for the syrup ($p < 0.0001$), while the overall acceptability was 95.3% vs. 80.7%, respectively [67]. About one-third of participants aged <6 months left a part of the

ODF sticking to the mother's breast or bottle. In another study in healthy newborns, vitamin D ODFs showed better compliance and higher caregiver preference compared to syrup (1 mL) [68]. However, concentrated liquid vitamin D preparations were not available at the time of the study which may limit findings [15].

8.4.4 Chewable Preparations

These formulations are intended to be chewed before being swallowed. In addition to chewable tablets, medicated chewing gums are another potential innovative drug delivery system [69]. Many over-the-counter vitamin preparations for children as available as "gummy" chewable dosage forms, although their close resemblance to confectionery may limit potential development and safe use [70]. Familiarity and experience with chewable formulations vary globally. A review in 2002 found more than 60 chewable tablet formulations approved for paediatric use in the USA, mostly vitamins/dietary supplements but also analgesic/cold preparations, anti-infectives, anticonvulsants, antacids, and anti-asthmatics [71].

Pharmaceutical development measures will include ensuring acceptable organoleptic properties and palatability, as well as ensuring dosage forms are easily crushed following mastication. For dosage forms to be chewed, the primary critical attributes to be included in acceptability testing include taste and aftertaste, mouth feel, smell, and time and effort needed to chew (duration of administration) [19]. Complexity of manufacture depends on the technology used and can include routine manufacturing processes with standard equipment (tablets) or specialist processes and equipment (deposited formulations and soft gels) [16].

Children's deciduous (or "milk") teeth usually start erupting around 6 months of age, and the complete set of 20 is usually present between two and two-and-a-half years. Many chewable preparations for children are licensed from the age of 2 years, and these formulations have been found to be safe, well-tolerated, and advantageous in children from this age [71]. In fact, mastication before swallowing is an innate default mechanism for articles which enter the mouth [2], thus these formulations may be preferable in children of this age, who might not fully understand the instruction to swallow or retain other dosage forms. The natural tendency for children, particularly around 2–3 years of age, to chew even mini-tablets has been shown [72, 73]. Conversely, as with (oro-)dispersible preparations, the consequences of swallowing chewable tablets intact should be investigated as highlighted in the EMA guideline [12]. However, as suggested WHO, it may even be preferable for these tablets to be formulated such that both methods of administration are possible [1]. The risk of choking should be carefully considered if children are unable to unwilling to take the tablets as intended [12].

In a study of children's perceptions and preferences, school children aged 6–11 years showed the most positive attitudes towards chewable formulations compared to other SODFs [53]. Mebendazole chewable tablets have been assessed in children aged 1–16 years in three studies [74–76]. In one study of children aged

2–10 years, 98% (n = 390/396) of children chewed the tablet and swallowed, while for others, the tablet was dispersed in water to facilitate swallowing (n = 2) or not taken (n = 4) [74]. Around half took the tablet with water, while 8% of children needed the tablet broken into half or quarters. In another study, gagging and difficulty swallowing were observed in children less than 3 years old [75].

8.4.5 Multiparticulate Technologies

Multiparticulate systems are versatile platform technologies with considerable promise for application in paediatric pharmaceutical development. A more detailed review can be found in Chap. 9 of this book. These dosage forms consist of multiple, small, discrete units, including powders, pellets, beads, and granules, which can be presented as dosage forms in themselves (e.g. "sprinkles") or further processed to produce other solid formulations (including tablets, capsules, (oro-)dispersible and chewable preparations). They are typically below 4 mm in diameter, and multiple units make up a single dose. This gives these technologies the potential to cater for the various doses and patient capabilities through infancy, childhood, and adolescence. It should be noted that currently, regulatory guidance only uses harmonised compendial terminology for dosage forms and routes of administration, thus pellets or mini-tablets are not referred to as specific examples [12].

Development and manufacturing complexity depend on the technology used [16]. For these dosage forms, the primary critical attributes to be included in acceptability testing include particle size, shape, texture (surface aspects), and hardness (grittiness); taste and aftertaste for uncoated multiparticulates; dose volume (quantity of multiparticulates that needs to be taken); need for a measuring/counting device; and ease of administration [19].

Multiparticulates can be presented as a unit dose package in sachets, stick-packs, or hard capsules, or as multidose presentations, although this is less common [77]. Provision of different sizes and strengths of the final product allows accurate and reliable dosing in the individualized manner that is often required across the paediatric population. However, this need for dose flexibility may result in the need for strengths and higher supply chain complexity and cost-of-goods when using unit-dose presentations [77]. Multidose presentation includes bottles and containers containing multiple doses, with doses achieved by volume using a scoop or syringe-like device. The design of product presentation, be it a package or device, requires careful consideration.

Multiparticulates can be administered directly in the mouth or mixed with soft foods and beverages. Indeed, appropriate compatibility studies will be required to assure that the API and formulation do not lead to adverse physical or chemical interactions with the foods or drinks that they can be mixed with; defining the compatibility studies needed to support administration with food is challenging [77]. Multiparticulates which are labelled for administration via sprinkling should have a target size of 2.5 mm with no more than 10% variation over this size to a maximum size of 2.8 mm, as recommended by the FDA [78]. This is to ensure adequate mouthfeel and reduce the risk of inadvertent chewing.

Anatomically, infants can start swallowing thick, semi-solid foods from 6 months, and this could include administration of powders and multiparticulates in these foods [79, 80]. Patients and caregivers should also be provided with other important information for use, including the type and quantity of foods or drinks which can be utilized, ensuring that volumes are appropriate to assure complete ingestion of the intended dose. When formulations are mixed with food and drink vehicles, acceptability and palatability need to be demonstrated to avoid potential risks of aversion, particularly in the case of breast milk.

The application of coatings can allow multiparticulates to be specifically functionalized and subsequently processing them into other solid formulations is feasible, though it may be challenging. Other challenges may be faced when taste-masking multiparticulates, such as ensuring uniform thickness of barrier coatings. Acceptability of lopinavir/ritonavir pellets was shown to decrease over time in infants in children with HIV with tablets becoming more preferable; however, this was attributable to the unpleasant taste of the drug which was perceived from the pellets [81]. Caregivers experienced problems with masking the unpleasant taste of the drug with food, leading to food refusal, needing to sweeten food with sugar or honey (which can be expensive in resource-limited settings), and concerns about not administering the whole dose [81]. Further, the detrimental effect of chewing to the coating and dosage form may be problematic in younger children. These platform technologies also have the potential to produce fixed dose combinations (FDCs), which combine multiple drugs into a single dosage form for convenient and reliable administration. These are especially advocated in resource-limited countries for conditions requiring multi-drug treatment, including tuberculosis and HIV.

The relationships between multiparticulate size, dose, patient age, and mode of administration (i.e. directly in the mouth or mixed in soft food or liquid) are complex and not fully understood [77]. The size and quantity of multiparticulates can affect the mouthfeel and perception of grittiness in the mouth, with those greater than ~200 μm in size being generally deemed to be perceived as gritty [77]. The acceptability of medicine-containing multiparticulates has been evaluated in various therapeutic areas, including hypercholesterolaemia, iron-deficiency anaemia, HIV, cystic fibrosis, and epilepsy; these findings have been well summarized by Litalien and colleagues [15]. Expert members of EuPFI have reported that in user studies (unpublished data) adolescents over 11 years prefer discrete medicines and are therefore much less accepting of multiparticulate formulations [77]. Reservations about these formulations in older children and adolescents have also been seen in survey studies [52, 53], which may be influenced by limited familiarity with this dosage form. However, the commercial availability of multiparticulates as paediatric formulations continues to increase [15, 47].

8.5 The Evolution of Mini-Tablets

Mini-tablets are a unique dosage form which affords the advantages of multiparticulates, with regard to ease of administration and dose flexibility, coupled with the established and cost-effective manufacturing technique of tableting. As such, research into the application of this dosage form for the paediatric population has shown much progress since the implementation of the EU regulation in 2007, mainly related to the potential use of mini-tablets in clinical practice and, to a lesser extent, the pharmaceutical formulation and production aspects specific to mini-tablets compared to conventionally-sized tablets. As mini-tablets could potentially overcome swallowing difficulties in patients, they may also be applicable for other populations, such as geriatric patients with dysphagia, or even for animals [82]. A comprehensive review by Meruva and colleagues outlines the current applications of mini-tablets, including marketed products, ongoing clinical trials, and manufacturing and packaging considerations [83].

8.5.1 Technical Aspects of Pharmaceutical Development

Mini-tablets may be formulated as immediate-release or modified-release products. In the former case, the small size of the mini-tablets enables easy swallowing, and the required dose is built up with multiple mini-tablets as necessary. In the latter case, individual sub-populations of mini-tablets are prepared with different matrices or coatings to give varying release profiles, and are often encapsulated in defined ratios to give the overall desired drug release profile. The potential for mucoadhesive or floating mini-tablets within the stomach to aid the control of drug release has also been investigated. Alternative routes of administration of mini-tablets have also been investigated, for example, sustained release mini-tablets for ocular use [84] and mucoadhesive tablets for vaginal use [85].

In all of the above situations, the formulation and process design of mini-tablets should follow the same patient-focused and quality-first approach as for any other tablet formulation, including assessment of the chemical and physical compatibility of the drug and the excipients. The use of processing methods such as direct compression (mixing, lubrication, compression) or wet granulation (mixing, granulation, drying, lubrication, compression) will need to be evaluated on a case-by-case basis, bearing in mind that direct compression tablets tend to be softer and more friable than wet granulation tablets and therefore less suited for abrasive further processes such as film coating.

However, there are several aspects of mini-tablet formulation and production whereby the size difference between mini-tablets and conventionally-sized tablets has a potential significant effect on mini-tablet behaviour. Although no formal pharmacopoeial definition of mini-tablets exists, the WHO describes mini-tablets as having a maximum diameter of 4 mm [1]; the same description is generally used in the literature, with many publications focusing on 2 mm diameter mini-tablets.

Powder flow and particle size distribution are arguably more important for mini-tablets than larger tablets. The narrow diameter of the die used in mini-tableting requires excellent powder flow for reproducible die filling and hence good tablet weight (and thereby drug content) uniformity. However, care must be taken to avoid flooding the die table with powder, particularly when running at slow speeds or with few punches. This then requires close control of powder flow into and out of the feedframe, which may lead to longer than normal powder residence time in the feedframe, with possible consequences of over-lubrication and segregation if the feedframe is continually stirred [86, 87].

The maximum particle size becomes increasingly important with smaller tablet diameters. A ratio of die diameter to maximum particle size of at least 3 has been recommended [88], with the depth of the die being similarly considered [72]. Particles which are too large may lead to high and unacceptable tablet weight variability as only a few comprise each mini-tablet. A secondary physical effect of poor tablet weight control is poor reproducibility of tablet strength, in terms of hardness (resistance to crushing) and friability. However, the minimum particle size is also important as particles which are too small risk falling in the void between the die and the punch; hence, an overall narrower particle size range is required for mini-tablets than larger ones. A sieve cut (e.g. 180–500 μm) can be used with 2 mm diameter punches to avoid these problems.

A related processing issue is powder segregation (or demixing) during the tableting process, potentially resulting in separation of formulation components. Given the extremely low mini-tablet weights, typically in the range of 10 mg for a 2 mm diameter mini-tablet, a small degree of powder segregation could lead to unacceptable drug content variability in the resultant mini-tablets. The segregation potential during prolonged residence times in a stirred feedframe also needs to be considered. To reduce such segregation issues, granulation is often considered as a formulation approach [89].

In addition, the small total weight of mini-tablets means that the drug loading for a mini-tablet formulation would typically be higher than for a larger tablet—e.g. consider a drug dose of 4 mg in a mini-tablet of weight 10 mg compared to the same dose in a more conventional tablet of weight 100 mg. In a mini-tablet formulation, any processing problems associated with this high drug loading (e.g. mixing, flow, compressibility) cannot simply be "diluted out" by increasing the content of excipients but need consideration at the formulation design stage.

The compaction process itself is obviously crucial to the production of mini-tablets and determining their physical properties. As the tablet's physical dimensions decrease, its surface-to-area ratio will increase, resulting in a greater proportion of the particles being in contact with the die wall and punch surfaces. During the compaction process, the stress experienced by the particles in contact with and just inside the die and punch surfaces will typically be higher than that experienced by the more internally situated particles, in effect leading to greater compression at the surfaces and giving rise to a tougher layer on the outside of the tablet. As mini-tablets have a greater surface-to-area ratio, this tougher surface layer would be expected to dominate the overall behaviour of the tablet, resulting in mini-tablets showing a greater tensile strength and hence physical resilience to further manipulation than larger tablets [90]. It is known that during compaction of larger tablets, particle movement can lead to areas of varying density being formed, with the lower density regions being weaker and therefore more likely to be associated with capping and lamination [91]. Given the smaller dimensions of mini-tablets, the decreased space for particle movement, and the greater influence of the tougher surface layer, it is unclear whether this effect will be seen, or to what extent, with mini-tablets.

It may be expected that mini-tablets will show a higher disintegration rate and faster drug dissolution profile than conventional tablets, due to their higher specific surface area, leading to greater interaction with the surrounding aqueous phase. However, this may be counteracted by the tougher surface compression layer of mini-tablets, as discussed above. In any case, all tablets must pass the necessary pharmacopoeial and industrial specifications; while this comparison is of intellectual interest, from a pragmatic perspective, a modicum of formulation optimization may be required to ensure appropriate dissolution profiles from tablets of all sizes.

A range of mini-tablet tooling is available, from single-tip punches used during research and development to multi-tip punches (up to 19 tips on a single punch) used during production. The small size of the mini-tablet punches means that they are very delicate and easily damaged—our experience is that a 2 mm single-tip punch can withstand only a maximum applied force of approximately 2 kN during compression. Hence, it is vital to understand the compactibility of the mini-tablet formulation and to optimize the powder or granule blend, as formulation issues cannot be overcome simply by increasing compaction force as may be the case for larger tablets. Care must be taken when compressing ungranulated powder mixes with a high content of brittle-facture material such as lactose monohydrate, or those materials that expand on compression, as this will lead to stronger tablets which may be more difficult to remove from the die, again leading to damage to the punches, particularly the lower punch.

Suitable lubrication is needed to ensure a low ejection force, prevent picking and sticking, and retain punch integrity. Proportionately, mini-tablets have a greater surface area in contact with the metal die and punch surfaces, so an increase in lubrication is required compared to larger tablets [86], but as mentioned above, there is a risk of over-lubrication due to long powder/granule residence times in a stirred feed-frame, with possible deleterious effects on disintegration times and drug dissolution profiles.

The standard pharmacopoeial specifications and methods for physical characterization of tablets may not be appropriate for mini-tablets. Pharmacopoeial specifications for uniformity of weight are dependent on the nominal weight of the tablets, with a greater variability allowed for smaller tablets. For tablets less than 80 mg in weight, a variability of ±10% is typically allowed. Even assuming a perfectly uniform distribution of the drug in the powder or granule used for compression, this variability in weight will lead to a similar variability in individual dose uniformity, which may pose more significance in a paediatric population than in adults. Any increase in the inherent variability of drug content in the powder or granule will result in greater variability in the product, potentially leading to a failure in dose uniformity testing. Specifications for weight uniformity and dose uniformity need, therefore, to be considered in conjunction.

The large volumes of dissolution media used in dissolution testing (typically 900 mL) would be even less applicable for dosage forms intended for children than those intended for adults, though this is a wider general issue for all paediatric formulations and not just mini-tablets. The standard mesh aperture for pharmacopeial disintegration equipment is ~2 mm which would let most mini-tablets pass through, while the effects on the disintegration testing process of reducing the mesh aperture are not known. Likewise, the pharmacopeial friability test is designed for standard-sized compacts. Typically, a composite sample of 10 tablets is tested if the individual tablet weight is more than 650 mg. For smaller tablets, a sample mass of 6.5 g is required. However, this would equate to 650 mini-tablets with a nominal weight of 10 mg. It remains to be established whether the standard dimensions and operation of the friability tester are appropriate for mini-tablets. Some authors have suggested the use of small glass beads to overcome the loss of stress inside the friabilator [88]. A suitable friability test will be important in determining whether mini-tablets are resilient enough to undergo a coating process.

The hardness (resistance to crushing) of mini-tablets is difficult to measure; the usual Schleuniger-type hardness testers are not able to register the small breaking force of a mini-tablet, often leading to non-pharmacopeial methods being employed instead, such as the use of texture analysers. In this case, however, the orientation of the tablet during the crushing process may be different: a texture analyser will crush the tablet "top to bottom", whereas the conventional pharmacopeial method uses diametral "side to side" crushing, hence results may not be exactly comparable.

The biopharmaceutical aspects of mini-tablet use need to be considered, from both a formulation development and a clinical use perspective. There is limited understanding of the gastrointestinal physiology of young children as it pertains to drug bioavailability and how this may vary with the age of the child. It may be expected, however, that gastric emptying times may be different for different formulations (solutions, disintegrating mini-tablets or multi-particulates, and non-disintegrating mini-tablets or multi-particulates), which may then affect the rate and extent of drug absorption. Designing different formulation types of the same drug and dose to be bioequivalent may be a challenge, therefore. However, rapidly dissolving levetiracetam mini-tablets were shown to be bioequivalent with the originator tablets [92].

Mini-tablets may be tricky to handle due to their small size, and there is a risk of inaccurate counting of multiple units; hence, to fully exploit their potential as a drug delivery vehicle for the disparate paediatric population (neonates to adolescents) with varying dosing requirements, accurate and flexible packaging and dispensing systems are required to ensure appropriate dosing and safe administration. Development in this area (e.g. filled capsules, sachets, blisters, home electronic dispensing systems, and manual dispensers) has proceeded alongside research into the clinical and pharmaceutical aspects of mini-tablets [93]. In this manner, dose control can be facilitated, which should lead to increased acceptance of mini-tablets as a dosage form.

8.5.2 *Emerging Evidence of Suitability*

In addition to developments in the pharmaceutical manufacture of mini-tablets, studies have also been undertaken to assess their acceptability and swallowability of this dosage form in the paediatric population. A selection of recent studies using placebo mini-tablets is summarized in Table 8.3. Many of the early exploratory studies involved administration of a single mini-tablet; however, considering the limited dose loading per individual mini-tablet, it is envisaged that in therapeutic cases, multiple mini-tablets will be required to provide the appropriate dose. Administration of multiple mini-tablets as a single dose has been investigated, as this may change organoleptic properties such as mouthfeel and may impact whether young children will successfully swallow all units and receive the intended therapeutic dose. Several studies have demonstrated the ability of infants to safely take multiple placebo mini-tablets, though in many cases, chewing before swallowing was observed.

While the pharmaceutical manufacturing ability of mini-tablets has been well demonstrated, evidence of end-user acceptability will continue to grow as the number of licenced products and those under clinical development (including via the paediatric-use marketing authorization (PUMA) route) increases. Marketed products using mini-tablets have been developed for various indications, including epilepsy (sodium valproate and levetiracetam), cystic fibrosis (ivacaftor, and umacaftor/ivacaftor), dyspnoea caused by respiratory conditions (procaterol), multiple sclerosis (dimethyl fumarate and diroximel fumarate), fungal infections (terbinafine), and insomnia and autism spectrum disorders (melatonin). Numerous clinical trials are also underway for other indications, including cardiac diseases (enalapril maleate), autoimmune diseases, solid tumours, and malaria (see Meruva et al. for a broader list as of 2024 [83]). This highlights how mini-tablets are continuing to become an important SODF for the paediatric population.

8.6 Conclusions

The significance of solid oral formulations as a critical drug delivery modality for children is well-supported by emerging evidence and current regulatory recommendations. Novel formulation technologies have widened the scope of potential paediatric drug delivery systems and the once recognized lack of evidence-based evidence of their suitability and acceptability across the population continues to improve. SODFs have important advantages to support their use as paediatric formulations,

Table 8.3 Studies evaluating the swallowability, acceptability, and/or palatability of placebo mini-tablets (MTs)

Study design	Formulations	Participants	Key findings	Reference
Exploratory single-dose study	1 MT (3 mm, uncoated) with water or drink of choice	$N = 100$ aged 2–6 years old	46% of 2-year-olds, 53% of 3-year-olds, 76% of 4-year-olds and 87% of 5–6-year-olds swallowed the MT whole	[72]
Open randomized cross-over trial	1 MT (2 mm, uncoated) placed on tongue and taken with three mouthfuls of a drink of choice. 3 mL 15% glucose syrup (without additional liquid)	$N = 60$ aged 0.5–6 years old	93% swallowed or chewed and then swallowed the MT vs. 78% who swallowed the syrup completely or with a small runlet. Chewing was observed particularly among children aged 2–4 years. 67% swallowed the MT without chewing	[73]
Randomized cross-over trial over consecutive days	1 MT (4 mm) 250 mg powder 2.5 mL suspension 2.5 mL syrup Chewing and co-administration/mixing with food or drinks was neither recommended nor forbidden	$N = 148$ aged 1–4 years old	MT had highest acceptability rating on 10 cm visual analogue scale from 0 "very unpleasant/ bothersome" to 10 "not at all unpleasant/bothersome". Mean number of fully swallowed intakes was significantly higher for MT. MT was preferred by children and parents	[94]
Open randomized cross-over trial	1 MT (2 mm, coated or uncoated) placed on tongue and taken with three mouthfuls of a drink of choice. 3 mL 15% glucose syrup (without additional liquid)	$N = 306$ aged 0.5–5 years old (stratified by age group)	Overall acceptability (swallowing with or without chewing) and swallowability (swallowing without chewing/liquid runlet) of uncoated and coated MTs were significantly higher than glucose syrup. Point estimates for overall acceptability were 78–100%, for uncoated MT, 84–100%, for coated MT, and 65–90% for the syrup	[95]

Open cross-over trial evaluating swallowability for up to 4 days	x5 or 10 MTs (2 and 3 mm, coated) administered in fruit jelly on spoon	*N* = 60 aged 2–3 years old	75% of 2-year-olds and 93% of 3-year-olds swallowed the MTs with or without chewing (50% and 64% respectively without chewing)	[96]
Open randomized cross-over trial	1 MT (2 mm, uncoated, fast-dissolving) placed in the child's cheek pouch followed by drink chosen by parent (breast milk, milk, tea, water, or maltodextrin) 0.5 mL 15% glucose syrup (without additional liquid)	*N* = 151 neonates aged 2–28 days	100% swallowed all or part of dose for both formulations. 82.2% completely swallowed MT vs. 72.2% for syrup	[97]
Open randomized 3-way cross-over single-dose trial	Group 1 25 or 100 MTs (2 mm uncoated) 5 mL 15% glucose syrup Group 2 100 or 400 MTs (2 mm uncoated) 10 mL 15% glucose syrup MTs administered with a soft food or a drink of child's choice on a teaspoon, or the entire dose of the higher number of MTs with the same vehicles	*N* = 372 in total Group 1: *n* = 186 aged 6–23 months Group 2: *n* = 186 aged 2–5 years	Group 1: Overall acceptability (swallowing with or without chewing/liquid runlet) was 87% for 25 MT and 95% for 100 MT vs. 79% for syrup. Swallowability (without chewing) was 81, 74, and 36%, respectively Group 2: Overall acceptability was 85% for 400 MTs, which was non-inferior to syrup (83%). Swallowability of MTs (16%) was inferior to syrup (31%) Acceptability of 100 MTs (73%) was inferior to syrup (83%). Swallowability was non-inferior (31% and 31%)	[98]

(continued)

Table 8.3 (continued)

Study design	Formulations	Participants	Key findings	Reference
Open randomized cross-over single-dose trial	7–31 MTs (2 mm, coated) 4–15 MTs (2.5 mm coated) Administered on teaspoon of soft food (yoghurt or fruit mousse) or with mouthful of drink (water, tea, milk, or juice) of choice	$N = 319$ aged 1 month to 6 years (stratified by age group)	Overall acceptability (swallowing with or without chewing) was 87.1%–95.8% for the 2 mm MTs and 90.3%–96.8% for the 2.5 mm MTs. Swallowing whole without chewing ranged from 56.9% to 87.5% with older children tending to chew the MTs. Palatability was observed as pleasant/neutral in 93.5%–100% of children	[99]
Open randomized parallel group single-dose trial	16 or 32 MTs (3 mm, coated) with soft-food or water Administered on spoonful of soft food (mashed banana, yoghurt, or applesauce) or water	$N = 300$ aged 2–7 years	Overall acceptability (swallowing with or without chewing) was 88% for 16 MTs with soft food, 84% for 32 MTs with soft food and 80% for both quantities in water. Swallowability (without chewing) was 53%, 46%, and 42%, respectively. Palatability was 82–90% for all administration modes	[100]

including ease of administration, dosing accuracy, safety, manufacturing ease, and cost. Provision of a suitable dosage form is an important factor which governs the age-appropriateness of paediatric formulations; it should enable both safe and accurate dose administration for all intended indications and settings.

Generally, limited information on formulation is published in paediatric studies and clinical trials [101, 102]. Studies addressing patient-related outcomes, including formulation acceptance in relation to adherence [103], are still lacking. Formulations need to be well accepted by both children and their caregivers to support patient adherence and consequently benefit therapeutic outcomes. The EMA PIP requires end-user acceptability of paediatric medicinal products to be assessed as an integral part of pharmaceutical development studies [12]. Assessing both children's ability and willingness to take these formulations, as well as considering the needs and preferences of caregivers, is vital. Furthermore, education about more novel SODFs and wider exposure of their use may support their acceptability among patients, caregivers, and healthcare professionals in diverse cultural settings. Until recently, liquid medicines have generally had precedence in paediatrics, and though the rationale for solid formulations is strong, evidence exploring and supporting their appropriate use in clinical practice continues to grow.

References

1. WHO Expert Committee on Specifications for Pharmaceutical Preparations & World Health Organization. Forty-sixth report of the WHO Expert Committee on Specifications for Pharmaceutical Preparations. WHO technical report series. Geneva: World Health Organization; 2012. p. 970.
2. Mrsny RJ. Oral drug delivery research in Europe. J Control Release. 2012;161(2):247–53.
3. Richey RH, et al. A systematic review of the use of dosage form manipulation to obtain required doses to inform use of manipulation in paediatric practice. Int J Pharm. 2017;518(1–2):155–66.
4. Yuliani SH, et al. Prevalence, risk, and challenges of extemporaneous preparation for pediatric patients in developing nations: A review. Pharmaceutics. 2023;15(3)
5. MacArthur RB, et al. How compounding pharmacies fill critical gaps in pediatric drug development processes: suggested regulatory changes to meet future challenges. Children (Basel). 2022;9(12):1885.
6. Richey RH, et al. Manipulation of drugs to achieve the required dose is intrinsic to paediatric practice but is not supported by guidelines or evidence. BMC Pediatr. 2013;13:81.
7. Binson G, et al. Accuracy of dose administered to children using off-labelled or unlicensed oral dosage forms. Pharmaceutics. 2021;13(7)
8. Ivanovska V, et al. Pediatric drug formulations: a review of challenges and progress. Pediatrics. 2014;134(2):361–72.
9. Batchelor HK, Marriott JF. Formulations for children: problems and solutions. Br J Clin Pharmacol. 2015;79(3):405–18.
10. Walsh J, et al. Path towards efficient paediatric formulation development based on partnering with clinical pharmacologists and clinicians, a conect4children expert group white paper. Br J Clin Pharmacol. 2022;88(12):5034–51.
11. Ernest TB, et al. Developing paediatric medicines: identifying the needs and recognizing the challenges. J Pharm Pharmacol. 2007;59(8):1043–55.

12. European Medicines Agency, Guideline on pharmaceutical development of medicines for paediatric use (EMA/CHMP/QWP/805880/2012 Rev. 2). 2013.
13. European Medicines Agency Committee for Medicinal Products for Human Use, Reflection paper: formulations of choice for the paediatric population. 2006.
14. Van Riet-Nales DA, et al. Paediatric drug development and formulation design-a European perspective. AAPS PharmSciTech. 2017;18(2):241–9.
15. Litalien C, et al. From paediatric formulations development to access: advances made and remaining challenges. Br J Clin Pharmacol. 2022;88(10):4349–83.
16. Walsh J, et al. Patient acceptability, safety and access: a balancing act for selecting age-appropriate oral dosage forms for paediatric and geriatric populations. Int J Pharm. 2018;536(2):547–62.
17. Meyers RS. The past, present, and future of oral dosage forms for children. J Pediatr Pharmacol Ther. 2024;29(1):22–31.
18. Khan D, et al. Paediatric specific dosage forms: patient and formulation considerations. Int J Pharm. 2022;616:121501.
19. Kozarewicz P. Regulatory perspectives on acceptability testing of dosage forms in children. Int J Pharm. 2014;469(2):245–8.
20. Liu F, et al. Patient-centred pharmaceutical design to improve acceptability of medicines: similarities and differences in paediatric and geriatric populations. Drugs. 2014;74(16):1871–89.
21. Liu F, et al. Formulation factors affecting acceptability of oral medicines in children. Int J Pharm. 2015;492(1–2):341–3.
22. Venables R, et al. Determination of formulation factors that affect oral medicines acceptability in a domiciliary paediatric population. Int J Pharm. 2015;480(1–2):55–62.
23. Baguley D, et al. Prescribing for children – taste and palatability affect adherence to antibiotics: a review. Arch Dis Child. 2012;97(3):293–7.
24. Nordenmalm S, et al. Children's views on taking medicines and participating in clinical trials. Arch Dis Child. 2019;104(9):900–5.
25. Walsh J, et al. Playing hide and seek with poorly tasting paediatric medicines: do not forget the excipients. Adv Drug Deliv Rev. 2014;73:14–33.
26. Cram A, et al. Challenges of developing palatable oral paediatric formulations. Int J Pharm. 2009;365(1–2):1–3.
27. Alessandrini E, et al. Colour of medicines and children's acceptability? A systematic literature review of children's perceptions about colours of oral dosage forms. Pharmaceutics. 2023;15(7)
28. Wang S. Formulations in paediatric investigation plans (PIPs): introduction to PIP quality section and regulatory framework. Int J Pharm. 2015;492(1–2):332–4.
29. Nellis G, et al. Potentially harmful excipients in neonatal medicines: a pan-European observational study. Arch Dis Child. 2015;100(7):694–9.
30. Ryu GS, Lee YJ. Analysis of liquid medication dose errors made by patients and caregivers using alternative measuring devices. J Manag Care Pharm. 2012;18(6):439–45.
31. Yin HS, et al. Liquid medication dosing errors in children: role of provider counseling strategies. Acad Pediatr. 2014;14(3):262–70.
32. Yin HS, et al. Evaluation of consistency in dosing directions and measuring devices for pediatric nonprescription liquid medications. JAMA. 2010;304(23):2595–602.
33. Craig SR, et al. Pediatric therapeutics and medicine administration in resource-poor settings: a review of barriers and an agenda for interdisciplinary approaches to improving outcomes. Soc Sci Med. 2009;69(11):1681–90.
34. Sosnik A, et al. Novel formulation and drug delivery strategies for the treatment of pediatric poverty-related diseases. Expert Opin Drug Deliv. 2012;9(3):303–23.
35. Gerrard SE, et al. Innovations in pediatric drug formulations and administration technologies for low resource settings. Pharmaceutics. 2019;11(10)
36. Orubu ES, Tuleu C. Medicines for children: flexible solid oral formulations. Bull World Health Organ. 2017;95(3):238–40.

37. Hill SR. Putting the priorities first: medicines for maternal and child health. Bull World Health Organ. 2012;90(3):236–8.
38. Lajoinie A, et al. Solid oral forms availability in children: a cost saving investigation. Br J Clin Pharmacol. 2014;78(5):1080–9.
39. Hori KS, Siu AM, Yamamoto LG. Replacing liquid antibiotics with cheaper pill equivalents: an opportunity for substantial savings. Clin Pediatr (Phila). 2018;57(11):1294–9.
40. Nunn T, Williams J. Formulation of medicines for children. Br J Clin Pharmacol. 2005;59(6):674–6.
41. Nahirya-Ntege P, et al. Young HIV-infected children and their adult caregivers prefer tablets to syrup antiretroviral medications in Africa. PLoS One. 2012;7(5):e36186.
42. Ansah EK, et al. Improving adherence to malaria treatment for children: the use of pre-packed chloroquine tablets vs. chloroquine syrup. Trop Med Int Health. 2001;6(7):496–504.
43. Bagenda A, et al. Adherence to tablet and liquid formulations of antiretroviral medication for paediatric HIV treatment at an urban clinic in Uganda. Ann Trop Paediatr. 2011;31(3):235–45.
44. Waning B, et al. The global pediatric antiretroviral market: analyses of product availability and utilization reveal challenges for development of pediatric formulations and HIV/AIDS treatment in children. BMC Pediatr. 2010;10:74.
45. Adams LV, et al. Children's medicines in Tanzania: a national survey of administration practices and preferences. PLoS One. 2013;8(3):e58303.
46. Strickley RG, et al. Pediatric drugs—a review of commercially available oral formulations. J Pharm Sci. 2008;97(5):1731–74.
47. Strickley RG. Pediatric oral formulations: an updated review of commercially available pediatric oral formulations since 2007. J Pharm Sci. 2019;108(4):1335–65.
48. Meltzer EO, Welch MJ, Ostrom NK. Pill swallowing ability and training in children 6 to 11 years of age. Clin Pediatr (Phila). 2006;45(8):725–33.
49. Yeung VW, Wong IC. When do children convert from liquid antiretroviral to solid formulations? Pharm World Sci. 2005;27(5):399–402.
50. Bergene EH, et al. Register-based study showed that the age when children were prescribed antibiotic tablets and capsules instead of liquids increased from 2004 to 2016. Acta Paediatr. 2019;108(4):699–706.
51. Bracken L, et al. Can children swallow tablets? Outcome data from a feasibility study to assess the acceptability of different-sized placebo tablets in children (creating acceptable tablets (CAT)). BMJ Open. 2020;10(10):e036508.
52. Alessandrini E, et al. Children's preferences for oral dosage forms and their involvement in formulation research via EPTRI (European Paediatric Translational Research Infrastructure). Pharmaceutics. 2021;13(5)
53. Ranmal SR, Cram A, Tuleu C. Age-appropriate and acceptable paediatric dosage forms: insights into end-user perceptions, preferences and practices from the Children's Acceptability of Oral Formulations (CALF) Study. Int J Pharm. 2016;514(1):296–307.
54. Ghuman JK, et al. Behavioral training for pill-swallowing difficulties in young children with autistic disorder. J Child Adolesc Psychopharmacol. 2004;14(4):601–11.
55. Garvie PA, Lensing S, Rai SN. Efficacy of a pill-swallowing training intervention to improve antiretroviral medication adherence in pediatric patients with HIV/AIDS. Pediatrics. 2007;119(4):e893–9.
56. Patel A, et al. Effectiveness of pediatric pill swallowing interventions: a systematic review. Pediatrics. 2015;135(5):883–9.
57. Rashed AN, et al. Feasibility of developing children's Pill School within a UK hospital. Arch Dis Child. 2021;106(7):705–8.
58. McCloskey AP, et al. KidzMed e-learning to upskill student pharmacists to teach pill swallowing to children. PLoS One. 2023;18(3):e0282070.
59. Abdulla S, et al. Early clinical development of artemether-lumefantrine dispersible tablet: palatability of three flavours and bioavailability in healthy subjects. Malar J. 2010;9(1):253.

60. Hamed K, Grueninger H. Coartem(®): a decade of patient-centric malaria management. Expert Rev Anti-Infect Ther. 2012;10(6):645–59.
61. Ogutu BR, et al. Efficacy and safety of artemether-lumefantrine and dihydroartemisinin-piperaquine in the treatment of uncomplicated Plasmodium falciparum malaria in Kenyan children aged less than five years: results of an open-label, randomized, single-centre study. Malar J. 2014;13:33.
62. Nasrin D, et al. Acceptability of and adherence to dispersible zinc tablet in the treatment of acute childhood diarrhoea. J Health Popul Nutr. 2005;23(3):215–21.
63. Wagner-Hattler L, et al. High acceptability of an orally dispersible tablet formulation by children. Children (Basel). 2021;8(3)
64. Lottmann H, et al. A randomised comparison of oral desmopressin lyophilisate (MELT) and tablet formulations in children and adolescents with primary nocturnal enuresis. Int J Clin Pract. 2007;61(9):1454–60.
65. Cohen IT, et al. Ondansetron oral disintegrating tablets: acceptability and efficacy in children undergoing adenotonsillectomy. Anesth Analg. 2005;101(1):59–63.
66. Orlu M, et al. Acceptability of orodispersible films for delivery of medicines to infants and preschool children. Drug Deliv. 2017;24(1):1243–8.
67. Klingmann V, et al. Acceptability of an orodispersible film compared to syrup in neonates and infants: a randomized controlled trial. Eur J Pharm Biopharm. 2020;151:239–45.
68. Rodd C, et al. Comparison of 2 vitamin D supplementation modalities in newborns: adherence and preference. Appl Physiol Nutr Metab. 2011;36(3):414–8.
69. Chaudhary SA, Shahiwala AF. Medicated chewing gum—a potential drug delivery system. Expert Opin Drug Deliv. 2010;7(7):871–85.
70. Lam HS, et al. Risk of vitamin a toxicity from candy-like chewable vitamin supplements for children. Pediatrics. 2006;118(2):820–4.
71. Michele TM, et al. Safety of chewable tablets for children. J Asthma. 2002;39(5):391–403.
72. Thomson SA, et al. Minitablets: new modality to deliver medicines to preschool-aged children. Pediatrics. 2009;123(2):e235–8.
73. Spomer N, et al. Acceptance of uncoated mini-tablets in young children: results from a prospective exploratory cross-over study. Arch Dis Child. 2012;97(3):283–6.
74. Friedman AJ, Ali SM, Albonico M. Safety of a new chewable formulation of Mebendazole for preventive chemotherapy interventions to treat young children in countries with moderate-to-high prevalence of soil transmitted helminth infections. J Trop Med. 2012;2012:590463.
75. Silber SA, et al. Efficacy and safety of a single-dose mebendazole 500 mg chewable, rapidly-disintegrating tablet for. Am J Trop Med Hyg. 2017;97(6):1851–6.
76. Palmeirim MS, et al. Efficacy, safety and acceptability of a new chewable formulation. EClinicalMedicine. 2020;27:100556.
77. Harris D, et al. Age-appropriate solid oral formulations for pediatric applications with a focus on multiparticulates and minitablets: summary of September 2019 EuPFI workshop. Eur J Pharm Biopharm. 2020;153:222–5.
78. Food and Drug Administration Center for Drug Evaluation and Research (CDER). Guidance for industry: size of beads in drug products labeled for sprinkle. 2012 [28 April 2024]; Available from: https://www.fda.gov/regulatory-information/search-fda-guidance-documents/size-beads-drug-products-labeled-sprinkle-rev1.
79. Van de Vijver E, et al. Treatment of infants and toddlers with cystic fibrosis-related pancreatic insufficiency and fat malabsorption with pancrelipase MT. J Pediatr Gastroenterol Nutr. 2011;53(1):61–4.
80. Suchdev PS, et al. Home fortification of foods with multiple micronutrient powders for health and nutrition in children under two years of age. Cochrane Database Syst Rev. 2020;2(2):CD008959.
81. Kekitiinwa A, et al. Acceptability of lopinavir/r pellets (minitabs), tablets and syrups in HIV-infected children. Antivir Ther. 2016;21(7):579–85.

82. Ekweremadu CS, et al. Development and evaluation of feline tailored amlodipine besylate mini-tablets using L-lysine as a candidate flavouring agent. Pharmaceutics. 2020;12(10)
83. Meruva S, et al. Current state of minitablet product design: a review. J Pharm Sci. 2024;113(5):1123–54.
84. Abd El-Gawad AEGH, et al. Formulation and evaluation of gel forming ocular minitablets containing piroxicam. Journal of Pharmaceutical Research International. 2012;2(3):141–67.
85. Hiorth M, Nilsen S, Tho I. Bioadhesive mini-tablets for vaginal drug delivery. Pharmaceutics. 2014;6(3):494–511.
86. Goh HP, Sia Heng PW, Liew CV. The effects of feed frame parameters and turret speed on mini-tablet compression. J Pharm Sci. 2019;108(3):1161–71.
87. Lura A, et al. Transfer and scale-up of the manufacturing of orodispersible mini-tablets from a compaction simulator to an industrial rotary tablet press. Int J Pharm. 2021;602:120636.
88. De Brabander C, et al. Matrix mini-tablets based on starch/microcrystalline wax mixtures. Int J Pharm. 2000;199(2):195–203.
89. Loo SJ, et al. Study of diminutive granules as feed powders for manufacturability of high drug load minitablets. Int J Pharm. 2023;638:122922.
90. Lennartz P, Mielck JB. Minitabletting: improving the compactability of paracetamol powder mixtures. Int J Pharm. 1998;173(1):75–85.
91. Eiliazadeh B, Pitt K, Briscoe B. Effects of punch geometry on powder movement during pharmaceutical tabletting processes. Int J Solids Struct. 2004;41(21):5967–77.
92. de Mey C, et al. Bioequivalence of a novel minitablet formulation of levetiracetam. Arzneimittelforschung. 2012;62(2):94–8.
93. Hejduk A, Lulek J. Dispensing of minitablets – has the problem been resolved? Int J Pharm. 2022;619:121666.
94. van Riet-Nales DA, et al. Acceptability of different oral formulations in infants and preschool children. Arch Dis Child. 2013;98(9):725–31.
95. Klingmann V, et al. Favorable acceptance of mini-tablets compared with syrup: a randomized controlled trial in infants and preschool children. J Pediatr. 2013;163(6):1728–1732.e1.
96. Kluk A, et al. Can preschool-aged children swallow several minitablets at a time? Results from a clinical pilot study. Int J Pharm. 2015;485(1–2):1–6.
97. Klingmann V, et al. Acceptability of uncoated mini-tablets in neonates—a randomized controlled trial. J Pediatr. 2015;167(4):893–896.e2.
98. Klingmann V, et al. Acceptability of multiple uncoated minitablets in infants and toddlers: a randomized controlled trial. J Pediatr. 2018;201:202–207.e1.
99. Münch J, et al. Evaluating the acceptability, swallowability, and palatability of film-coated mini-tablet formulation in young children: results from an open-label, single-dose, cross-over study. Pharmaceutics. 2023;15(6)
100. Münch J, et al. Acceptability, swallowability, palatability, and safety of multiple film-coated mini-tablets in children aged ≥2–<7 years: results of an open-label randomised study. Pharmaceutics. 2023;15(2)
101. Standing JF, Khaki ZF, Wong IC. Poor formulation information in published pediatric drug trials. Pediatrics. 2005;116(4):e559–62.
102. Pandit S, et al. Inappropriate oral formulations and information in paediatric trials. Arch Dis Child. 2010;95(9):754–6.
103. van Riet-Nales DA, et al. Effects of the pharmaceutical technologic aspects of oral pediatric drugs on patient-related outcomes: a systematic literature review. Clin Ther. 2010;32(5):924–38.

Chapter 9
Micropellets: A Modern Multiparticulate Technology Platform for Pediatric Medicines

Norbert Pöllinger

Abstract Multiparticulates like pellets and micropellets are an ideal approach to provide oral solid dosage forms, which are most suitable for the pediatric and the geriatric population. Micropellets are a versatile, integrated technology platform allowing a wide variety of oral drug products. A broad range of dosage strengths can be achieved when different quantities of a micropellet product are filled into capsules, stickpacks, or sachets. Micropellets can be administered as such into the mouth directly, sprinkled on food, or dispersed in a liquid.

With one basic micropellet core formulation concept, different drug release profiles, such as immediate release, controlled, or modified release, can be realized. Established technologies with proven scale-up concepts are capable of providing modern micropellet-based medicines to developing countries at a viable cost. Compared to micropellet technologies, the production of mini-tablets is technically more complicated and challenging. Simpler containment concepts are applicable to very effective "high potent" drug compounds with the micropellet concept.

Intelligent formulation and development concepts for formulation and clinical development would ideally go with the final market formulations from the very beginning. This strategy can avoid preliminary simple enabling formulations and reduce bioavailability bridging studies to the minimum.

Keywords Multiparticulates · Pellets and micropellets · Micropellets and minitablets · Micropellet manufacturing technologies · Optimized API solubility and bioavailability · Personalized medicines

N. Pöllinger (✉)
Glatt Pharmaceutical Services GmbH & Co. KG, Binzen, Germany
e-mail: Norbert.Poellinger@glatt.com

H. Batchelor, K. Rose (eds.), *Pediatric Formulations*, AAPS Advances in the Pharmaceutical Sciences Series 58, https://doi.org/10.1007/978-3-031-77240-5_9

9.1 Introduction

Only a fraction of our licensed drug products are approved for therapy in children. In particular, for very young children and rare diseases in childhood, clinically tested and approved drug products are missing. Often, children must be treated off-label with drug products and formulations that are not approved or suitable for their age group [1]. Depending on the pediatric age groups, adult formulations may not be suitable for use in children, and new pediatric formulations must be specifically developed for clinical development and for the market [2]. The physical, metabolic, and psychological processes inherent to growth from birth to adulthood reveal that children cannot be regarded as small adults, nor can they be regarded as a homogeneous group in themselves. As a consequence, clinical trials in adults are not necessarily predictive for children and vice versa. Thus, in many cases, clinical trials will be needed in children of different ages in order to demonstrate that a pediatric medicine is safe and effective in all of the target groups for which the medicine is being developed [3].

Extensive variability in dose and dosing regimens used to treat pediatric patients, based on maturational and non-maturational differences between individuals, is part of the essence of pediatric clinical pharmacology today. Consequently, the pediatric community is in need of drug formulations tailored to the specific needs of neonates, infants, children, and adolescents. This must include valid data on product stability, palatability, and compatibility. Most of the time, children are still treated with medicines that were neither designed or developed nor evaluated specifically for use by children. As a consequence, there is a risk of suboptimal (too low, too high, or too variable) dosing and side effects from potentially toxic ingredients, including excipients [4].

The EU pediatric regulation mandates a pediatric investigation plan (PIP) for any new medicine at the end of human pharmacokinetic studies or after proof of concept. The European Medicines Agency (EMA) will not validate a submission without an approved PIP, which must address the potential future use of the drug for all pediatric groups from preterm newborns to adolescents. The US FDA would review the pediatric plan (PP), but pediatric discussions will not impede adult registration [5].

Clinical testing in children is subject to strict regulations. As children are not a very uniform population, they are considered as a "moving target" divided into five different age groups: pre-term newborn infants, term newborn infants, infants and toddlers, children, and adolescents [6, 7].

Potential and promising formulation approaches for oral solid pediatric dosage forms are also considered, reflecting the need for different and dedicated drug products for children and adults, including the potential need for bridging of formulations used during clinical development.

	neutral tasting API	inconvenient tasting API	very bad tasting API
effective taste masking required ?	not required	required	**absolutely required**
Taste masking approaches			
flavors	not required	√	na
sweeteners	not required	√	na
complexation	not required	√	to be evaluated
salt formation	not required	√	to be evaluated
cyclodextrins	not required	√	to be evaluated
taste concealing with coating	not required	(√)	**√**
Oral solid dosage forms			
powders	√	√	na
granules	√	√	na
micropellets (coated)	micropellets	micropellets	**coated micropellets**
mini-tablets (coated)	mini-tablets	mini-tablets	**coated mini-tablets**
Oral liquid dosage forms			
solution	API in solution	(API in solution)	na
suspension	API in suspension	(API in suspension)	suspension with **coated micropellets**

* single-use oral liquids with coated micropellets from sachet / stickpack / dispersible tablet / capsule
* multiple-dose oral liquids with coated micropellets (dry suspension, ready-to-use suspension)

Fig. 9.1 Pediatric formulation concepts considering the taste of API

9.2 Oral Pediatric Formulations: General Considerations

The oral route is the preferred route of administering a medicine: oral dosage forms are well accepted by all patient age groups if administered in a suitable form. To ensure that a medicine is accepted by the young patients, palatability has become an integral part of pediatric formulation development (a quality part of PIP [8]). Bad-tasting APIs must be effectively concealed with feasible technologies (Fig. 9.1).

Moreover, the size and form of a pediatric medicine must be adequate to allow easy swallowing even by newborn children. Normal-sized tablets or capsules are not appropriate: their acceptability for very young patients is very limited [9]. Splitting of tablets is not an option as it often is not precise enough and moreover not applicable to coated tablets, as functional coatings providing, e.g., taste masking or controlled drug release would be destroyed.

On the other hand, micropellets <500 µm with the API enclosed in a taste-masked form are a very promising pediatric formulation approach (Figs. 9.1 and 9.2): they can be administered as such as well as an oral liquid. Mini-tablets with a diameter of, e.g., 2 mm are larger-sized multiparticulates. Randomized controlled trials showed the acceptability of this oral solid dosage form in the pediatric population and even in neonates [10, 11].

9.3 Pediatric Drug Products with Micropellets: Administering Solid and Liquid Oral Dosage Forms

Oral drug delivery remains the most widely accepted and preferred route of administration in pediatric and adult populations alike, both from a manufacturing and end user perspective [12, 13]. While liquid medications have traditionally been the

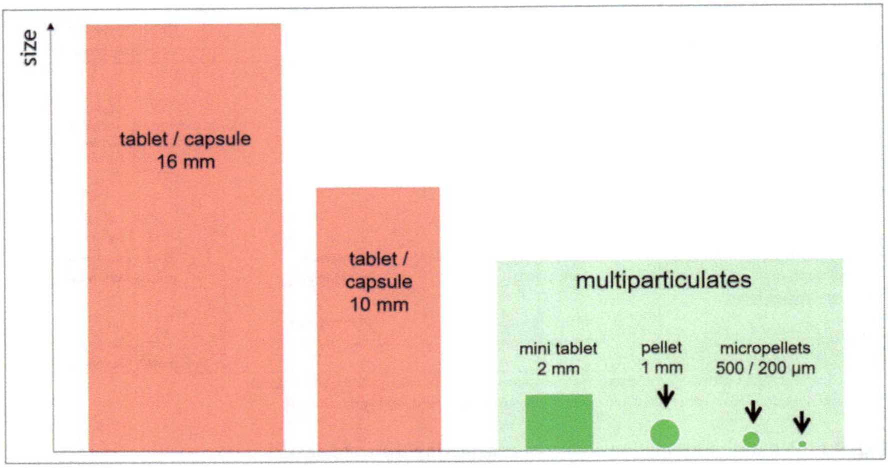

Fig. 9.2 Size of various oral solid drug products

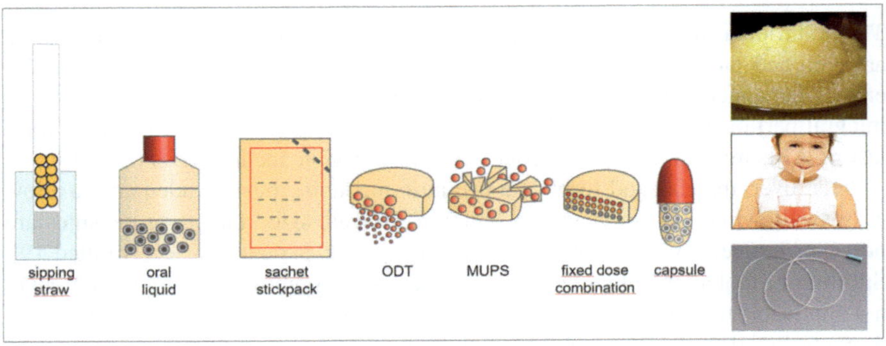

Fig. 9.3 Various drug products with one micropellet formulation

preferred choice in pediatric care, there is substantial interest in solid oral dosage forms for this population. Tablets and capsules may pose challenges for young children who cannot swallow them whole. Modern solid formulations, like micropellets, offer a promising alternative. Micropellets not only eliminate the need for swallowing large units intact but also allow for flexible and individualized drug delivery. Different drug products and different dose strengths can be made available easily with multiparticulates [14] (see Figs. 9.3 and 9.4). Formulation development is a global concern, particularly in resource-limited settings. The World Health Organization (WHO) advocates for flexible solid oral dosage forms that can be administered in various ways, including without water, to address diverse healthcare needs.

For infants, micropellets can be incorporated into semi-solid foods from as early as 6 months, ensuring compatibility and acceptability. However, studies must

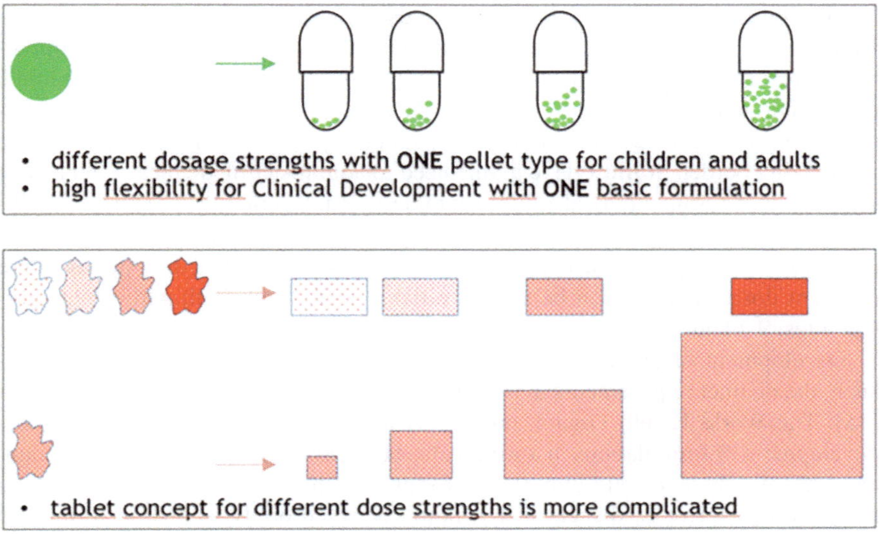

Fig. 9.4 Dose flexibility with micropellets—compared with the tablet concept

confirm compatibility with food and drinks to prevent aversion, especially in the case of breast milk. Addressing these considerations is crucial for developing safe and effective pediatric formulations.

Oral drug delivery remains the most widely accepted and preferred route of administration in pediatric populations.

While liquid medicines have historically been considered the "gold standard" in pediatrics, the emergence of innovative drug delivery technologies has led to a paradigm shift toward research and development into solid oral dosage forms for use across this heterogeneous population. Provision of a suitable dosage form is an important factor that governs the age-appropriateness of pediatric formulations. It should enable both safe and accurate dose administration for all intended indications and settings. Formulations need to be well accepted by both children and their caregivers to support patient adherence and consequently benefit therapeutic outcomes. Given the extensive physiological and developmental differences across this group (ranging from neonates to adolescents), it is unlikely that one single formulation could fulfill all requisites and be deemed appropriate for all intended ages. Rather than a "one-size-fits-all" approach, investigators aim to tailor each formulation to cover the widest age range, ensuring pharmaceutical development strategies are effective and economical whilst providing authorized products for each target subset [14].

The use of tablets and capsules may be limited in the pediatric population due to the inability of very young children to swallow monolithic dosage forms intact. More contemporary solid formulations include multiparticulate forms such as micropellets. These negate the need to swallow large units intact and can potentially provide a flexible and individualized approach to drug delivery, with prime application in pediatrics [14]. Micropellets also offer advantages in terms of stability, excipient use, palatability, and functionality. Moreover, they provide the opportunity for the development of functionalized formulations such as modified, prolonged, and delayed-release systems. This not only gives the potential for targeted drug delivery but can also benefit patients by reducing dose frequency and minimizing burden of lifestyle, e.g., the need to administer medicines at nursery or school. Solid formulations also offer complete and accurate dose delivery, facilitated by individual, uniform dose units or packages (e.g., sachets, stickpacks, or capsules for micropellets) which are easy to administer. Different drug products and different dose strengths can be made available easily with multiparticulates [14] (see Figs. 9.3 and 9.4).

Development of appropriate formulations is a global health challenge, and as such, the distinct requirements in resource-limited settings must also be considered [15]. The World Health Organization (WHO) has recommended prioritizing the development of formulations that would also be suitable for use in developing countries, namely flexible solid oral dosage forms, which can be administered in more than one manner (e.g., dispersed in a liquid or taken orally as a solid form even without water) [12]. Also in relation to resource-limited settings, solid formulations have the added advantage of superior stability and having low bulk and weight, thus being easy to transport and store [14]. Micropellets fulfill those requirements in an excellent manner.

Anatomically, infants can start swallowing thick, semi-solid foods from 6 months, and this could include administration of micropellets in these foods. Appropriate studies must demonstrate that physical and chemical compatibility of the drug product with food or drinks they are mixed with is guaranteed. Acceptability and palatability need to be demonstrated to avoid potential risks of aversion, particularly in the case of breast milk [14, 16].

9.3.1 Various Drug Products with Micropellets

9.3.1.1 Capsules, Sachets, Stickpacks with Micropellets (see Fig. 9.3)

Micropellets can easily be filled into capsules, sachets, or stickpacks. Different micropellet qualities can be filled into one packaging unit, e.g., micropellets with different APIs or drug release profiles. Based on the coated micropellet concept, chemically incompatible APIs can be combined in one administration form.

Different dosage strengths are easily realized by filling different micropellet quantities into one unit. Individualized dosing strengths can be provided for dose titration in clinical dose finding studies and in the initial treatment phase, but also for a gradual withdrawal of a medicine when an exit from therapy is intended [17]. With micropellets, dose flexibility is much simpler to achieve than with the tablet or mini-tablet concept (Fig. 9.4).

The content of a single unit can be directly administered to the patient's mouth. Alternatively, micropellets are sprinkled on a small portion of soft food or small volume of a beverage. For this application, the compatibility and stability of the sprinkled products with soft food or beverage must be proven.

Filling micropellets into capsules is an economic standard "packaging" technology. Very large capsule sizes can be used as the capsules are serving as primary packaging material only and must not be swallowed. Compared with capsule filling, sachet or stickpack production is more challenging from a technical and economic point of view.

Hydrocortisone taste masked pellets are filled into capsules size 00 elongated (Alkindi®, Diurnal Ltd., UK). Excellent content uniformity of capsules with different strengths from 0.5 mg to 5 mg is achieved [18, 19]. Sprinkle capsules are a new generation of capsule shells having an innovative closure that needs less force to open—opening of capsules is made easier and safer for parents and caregivers [20].

9.3.1.2 Tablets with Micropellets (ODT, MUPS, Fixed Dose Combinations) (see Fig. 9.3)

Micropellets could be processed to different types of tablets, such as orally disintegrating tablets (ODT), MUPS tablets, or also fixed dosed combination tablets.

Compared with simple filling of micropellets into capsules, sachets, or stickpacks, compaction of micropellets into tablets is technically more complicated. The compression of micropellets goes along with considerable challenges, referring to tablet weight variation and segregation phenomena. De-mixing is usually due to differences in size, shape, surface, and density between micropellets and extragranular tableting excipients. A threshold of at least 50% w/w pellet content should be attained in any tableting blend to avoid segregation [21]. The biggest challenge when micropellets are compressed into tablets often results from the inevitable deformation of the particles during compression: film coating can be damaged, resulting in a subsequent reduction or even loss of the taste-masking properties and a modification of the drug release properties.

Considering the potential risks and challenges with micropellets compressed to tablets, this technical option cannot be recommended as a top priority formulation approach.

9.3.1.3 Oral Liquids with Micropellets

With micropellets <500 μm, oral suspensions can be easily prepared. In contrast, 2 mm mini-tablets tend to fast sedimentation and are therefore almost not suitable for oral liquids.

9.3.1.3.1 Multiple-Dose Oral Liquids (Ready-to-Use Suspensions, Dry Suspensions) (see Fig. 9.3)

A blend of micropellets and additional suspension stabilizing excipients, and eventually flavors, is filled as a dry suspension form into bottles. Prior to the first administration, the dry suspension is transferred into a suspension by adding an appropriate dispersion medium, usually water. Sedimented particles must be easily redispersible when the suspension is re-suspended prior to each administration and must not form a solid cake. Taste masking and drug release must be guaranteed over the whole in use time of the oral liquid, which is usually 1 to 2 weeks. Antimicrobial compounds may be required.

In order to avoid any instability issues associated with a ready-to-use suspension, the dry suspension concept is often preferred. The chemical and physical stability of a suspension prepared from a dry suspension prior to first use usually must cover a 1–2 week time period only, which is much easier to fulfill than a 3–5 year expiry date of a ready-to-use liquid suspension preparation.

9.3.1.3.2 Single-Dose Oral Liquids (Sachet, Stickpack, Drinking Straw) (see Fig. 9.3)

Sachets and stickpacks are monodose packs containing the micropellets and further additives. To physically stabilize the resulting aqueous dispersion, density-increasing compounds such as sugars and immediately swelling viscosity-increasing water-soluble polymers like acacia gum may be included. Antimicrobial compounds are not required, as the liquid prepared from a sachet or stickpack should be administered immediately after preparation.

A defined single dose of micropellets can also be filled into drinking straws. The young patient can enjoy a favorite drink with the straw while taking the exact amount of medicine prescribed. A control filter element in the drinking straw ensures that the complete dose is administered [22].

9.3.1.4 Nasogastric Tubes (see Fig. 9.3)

Children not able to swallow may have to be fed via nasogastric tubes, which can be used for liquid medication as well [23]. Micropellets <500 μm can be administered easily with a syringe through nasogastric tubes, which usually have a diameter of 5–16 Charriere = ~ 1.7–5 mm [24].

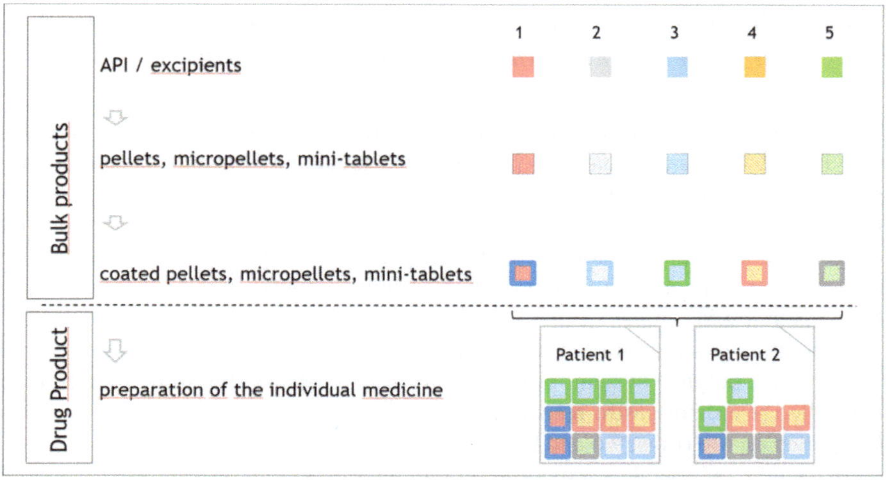

Fig. 9.5 Individual precision polymedicine with multiparticulates

9.3.1.5 Personalized Pediatric Medicine (Individual Precision Polymedicine)

With the micropellet concept, individually composed and dosed medication can be provided for children—but also for older adults (Fig. 9.5). Different types of micropellets containing different APIs, drug doses, and drug release profiles can be filled into capsules, sachets, stickpacks, or drinking straws [25]. With this concept, patient compliance and ease of medication can be improved considerably.

9.4 Micropellets: A Versatile Multiparticulate Concept for Oral Pediatric Dosage Forms

The micropellet concept is a versatile platform technology with considerable promise for application in pediatric pharmaceutical development. This technology has the potential to cater for the various doses and patient capabilities through infancy, childhood, and adolescence [14]. The very unique feature of the API containing core micropellets is that they can be processed to very different final micropellet qualities in terms of taste concealing and drug release. Moreover, aspects like optimizing drug solubility, release, and bioavailability as well as smart formulation concepts for "high potent" compounds are covered.

9.4.1 *Multipurpose Coating Applications on Core Micropellets for Taste Masking and Drug Release Control*

Palatable taste is a basic requirement of pediatric medicines. Powders and granules may be a convenient option to prepare pediatric medicines when the taste of a drug is acceptable, and palatability is good enough with simple taste "adjusting" operations, such as adding flavors and sweeteners. For bad-tasting APIs, the micropellet concept is a preferable option: the bad-tasting drug is completely encapsulated with a taste-concealing coating on top of the core micropellets containing the API (see Figs. 9.1 and 9.5).

For optimized and tailored drug release, the core micropellets are film-coated with feasible polymer systems providing any drug release profile from immediate to controlled release or gastro-resistance. If required, a final taste-concealing coat can be applied on top (Figs. 9.6 and 9.7).

In the end, micropellets provide several biopharmaceutical advantages compared with single-unit forms: the variability of the gastric emptying and dependency on the nutritional state (fasted, fed) is reduced. Application of multiparticulates minimizes the risk of high local concentrations within the gastrointestinal tract going along with mucosa irritations. The intra- and interindividual variability of blood levels is known to be reduced, as well as the dose dumping risk. A controlled onset time of drug release can be achieved. In addition, selecting feasible coating systems, drug delivery to distal sites within the GI tract is possible, e.g., to the colon for treatment of local mucosa irritations and inflammations [26].

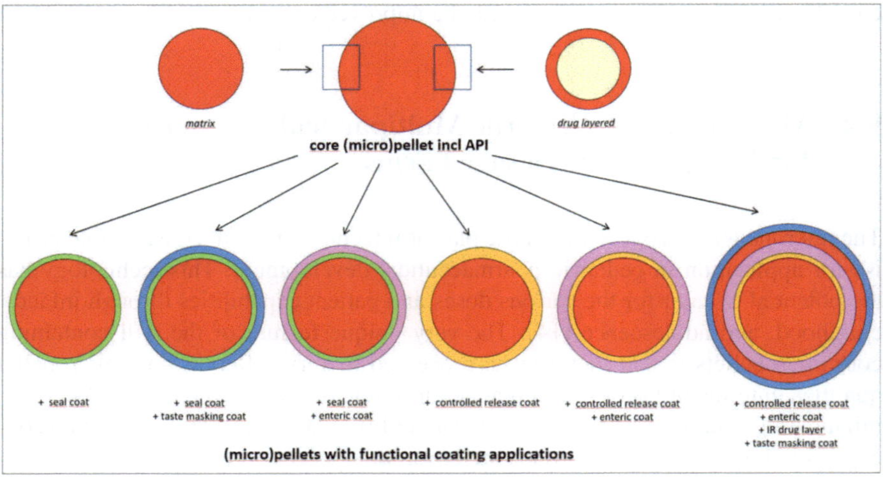

Fig. 9.6 Formulation and drug release variations with one basic drug core micropellet (examples)

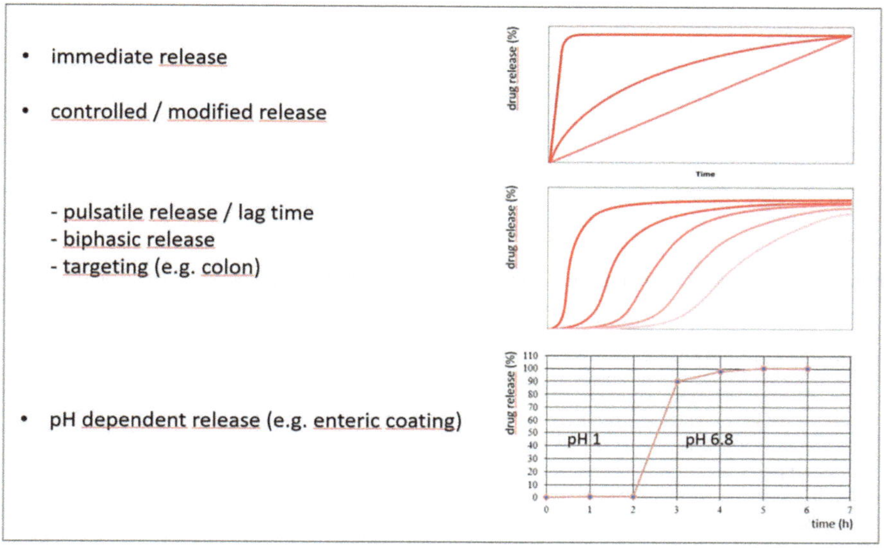

Fig. 9.7 Tailored drug release profiles with micropellets

9.4.2 Optimizing the Solubility and Bioavailability of Poorly Soluble Drugs with Appropriate Micropellet Core Technology Platforms

Many pharmaceutical drug compounds are poorly water-soluble or almost insoluble in water. Nowadays, around 70% of newly discovered drug candidates and approximately 40% of marketed oral drugs are categorized as practically insoluble [27]. Drug solubility is one of the most important parameters to achieve therapeutic concentration. Concepts for optimization of drug release and bioavailability of poorly soluble or almost water-insoluble active compounds can effectively be implemented in micropellets by applying different techniques.

9.4.2.1 Microenvironmental pH Modulation

Microenvironmental pH modulation is required for many basic drugs, with the aid of weak acids such as tartaric acid. Due to the high hygroscopicity of weak acids, a probable degeneration of the drug candidate by hydrolysis must be considered when the drug is in permanent direct contact with a weak acid, e.g., in matrix tablets. Micropellets with a seal coat layer separating the sensitive drug compound from the acid core are a promising technical option to achieve a product with long-term stability (Fig. 9.8).

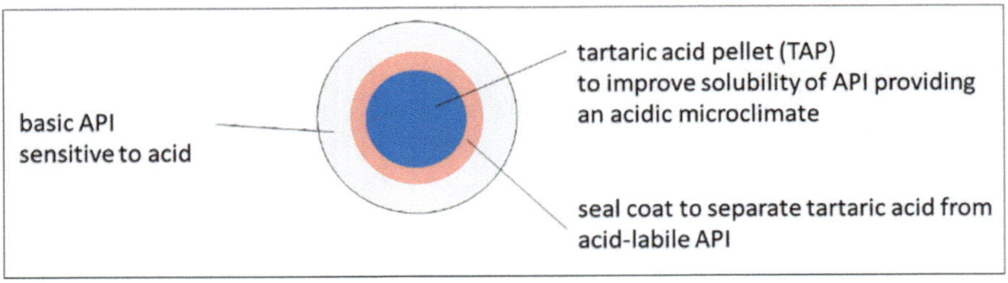

Fig. 9.8 Microenvironmental pH modulation of a basic API (Dabigatran) sensitive to acid [28, 29]

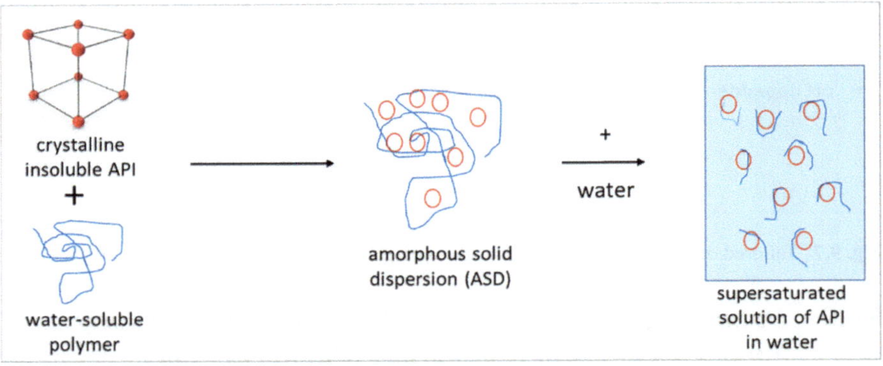

Fig. 9.9 From crystalline API to Amorphous Solid Dispersion (ASD)

9.4.2.2 Amorphous Solid Dispersions (ASD)

A common approach to improve the solubility of poorly water-soluble compounds is to transfer them into Amorphous Solid Dispersions (ASD) with suitable polymer candidates [30]. An ASD is a solid dispersion in which a previously crystalline active ingredient is dispersed within an excipient matrix in an amorphous form (Fig. 9.9).

With the drug in an amorphous state, no energy is required to break a crystal lattice in order to dissolve the compound. For this reason, relative to the crystalline form, the amorphous form of many poorly water-soluble drugs can achieve substantially higher apparent water solubility and markedly faster dissolution (Fig. 9.10).

ASD is also known to result in higher membrane flux due to a higher supersaturation and thus improve bioavailability. ASD has higher wettability than pure lipophilic drugs due to the presence of hydrophilic polymers. In addition to functional polymers, an ASD can include surfactants in order to further enhance drug release and stability [32].

Different types of ASD micropellets can be prepared in addition to ASD granules (which are usually further processed to tablets): ASD matrix micropellets and ASD-layered micropellets (Fig. 9.11). The crystalline API and the polymer are dissolved

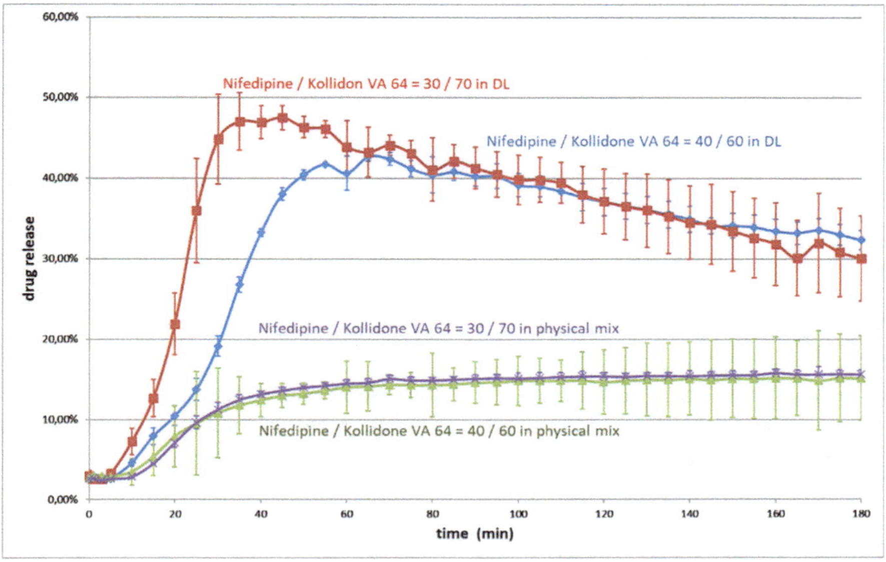

Fig. 9.10 In vitro release of Nifedipine from ASD layered pellets with different drug loads (30% vs 40%), (USP paddle, pH 6.8) [31]

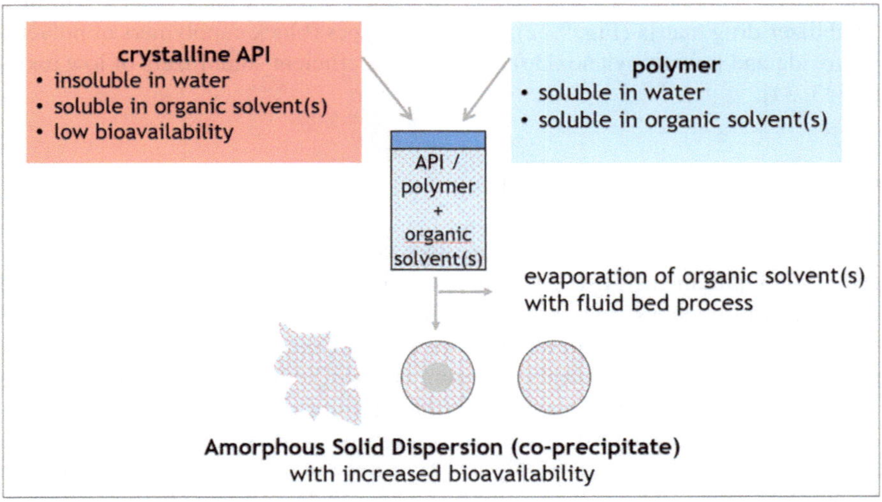

Fig. 9.11 Drug-layered and matrix micropellets including the drug compound as an Amorphous Solid Dispersion (ASD)

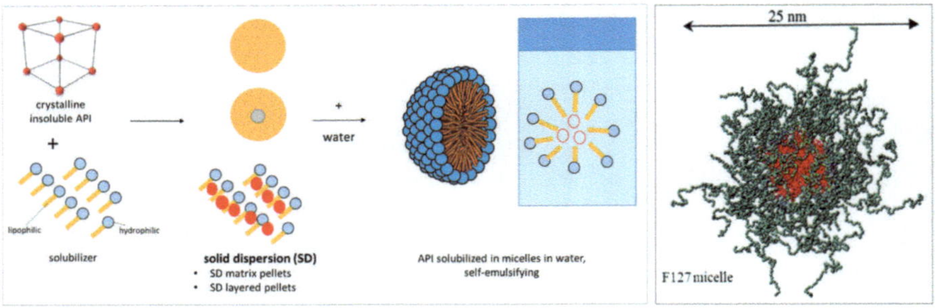

Fig. 9.12 Solid dispersion (SD) matrix and layered pellets Pluronic 127 micelles [33]

in a feasible organic solvent (e.g., ethanol, isopropanol, acetone) or a mixture thereof. The ASD solution is sprayed onto carrier material (powders, starter beads) in order to provide a solid amorphous solid dispersion preparation.

9.4.2.3 Solid Dispersions Including Solubilizer(s)

Solid dispersions consisting of a poorly water-soluble or water-insoluble drug and O/W surfactant(s) are also capable of improving drug solubility and bioavailability. The active ingredient is dissolved or dispersed with the surfactant to form a solubilizer-drug matrix (Fig. 9.12). Poloxamer types (block copolymers of polyethylenoxide and polypropylenoxide) are known as efficient solubilizers of low toxicity [33, 34].

9.4.2.4 "High Potent" Micropellets

Handling of "high-potent" active compounds is a challenge in terms of work safety and GMP requirements. Also, NCEs are not fully characterized toxicologically, and are often considered as "high potent" compounds. Today, the term "high potent" is replaced by the quantitative description with OEL (Occupational Exposure Limit in $\mu g/m^3$) and PDE (Permitted Daily Exposure in mg). Qualitative descriptions such as cytotoxic, hormone, or carcinogen, therefore, have become irrelevant. OEL values $<100\ \mu g/m^3$, referring to OEB (Occupational Safety Band) 3–5, are considered to be critical in terms of work safety (Fig. 9.13)—contacts of operating staff with such products should be reduced to the minimum, introducing the most feasible platform technologies and containment.

PDE (Permitted Daily Exposure in mg) value refers to a potential cross-contamination risk for products manufactured in multipurpose units. PDE value is an important indicator when the maximum allowed carryover quantities after cleaning of a processing unit must be defined. The production of a product can only take place in a multipurpose unit when a cleaning validation study, performed with

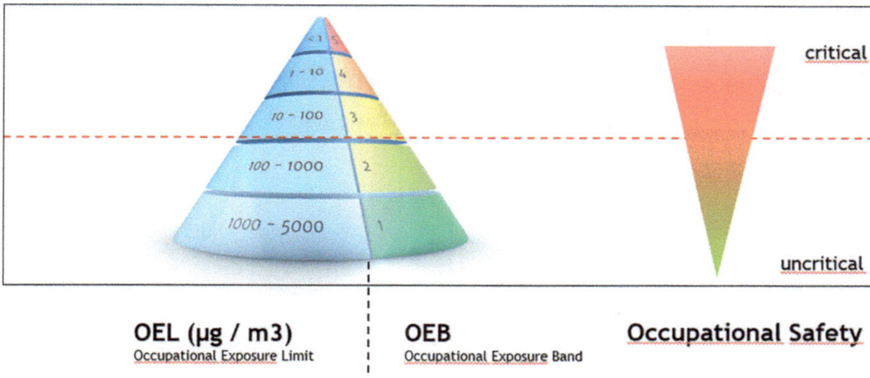

Fig. 9.13 OEL and OEB: Health/Safety/Environment (HSE) criteria

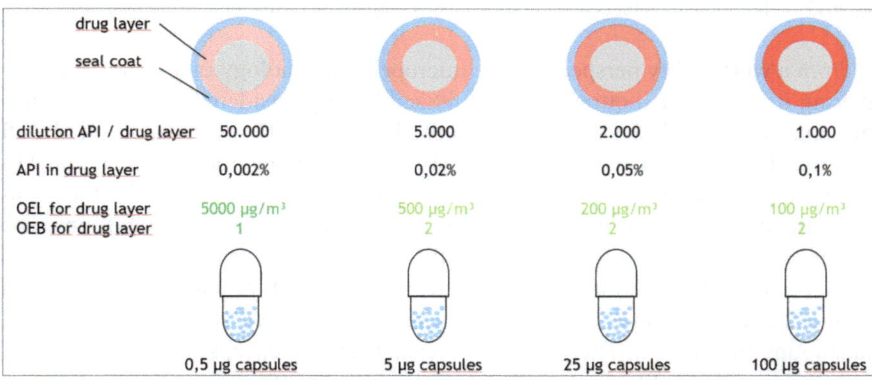

Fig. 9.14 Low-dosed API (Thyroxin, T4) with OEL 0,1 µg/m^3 = OEB 5 formulated to drug-layered pellets 0.5–100 µg/capsule (pellet fill weight per capsule: 150 mg, capsule size: 3)

validated analytical methods, can prove that the maximum carryover limits are kept. Cleaning validation studies have to be performed for every sub-unit of a production line, as defined in a risk assessment.

The micropellet technology platform is a promising option to process compounds with an extremely low OEL value, such as Thyroid hormone Thyroxin T4 (OEL: 0,1 µg/m^3 = OEB 5) applying established fluid bed technology: for drug layering on starter beads, the API is diluted extensively, reducing the OEB from 5 to 1, resp. 2 (Fig. 9.14).

Today, Levothyroxin (L-Thyroxin) tablets with dosage strengths of 25, 50, 75, 88, 100, 112, 125, 137, 150, 175, and 200 µg are commercially available. With Levothyroxin micropellets, an even more individualized dosing is easily possible, filling individual micropellet quantities into capsules. To treat children, sprinkling capsules would be a feasible dosage form to allow the most convenient application.

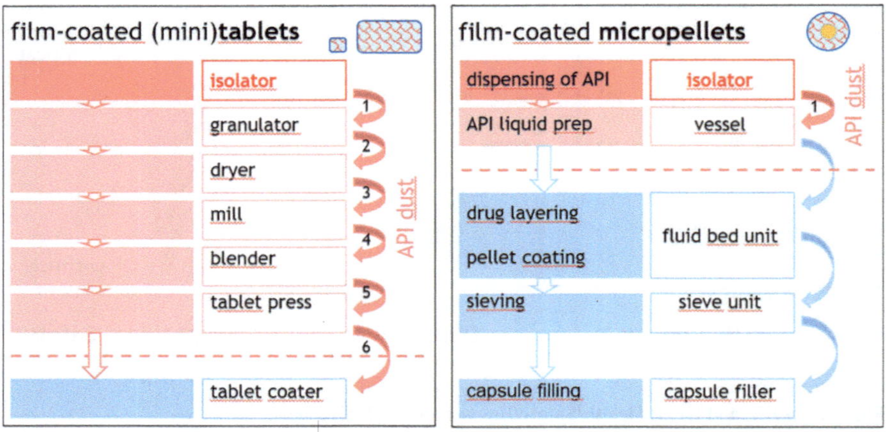

Fig. 9.15 "High potent" tablets and micropellets—comparison of manufacturing processes

From a work safety perspective, the micropellet technology platform using drug layering in a Wurster (bottom-spray) fluidized bed is considered to be a very suitable approach for "low OEL" compounds (Fig. 9.15). Dispensing of small API quantities is performed in a safety bench or isolator (e.g., flexible isolator). The API can be transferred directly into a liquid concentrate inside the isolator. Further dilution to the final drug layering liquid can be arranged under standard conditions. The drug layering liquid is directly fed into the fluidized bed process and applied on starter beads, followed by a seal coating step (Fig. 9.16). If required, any functional coating could be sprayed on the drug-layered/seal-coated micropellets without interrupting the process. The coated micropellets can be encapsulated using a standard capsule filler or filled into sachets, stickpacks, or bottles.

In total, producing "high potent" micropellets allows minimizing the handling of API powders and product transfers going along with a containment concept which is considerably simplified compared with a tablet manufacturing line. Less equipment must be cleaned and cleaning validated compared with the more diversified tablet manufacturing concept.

9.5 Micropellet Manufacturing Technologies

Micropellets consist of a core containing the API, which is usually covered with one or more functional coating(s) and eventually with an additional immediate-release drug layer (Fig. 9.6). The manufacturing procedures for the core micropellets and the coated micropellets are described in this chapter. The technologies presented are established industrial technologies being available from lab scale to commercial scale. The scale-up of (micro)pelletization and (micro)pellet coating processes are established state-of-the-art operations.

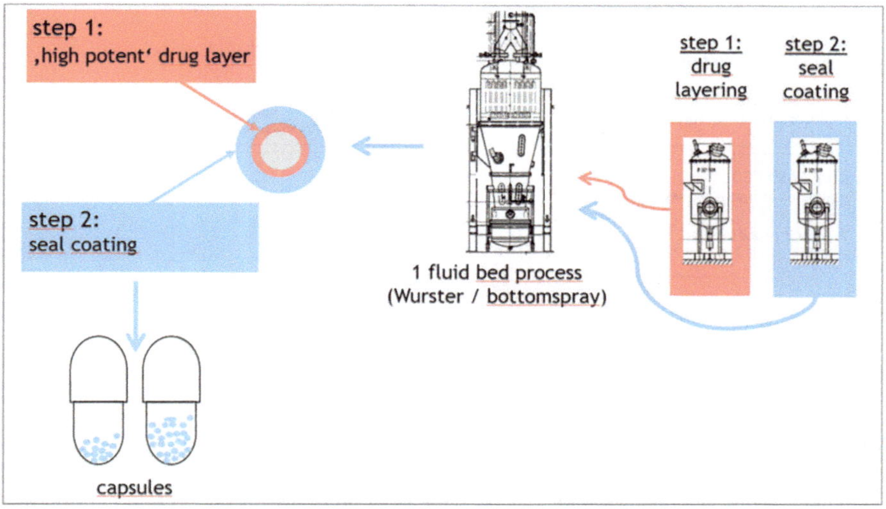

Fig. 9.16 Safe handling of "high potent" APIs with seal-coated (micro)pellets

9.5.1 Manufacturing Technologies for Core Micropellets (Fig. 9.17)

9.5.1.1 Manufacture of Drug/ASD/SD Layered Core Micropellets

9.5.1.1.1 Drug/ASD/SD Layered Core Micropellets with Wurster (Bottomspray) Fluidized Bed Technology (see Fig. 9.16)

The Wurster (bottomspray) fluidized bed technology is the most frequently applied and preferred technology for drug layering. In a batch process, an API containing liquid is layered on starter pellets having a size of 100 μm or larger. As a prerequisite for the process, the API must be chemically stable in the drug layering liquid, which can be either a solution or—to make the process even more efficient—a dispersion of the API in a water-based or organic solvent-based liquid.

The Wurster partition divides the fluidized bed into zones of differing airflow: the "up-bed" and the "down-bed" zone. This particular configuration generates a defined and controlled circulation of all particles in the process. The rising stream of particles in the "up-bed" is concurrently sprayed with the drug layering liquid by means of binary high-speed nozzles. The resulting fluidization pattern causes the particles to be individualized and highly scattered when they pass the spray nozzle(s) zone layering them with drug layering liquid. The risk of unwanted particle agglomeration can be mostly ruled out when feasible process parameters are chosen.

The API content in the drug-layered core pellets can vary from <0.1% up to 70%. With one drug-layering step, a drug load of up to 60% is usually achievable. For a higher drug load of up to 70%, a batch split must be applied, most likely. Very low drug

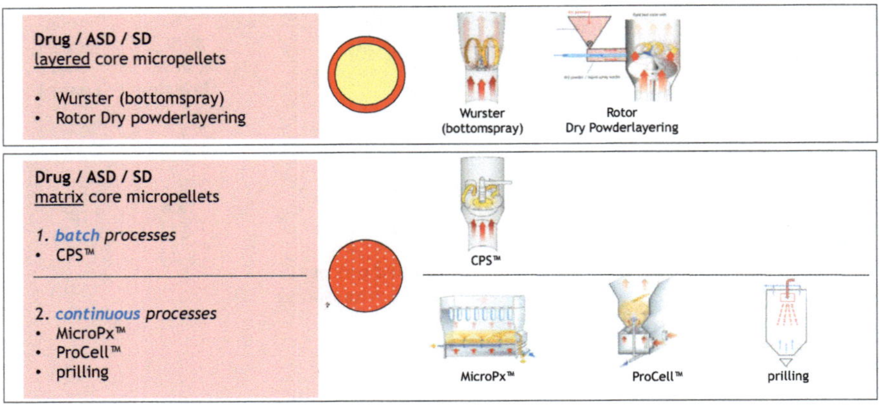

Fig. 9.17 Core micropellet manufacturing technologies

loads are safely achievable with the Wurster (bottomspray) fluidized bed technology, resulting in excellent content uniformity of the drug-loaded micropellets. For low- and medium-dosed APIs, a minimum core particle size of ~200 μm is achieved. Bulk uniformity (BU) and content uniformity (CU) of drug-layered pellets/drug products are the most important quality parameters. The Wurster (bottomspray) fluidized bed technology provides excellent homogeneity of drug distribution on the micropellets [26].

A Cetirizine drug layering liquid, including a binder, was layered on 70 kg of 200–400 μm starter beads using an 18″ pilot-scale high-speed (HS) fluidized bed Wurster system. For three batches, practical yields from 98.6% to 100% were found. Levocetirizine immediate release micropellets (assay: 0.3% w/w) show an RSD of 0.8–1.3% for bulk uniformity; also, content uniformity of stickpacks was excellent (Fig. 9.18, Internal Glatt documents which have not been released to the public).

Also, pediatric Alkindi® 0.5 mg capsules containing Hydrocortisone micropellets are characterized by an excellent content uniformity. Before this pediatric drug product was made commercially available, handmade capsules were prepared from crushed (Fig. 9.19).

The same Wurster (bottomspray) fluidized bed layering technology is used to deposit ASD (Amorphous Solid Dispersion) and SD (solid dispersion including emulsifier/s) on starter pellets.

9.5.1.1.2 Drug-Layered Core Micropellets with Fluidized Bed Dry Powder-Layering Technology (see Fig. 9.17)

The fluidized bed dry powder-layering technology is mainly applied when a high-dose and moisture-sensitive API must be layered on starter beads. The API is not processed from a liquid but as a powder blend which may contain further excipients, such as glidant and binder.

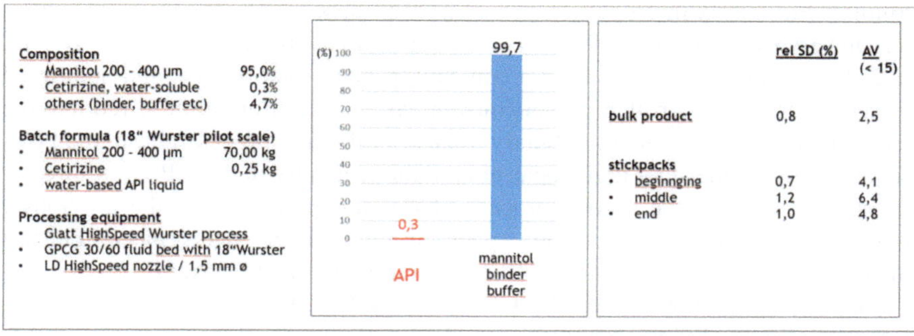

Fig. 9.18 Drug-layered Cetirizine micropellets

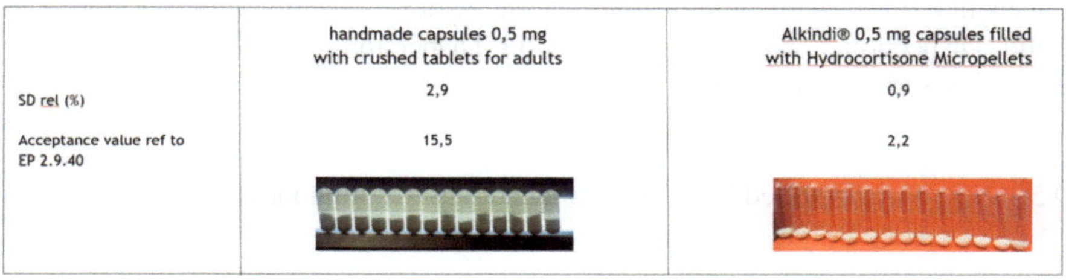

Fig. 9.19 Alkindi® (Hydrocortisone) 0.5 mg capsules (Diurnal Ltd., UK): content uniformity

For drug-layering of a prazole compound, 20 kg of starter beads are layered with 65 kg of API powder blend using a 1000 mm fluidized bed rotor system. The fully automatic production of 85 kg drug-layered pellets, applying liquid spray rates of 220 g/min and powder feed rates of 1200 g/min, takes 1 h [26].

9.5.1.2 Matrix-Type Core Micropellets

As an alternative to the described layered formulations, matrix-type core micropellets can be produced without starter pellets. Different fluidized bed technologies are in place to manufacture matrix-type core micropellets (Fig. 9.17).

The widely used extrusion/spheronization technology is not an option to produce micropellets as the lower limit in particle size achievable is ~700 µm; micropellets in a size range of 100–500 µm cannot be produced with extrusion and spheronization due to technical limitations of the extrusion tools.

9.5.1.2.1 Fluidized Bed CPS™ Technology (Batch Process) (Fig. 9.17)

The Glatt CPS™ fluid bed technology is the next generation of the rotor fluid bed technology; the advanced batch process is used to directly transfer powders into pellets and micropellets (direct pelletization). Extremely low-dosed drugs can be formulated to matrix micropellets as well as high-dosed APIs: the drug concentration can vary from <1% up to 90% w/w. Compared to the fluidized bed rotor system, the CPS™ technology works with a conically shaped rotating disc and additional devices ensuring a controlled particle movement. Pelletization liquid is fed into the process using a tangential spray binary nozzle. In the first process phase, the powders are wetted either in a high-shear granulator or ideally in the CPS unit directly to form granules. In the second process phase, spheronization of the previously irregularly shaped aggregates into spherical micropellets takes place. Particle growth is controlled with inline or offline optical PAT camera system.

Starter beads are not required for CPS direct pelletization technology. Typically, microcrystalline cellulose powder is used as a basic excipient; moreover, other functional compounds like polymers, disintegrants, and solubilizers can be part of the matrix pellet formulations in combination with the API to provide particular effects in terms of API solubility and release. As a pelletization liquid, Amorphous Solid Dispersion (ASD) liquids or Solid Dispersion (SD) liquids containing emulsifier/s are also possible.

9.5.1.2.2 Fluidized Bed MicroPx™ Technology (Continuous Process) (Fig. 9.17)

The Glatt MicroPx™ technology is a continuous fluidized bed agglomeration process that produces matrix-type micropellets in the size range of 100–500 μm (or larger pellets), which have a high drug load of typically up to 95%. Functional pharmaceutical excipients like emulsifiers can be integrated into the micropellet matrix, providing a Solid Dispersion matrix. Also, Amorphous Solid Dispersion (ASD) liquids can be directly processed to micropellets with the MicroPx™ technology.

Typically, all micropellet components like API, pharmaceutical binder, and other functional ingredients like emulsifiers or polymers for ASD formation are contained in a water- or organic solvent(s)-based liquid, which is fed into the MicroPx™ process via bottom-spray binary nozzles. The fluidization gas is supplied into the apparatus evenly distributed over a gas distributor plate, which can be round-shaped (lab units) or rectangular-shaped (pilot and commercial units) [35]. The pelletization liquid can be a solution, suspension, or emulsion. Also, a melt consisting of a molten emulsifier containing the API in a dispersed or dissolved form can be processed.

The pelletization process starts with spraying the API-containing liquid into the empty MicroPx™ fluidized bed unit. Initially, powder is generated by spray-drying or spray-solidification. Stepwise, the powders are agglomerated to seeds, which are continuously layered with API-containing droplets from the bottomspray guns, ending up in onion-like structured spherical micropellets. An online acting zig-zag-formed air sifter provides micropellets with a narrow particle size without an additional sieving step. Alternatively, a sieve-mill circle can be used for the separation of well-sized particles from the ongoing continuous process, whereas oversized

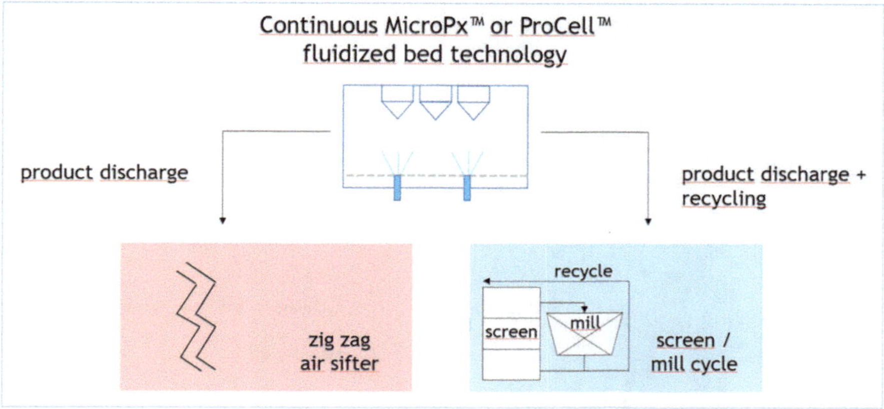

Fig. 9.20 Continuous MicroPx and ProCell pelletization processes including zig-zag air sifter of screen/mill cycle

particles are milled and fed back into the process together with the fines [26] (Fig. 9.20).

9.5.1.2.3 Spouted Bed ProCell™ Technology (Continuous Process) (Fig. 9.21)

The Glatt ProCell™ technology is a spouted-bed type direct pelletization process, which could be used as an alternative to the MicroPx™ fluid bed process.

The main difference between spouted beds and conventional fluidized beds is the variable cross-sectional area of the apparatus as a function of the height. Contrary to a conventional fluid bed, the process gas is not supplied into the apparatus evenly distributed over a gas distributor plate, but enters the process chamber through narrow slits as a gas jet with high velocity. Particles located in the process chamber are entrained by the gas jet to the top of the apparatus and finally reach the conical-shaped annulus zone, where they slide back downwards due to gravity. Subsequently, the particles enter the jet zone again, where they are conveyed upwards once again. Thus, a characteristic continuous particle circulation is affected, and good contact between the arising solid microparticles and liquid droplets generated by binary nozzle(s) is achieved. ProCell™ can be equipped with the same inline particle fractionation systems as the MicroPx™ technology: zig-zag air sifter or sieve-mill circle [26].

The ProCell™ spouted bed technology is predominantly used for solid particle systems that are difficult to fluidize with the more conventional setup of MicroPx™ continuous technology. Moreover, it is advantageous for temperature-sensitive products: due to the particular fluidization characteristics of the spouted bed and the high gas velocities coming from the narrow inlet gas slits, high inlet air temperatures can be applied without damaging temperature-sensitive compounds.

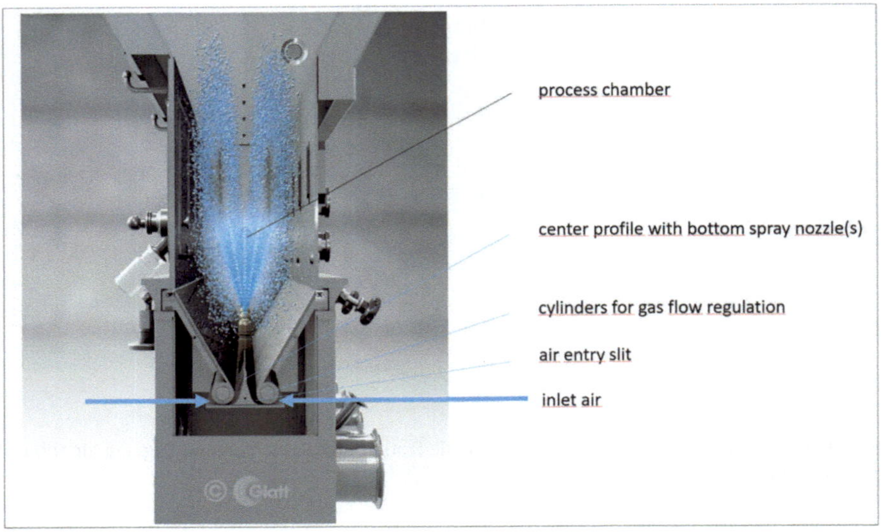

Fig. 9.21 ProCell™ spouted bed technology

9.5.1.2.4 Continuous Prilling Technology (Fig. 9.17)

Prilling (syn.: spray cooling, spray solidification, chilling) is a process in which a melt of a lipophilic material (e.g., wax) or a hydrophilic material (e.g., polyethylene glycol, emulsifier, etc) is solidified in a stream of cold gas. In spray cooling, an ambient temperature air stream is utilized to cool a molten material having a melting point of >45 °C. In fact, the prilling process essentially consists of two operations: firstly producing liquid droplets and secondly solidifying them individually by cooling them as they fall through a rising cool air stream. As agglomeration does not take place, the size distribution of the drops determines that of the product.

A variety of atomization techniques is available, including pressure spray guns, vibrating spray guns, and spinning disc atomizers. Microspheres are the most common encapsulation morphology prepared with this technique, with an active ingredient homogenously dispersed throughout the encapsulating matrix [26].

A Short Recommendation of Core Pelletization Technologies

- Fluidized bed Wurster (bottomspray) drug-layering approach is the no. 1 technology when it comes to transferring water/solvent stable APIs into micropellets with a drug-load range of <0.1–70%. It is a highly variable batch process technology that can be applied in a fine-tuned, optimized way for different particle sizes, API qualities, etc., in order to achieve the best product quality together with optimal process efficacy.
- Continuous MicroPx™ and ProCell™ processes are unique pelletization technologies to produce 200–400 μm sized micropellets with an API load of usually 90–95%. These quality attributes cannot be reached with other currently available technologies.

- Batch-wise CPS™ technology is considered a niche alternative; from a technical point of view, the physico-chemical properties of an API play a more important role than for drug layering with the Wurster technology.

9.5.2 Coating of Core Micropellets with Wurster (Bottomspray) Fluidized Bed Technology

Very often, core multiparticulates are coated with one or more functional films in order to achieve any kind of controlled drug release, drug targeting, or taste masking. Spherical core micropellets are an ideal substrate for fluidized bed coating with the Wurster (bottomspray) technology. Due to their spherical shape and smooth surface, less coating material is required than for the coating of powders or irregular-shaped granules. This is advantageous for pediatric medicines for which as less as possible excipients should be used.

The Wurster (bottomspray) fluidized bed technology (Fig. 9.17) is most frequently applied for all drug layering and coating applications on multiparticulates. The process technology for drug layering and functional coating applications is identical.

The parameters to be selected for a particular process depend mainly on the coating liquid properties and on the size of the starter cores. Minimum film forming temperature MFT of the coating polymer(s) is of highest importance for the formation of a uniform and dense film from an aqueous dispersion. As stability issues with respect to the in vitro dissolution profile were frequently experienced when aqueous polymer dispersions were processed, a revival of organic solvent-based coating liquids took place in the near past. With feasible equipment, configuration, and processing know-how, very small starter pellets, e.g., 100 μm in size, can be optimally layered with drug and finally coated with functional films.

Scale-up of the process from lab scale to pilot and commercial scale has been a well-proven exercise worldwide for decades.

Compared with other fluid bed drug layering and coating technologies, the Wurster (bottomspray) fluidized bed technology is characterized by an extraordinary process efficacy. The basis for the high efficacy is the principle of singularization of the particles when they pass the bottomspray type spray gun(s), allowing for high spray rates without the risk for particle agglomeration (Fig. 9.22).

Underbed spraying, on the other hand, is going at a higher risk for undesired particle agglomeration, in particular when very small particles are processed: a singularization of the particles passing the spray guns does not take place. In order to avoid particle agglomeration, the process with underbed spraying must be performed at a reduced speed in order to avoid agglomeration. As a consequence, the HighSpeed Wurster coating process was found to be 33% faster for a 220 kg batch size than the underbed tangential spraying process. With the HighSpeed Wurster coating technology, 9.2 kg Omeprazole pellets/h were achieved, while with the underbed tangential spray technology, the production rate is 6.11 kg/h [36].

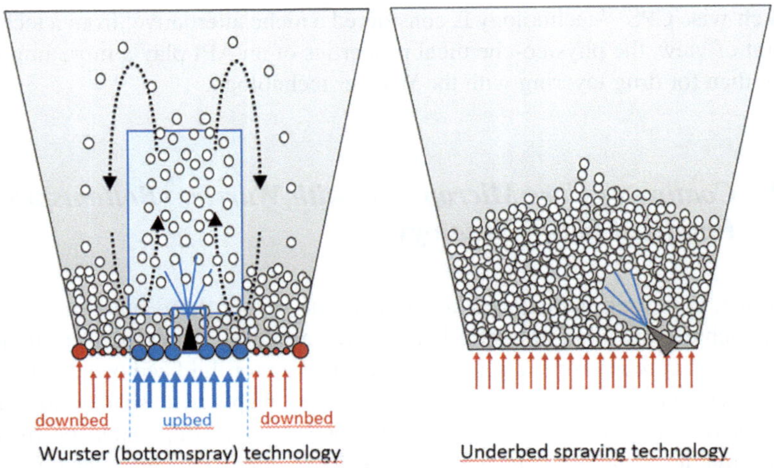

Fig. 9.22 Comparison of Wurster (bottomspray) and underbed spraying fluidized bed technology

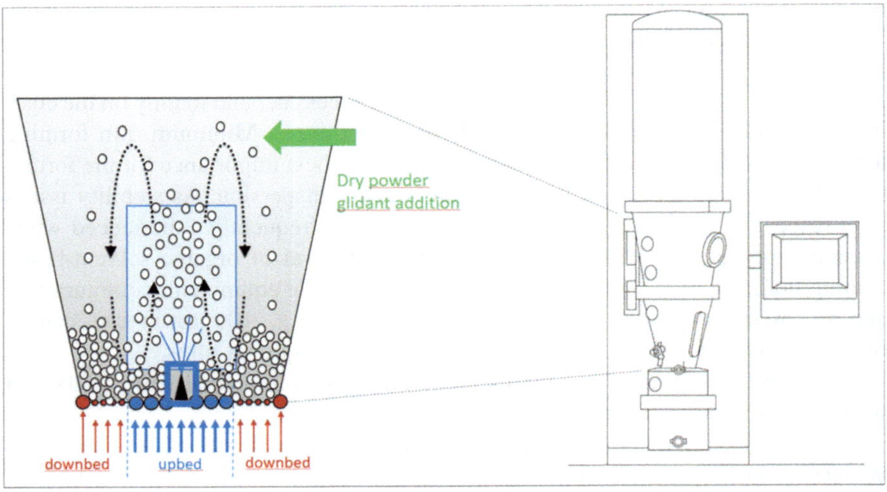

Fig. 9.23 Schematic illustration of Wurster fluidized bed coating process with powdered glidant addition [38]

An innovative concept in coating sustained-release microparticles with scalable Wurster (bottomspray) fluidized bed uses the addition of a glidant to the process, with the aim of reducing agglomeration tendency and increasing the yields (Fig. 9.23). Microcrystalline cellulose cores with a particle size <150 μm (Cellets® 90 and 100) were coated with Eudragit NM 30 D and Eudragit RS/RL 30 D aqueous dispersions using a fluidized bed Wurster coater. A dry powder glidant (magnesium stearate or silicon dioxide Aerosil 200) was periodically added (every 15 or 30 min, at 0.1% w/w based on starting cores for each addition) to the coating chamber through an external feeding port. With this novel approach, the high efficiency of the Wurster (bottomspray) fluidized bed process was further improved. In addition, powder rheology tests showed that dry powder glidants increased the tapped density and decreased the cohesive index of coated microparticles.

Reproducible microencapsulation of highly water-soluble metoprolol succinate was achieved, yielding coated microparticles less than 200 μm in size, with 20 h sustained release, suitable for use in liquid suspensions. The robust, scalable Wurster technology, including the novel in-process addition of a glidant, offers an important solution to the challenges of formulating sustained-release, but also taste-masked dosage forms suitable for children and older people with swallowing difficulties [37, 38].

Other fluidized bed coating technologies, such as tangential spray, CPS™, or rotor process, are recommended as fall-back options when, e.g., Wurster process, referring patent hurdles must be overcome. The topspray fluidized bed technology is an excellent technology for particle agglomeration, but not recommended for drug layering and coating applications.

9.6 Manufacturing Taste Masked Micropellets: Aspects to Consider and Taste Evaluation Approaches

To achieve effective taste masking, migration of the active compound into the taste masking film must be absolutely avoided, as this would compromise the taste quality aimed at. For this reason, a seal coat is frequently applied on top of the core micropellets before the taste-concealing film is processed (Fig. 9.6). By this means, a physicochemical barrier is introduced, assuring the proper functionality of the taste masking film coat. Figure 9.24 presents EDX evaluation results demonstrating that no API migrated into the taste masking film.

Usually, water-soluble polymers, such as HPMC, HPC, or PVP, are used for seal coating applications. Depending on the solubility of the API concerned, processing of the seal coat either from aqueous or organic solvent-based liquids must be

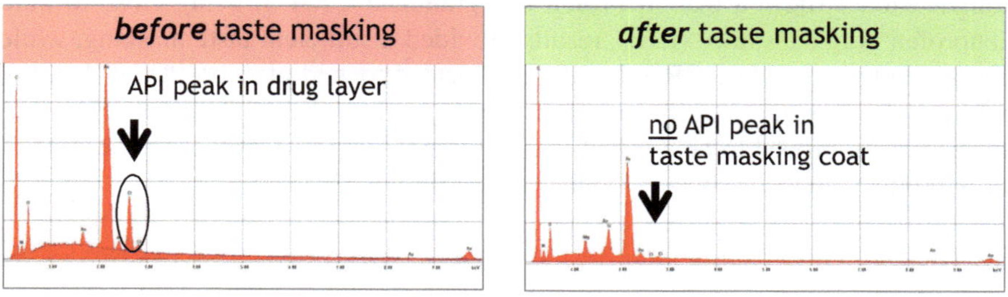

Fig. 9.24 Evaluation of API migration in taste masked micropellets with EDX (energy dispersive x-ray technology) (Glatt GmbH, 79589 Binzen/Germany, 2014)

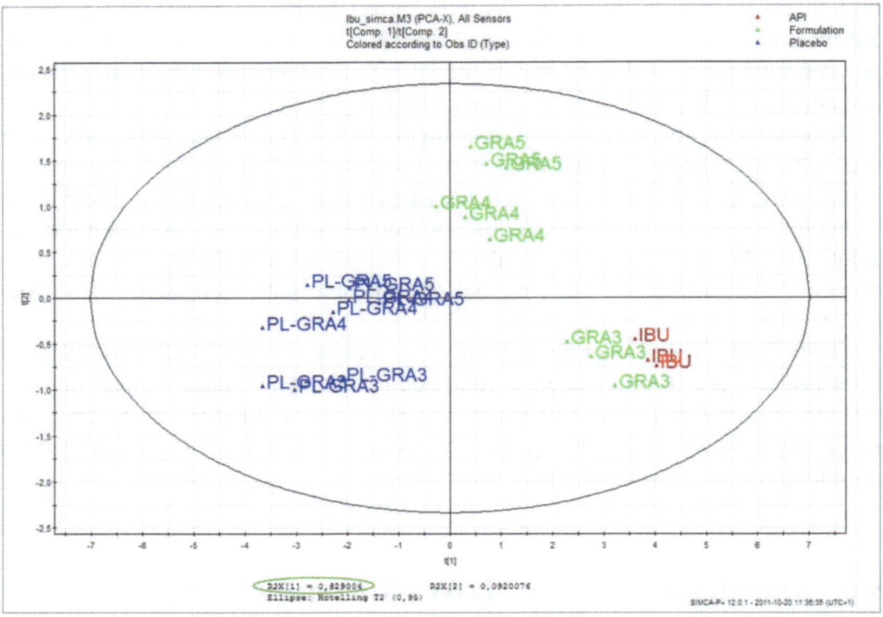

Fig. 9.25 E-tongue results including multivariate analysis in development of taste masked Ibuprofen [40]

considered. In order to end up with an optimal result, the application of a seal coat on a highly water-soluble API would be processed from an organic solvent(s) system in which the API is less soluble than in water. Antitacking agents such as silicon dioxide or talc facilitate the seal coating process of the core micropellets. Plasticizers, such as polyethyleneglycol (PEG), could be added to increase the flexibility of the seal coat film.

The quality of the taste masking can be tested in vitro (electronic tongue model, in vitro drug release studies, etc.) or with in vivo methods (rat taste panels with BATA model, human taste panel) [39].

Evaluation of different taste-masked Ibuprofen granulates with an electronic tongue model showed that Ibuprofen test granulate GRA 3, being close to pure Ibuprofen drug substance (IBU), results provided insufficient taste masking, while Ibuprofen test granulate GRA 4 and 5 are located toward placebo PL-GRA 3,4,5, with results showing a sufficient taste masking effect (Fig. 9.24) [40]. The electronic tongue proved to be a valuable tool for assessing and predicting the taste of drug products in the early development phase (Fig. 9.25).

9.7 Micropellets and Mini-Tablets: Comparing Two Pediatric Drug Product Concepts

Mini-tablets are a dosage form that affords the advantages of multiparticulates, with regard to ease of administration and dose flexibility. The World Health Organization (WHO) defines mini-tablets as no bigger than 4 mm in diameter [12]. Many of the manufacturing unit operations used to produce mini-tablet-based dosage forms are the same as those employed in manufacturing of larger tablet images. However, due to their unique size, some adjustments to the manufacturing processes and testing approaches may be required.

Biconvex mini-tablets, being nearly spherical, present specific challenges relating to their compaction. The narrow diameter of the die used in mini-tableting necessitates excellent flow to achieve tablets within the desired weight range. The size of the tablet punch can dictate the force used to compact the powder to a tablet; low forces required for small punches can be challenging for routine use [14].

Mini-tablets are usually manufactured by compression molding using a multiple-tip tool in a rotary tableting machine. Owing to their special structure, ensuring uniformity is a very important challenge in the manufacturability of mini-tablets using the multiple-tip tool. Further, as the rotational speed of the rotating disc in the rotary tableting machine is an important parameter directly related to the manufacturing capacity of mini-tablets, optimization of this parameter according to powder characteristics is required [41].

Sticking of powders to punches was observed when a powder blend should be directly compressed into mini-tablets. Various lubricants, including magnesium stearate, sodium stearyl fumarate, talc, and combinations thereof, were tested to minimise the sticking tendency. However, there were signs of sticking on the upper and lower punches depending on the formulation, and sometimes whole parts of the mini-tablets remained on the punches. In some cases, the mini-tablets could not even be removed from the dies without causing any damage to the mini-tablets. Environmental moisture fluctuations were not responsible, as tableting was carried out in a climatized room at 21 °C and 45% r.h. 2% (*w/w*) magnesium stearate, and 1% (*w/w*) talc proved to work on the compaction simulator regarding a reduction in the sticking tendency. However, the high amount of lubricants used in the formulation, resulting possibly in overlubrication, could have been responsible for a reduced tensile strength of the mini-tablets. Reducing the magnesium stearate quantity from 2% to 1% was not successful, as high ejection forces were the limiting factor, demonstrating insufficient lubrication. Batch size increase led to compaction of the tablet blend in the hopper of the tablet press above the feed mechanism, which made further direct tableting no longer feasible, as the dies were no longer filled accurately. Finally, an intermediate dry-granulation step was introduced instead of direct compression [42].

Manufacturing of 1–3 mm *mini-tablets* is a sophisticated and challenging process considering the physical sensitivity of multi-tip punches, the achievable tensile strength, the drug load, and the bulk uniformity of mini-tablets, the content uniformity of the drug product, respectively. Direct compression is the preferred process variant; however, in most cases, a granulation step is required in order to provide the required excellent flow properties of the tablet blend and mini-tablet quality. Bulk and content uniformity are critical quality parameters; to fulfill the requirements, the quality of the tablet blends as well as the tableting parameters, including the tablet press speed, must be evaluated in depth and optimized. Using bi-convex multi-tip punches, mini-tablets with a high sphericity can be manufactured.

Frequently, relatively high lubricant concentrations are used to ensure even mini-tableting, having on the other hand an impact on the tensile strength of the resulting mini-tablets. From this point of view, it is unlikely that identical tablet blends may be used for normal-sized tablets for adults and mini-tablets for pediatrics. Adequate tensile strength is an important prerequisite for coating of mini-tablets, either with fluidized bed technology (Wurster/bottomspray) or drum coater.

In total, a number of unit operations—blending, granulation, milling of the granules, tableting, and coating—are necessary to produce coated mini-tablets. The mini-tableting step is considered to be most critical. Quality of resulting mini-tablets depends to a high degree on the quality of the tablet blend (powders, granules). When solvents must be completely avoided due to the API's properties, mini-tablets made with direct compression or dry granulation are a promising approach (Fig. 9.26).

Dispensing of mini-tablets, e.g., counting into capsules, can be a technical challenge due to the cap-wall form of the particles, resulting in non-ideal behavior in the dispensing unit.

Micropellets, on the other hand, can be produced with very straightforward and well-established processes (Fig. 9.27). Drug layering on starter beads with fluid bed Wurster (bottomspray) fluidized bed technology is a worldwide established and proven standard technology in commercial pharmaceutical production up to batch sizes of 600–800 kg. Moisture-sensitive compounds can be processed using the

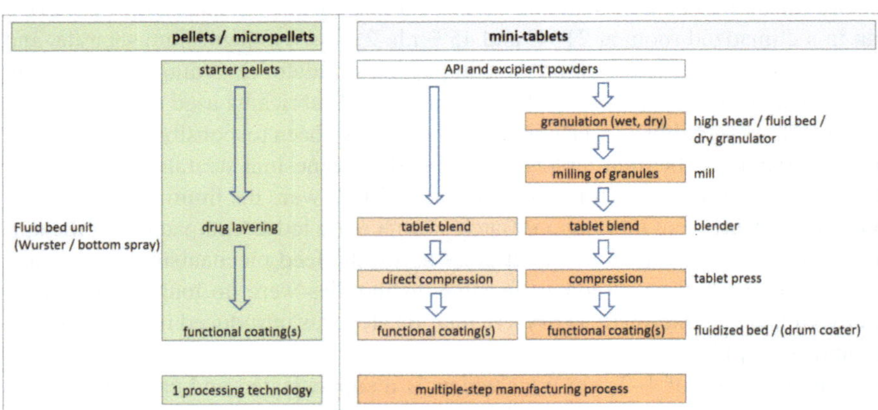

Fig. 9.26 Manufacturing micropellets and mini-tablets

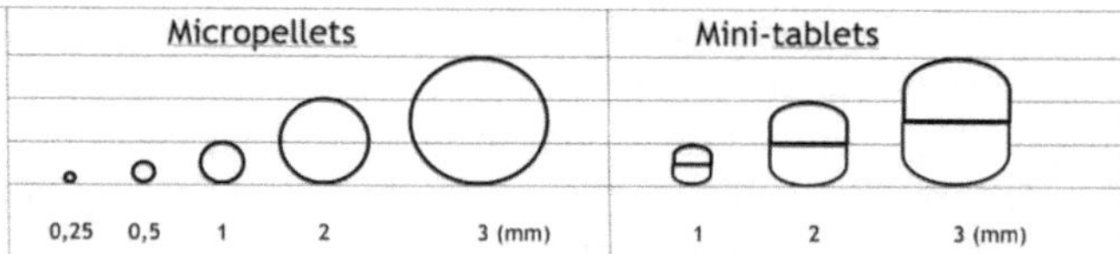

	Micropellets	Mini-tablets
• geometry • particle size	spherical (micro)pellets < 500 µm - 3 mm	bi-convex 1 - 3 mm
• drug load • content uniformity	drug layered (micro)pellets: up to 70% matrix (micro)pellets: up to 95% excellent / standard	depending on API properties challenging
• functional coating applications	standard with Wurster (bottomspray) process ideal with spherical cores	Wurster (bottomspray) or drum coater challenging with bi-convex cores
• liquid medication possible	with micropellets < 500 µm	no
• manufacturing process • commercial production • dispensing into capsules	straightforward established standard	complex to be developed complex
• high potent handling	smart and less complex	complex

Fig. 9.27 Comparing micropellets and mini-tablets

dry-powder layering fluidized bed rotor technology. In addition, different matrix micro-pelletization technologies are available, allowing the selection and nomination of the most feasible process for any API. With extremely low as well as with high drug loads, excellent bulk and content uniformity is achieved: Wurster (bottomspray) fluidized bed process enables core micropellet drug loads from <0,1% up to 70%. With matrix pelletization technologies such as continuous MicroPx™ and ProCell™ technology, core micropellet drug loads of 95% are regularly realized. The resulting core micropellets are ideally spherical, physically robust, and are a perfect substrate for homogenous coating applications which—when the Wurster fluidized bed technology is used—can be applied directly after drug layering without interrupting the manufacturing procedure.

Following this production strategy, coated micropellets are manufactured with one unit operation only. This scenery also facilitates the safe handling and processing of "highly active" compounds as well as not fully characterized NCEs. Using the same technologies, larger pellets up to a size range of 3 mm in diameter can be produced if needed.

Dispensing of spherical pellets into capsules, sachets, stickpacks, and bottles has been a long-time proven standard technology.

9.8 Conclusions and Perspective

Multiparticulates like pellets and micropellets are a promising platform technology for our society: one concept for the different age groups from preterm babies to the elderly, one basic type to be extrapolated to different drug release and drug targeting options. Why not start clinical development directly with an immediate release (micro)pellet formulation and to stepwise expand it to different drug products without starting from the beginning again and again—as it is often the case when clinical testing starts with tablets. A very precise and narrow dosing regimen is easily possible—without development of x different tablet dosage strengths. A solid dosage form development strategy for clinical testing and the market for pediatrics, adults, geriatrics, and intensive care patients—with (micro)pellets, an overall integrated concept is possible.

A future step could be: Individual Precision Polymedicine with multiparticulates, considering individual, tailor-made, convenient dosage forms and patient profiles, pharmacogenomic patient profiles, physiological poly-drug-drug interactions, and optimal doses and treatment regimen. The result could be a patient-oriented, convenient, safe, and effective dosage form, a better self-management of patients with reduced medication errors and non-adherence to therapy—all in all a benefit for efficacy, safety, and compliance.

At last, processing of "high potent" drugs is facilitated and safer with a drug-layering approach from a liquid followed by a seal-coating step in a fluid bed unit rather than with potentially dusty processing steps such as sieving, milling, and tableting.

References

1. ZAK® Initiative Arzneimittel für Kinder e.V. (IKAM) auf www.zak-arzneimittel.de
2. Ricci B. Bridging studies in support of pediatric formulation development. In: EuPFI 4th conference, Prague; 2012.
3. Guideline of pharmaceutical development of medicines for pediatric use; EMA/CHMP/QWP805880/2012, p. 4/24.
4. Allegaert K, De Cok P, van den Anker JN. The clinical relevance of pediatric formulations. In: Bar-Shalom D, Rose K, editors. Pediatric formulations. AAPS Springer; 2014. p. 69.
5. Rose K. A paediatric investigation plan case study. Pharm Med. 2012;26(5):287–95.
6. Blasius H. Arzneimittelentwicklung, 2. Teil. DAZ. 2014;154(26):64–70.
7. t'Jong G. Pediatric development: physiology, enzymes, drug metabolism, pharmacokinetics and pharmacodynamics. In: Bar-Shalom D, Rose K, editors. Pediatric formulations. AAPS Springer; 2014. p. 9–23.
8. Guideline on pharmaceutical development of medicines for pediatric use: EMA/CHMP/QWP805880/2012.
9. Sam T. Exploring a risk management: approach towards developing the optimal paediatric formulation—applying quality by design and multi-criteria decision making. In: EuPFI 3rd conference, Strasbourg; 2011.
10. Klingmann V, Spomer N, Stoltenberg I, Lerch C, Meissner T, Breitkreutz J. Acceptance of drug-free minitablets in young children. In: EuPFI 3rd conference, Strasbourg; 2011.
11. Klingmann V, Seitz A, Meissner T, Breitkreutz J, Moeltner A, Bosse HM. Acceptability of uncoated mini-tablets in neonates—a randomized controlled trial. J Pediatr. 2015;167(4):893–896.e2.
12. World Health Organization (WHO). Annex 5 development of paediatric medicines: points to consider in formulation. In: Forty-sixth Rep. WHO Expert Comm. Specif. Pharm. Prep. Geneva: World Health Organization; 2012. p. 235.
13. Mrsny RJ. Oral drug delivery research in Europe. J Control Release. 2012;161:247–53. https://doi.org/10.1016/j.jconrel.2012.01.017.

14. Ranmal SR, Barker SA, Tuleu C. Paediatric solid formulations. In: Bar-Shalom D, Rose K, editors. Pediatric formulations. AAPS Springer; 2014. p. 153.
15. Craig SR, Adams LV, Spielberg SP, Campbell B. Pediatric therapeutics and medicine administration in resource-poor settings: a review of barriers and an agenda for interdisciplinary approaches to improving outcomes. Soc Sci Med. 2009;69:1681–90.
16. Alkindi Sprinkle (hydrocortisone) oral granules. Prescribing Information Medication Guide & IFU. https://www.alkindi.sprinkle.com/patient/taking-alkindi-sprinkle. 2022.
17. Lubahn B. Der schwere Ausstieg. DIE ZEIT. 2019;53:38.
18. Grave A, Poellinger N, Whitaker M. Development of taste masked micropellets for pediatric use. In: 9th World meeting on pharmaceutics, biopharmaceutics, Lisbon; 2014.
19. Poellinger N, Grave A. TAIN—a collaborative approach to meet patient needs in a rare disease. In: 6th conference of the EuPFI, Athens; 2014.
20. Capsugel, Coni-Snap Sprinkle Caps. 2015.
21. Bhad ME, Abdul S, Jaiswal SB, Chandewar AV, Jain SM, Sarkarkar DM. MUPS tablets—a brief review. Int J Pharm Tech Res. 2010;2:847–55.
22. Timpe C, Dischinger A. Drug delivery strategies for oral pediatric and geriatric products. J Asia's Pharm Biopharm Ind. 2014;
23. Arzneimittelapplikation per Sonde. DAZ. 2006;29:80.
24. Schmitt V, Lipp H-P. Orale Liquida für die Pädiatrie—Entwicklung und Optimierung kindgerechteZubereitungen. DAZ. 2009;44:54.
25. Stegemann S. Maßgeschneiderte Arzneiformen. DAZ. 2014;46:60–5.
26. Poellinger N. Multiparticulates—modern oral drug products for our society. Part 1,1. Pharm Ind. 2022;84(12):1440–50.
27. Kawabata Y, Wada K, Nakatami N, et al. Formulation design for poorly water-soluble drugs based biopharmaceutical classification systems: basic approach and practical applications. Int J Pharm. 2011;420(1):1–10.
28. Weitz J, Earl KM, Leblanc K, et al. Establishing therapeutic evidence of complex pharmaceuticals: the case of dabigatran. Can J Cardiol. 2018;34(9):1116–9.
29. Pradaxa Oral Pellets: package insert/prescribing information. https://www.drugs.com/pro/pradaxa-oral-pellets.html.
30. Vasconcelos T, Sarmento B, Costa P. Solid dispersions as strategy to improve oral bioavailability of poor water soluble drugs. Drug Discov Today. 2007;12(23):1068–75.
31. Neuwirth M. Versuchsbericht: Entwicklung einer Methode zur kontinuierlichen Produktion einer Amorphen Solid Dispersion (ASD) in der Wirbelschicht. Bonn: Glatt Pharmaceutical Services/Univ; 2017.
32. Bhujbal SV, Mitra B, Jain U, et al. Pharmaceutical solid dispersion: a review of manufacturing strategies. Acta Pharm Sin. 2021;11(8):2505–36.
33. Bodratti A, Alexandridis PJ. Formulation of poloxamers for drug delivery. J Funct Biomater. 2018;9(1):11.
34. Poellinger N. Multiparticulates—modern oral drug products for our society. Part 1,2. Pharm Ind. 2022;85(2):187–97.
35. Uhlemann H, Mörl L. Wirbelschicht-Granulation. Springer-Verlag; 2000.
36. Merimeche P. Fluid bed technology: comparison of the conventional coating technology with the Hüttlin technology at the example of the production of omeprazole pellets. In: Int Seminar "Innovative Technologies and Equipment for Pharmaceutical Industry", Moscow; 2005.
37. Mohylnuk V, Patel K, Scott N, et al. Wurster fluidized bed coating of microparticles: towards scalable production of oral sustained-release medicines for patients with swallowing difficulties. AAPS PharmSciTech. 2020;21:3.
38. Fang L. Method for coating particles. WO2019073257(A1). 2019.
39. Soto J. Assessing the aversive taste of medicines: a comparison between rate taste panels (via the BATA model)and human taste panels. In: 7th conference of the EuPFI, Brussels.
40. Guhmann M, Preis M, Gerber F, Poellinger N, Breitkreutz J, Weitschies W. PBP World Meeting Lisbon; 2014.

41. Kurashima H, Uchida S, Kashiwagura Y, Tanaka S, Namikia N. Evaluation of weight variation in mini-tablets manufactured by a multiple-tip tool. Chem Pharm Bull. 2020;68:981–8.
42. Elezai V, Lura A, Canha L, Breitkreutz J. Pharmaceutical development of film-coated mini-tablets with losartan potassium for epidermolysis bullosa. Pharmaceutics. 2022;14(3):570. https://doi.org/10.3390/pharmaceutics14030570.

Chapter 10
Oral Semisolid Formulations

Carsten Timpe

Abstract Oral semisolid dosage forms for children are pharmaceutical formulations designed to administer medication in a convenient and easily ingestible form. These formulations are particularly beneficial for infants and young children who may have difficulty swallowing solid tablets or capsules. Semisolid formulations can be manufactured to provide good palatability by balancing their physical/ mechanical properties, providing a pleasant taste and texture that children may find more enjoyable compared to traditional dosage forms. However, as some semisolid formulations can resemble regular candies, they may be tempting for children to consume in excess. This chapter provides an overview of semisolid oral formulations with many examples of research on suitable products. Overall, they offer improved palatability, accurate dosing, and ease of administration, ensuring that children receive their prescribed medication in a safe and effective manner.

Keywords Jellies · Gummies · Soft-pudding · Swallowability · Multiparticulates · Texture · Adhesiveness · Cohesiveness

10.1 Introduction

A successful medical treatment, especially of young children, requires development of age-appropriate formulations which should be specifically designed, developed, and evaluated for its pediatric use [1, 2]. Testing of the acceptance of such pediatric formulations is nowadays an expectation of regulatory bodies (e.g., EMA, FDA, MHRA) when pharmaceutical companies submit their pediatric investigation plans (called "PIP" in the EU at the end of phase 1, pediatric submission plan "PSP" in the USA at the end of phase 2). Acceptability is defined as "the overall ability and

---

C. Timpe (✉)
Pediatric Consultant, Bad Bramstedt, Germany

H. Batchelor, K. Rose (eds.), *Pediatric Formulations*, AAPS Advances in the Pharmaceutical Sciences Series 58, https://doi.org/10.1007/978-3-031-77240-5_10

willingness of the patient to use and its caregiver to administer the medicine as intended" [3].

Typically, 25–35% of developing children show signs of feeding disorders, and 33–80% of children with developmental delays have swallowing issues [4]. The authors of [4] concluded "that complex pediatric feeding problems are biobehavioral conditions in which biological and behavioral aspects mutually interact, and both need to be addressed to achieve normal feeding." In a more recent paper, children's preferences for oral dosage forms in different European countries have been analyzed. The average age of the population surveyed was 12.3 years with 4.8 years standard deviation (SD) [5]. More than 1100 respondents were involved, and it was found that conventional dosage forms, i.e., liquid (35%), tablets (19%), and capsules (14%), were the most popular. Children below the age of twelve years preferred liquid dosage forms, and this preference was also found for healthy and rarely sick children. Monolithic dosage forms were chosen by adolescents and children with chronic diseases. Overall, the authors stated that there was a lack of familiarity with more novel dosage forms. Here, they listed orodispersible films and granules as examples.

Dosage forms that would not fit within these typical categories for developing pediatric dosage forms, for example, oral gels ("jellies"), gummies, or pudding-like products, are mostly not considered by global pharma companies for novel medicines. Although there might be high acceptance by children and caregivers, there might also be multiple reasons for this reluctance, which will be discussed under different aspects in more detail below. Despite a general neglect by larger parts of the pharmaceutical industrial community, there are indications of interest in these dosage forms that are going to increase. Several papers have recently been published in that field, e.g., a paper about the preparation and in vivo animal testing of Vardenafil jellies as potential treatment against irritative bowel disease in children in 2021 [6]. Another paper describes easy-to-swallow "instant" jelly formulations for sustained release Gliclazide delivery [7]. In Japan, several jelly-based drug products have already been successfully launched for adults, e.g., Alendronate oral jelly 35 mg (BonalonVR Oral Jelly 35 mg; Teijin Pharma Limited, Tokyo, Japan), approved in 2012 by the Japanese health authorities for osteoporosis. In addition, calcium gluconate, amlodipine, acyclovir, donepezil hydrochloride, sildenafil, and tadalafil are also available as oral jelly products in Japan [8].

10.2 Classification of Oral Semisolid Formulations

10.2.1 Physical Characterization Aspects

In general, the term "semisolid formulations" is associated with drug product formulation systems like ointments or gels for local delivery to the oral mucosa, vaginal, or nasal. This chapter focuses only on semisolid preparations that are intended for oral drug delivery to children. One of the physical characteristics is that they are much softer than solid oral products like tablets or mini-tablets.

Typically, the deformation of a pure solid is described by Hooke's law, and then the inherent stiffness of the material can be described by Young's modulus. The Young's modulus E is the ratio of the stress-induced deformations in a material under an applied strain. For very small deformations, strain varies linearly with stress. The relationship between stress and strain can be graphed, and then the slope of the line gives the Young's modulus, E (proportionality constant between stress and strain in Hooke's Law).

In case of semisolid formulations (like gels or jellies), the system comprises elastic and viscous components (e.g., a solid polymer like carrageenan or gelatin, and a liquid, e.g., liquid syrup). A number of soft semisolid materials also fall under the category of structured fluids since they have a multiphase structure and exhibit complex flow behavior. These components can be characterized, for instance, by their rheological properties: While shear-thinning behavior is common with increasing shear rates for such semisolid systems, strain and frequency sweep experiments can further help to characterize under which strain conditions the gel network structure would disrupt. Loss G' (the elastic part of the system) and storage modulus G'' (the viscous component) could be indicative of the mechanical stability: If $G' > G''$, the material is highly structured, and increasing the strain above a critical value could disrupt its structure [9].

As explained, semisolid formulation systems have a certain gel strength, which is the maximum stress before breakage of the system. Elasticity and rupture forces are important parameters that can be measured technically with texture analysis in addition to rheology tests described above. Another key parameter of such gel structures is the extensibility, which is represented by the maximum strain value before breakage. Figure 10.1 shows a typical stress/strain curve with gel rupture point. The ratio of the shear stress to strain is the shear modulus (for pure solids, it is the Young modulus E) and is a parameter that quantifies the ability of the gel to resist deformation.

10.3 Gummies, Jellies, and Soft-Pudding

Oral semisolid dosage forms could be categorized according to their different physical characteristics (see chapter 10.2.1) into (a) very stiff, hardly chewable gummies, (b) soft chewable gels ("jellies"), including squeezable gels and new energy supplements, and (c) soft-pudding types with low mechanical strength, which can often be used to sprinkle multiparticulates (mini-tablets, pellets) and improve swallowability [10]. The jelly-type dosage form in particular has received increased interest for pediatric applications and will therefore represent the major part of this chapter.

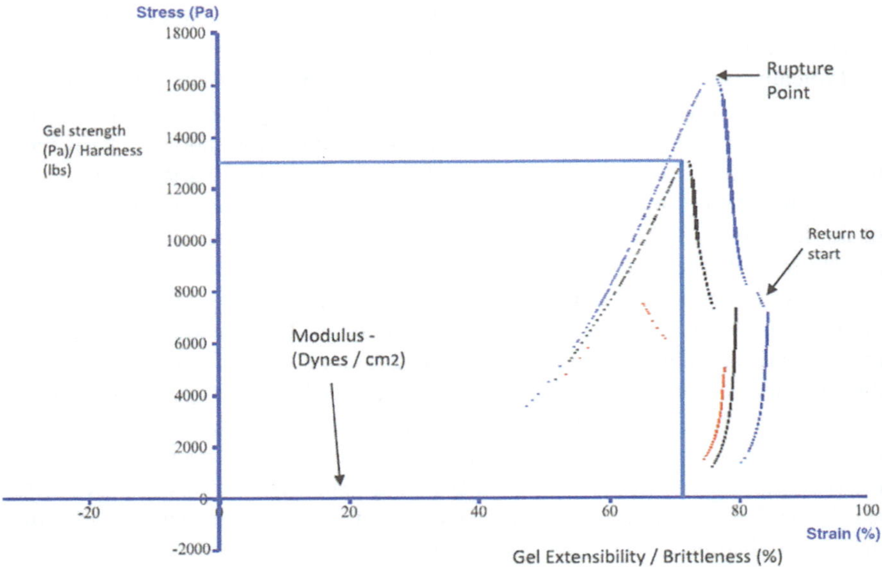

Fig. 10.1 Sample texture analysis graph of gel dosage form stress-strain curve for three different gel samples (shear modulus is the slope at the very beginning of the stress-strain curve) (taken from [10])

10.3.1 Pastilles or Gummies as Pediatric Dosage Form

Pastilles or gummies are typically manufactured based on gelling agents combined with sugars or sugar substitutes. For example, liquid syrup is deposited into previously formed starch molds and will then obtain the final consistency in a drying step. After the separation from the starch, the pastilles or gummies are either coated with glazing agents or sugar. This semisolid dosage form can be used for many different indications from pharmaceuticals for the treatment of the oral cavity, medical devices with herbal extracts to food supplements with vitamins and minerals [11]. While gummies have been a traditional approach for more than 100 years, based on confectionery food technology started with the Haribo company in Germany in the early 1900s, more recently, these products could now also be modified by using modern 3D printing technologies to develop patient-centric and safe dosage forms [12]:

E.g. different thermo-reversible inks were prepared by combining different hydrocolloids (carrageenan, xanthan gum, gelatin) and liquid sweetener, strawberry flavor, and some food coloring. Compositions differed mainly in the presence or absence of corn starch and ranitidine (as model drug) among all the ingredients. 3D printing occurred via a syringe-based extrusion mechanism. Mass uniformity and drug release of ranitidine measured with UV-vis spectrophotometry, showed successful results both in dose accuracy, uniformity of drug content, and dissolution. A

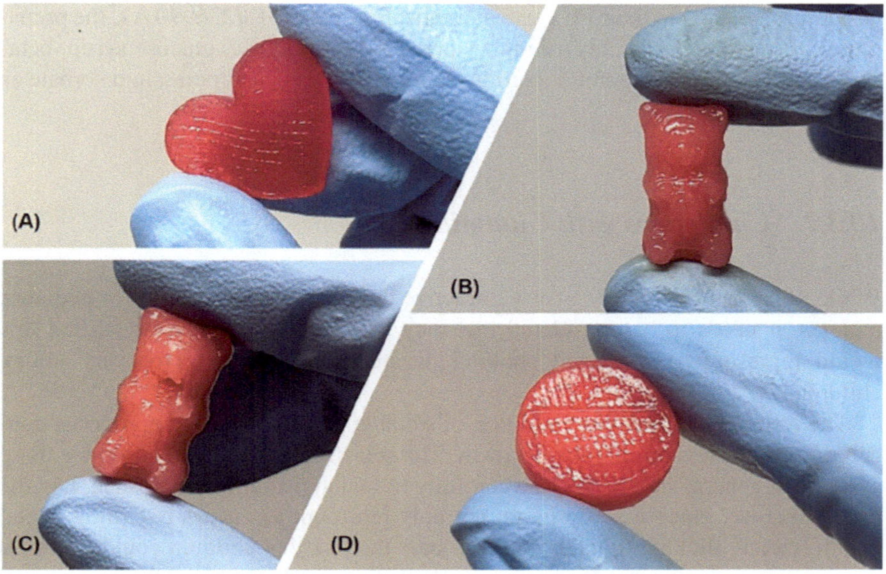

Fig. 10.2 3D printed gummies (**a**) heart, (**b**) gummy bear with 80% infill density, (**c**) gummy bear with 65% infill density, (**d**) disk. Figures (**b**) and (**c**) have an identical appearance but differ in weight and mechanical strength (taken from [12])

bright and uniformly colored appearance was achieved combined with a pleasant aroma (see Fig. 10.2).

Ink printability was checked with rheological tests (shear-viscosity, amplitude, and frequency sweep analysis): Frequency sweeps depicted that these inks had viscoelastic properties with a typical weak-gel behavior, since storage modulus (G') values were higher than loss modulus (G'') values, regardless of the formulation chosen or the temperature at which the measurements were made.

Historically, these gummy forms contain gelatin, natural or artificial sweeteners, flavors, colorants, citric acid as acidulent, and as anti-sticking agents, e.g., beeswax, coconut oil, or starch [10].

More recently, the FDA has granted a US company the first Investigational New Drug Application (IND) for an allergy cetirizine-based gummy medication, Ceteric™ Allergy Gummy, IND 140312 [13]. The company has developed a portfolio of gummy drugs using their proprietary "API-Complexing technology and sugar-free GUMMY ZERO® matrix" for multiple indications like pain, fever, cough, nausea, insomnia, inflammatory and bacterial infections besides allergy [14]. To get a regular drug approval by the FDA, the company has planned to conduct a randomized (open-label, single-dose, five-period crossover) relative bioavailability study to evaluate cetirizine HCl Gummy 10 mg® and cetirizine HCl Oral Tablets 10 mg® administered in a total of 30 healthy adult male and female subjects. It is also planned to include in the study the swallowing and chewing of these gummies with and without 240 ml of room temperature water [15]. The company has described, e.g., in their PCT patent application WO 2018/236990 A1, the preparation of cetirizine gummies using, besides the drug, sucrose, glucose syrup, beta-cyclodextrin, gelatin, mannitol, taurine, sodium benzoate, and potassium sorbate as preservatives.[1]

10.3.2 Challenges with Gummies

While gummies seem to be a very attractive dosage form, especially for pediatric patients, it needs to be considered that they should not be oversized so that this would represent a significant choking hazard risk, especially for very young children.

The water content requires addition of suitable preservatives in appropriate concentrations to prevent microbial growth. The selection of excipients used for these systems in general should be done carefully to match pediatric acceptance, i.e., the use of sucrose, glucose syrup might not only represent a concern regarding dental health (caries, sticking to teeth) but also bear the risk that children could overdose these drug products since gummies can look like candy. Further discussions and experiences with marketed gummy drug products for children could be required to optimize such dosage forms. Probably modern 3D printing technologies could help to support these products while avoiding the use of some rather unacceptable excipients for pediatric applications.

10.4 Jellies

Gels or jellies are semisolid, gel-network structured products formed from dry powders of hydrocolloid polymers via a certain degree of crosslinking or association when these are hydrated or dissolved in an aqueous medium. They are either transparent or translucent and should be softer compared to gummies, although there might be an overlap. Due to the soft texture and the taste, odor, and flavor (when flavors and sweeteners are added for commercial products), these dosage forms might represent a very attractive drug delivery system not only for elderly patients suffering from dysphagia but also for young children who often do not like to swallow tablets or capsules. As already mentioned, jellies are nowadays quite common in Japan [16]. In 2011, medicated jellies had already been listed in the Japanese Pharmacopeia, 16th Edition, and in a recent update [17].

---

[1] https://patents.google.com/patent/WO2018236990A1/en (accessed June 20, 2022)

10.4.1 "Instant" Jellies

10.4.1.1 Giclazide Jellies

F. Liu and co-workers describe in their paper the preparation and characterization of Giclazide "instant" jellies for sustained Giclazide, a sulfonylurea for type 2 diabetes [7]. The concept of their system is the incorporation of sustained release Giclazide microparticles (prepared by layering of MCC pellets in a Wurster column together with a polymethacrylate polymer) into different types of jellies (polymers sodium alginate or polymer mixture with guar gum or low acyl gellan gum, 50:50) by mixing the required quantity (142 mg, equivalent to 30 mg gliclazide) of the particles with the jelly polymers. As gel formation should occur at room temperature without any heating, calcium ions in the form of dicalcium phosphate dihydrate (0.1 to 1% w/v) in 20 mL of deionized water were added to the blend, using a spatula until no dry powder could be seen. The different jelly types were fully characterized with regards to their texture (gel strength, firmness, cohesiveness, and adhesion forces were measured using a texture analyzer), rheological properties (using oscillatory viscosity measurements, i.e., determining loss G' and storage G'' moduli; see above), appearance, in vitro swallowability (using as a simulator a "Cambridge Throat" [18]), and in vitro dissolution tests for the sustained release behavior of the medication.

One of the objectives of this research was to match properties of commercialized jellies in Japan and "ready-to-eat" jellies in the UK. Two compositions were identified that were comparable with the commercial jellies in the UK with regard to appearance, rheological, and textural properties. In vitro swallowability data showed acceptable oral transit times and cohesive bolus flow. Incorporation of the Giclazide microparticles into these hydrocolloid jelly mixtures further reduced the drug release rates compared to those from coated particles without jelly. This should be taken into consideration when developing such products for the pharmaceutical industry.

10.4.1.2 Paracetamol Jellies

Another example of an "instant" jelly with paracetamol was published in 2020 by Abd Almonem Doolaanea from Malaysia [19]. The focus of the authors was to aid the swallowability and pediatric patient acceptability of paracetamol multiparticulates with a soft and smooth jelly dosage form to overcome the grittiness and rough mouthfeel perception associated with multiparticulate dosage forms. The authors also state that jellies would be more advantageous compared to liquid dosage forms with regard to their high viscosity, improving taste perception by children, and avoiding fast sedimentation of particles during swallowing. Appropriate packaging of jellies for light protection and patient convenience in terms of spillages during travel should be addressed for these dosage forms. Addition of preservatives for microbial stability is another important aspect and ease of preparation without using any heat for the jelly preparation from the packaged dry powder preparation [17].

For the manufacture of the paracetamol jellies, carrageenan has been used as a hydrocolloid polymer. There are three types of carrageenan—kappa κ, iota I, and lambda λ—which vary in sulfation, affecting their functional properties. Most suitable to form viscoelastic jellies is the iota I carrageenan, whereas the kappa κ forms rigid and brittle jellies. The lambda λ carrageenan gives only weak jellies and is used as a thickener in foods [19].

Preparation of the paracetamol multiparticulate beads with approximately 1.4 mm diameter was done by electro-spraying: Paracetamol drug substance was mixed with sodium alginate solution to form a suspension, which was then extruded through a plastic syringe with the needle as positive electrode and low molecular weight chitosan as negative electrode. A calcium chloride solution with paracetamol was used as the collector solution. Beads were dried in an oven at 50 °C. For the manufacture of the jellies, the electro-spray coated beads were mixed with several gelling polymers (e.g., gelatin, different carrageenan types, low acyl gellan gum), and then water was added prior to use. Finally, the iota I carrageenan was selected, and different jelly compositions were produced, including, e.g., sorbitol as a sweetener and citric acid/sodium citrate added as a buffering agent.

Paracetamol jellies were prepared as a single dose in a sachet and could be reconstituted by adding 20 mL of water. Spherical shapes with an appropriate size (2.5–2.8 mm is the acceptable limit for sprinkled multiparticle according to FDA [20]) would be clearly preferable according to the authors to avoid chewing. Figure 10.3 shows the dry powder instant jelly in a sachet and the jelly after addition of water.

A thorough analytical characterization of the paracetamol jellies by the authors included differential scanning calorimetry to study potential interactions between

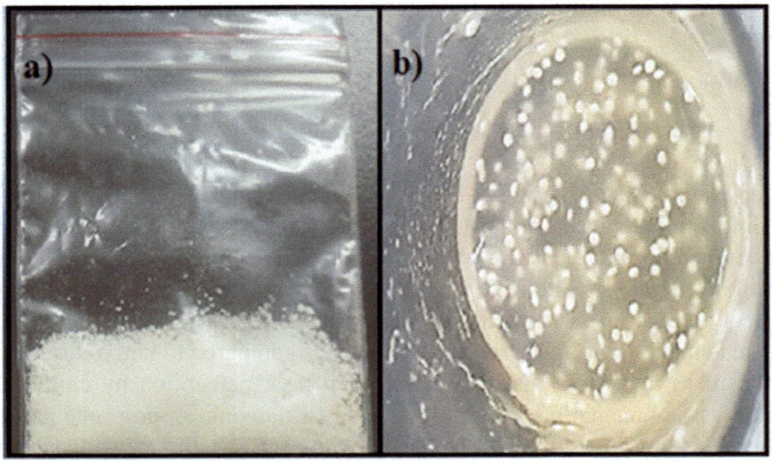

Fig. 10.3 (**a**) Paracetamol instant jelly sachet, and (**b**) paracetamol instant jelly (taken from [19])

the drug paracetamol and the polymer (carrageenan) used. The absence of such interactions was also confirmed by FTIR analysis. Texture analysis of the jellies (strength/hardness/gumminess, cohesiveness, adhesiveness) depicted for the hardness values similar to the majority of tested marketed products.

Like for the Giclazide instant jellies, the authors conducted oscillatory viscosity measurements and found for their paracetamol instant jellies a higher elastic or storage modulus ($G' = 2044$) than the loss modulus ($G'' = 380.3$). Shear-thinning behavior of these jellies is expected to be easiery and safer compared to Newtonian fluids [21].

The moisture content was approximately 2.6% (w/w), which was considered low and would have a significant impact on the dry powder stability of the instant paracetamol jellies. Additionally, packaging in tightly sealed sachets or the addition of desiccant might further contribute to a satisfactory long-term shelf-life of such products.

In vitro drug release of the paracetamol jellies was clearly inferior to a marketed suspension due to the gel structure slowing down the dissolution rate. Nevertheless, the time to release 50% of the drug was quite comparable to chewable paracetamol tablets, with a low difference of 5 min compared to the delay of 16 min when compared against the commercial suspension.

10.5 Palatability of Jellies

In terms of an optimum with respect to mechanical strength/hardness (gumminess), cohesiveness, and adhesiveness of these products, the softer jellies are easier to swallow than those with a high gumminess, low cohesiveness (risk of scattered particulates), and high adhesiveness (lower chewing frequencies if jelly sticks to teeth). These factors would represent higher risks for aspiration, especially in pediatric patients with dysphagia and higher efforts in terms of chewing time, muscle activity, and jaw movements [22, 23].

Figure 10.4 illustrates these rheological and texture-related aspects of jellies with regard to pediatric palatability

For the paracetamol jellies, the authors conducted a palatability (taste, smell) study with 12 adult volunteers and a ranking from 1 as most negative to 5 representing a most positive outcome, with 3 as the neutral rating [19].

Regarding smell, a commercial paracetamol suspension was rated as the highest score due to addition of flavoring agents, while the jelly formulation was assessed as neutral due to the absence of flavors. With respect to taste masking and texture mouthfeel, the jelly formulations were rated as acceptable as the commercial paracetamol suspension. The bitter aftertaste of paracetamol was found in paracetamol powder and the marketed suspension product, with no special observations for the jellies. Overall, the formulated paracetamol jellies depicted equal or superior palatability compared to the commercial suspension, which contains a significant amount of sucrose (which may be associated with dental caries issues!).

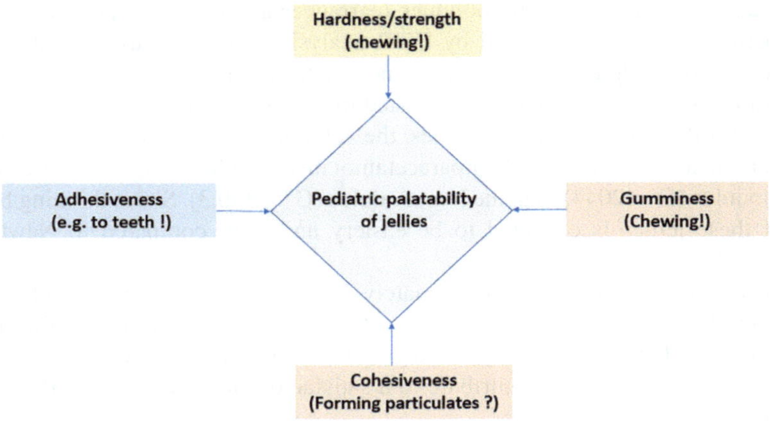

Fig. 10.4 Texture-related physical parameters of jellies influencing pediatric palatability

10.6 Patented "Instant" Jelly Technologies

The swallowability problems for patients suffering from dysphagia syndromes (elderly, young children) have been recently addressed by a few technologies that try to combine the benefits of an instant jelly with the advantages of multiparticulates, e.g., the Parvulet™ technology [24]. The commercial product is available in the form of tablets or dispersible granules and can be transformed into a semisolid jelly-like form in which multiparticulates (e.g., minitablets, pellets) could be sprinkled to ease the swallowing of the solid form. Another pellet-based product that swells with water and forms a soft pulp has also been reported [7].

10.7 Excipients Typically Used for Jelly Preparation

In a review article published in 2018 by G. Kauer et al. about oral jellies for pediatrics, a list of typical excipients used for jelly preparation has been included [25]. Table 10.1 lists these with a main focus on gelling agents and other excipients, with additional information.

Selection of these excipients listed in Table 10.1 for development of a jelly for pediatric purposes should in general be aligned with appropriate information about the acceptability in the targeted age groups from multiple databases [36].

Table 10.1 List of excipients used for preparation of jellies [25]

Function	Excipient	Source	Comment
Gelling agent (hydrocolloid)	Pectin	Pectin is a structural acidic heteropolysaccharide contained in the primary and middle lamella and cell walls of terrestrial plants [26]	Main component is galacturonic acid, a sugar acid derived from galactose
Gelling agent (hydrocolloid)	Tragacanth	Gum tragacanth is a viscous, odorless, tasteless, water-soluble mixture of polysaccharides obtained from sap that is drained from the root of middle Eastern legumes of the genus *Astragalus* and dried [27]	Tragacanth contains 60–70% water-soluble Tragacanthin (*Tragacanthin acid or Traganthin*) and 30–40% water insoluble, but swellable Bassorin (also *Traganthin acid*). Tragacanthin contains galacturonic acid and D-xylose, Bassorin L-arabinose, D-galactose, L-rhamnose, and the methylester of D-galacturonic acid
Gelling agent (hydrocolloid)	Gellan	Water-soluble anionic polysaccharide produced by the bacterium Sphingomonas elodea. Substitute for agar [28]	Tetrasaccharide, which consists of two residues of D-glucose and one of each residue of L-rhamnose and D-glucuronic acid
Gelling agent (hydrocolloid)	Carrageenan	Extracted from red edible seaweeds [29]	Carrageenans contain 15–40% ester-sulfate content, which makes them anionic polysaccharides (kappa-carrageenan has one sulfate group per disaccharide, iota-carrageenan has two, and lambda-carrageenan has three)
Gelling agent (hydrocolloid)	Hydroxypropyl methylcellulose (HPMC)	Semisynthetic cellulose polymer	Methylcellulose substituted with propylene oxide
Gelling agent (hydrocolloid)	Gelatin	Commonly derived from collagen taken from animal body parts	Mixture of animal-derived proteins
Gelling agent (hydrocolloid)	Modified starches	Prepared by physically, enzymatically, or chemically treating native starch to change its properties [30]	Chemical modifications, e.g., by addition of acids or bases. Physical modifications via thermal treatment

(continued)

Table 10.1 (continued)

Function	Excipient	Source	Comment
Gelling agent (hydrocolloid)	Carboxymethyl-cellulose	Semisynthetic cellulose polymer	Cellulose derivative with carboxymethyl groups (-CH_2-COOH) bound to some of the hydroxyl groups of the glucopyranose monomers that make up the cellulose backbone [31]
Gelling agent (hydrocolloid)	Agar	Jelly-like substance consisting of polysaccharides obtained from the cell walls of some species of red algae	Mixture of the linear polysaccharide agarose and a heterogeneous mixture of smaller molecules called agaropectin [32]
Gelling agent (hydrocolloid)	Xanthan gum	Natural polysaccharide prepared via fermentation from bacteria of the genus *Xanthomonas* from sugar-containing substrates (E 415) [33]	Long-chain polysaccharide having D-glucose, D-mannose, and D-glucuronic acid as building blocks in a molecular ratio of 3:3:2 with a high number of trisaccharide side chains. Xanthan gum has an average molecular weight of about 2000 kDa [34]
Gelling agent (hydrocolloid)	Pullulan	Polysaccharide polymer produced from starch by the fungus *Aureobasidium pullulans* (E 1204)	Consisting of maltotriose units, also known as α-1,4-;α-1,6-glucan. Three glucose units in maltotriose are connected by an α-1,4 glycosidic bond; consecutive maltotriose units are connected to each other by an α-1,6 glycosidic bond [35]
Solvents	Propylene glycol, glycerine	Chemical synthesis	Facilitate dissolution and dispersion of the gum Cave: Pediatric toxicity in young children!
Preservatives	Benzoic acid, sodium benzoate, sorbic acid, potassium sorbate, parabenes	Chemical synthesis	Consider pediatric acceptance (e.g., parabens)
Emulsifiers	Sorbitan monooleate, polyoxyethylene monooleate	Chemical synthesis	
Sweeteners	Sucralose, acesulfame-K, aspartame, sugar alcohols like sorbitol, xylitol, mannitol, sucrose, fructose, maltose	Chemical synthesis	
Flavors	Strawberry, vanilla, chocolate	Chemical synthesis	

10.7.1 Preparation of Jellies

Jellies typically are prepared by dispersing accurately weighed amounts of the gelling polymers into water, e.g., heated up to a temperature of 90 °C [37]. The dispersion should be continuously stirred until the polymer is dissolved, followed by permanent stirring by addition of other excipients like sweetener, pH modifiers like citric acid, and preservatives. The final weight will be adjusted, and the mixture transferred to molds and allowed to cool.

10.8 Conclusion

Gummies and jellies—while still neglected by the majority of the pharmaceutical industry—could become an attractive alternative as a pediatric semisolid dosage form, as the acceptance of these forms by children and caregivers might be high with regard to swallowability (if the forms are not oversized), taste, smell, and color. Soft-puddings can often be used to sprinkle multiparticulates (mini-tablets, pellets) and improve swallowability.

Challenges for the industry might be to balance between this attractiveness and the risk for abuse of these dosage forms as they might look like confectionery. In addition, the requirement to add preservatives would represent a drawback in terms of excipients'safety and acceptance by regulatory bodies in different countries (e.g., parabens discussion). The use of polymers from natural sources might represent another issue in terms of maintaining a stable product quality. The manufacture of these dosage forms also requires specific manufacturing and packaging equipment, which might not be available at many pharma companies.

Despite these downsides, Japan seems to be open to approving more of such products also for adult patients, and recent activities in the USA seem to move in the same direction. Especially the option to develop "instant" jellies for children could be the ideal approach to improve swallowability and acceptance as the jelly is prepared only when drug administration is required and insofar preservation could be avoided if the in-use time would be rather short (e.g., not more than two hours) and manufacturing of dry powders should be possible in multiple pharmaceutical facilities around the globe. The availability of certain patented technologies (e.g., [24]) should now further stimulate the possibility of also including these dosage forms and new ways of a more child-friendly administration into the concepts of drug developing pharma companies. In addition, the progress regarding 3D printing technology might also further contribute to better acceptance of these dosage forms globally.

References

1. Ternik R, Liu F, Bartlett JA, Khong YM, Thiam Tan DC, Dixit T, Wang S, Galella EA, Gao Z, Klein S. Assessment of swallowability and palatability of oral dosage fo in children: report from an M-CERSI pediatric formulation workshop. Int J Pharm. 2018;536(2):570–81. https://doi.org/10.1016/j.ijpharm.
2. Mistry P, Batchelor H. SPaeDD-UK project (smart paediatric drug development—UK). Evidence of acceptability of oral paediatric medicines: a review. J Pharm Pharmacol. 2017;69(4):361–76. https://doi.org/10.1111/jphp.12610.
3. Kozarewicz P. Regulatory perspectives on acceptability testing of dosage forms in children. Int J Pharm. 2014;469(2):245–8. https://doi.org/10.1016/j.ijpharm.2014.03.057.
4. Burklow KA, Phelps AN, Schultz JR, McConnell K, Rudolph C. Classifying complex pediatric feeding disorders. J Pediatr Gastroenterol Nutr. 1998;27(2):143–7. https://doi.org/10.1097/00005176-199808000-00003. PMID: 9702643
5. Alessandrini E, Brako F, Scarpa M, Lupo M, Bonifazi D, Pignataro V, Cavallo M, Cullufe O, Enache C, Nafria B, et al. Children's preferences for Oral dosage forms and their involvement in formulation research via EPTRI (European Paediatric translational research infrastructure). Pharmaceutics. 2021;13:730. https://doi.org/10.3390/pharmaceutics13050730.
6. Gomaa E, Ayoub MM. Vardenafil oral jellies as a potential approach for management of pediatric irritable bowel syndrome. Saudi Pharm J. 2021;29(9):955–62. https://doi.org/10.1016/j.jsps.2021.07.020.
7. Patel S, Scott N, Patel K, Mohylyuk V, McAuley WJ, Liu F. Easy to swallow "instant" jelly formulations for sustained release Gliclazide delivery. J Pharm Sci. 2020;109(8):2474–84. https://doi.org/10.1016/j.xphs.2020.04.018.
8. Imai K. Alendronate sodium hydrate (oral jelly) for the treatment of osteoporosis: review of a novel, easy to swallow formulation. Clin Interv Aging. 2013;8:681–8. https://doi.org/10.2147/CIA.S37199.
9. TA instruments, Frank A., Understanding Rheology of Structured Fluids, document AAN016.
10. Jacobs IC. Semi-solid formulations. In: Pediatric formulations. New York: Springer; 2014. p. 171–9.
11. https://www.cphi-online.com/pastilles-and-gummies-the-most-versatile-solid-prod479120.html, Accessed on 14 June 2022.
12. Herrada-Manchón H, Rodríguez-González D, Alejandro Fernández M, Suñé-Pou M, Pérez-Lozano P, García-Montoya E, Aguilar E. 3D printed gummies: personalized drug dosage in a safe and appealing way. Int J Pharm. 2020;587:119687. https://doi.org/10.1016/j.ijpharm.2020.119687.
13. https://www.formularywatch.com/view/is-medicine-delivered-via-gummies-the-future-. Accessed 20 June 2022.
14. https://www.gbi100.com/en/exhibition/463. Accessed 20 June 2022.
15. https://clinicaltrials.gov/ProvidedDocs/21/NCT04071821/Prot_000.pdf, Accessed 20 June 2022.
16. Ryukakusan Co. Ltd. Swallowing aid jelly. Available at: https://www.ryukakusan.co.jp/productjelly/en. Accessed 21 June 2022.
17. Kakino Y, Hishikawa Y, Onodera R, Tahara K, Takeuchi H. Gelation factors of pectin for development of a powder form of gel, dry jelly, as a novel dosage form. Chem Pharm Bull (Tokyo). 2017;65(11):1035–44. https://doi.org/10.1248/cpb.c17-00447.
18. Patel S, McAuley WJ, Cook MT, Sun Y, Hamdy S, Liu F. The swallowing characteristics of thickeners, jellies and yogurt as observed using an in vitro model. Dysphagia. 2019; https://doi.org/10.1007/s00455-019-10074-1.
19. Almurisi SH, Doolaanea AA, Akkawi ME, Chatterjee B, Ahmed Saeed Aljapairai K, Islam Sarker MZ. Formulation development of paracetamol instant jelly for pediatric use. Drug Dev Ind Pharm. 2020;46(8):1373–83. https://doi.org/10.1080/03639045.2020.1791165.

20. Nagavelli LR, Lionberger RA, Sayeed VA, Yu L, Allgire J, Smith A, Wokovich A, Westenberger BJ, Buhse L. Analysis of bead sizes for MR capsules labeled for sprinkle. AAPS Pharm Sci Tech. 2010;11(4):1508–10. https://doi.org/10.1208/s12249-010-9529-2.
21. O'Leary M, Hanson B, Smith C. Viscosity and non-Newtonian features of thickened fluids used for dysphagia therapy. J Food Sci. 2010;75(6):E330–8. https://doi.org/10.1111/j.1750-3841.2010.01673.x.
22. Momosaki R, Abo M, Kobayashi K. Swallowing analysis for semisolid food texture in post-stroke dysphagic patients. JStroke Cerebrovasc Dis. 2013;22:267–70.
23. Momosaki R, Abo M, Kobayashi K. Swallowing analysis for semisolid food texture in poststroke dysphagic patients. J Stroke Cerebrovasc Dis. 2013;22(3):267–70. https://doi.org/10.1016/j.jstrokecerebrovasdis.2011.08.009. Epub 2011 Oct 2
24. Adare Pharmaceuticals. Parvulet™ taste masking technology. 2019. Available at: https://adarepharmasolutions.com/technologies/parvulet/. Accessed 27 June 2022
25. Kaur G, Gananarajan G. A review article on oral jellies for pediatrics. Indo Am J Pharm Sci. 2018;05(01):444–52. https://doi.org/10.5281/zenodo.1161151.
26. https://en.wikipedia.org/wiki/Pectin. Accessed 27 June 2022.
27. https://de.wikipedia.org/wiki/Traganth. Accessed 27 June 2022.
28. https://en.wikipedia.org/wiki/Gellan_gum. Accessed 27 June 2022.
29. https://en.wikipedia.org/wiki/Carrageenan. Accessed 27 June 2022.
30. https://en.wikipedia.org/wiki/Modified_starch. Accessed 27 June 2022.
31. https://en.wikipedia.org/wiki/Carboxymethyl_cellulose. Accessed 27 June 2022.
32. https://en.wikipedia.org/wiki/Agar. Accessed 4 July 2022.
33. https://en.wikipedia.org/wiki/Xanthan_gum. Accessed 4 July 2022.
34. Jindal N, Khattar JS. Chapter 4—Microbial polysaccharides in food industry. In: Grumezescu AM, Holban AM, editors. Handbook of food bioengineering, biopolymers for food design. Academic Press; 2018. p. 95–123. ISBN 9780128114490.
35. https://en.wikipedia.org/wiki/Pullulan. Accessed 4 July 2022.
36. Reflection Paper: Formulations of Choice for the Paediatric Population, EMEA/CHMP/PEG/194810/2005.
37. Sunil S, Sharma UK, Arathy SA. Pharmaceutical Jellies: a novel way of drug delivery. J Pharm Sci & Res. 2020;12(7):904–9.

Chapter 11
Buccal/Sublingual Drug Delivery for the Paediatric Population

Darragh Murnane and Marc B. Brown

Abstract Oral mucosal drug delivery has emerged as a potential route of drug administration which represents a strategy of maximising the availability of the oral route for easy administration without the need to swallow a dosage form whilst offering the possibility of local delivery and direct access to the systemic circulation thus avoiding first pass metabolism. However, changes in anatomical, physiological and biochemical functions that occur from birth to adolescence are required to be considered when designing such formulations. Currently marketed formulations include liquids, semi-solids and solid dosage forms; however, there is significant research in developing alternative dosage forms for oral mucosal delivery such as nanotablets, films, microparticles and mouth sprays which is partly due to problems (e.g., posology or excipients present in formulation) associated with using currently marketed products in children. Despite the development of novel technologies, including NanoTabs and mucosal sprays, there still remain relatively few products approved for use in the paediatric population. This is likely explained by the fact that alongside the normal problems that need to be overcome during product development, additional issues including compliance, palatability, paediatric safety, the needs and capabilities of carers and the implications of global paediatric regulatory guidance need to be considered.

Keywords Dosage form · Solid dosage form · Pass metabolism · Fentanyl citrate · Buccal tablet

D. Murnane (✉)
School of Health, Medicine and Life Sciences, University of Hertfordshire, Hatfield, UK

Informix Pharma Ltd., Bury St. Edmunds, U.K.
e-mail: d.murnane@herts.ac.uk

M. B. Brown
School of Health, Medicine and Life Sciences, University of Hertfordshire, Hatfield, UK

MLBT Investments Ltd., Aylesbury, UK

H. Batchelor, K. Rose (eds.), *Pediatric Formulations*, AAPS Advances in the Pharmaceutical Sciences Series 58, https://doi.org/10.1007/978-3-031-77240-5_11

11.1 Introduction

Legislation in the European Union (EU) and the United States (US) has led pharmaceutical companies to consider the paediatric population during the early stages of product development, including the potential routes of administration of medicines appropriate for children [1–3]. Oral administration is the drug delivery route of choice in children with the exception of neonates where the parenteral route of administration is preferred [4]. However, there is increasing evidence of limitations associated with such enteric oral delivery in children, including gastric intolerance, enzymatic degradation and first-pass metabolism. A particular barrier to enteric oral delivery is the presentation of some age-related disease conditions that affect swallowing (e.g., dysphagia). Thus, oral mucosal drug delivery represents a strategy of maximising the availability of the oral route for administration without the need to swallow a dosage form whilst offering the possibility of local delivery and direct access to the systemic circulation, thus avoiding first pass metabolism [5]. However, there are many changes in the anatomical, physiological and biochemical functions of the oral mucosa that occur from birth to adolescence that influence the pharmacokinetics and pharmacodynamics of a drug that need to be considered when designing such formulations for the paediatric population [6].

11.2 The Physiology of the Oral Mucosa and Its Implications for Paediatric Drug Delivery

The physiological development of the buccal/sublingual mucosa in children is difficult to predict, and extrapolation of safety and efficacy data derived from an adult trial may not be appropriate for children of different age groups. The permeability of the buccal/sublingual mucosa of children is often greater than that of an adult mucosa for certain drugs due to the differences in physiology as well as factors such as salivary flow, enzymatic environment and the thickness of the mucosa [7]. In a study by Sonesson et al. [8] the rate of buccal salivary secretion in young children of 3 years of age was found to be lower than in adults as a result of underdeveloped salivary glands. This suggests that mucoadhesive formulations may be retained for a longer period of time at the site of application, resulting in an increased concentration of drug being absorbed across the mucosa and subsequently a higher drug plasma concentration. In another study [7], the buccal administration of lidocaine patches in children of 2–7 years old produced a four to five times higher maximum plasma drug concentration (C_{max}) in blood compared to an adult population at 9 min. In addition the time to attain maximum plasma concentration after patch application was comparable in children and adults, despite the patch application time being shorter in the former (5 min) compared to the latter (15 min). Obviously the relatively lower body mass and blood volume in children is thought to be one of the reasons for the higher C_{max} observed compared to adults. However, other investigators have reported a trend towards decreased permeability to water with age, especially in the sublingual mucosa of females [9]. The slight thinning of the epithelium with a concomitant flattening of the epithelial–connective tissue interface resulting in a more compact structure of the mucosa may be the reason behind this phenomenon [10]. Such differences need to be considered when evaluating the potential for efficacy and toxicity of a buccally/sublingually applied medicine in children.

11.3 A Review of the Currently Marketed Oral Mucosal Products for Drug Delivery in Paediatrics

The availability of approved products for paediatric use via the buccal/sublingual mucosa varies depending on the regulatory market area. Despite several products being marketed in the United States for buccal/sublingual mucosal delivery in adults, none are approved for a paediatric indication [11]. However, some products have been approved for paediatric indications in the United Kingdom and are summarised in Table 11.1. Such approved products span a range of dosage forms that are used for the treatment of systemic conditions as well as in the topical therapy of localised oral diseases. The marketed dosage form options include liquids, semi-solids and solid dosage forms, although currently there is significant research in developing alternative dosage forms for oral mucosal delivery such as nanotablets, films, microparticles and mouth sprays [5]. Liquid formulations including solutions, suspensions and emulsions are easy to administer, and the dosage can be easily measured out. However, such dosage forms can present some issues with regard to short-term stability and are generally bulky in nature, both of which can adversely influence the supply chain and increase cost, thus limiting their wider acceptability by both healthcare professionals and the pharmaceutical industry. Additionally liquid formulations typically require the use of preservatives and antioxidants to avoid microbial growth or spoilage for products intended for multiple dosing. However, the use of certain preservatives (e.g., Thiomersal) is not always desirable for paediatric therapies as they have reported toxicity in this lower age group population [12]. Other disadvantages can include that the residence time of the formulation in the oral cavity can be low [13] and the administered dose is often either swallowed or expectorated, both of which can lead to poor drug delivery at or to the mucosal membrane.

Semi-solid dosage forms such as gels, pastes and ointments are an alternative to solutions, especially where buccal residence of the drug is required for a prolonged period of time. Currently, gels are only the semi-solid formulation approved to be used in paediatrics via buccal/sublingual route (Table 11.1) and are intended typically for topical oral treatment (e.g., teething and anti-fungal gels). Solid dosage forms such as lozenges, tablets, capsules and films/wafers appear to be the obvious choice for development because of the improved drug stability, ease of manufacture

Table 11.1 Summary of oral mucosally applied products available in the United Kingdom for paediatric populations

Name of the drug	Brand name	Name of company	Approved use	Approved age range	Comments	References
Buprenorphine hydrochloride	Temgesic sublingual tablets 200, 400 μg	Reckitt Benckiser Pharmaceutical Ltd	Treatment of moderate to severe pain	6 years or more	According to the SmPC, the dose for children under 12 years of age is based on weight and is 1/2 tablet for children weighing from 16 to 25 kg, 1/2 to 1 tablet for 25–37.5 kg and 1 to 1 ½ tablets for 37.5–50 kg and should be given every 6–8 h. Hence the patient needs the dosages in fractions of tablets. As per the product description provided within the SmPC and patient information leaflet (PIL), Temgesic tablets are "round white tablets, marked with an L on one side and a sword on the other". It means that dividing the tablets into two halves may not be accurate as the formulation lacks the break line and thus could increase the chance of dosing error. This can impact the effectiveness of treatment. The physical and chemical stability of the formulation upon adaptation needs to be considered	[14]
	Tephine sublingual tablets 200, 400 μg	Sandoz Limited	Treatment of severe pain following surgery or injury, myocardial infarction and severe pain associated with cancer	2 years or more or children weighing more than 16 kg	The recommended dose for Tephine sublingual tablet in a patient weighing 16–35 kg is 100 μg, but according to the SmPC [15] the Tephine is not divisible; hence, it is of limited use in patients weighing more than 16 kg	[15]

Midazolam hydrochloride	Buccolam oromucosal solution	ViroPharma Ltd	Treatment of prolonged, acute, convulsive seizures	3 months to <18 years	To be used by parents/carers in patients with confirmed epilepsy. Administration in infants must be carried out in a hospital setting where therapeutic monitoring is possible. The advantage of this formulation is that it is colour coded and pre-filled according to the strength of the formulation to reduce chances of under- or over-dosing	[16]
Miconazole	Daktarin oral gel, Daktarin sugar-free 2% oral gel	Janseen-Cilag Ltd McNeil products Ltd	Oral treatment and prevention of fungal infections of the oropharynx and of superinfections due to Gram-positive bacteria	4 months or more	Despite being approved for the same indication, the products have a different posology and method of administration, and hence there is an increased chance of error on switching the patient from a normal oral gel to the sugar-free formulation	[17, 18]
					For example, in the case of Daktarin oral gel, it is recommended to use one spoonful (5 ml) four times a day in children from the age of 6 years and over, one spoonful two times a day in children from the age of 2 to 6 years and half a spoonful (2.5 ml) should be used in children from the age of 4 months to 2 years	
					In the case of Daktarin sugar-free 2% oral gel, it is recommended to apply a small amount of gel directly to the affected area with a clean finger four times a day in children from the age of 6 years and twice a day in children from the age of 4 months to 6 months with no indication of the quantity of the gel to be used despite such a huge difference in the child's age	[16]

(continued)

Table 11.1 (continued)

Name of the drug	Brand name	Name of company	Approved use	Approved age range	Comments	References
Lidocaine hydrochloride, chlorocresol and cetylpyridinium chloride	Anbesol teething gel	Alliance Pharmaceutical	Temporary relief of pain caused by recurrent mouth ulcers, denture irritation and teething	No age restriction	Anbesol teething gel contains 66.605% ethanol (alcohol), i.e. up to 0.754 ml per dose, which is equivalent to 15.1 ml beer or 6.3 ml wine per dose which may be harmful to children who may not have yet developed tolerance for alcohol or in children suffering from liver disease or epilepsy	[19]
Dextromethorphan	Boot dry cough relief lozenges	The Boots Company PLC	Cough suppressant for the relief of acute non-productive (dry, tickly) cough associated with respiratory tract infection		Each lozenge contains 1.4 g of sucrose and 1.1 g of glucose. According to the normal dosing schedule, the patient is likely to take 9.8 g of sucrose (37.93 calories) and 7.7 g of glucose (30.8 calories) every day which may restrict use in patients suffering from type II diabetes or those with poor glucose control	[20]

and their less bulky nature, all of which increase paediatric patient compliance and decrease the cost of goods. As such lozenges and tablets are the most commonly approved products for the buccal/sublingual route in paediatric patients (Table 11.1). However, where prolonged contact with the mucosa is required semi-solid formulations may offer greater paediatric acceptability than solid dosage forms as the former can be spread evenly and thinly over the mucosa rather than having to be deliberately retained and thus obstructing swallowing, eating and drinking.

11.4 Specific Difficulties with Using Marketed Oral Mucosal Products in Paediatric Therapy

Despite there being several marketed products demonstrating the applicability of oral mucosal delivery, several complications adversely influence the use of such commercialised medicines in many child treatment groups. For example, issues surround the posology, dose and excipients of such formulations making them difficult for use in children (Table 11.1). For example, Temgesic sublingual tablets need to be split in order to administer the required dose for children [14], yet there is an absence of a break line which would make dosage division easier and more accurate (Table 11.1). Tephine sublingual tablets are not available in a strength suitable (or even manually divisible) for children weighing 16–35 kg [15]. Another example is the case of Daktarin sugar-free gel where it is recommended to apply a small but undefined amount of gel directly to the affected area four times a day in children above 6 years of age and twice a day in children between 4 months and 6 years [18]. Also the presence of a high quantity of alcohol and sugar in Anbesol teething gel and Boot's dry cough relief lozenges, respectively, may restrict their use in certain paediatric populations such as those children who are either intolerant to alcohol or suffering from liver disease or epilepsy, or who are suffering with diabetes, respectively [19, 20].

Glyceryl trinitrate (GTN) provides an excellent example of an active pharmaceutical ingredient benefitting from the avoidance of first-pass metabolism when administered via the sublingual route and was, in fact, one of the first drugs successfully developed to be delivered via the buccal/sublingual mucosa in adults. Currently, an injectable formulation is available for use in children to control hypertension as well as in heart failure [21]. However, although conventional and sustained release buccal tablets and spray formulations have been developed for adults none are approved for use in children. In addition, although fentanyl has been formulated in lozenges, buccal tablets and recently in a thin film to avoid the extensive first-pass metabolism when administered for oral enteric delivery, none are available for children less than 12 [22]. This is despite the same drug being used in injectable form and as transdermal patches for pre-operative analgesia and severe chronic pain, respectively, in children [23]. However, a study by Camacho Parreño et al. [24] reported off-label use of an oral transmucosal fentanyl citrate lozenge formulation for analgesia for bone fracture alignment in children. The author emphasised that transmucosal fentanyl could be used as an alternative to other pain medications but that more studies would be required to establish effective and safe use of fentanyl in children. The obvious problem being whether such transmucosal medications can be manipulated to get the reduced dose required for paediatrics.

11.5 Current Research Trends in Oral Mucosal Drug Delivery

Previous research into the improvement of drug delivery via the buccal/sublingual mucosa has resulted in the development of several novel dosage forms like solutions, lyophilised and bioadhesive tablets and lozenges, chewing gum, solution sprays, laminated systems, patches, hydrogels, adhesive films, hollow fibres and microspheres [5]. Formulators have used creative approaches that incorporate a combination of strategies to create a balance between patient convenience and clinical benefits [5], but not all of these may be appropriate for use in the paediatric population. For example, bioadhesive formulations need to be kept within the mouth for longer period of time compared to conventional formulations to achieve the desired therapeutic outcome, but children may not be compliant for such extended periods.

Palatability including texture, flavour and taste plays a major role in the compliance and concordance of children to treatment. Individuals differ in their experiences when tasting the same product because genetics plays a significant role in perception of palatability [25]. Taste in children is an even bigger problem because the sense of taste develops as the child grows [26]. It is reported that children have a stronger liking for sweet [27] and salty tastes [28] compared to adults, but yet they seem to be more sensitive to a bitter taste [29, 30], and although adults may think that the worse a medication tastes, the better it works, this perception does not work in children. Several approaches have been explored to mask the taste of drugs such as the addition of chemicals (flavours, sweeteners, pH modifiers), encapsulation of the drug (coating or matrix formation using polymer, or complexation with cyclodextrins) and the development of a novel formulation (liposomes, double emulsion) [31]. However, such approaches have very limited application in formulations designed to be delivered via the buccal/sublingual mucosa because any such alteration may impact drug release at the site of absorption. It is also essential to consider that the excipients included to increase palatability may not be approved for use in children [32]. Hence more research is needed to address the issue of taste masking without compromising formulation performance.

The size and shape of the formulation should also be considered carefully for oral transmucosal administration for each target age group depending on the stage of childhood development. However, in a study by Spomer et al., it was found that the overall acceptance of 2-mm uncoated mini-tablets was at least equal to or even better than that of sweet tasting syrup, even for children aged 2 years or less. Almost 40% of the children aged 1–2 years refused the liquid formulation, but only 10% refused the mini-tablet [33].

While traditionally employed for enteric delivery, multi-particulate delivery systems have also been explored for buccal/sublingual administration [34]. The benefit of these drug delivery systems is the flexibility provided to adapt the dose when required for a posology determined by bodyweight or age [35]. The other benefits from a formulation perspective are the potential for enhanced retention at site of administration via mucoadhesion and tailoring drug release. Striant® (Testosterone) and Onsolis® (Fentanyl citrate) are currently approved products for use in adults utilising mucoadhesion strategies to prolong the residence time of the formulation with the oral mucosa [36, 37]. These formulations have used polycarbophil, hypromellose and carbopol 934P, carboxymethyl cellulose and hydroxyethyl cellulose as mucoadhesive polymers [36, 37], and this strategy can be extended to develop suitable paediatric mucoadhesive formulations. The multi-particulate dosage form is usually placed directly in a patient's cheek or under the tongue and can be supplied in a bottle (with appropriate dosing scoop) or a pre-packed sachet or capsule which allows for more accurate dosing.

Thin film dosage forms and lyophilised wafers fall under the category of orodispersible dosage forms which usually melt or disperse/dissolve within a matter of seconds when placed under the tongue or beneath the cheek. Thin films hold great potential for the paediatric population as they are easy to administer, are difficult to spit out due to rapid dispersion and can provide a range of dosages appropriate for use in children [38]. These formulations are usually compact (the size of postage stamp) and hence do not require bulk packaging and can be packed in a sachet to provide a unit dose. The films can be easily manipulated without altering the quality of formulation by cutting into the desired size, if required, for children of different ages. However, taste is a challenge for such formulations as they have very limited capacity to hold flavourings to mask the taste of the drug or film-forming agents without them crystallising upon film-drying. Potentially this could be overcome by incorporating a drug coated with polymer using particle coating technology [39].

Oral mucosal sprays have been extensively explored in adults as a mean of directly delivering the drug in a liquid formulation to the oral mucosa [40]. Such an approach has potential for use in children, but issues including the smaller size of the oral cavity available for spray application and local irritation at the site of application along with difficulty in modifying the dose must be considered. Nevertheless, advantages such as rapid application to the desired site, negligible chances of spitting out the formulation, mucoadhesion, dose manipulation and accurate and controlled dosing may outweigh these concerns.

AcelRX Pharmaceutical Inc. has filed a patent (US 20090010992) in the United States for "Drug formulations for oral transmucosal drug delivery to paediatric patients" [41]. The invention describes the development and evaluation of small volume bioadhesive dosage forms called "NanoTabs" containing the opioid sufentanil, for the oral transmucosal delivery route with the help of a dosing applicator or device. The NanoTabs formulation, which is administered with an applicator, offers a practical solution where (1) intravenous (IV) access is not available; (2) initiation of IV access requires sedation; (3) the patient cannot or will not swallow pills; (4) pre-procedural sedation or a relatively rapid onset of action is required which cannot be achieved by oral GI administration and (5) for paediatric patients who are not able to use the other non-invasive routes of administration effectively [41]. The NanoTab ranges between 8 and 15 mg in weight, with an active ingredient in the range of 0.001–10 mg per NanoTab. The formulation is currently under phase III trial evaluation for use in post-operative patient-controlled analgesia and is reportedly likely to overcome the programming and delivering errors associated with traditional intravenous patient-controlled analgesia which has led to patient deaths [42].

Although not approved for use in children as yet, Schechter et al. [43] have evaluated the potential of fentanyl citrate-based lozenges applied for oral transmucosal delivery for use in painful procedures such as bone marrow aspiration or lumbar puncture in 48 children whose ages ranged from 3 to 18 years. The study concluded that such delivery resulted in the children suffering less procedural pain when compared to children who were on placebo. However, significant incidence of nausea and vomiting may ultimately limit their paediatric use.

Recently, Gerrard et al. proposed a new concept "nipple shield delivery system (NSDS)" to deliver an anti-HIV drug to prevent mother-to-baby transmission of the disease. The approach to deliver the drug to infants incorporates a drug-loaded insert into an NSDS. The drug is released directly into milk during breastfeeding [44]. Though the goal of this research is to deliver the drug orally through the breastfeeding mechanism, it can be extended to target the oral mucosa by incorporating mucoadhesive properties in the formulation [45]. LMA international, a device manufacturing company, developed the LMA MADdy™ paediatric laryngo-tracheal mucosal atomisation device which consists of a small atomising tip at the end of a flexible applicator that is partially concealed by a colourful, child-friendly blowfish and used for dispensing topical medications to the nose, mouth, throat, hypopharynx, larynx and trachea in a fine, gentle mist. It is not a pre-filled device and hence gives flexibility to adapt the dose [46].

11.6 Conclusions

Oral mucosal drug delivery is a suitable alternative to enteric drug delivery for paediatric patients and offers the potential avoidance of first-pass metabolism along with an easy and accessible site to deliver the drug for local or systematic administration. Nevertheless, despite these advantages and the development of novel technologies including NanoTabs and mucosal sprays, there still remain relatively few products approved for use in the paediatric population. This is likely explained by the fact that alongside the normal problems that need to be overcome during product development, additional issues including compliance, palatability, paediatric safety, the needs and capabilities of carers and the implications of global paediatric regulatory guidance [47, 48] need to be considered when developing or applying such adult formulations for use in children.

References

1. Breitkreutz J. European perspectives on pediatric formulations. Clin Ther. 2008;30:2146–54.
2. Dunne J. The European regulations on medicines for paediatric use. Paediatr Respir Rev. 2007;8:177–83.
3. Rodriguez W, Selen A, Avant D, et al. Improving pediatric dosing through pediatric initiatives: what we have learned. Pediatrics. 2008;121:530–9.
4. Committee for Medicinal Products for Human Use (CHMP) Reflection paper: formulations of choice for the paediatric population. http://www.ema.europa.eu/docs/en_GB/document_library/Scientific_guideline/2009/09/WC500003782.pdf. Accessed 2 Dec 2011.
5. Patel VF, Lui F, Brown MB. Advances in oral transmucosal drug delivery. J Control Release. 2011;153:106–16.
6. Costello I. Paediatric pharmacokinetics and pharmacodynamics. In: Florence AT, Moffat AC, editors. Paediatric drug handling. 1st ed. London: Pharm Press; 2007.
7. Leopold A, Wilson S, Weaver JS, Moursi AM. Pharmacokinetics of lidocaine delivered from a transmucosal patch in children. Anesth Prog. 2002;49:82–7.
8. Sonesson M, Eliasson L, Matsson L. Minor salivary gland secretion in children and adults. Arch Oral Biol. 2003;48:535–9.
9. Williams DM, Cruchley AT. Structural aspects of aging in the oral mucosa. In: Squier CA, Hill MW, editors. The effect of aging in oral mucosa and skin. Boca Raton: CRC Press; 1994. p. 65–74.
10. Squier CA, Wertz PW, Williams DM, Cruchley AT. Permeability of oral mucosa and skin with age. In: Squier CA, Hill MW, editors. The effect of aging in oral mucosa and skin. Boca Raton: CRC Press; 1994. p. 91–8.
11. http://www.medilexicon.com/drugs-list/pediatrics.php. Accessed 15 July 2012.
12. http://www.fda.gov/BiologicsBloodVaccines/SafetyAvailability/VaccineSafety/UCM096228. Accessed 14 Feb 2013.
13. Kaye JL. Review of paediatric gastrointestinal physiology data relevant to oral drug delivery. Int J Clin Pharm. 2011;33:20–4.
14. Temgesic. www.medicines.org.uk. Accessed 14 June 2012.
15. Tephine. www.medicines.org.uk. Accessed 14 June 2012.
16. Buccolam. www.medicines.org.uk. Accessed 14 June 2012.
17. Daktarin oral gel. www.medicines.org.uk. Accessed 22 June 2012.
18. Daktarin Sugar free 2% oral gel. www.medicines.org.uk. Accessed 22 June 2012.
19. Anbesol Teething gel. www.medicines.org.uk. Accessed 14 June 2012.
20. Boots Dry Cough Relief Lozenges. www.medicines.org.uk. Accessed 14 June 2012.
21. Glyceryl Trinitrate, Paediatric Formulary Committee. BNF for Children (online). BMJ Group, Pharmaceutical Press, and RCPCH Publications, London. http://www.medicinescomplete.com. Accessed 18 Oct 2012.
22. Actiq. www.medicines.org.uk. Accessed 14 June 2012.
23. Fentanyl, Paediatric Formulary Committee. BNF for Children (online). BMJ Group, Pharmaceutical Press, and RCPCH Publications, London. http://www.medicinescomplete.com. Accessed 18 Oct 2012.
24. Camacho Parreño S, Prieto MF, Santana VM, Sanchez MTS. Off-label oral transmucosal fentanyl citrate analgesia for bone fracture alignment in children. Eur J Hosp Pharm. 2012;19:103.
25. Bartoshuk LM. Comparing sensory experiences across individuals: recent psychophysical advances illuminate genetic variation in taste perception. Chem Senses. 2000;25:447–60.

26. Birch LL, Fisher JO. Development of eating behaviors among children and adolescents. Pediatrics. 1998;1(Suppl 2):539–49.
27. Beauchamp GK, Cowart BJ. Development of sweet taste. In: Dobblng J, editor. Sweetness. New York: Springer; 1987.
28. Beauchamp GK, Moran M. Acceptance of sweet and salty tastes in 2-year-old children. Appetite. 1984;5:291–305.
29. KaJulra H, Cowart BJ, Beauchamp GK. Early developmental change in bitter taste responses in human infants. Dev Psychobiol. 1992;25:375–86.
30. Mennella JA, Peplno MY, Reed DR. Genetic and environmental determinants of bitter perception and sweet preferences. Pediatrics. 2005;115:e216–22.
31. Bar-Shalom D. Pediatric formulations – are we living in the past? Am Pharm Rev. 2011. http://www.americanpharmaceuticalreview.com. Accessed 27 Sept 2012.
32. Shojaei AH. Buccal mucosa as a route for systemic drug delivery: a review. J Pharm Pharm Sci. 1998;1:15–30.
33. Spomer N, Klingmann V, Stoltenberg I, Lerch C, Meissner T, Breitkreutz J. Acceptance of uncoated mini-tablets in young children: results from a prospective exploratory cross-over study. Arch Dis Child. 2012;97:283–6.
34. Limbachiya S, Nagar H, Dodiya H, Mehta D. Development and characterization of fast dissolving mucoadhesive microsphere of nebivolol hydrochloride using modified methacrylate polymer. Int J Pharm Sci. 2013;3:136–41.
35. De Matas M, Okwelogu C, York P. Paediatric dosage forms: proposal of a versatile platform technology. Chim Oggi Chem. 2009;27:20–4.
36. Onsolis. http://www.onsolis.com/pdf/onsolis_pi.pdf. Accessed 18 Feb 2013.
37. Striant. http://www.actientpharma.com/filebin/pdf/products/58816-10_Striant_full_PI_4pager.pdf. Accessed 18 Feb 2013.
38. Dixit RP, Puthli SP. Oral strip technology: overview and future potential. J Control Release. 2009;139:94–107.
39. http://www.catalent.com/index.php/news-events/news/Catalent-Licenses-Technology-with-NJIT. Accessed 4 March 2013.
40. Novamist Technology. http://www.novadel.com/technology/index.htm. Accessed 18 Oct 2012.
41. Palmer P, Schreck T, Tzennis S, Hamel L, Poutiatine AI. Drug formulations for oral transmucosal administration in pediatric patients. US 20090010992. 2009. www.uspto.gov. Accessed 24 July 2012.
42. ARX-01 Sufentanil NanoTab PCA System. http://www.acelrx.com/pipeline/index.html. Accessed 24 July 2012.
43. Schechter NL, Weisman SJ, Rosenblum M, Bernstein B, Conard PL. The use of oral transmucosal fentanyl citrate for painful procedures in children. Pediatrics. 1995;95:335–9.
44. Gerrard SE, Baniecki ML, Sokal DC, Morris MK, Urdaneta-Hartmann S, Krebs FC, Wigdahl B, Abrams BF, Hanson CV, Slater NK, Edwards AD. A nipple shield delivery system for oral drug delivery to breastfeeding infants: microbicide delivery to inactivate HIV. Int J Pharm. 2012;434:224–34.
45. http://justmilk.org/index.html. Accessed 5 Oct 2012.
46. http://www.lmana.com/pwpcontrol.php?pwpID=6358. Accessed 5 Aug 2012.
47. European Medicine Agency. Draft guideline on pharmaceutical development of medicines for paediatric use. Committee for Medicinal Products for Human Use (CHMP); 2011. http://www.ema.europa.eu. Accessed 19 July 2012.
48. World Health Organization. Report of the informal expert meeting on dosage forms of medicines for children. December 2008. http://www.who.int/selection_medicines/committees/expert/17/application/paediatric/Dosage_form_reportDEC2008.pdf. Accessed 5 Aug 2012.

Chapter 12
Topical and Transdermal Drug Delivery

William J. McAuley and Marc Brown

Abstract Formulations designed to be applied to the skin are commonly used for children, particularly for the treatment of skin diseases. There are a number of important issues with regard to the use of these types of medicines in the paediatric population in comparison to adults; for example children have a greater risk of experiencing systemic side effects as a result of treatment. This chapter discusses the roles of the different formulations used for topical and transdermal treatment, their design and considerations specifically relevant for the development of these medicines for children.

Keywords Skin · Topical · Formulation · Transdermal · Children

12.1 Introduction

The topical and transdermal administration of medicines are considered as two separate treatment modalities. For topical treatment the purpose is to apply a formulation to the skin to treat the skin or other localised tissues, whereas with transdermal drug delivery, the drug is applied to the skin with the intention that it is absorbed into the systemic circulation to have an effect elsewhere in the body. Topical treatment of skin diseases is common and offers the possibility of obtaining high, localised drug concentrations in the affected tissue. This can, in many cases, provide suitable therapeutic treatment and reduce systemic exposure to the drug in comparison to other routes of delivery, thereby minimising side effects and toxicity.

W. J. McAuley (✉)
Centre for Research in Topical Drug Delivery and Toxicology, School of Life and Medical Sciences, University of Hertfordshire, Hatfield, UK
e-mail: w.j.mcauley@herts.ac.uk

M. Brown
MedPharm Ltd., Guildford, UK

H. Batchelor, K. Rose (eds.), *Pediatric Formulations*, AAPS Advances in the Pharmaceutical Sciences Series 58, https://doi.org/10.1007/978-3-031-77240-5_12

Transdermal formulations for children are less common than topical treatments. However, for suitable drug candidates they offer advantages including the provision of the controlled, prolonged release of a drug. This minimises dosing frequency, improving patient compliance and thus therapeutic treatment for all patient groups, including children.

12.2 Skin Conditions in Children

The impact of skin and subcutaneous tissue conditions is significant, representing the fourth leading cause of non-fatal global burden of disease, and the topical route of administration is of particular importance as many of these conditions are treated topically [1]. Skin conditions in children are common and differ in incidence to those of the adult population. Some conditions such as seborrheic dermatitis and Mongolian spots are quite common in neonates but may not require any treatment; however, others can have a significant impact on an individual's quality of life and are more prevalent in children. For example, atopic dermatitis affects between 11% and 20% of children in the UK but only 5–10% of adults [2]. Other skin conditions such as acne, viral and bacterial skin infections and skin infestations such as scabies also have bigger impacts on the paediatric population [1]. Acne is particularly common, for example, affecting between 42% and 73% of adolescents in Europe, with studies elsewhere having reported prevalence of up to 92–95% in adolescents [3, 4]. In contrast the prevalence of clinical adult acne has been reported as being as low as 3% in men and 12% in women [5]. Some conditions such as napkin/diaper dermatitis (nappy rash) are predominantly thought of as affecting paediatric populations, though they also affect a large number of the adult population, affecting individuals of any age who wear napkins or incontinence pads [6]. The prevalence in infants has been estimated to be between 7% and 35% [7], and a study in the UK demonstrated a prevalence of 25% in the first 4 weeks of life alone [8]. In contrast to the above conditions, skin disease burden from conditions such as psoriasis, fungal infection and alopecia areata are higher in adults [1]. The prevalence of psoriasis has been seen to increase through childhood, the incidence increasing tenfold from 0.12% at age 1 to 1.2% at age 18, which is still considerably lower than the 2.5% incidence across the whole population [9].

12.3 Topical Products Used for Children

A large array of topical products and formulation types are available for children, including for example emollients, anti-inflammatories, anti-infectives and acne and psoriasis treatments. Common formulation types include semisolids such as creams, ointments and gels, more fluid formulations such as shampoos and lotions with other formulation types such as sprays, powders and foams also being available.

This wide range of products is perhaps unsurprising given the considerable burden of skin disease and the desirability of treating these conditions topically if possible. Additionally, the use of certain other products is particularly common in children, such as the use of local anaesthetics prior to minor skin procedures, including venepuncture or venous cannulation. However, few topical products are marketed exclusively for children with some products available for adults not being licensed for paediatric use. The ages of children for which different products are licensed vary considerably; in some cases no age range is specified, whilst in others a minimum age is quoted, with ages of 1, 2 and 12 being relatively common. Often new products are initially licensed to older age groups, with subsequent approval for younger patients following supporting clinical trials. For example, Eucrisa, which contains the phosphodiesterase 4 inhibitor, crisaborole, is used for atopic dermatitis. It was first approved in 2016 for patients aged 2 years and older and was subsequently lowered for patients 3 months and over in 2020. Enstilar, a foam used for psoriasis, was initially approved for adults (18 years and older) before being lowered to patients 12 years and older also in 2020. Opzelura, which contains the JAK inhibitor ruxolitinib, used for atopic dermatitis, was approved in 2021 and is marketed for patients 12 years and older, with clinical trials currently ongoing in patients between 2 and 12 years old [10]. Treatments for acne are usually marketed for 12 years and above, which reflects the need of the population, though a few products such as Amzeeq, which contains the topical antibiotic minocycline, are licensed for those over 9 years old [11].

12.4 Skin Structure

In designing topical/transdermal formulations it is important to have an awareness of skin anatomy and physiology, as this is useful for consideration of the site of drug action, skin condition, the delivery of a drug and the overall formulation development process. The skin presents a considerable barrier to the absorption of drugs, and it has been estimated that typically only a few per cent of a topically applied dose is bioavailable [12]. However, a variety of different factors affect the skin's permeability to drugs, and again appreciation of these can help inform the topical/transdermal formulation development process.

Figure 12.1 is a schematic cross-section of the skin showing the three macroscopic layers, the hypodermis or subcutaneous tissue, the dermis and the epidermis.

The subcutaneous tissue contains the main blood vessels to the skin and has thermal, mechanical support and energy storage functions. It consists of fat cells arranged in lobules connected with elastin and collagen. This layer also contains the main blood and lymphatic vessels and nerve fibres which link to the skin tissue. The next layer, the dermis, makes up the majority of the actual skin thickness and is approximately 2–5 mm thick. It contains connective tissue that provides elasticity and flexibility to the skin, with cells being sparsely distributed throughout. It also contains an extensive vasculature, which supplies the skin with required nutrients

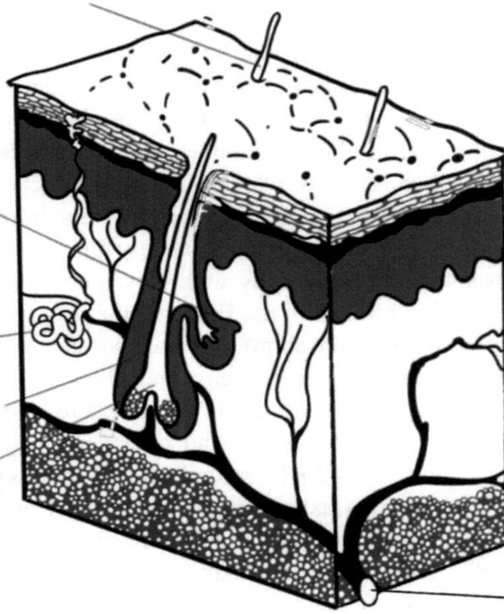

Fig. 12.1 Schematic illustration of a cross-section of the skin showing the three macroscopic layers, the subcutaneous tissue, the dermis and the epidermis

and removes waste products, and nerves and lymphatic vessels are also present. The blood vessels in the dermis have an additional important role in heat regulation and are the site of access of drugs applied to the skin into the systemic circulation. Skin appendages including hair follicles with associated sebaceous glands and eccrine and apocrine sweat glands originate in the dermis. The sebaceous gland secretes sebum, a waxy/oily substance, which forms a thin discontinuous layer over the skin surface and predominantly consists of a mixture of triglycerides, fatty acids, wax esters, squalene and cholesterol. Eccrine glands, present over most of the body, secrete sweat, a dilute salt solution, and primarily are involved in temperature regulation. In contrast apocrine sweat glands develop during puberty and are mainly localised to the axilla and anogenital regions. They produce a secretion that largely consists of proteins, lipoproteins and lipids. It is the secretion from the apocrine sweat glands that is associated with body odour, produced by the action of bacteria on the secretion.

The outermost of the three macroscopic layers is the epidermis. This layer is much thinner than the dermis, and although the actual thickness depends on body region, it is typically 80 μm thick. The layer lacks blood vessels and consists primarily of keratinocytes, though other cells are also present, including melanocytes, which produce melanin and protect the body from UV radiation, and Langerhans cells, part of the immune system. The epidermis is continually being renewed, with cells undergoing a complex differentiation process so that cells which are sloughed

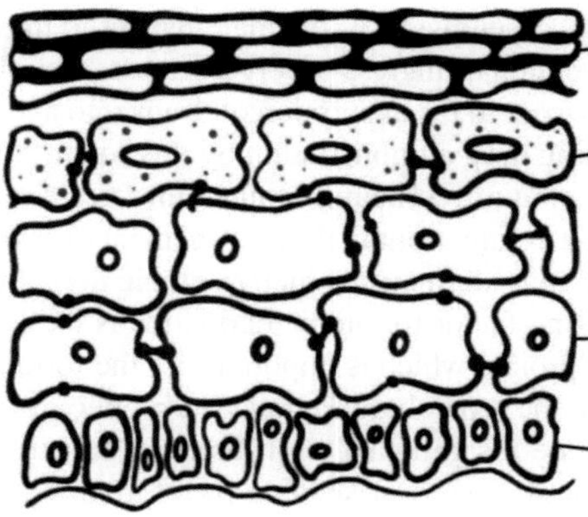

Fig. 12.2 Schematic representation of the strata of epidermis

off from the surface in a process known as desquamation are replaced by cells from below. Histological examination allows subdivision of the epidermis into four further layers, which represent stages in the differentiation of keratinocytes. A magnified schematic illustration showing these layers can be seen in Fig. 12.2.

The keratinocytes of the basal layer (stratum basale) are viable cells connected to the basement membrane and able to undergo cell division, with the newly produced cells moving further into the epidermis and towards the surface of the skin. As the cells migrate through the epidermis they differentiate; the next histological layer is known as the stratum spinosum and comprises two to six layers of keratinocytes. The cells change from being columnar in shape to polygonal and have a spiny appearance, which is a result of connections appearing between the cells known as desmosomes. These desmosomes connect the cells, providing structural integrity to the epidermis. The cells also start to produce keratin filaments. With further differentiation granules become visible in the cytoplasm of the next layer, the stratum granulosum. These granules contain precursors of the (intercellular) lipids, which exist between the cells in the outermost layer, the stratum corneum. Within the stratum granulosum the keratinocytes continue to produce keratin, their organelles break down, and the cells begin to flatten. The final layer, the stratum corneum, is an example of a stratified squamous epithelium. By the time that the keratinocytes have reached this layer, all of the cell organelles have degraded, and the keratinocytes consist primarily of keratin, held within a proteinaceous envelope. As they are fully keratinised or cornified at this stage, the cells are commonly known as corneocytes. The cells are flattened, for example, they may be 1–2 μm deep but 50 μm wide, although their dimensions depend to some extent on body region. The cells are imbedded in a structured intercellular lipid matrix, which was produced in the stratum granulosum and extruded into the intercellular spaces. The stratum corneum is relatively thin, typically 10–20 μm thick, though again this varies according to body region. A common analogy used to describe the stratum corneum is the "bricks and mortar" model, with the corneocytes being represented by bricks and the intercellular lipid by the mortar.

The stratum corneum acts as the main barrier to the loss of water from the body and to the prevention of ingress of foreign substances. Thus, the layer is also generally the main barrier to the absorption of drugs into and across skin and is important for understanding drug administration by this route. Additionally, the correct functioning of the stratum corneum, maintaining correct skin barrier properties, has been recognised as important in the management of several skin diseases, in particular for inflammatory dermatoses such as atopic dermatitis and psoriasis. For example a predisposing factor for atopic dermatitis is the loss of function gene variants for filaggrin, a protein which is important for the formation of the stratum corneum [13]. Moreover, the proteolytic degradation products of filaggrin, present in the stratum corneum, are a significant contributor to natural moisturising factor. Natural moisturising factor assists in keeping the stratum corneum hydrated and flexible, and decreased levels are typically found in patients with atopic dermatitis. Additionally changes in the stratum corneum intercellular lipid composition and packing, which are believed to be linked to the increased transepidermal water loss in the condition, have been observed in patients with atopic dermatitis [14].

12.5 Skin in the Paediatric Population

Development of the stratum corneum and skin barrier formation in the foetus starts at approximately 20 weeks gestation. The development initially occurs at specific sites, typically around hair follicles, which then increase in size to eventually cover the entire body [15]. The skin barrier function gradually improves with time with full maturation occurring a few weeks before birth (around week 34). The reduced skin barrier function of preterm infants has been exploited to deliver theophylline transdermally to preterm infants [16], although no products are marketed for this purpose. The skin barrier function from around weeks 30–32 has been shown to be similar to that of an adult using transepidermal water loss (TEWL) measurements, i.e. the determination of the amount of water evaporating across the skin [17] and drug delivery across full-term infant skin has been assessed as being similar to that of adult skin [18]. Nonetheless, more recent work has described continuing development of the skin barrier in childhood with reductions in TEWL, signifying improvements in the skin barrier along with changes in other skin parameters such as skin water holding capacity occurring between birth and 4 years old for children across different geographic locations and ethnic backgrounds [19]. These data indicate that the skin barrier of children is not identical to adults and may be influenced by body site and exposure to the environment. Differences in skin parameters, including stratum corneum thickness and corneocyte size, have been observed between children and adults and may contribute to these functional differences [20]. These differences in skin barrier formulation do not typically inform the development of topical formulations for children in terms of drug delivery but may do in relation to the potential likelihood of causing irritation, and it is accepted that the skin of children may have an increased irritant response in comparison to adults

[21]. Moreover, it is also accepted that paediatric patients may be at greater risk of systemic side effects following topical treatment. This is primarily because the surface area to body mass ratio of a child is considerable larger than that of the adult meaning that the systemic exposure of a child to a topically applied treatment can be considerably greater than that of an adult. Systemic side effects in children from topically applied medicines have been observed with particular treatments. Examples include adrenal suppression in children who have been treated with topical corticosteriods and neurotoxicity following the topical treatment of scabies [22, 23]. In practical terms to manage this risk with treatments such as topical corticosteroids, treatment is often advised to be limited to relatively short time periods with continuous therapy being avoided where possible.

12.6 Skin Factors Affecting Topical Absorption

In addition to the effect of age on the barrier function of skin, there is considerable variation in skin drug absorption across skin from different regions of the body. For example, higher drug absorption occurs from the face, neck or genitals, greater in comparison to the trunk, legs or arms [24]. The presence of hair follicles can have a significant effect, particularly at body sites where they are at high density, such as the scalp. Their presence may increase absorption, though the effect may vary with the particular type of drug characteristics and be larger for hydrophilic drugs [25]. Regional variation in skin permeability has been exploited with testosterone-containing transdermal patches having been designed for application to the scrotum and scopolamine-containing patches being applied behind the ear, both of these body sites showing relatively high drug absorption [24]. Typically, increased drug absorption is expected across diseased skin; examples include nappy rash as well as lesions in eczema or psoriasis [26]. This may be beneficial, enabling improved delivery of a particular drug and thus therapeutic treatment. However, care must also be taken, particularly with conditions such as nappy rash that can involve a relatively large area. The increased absorption because of the damaged skin barrier and the inherently lower skin barrier function of the region may lead to systemic side effects [27]. Potentially the increased drug absorption across diseased skin may make it possible to deliver drug molecules that are ordinarily very difficult to get across skin, for particular skin conditions [28]. The decreased barrier function of diseased skin also has implications for treatment as the skin's barrier function will recover with improvement in the condition, reducing drug absorption.

12.6.1 Drug Absorption Across the Skin

There are three main routes that drugs can traverse the stratum corneum, the main barrier to skin absorption. These can be visualised with reference to Figs. 12.1 and 12.2. The transappendageal route involves drugs being absorbed through hair follicles and other appendages, effectively bypassing the stratum corneum. Overall drug absorption via this route is believed to be low with the available surface area having been estimated to be approximately 0.1% of the total available surface area. Nonetheless, it is important when delivering drugs to the appendages, e.g. for acne treatment, and as mentioned previously it may be important for hydrophilic drugs and at particular body regions, such as the scalp. In contrast the transcellular route which involves drug diffusing into the stratum corneum's corneocytes, across the intercellular lipid spaces and back into deeper corneocytes continually until it reaches the stratum granulosum, has the largest available surface area and a relatively short diffusional path length. Although this suggests that this route would be the most likely for drug permeation, the multiple partitioning steps from cells to the intercellular lipid spaces and back into the corneocytes are thought to be unfavourable. Instead, it is the intercellular pathway, where the drug diffuses through the intercellular 'mortar', that is thought to be the most likely route for drug transport across the stratum corneum. This route only offers approximately 1% of the available surface area and, in contrast to the transcellular route, is much more tortuous; the increased diffusional path length contributes to the skin's barrier function.

12.7 Formulation Factors Affecting Drug Transport Across the Skin

Although the skin is an effective barrier to the absorption of many chemicals, the characteristics of individual permeants affect this with particular drug molecules showing a much greater propensity to cross it than others. Drugs that have a low molecular weight, that are lipophilic but also have some hydrophilic character, enabling absorption both into the relatively lipophilic stratum corneum and the deeper, hydrophilic skin layers, penetrate the skin best [29, 30]. Diffusion of drug molecules across skin can successfully be described through the application of Fick's laws of diffusion. The simplest case is that where there is a constant concentration gradient of drug across the skin, such as occurs when an infinite dose is applied to the membrane and sink conditions existing on the other side. A common form of Fick's first law can then be applied (Eq. 12.1).

$$J = \frac{KDC_{\text{app}}}{h} \tag{12.1}$$

where J is the rate of drug transport across the skin (flux), D is the diffusion coefficient of the drug in the skin, K is the partition coefficient of the drug from the formulation to the superficial layer of the stratum corneum, C_{app} is the applied concentration and h is the diffusional path length. The description of Fick's law in terms of concentration works well when a single vehicle is used, for example in the case of a simple aqueous drug solution, as the concentration of the drug is increased, a proportional increase in drug flux is expected. In this case the concentration of the drug in the vehicle is representative of its thermodynamic activity. A modified form of Eq. 12.1 incorporating thermodynamic activity is given below (Eq. 12.2), where α is the thermodynamic activity of the drug in the formulation and y is the thermodynamic activity of the drug in the stratum corneum.

$$J = \frac{\alpha D}{yh} \tag{12.2}$$

Equation 12.2 indicates that drug flux across the skin does not necessarily correlate with the concentration of the drug in the formulation, and instead it is the thermodynamic activity of the drug that is important. To envisage this it is perhaps easiest to consider the saturation of the drug in the formulation instead of its concentration. The solubility limit of the drug in a particular formulation is the concentration at which the formulation is fully saturated with the drug. The drug concentration at which this occurs varies depending on the particular formulation; however, regardless of the actual concentration, at the solubility limit, the drug is at its maximum (stable) thermodynamic activity which has a value of 1. Changes in formulation may alter the drug's solubility in the vehicle; thus the thermodynamic activity of the drug may be altered and delivery into the skin, even though the drug may be at the same concentration in the two different formulations [31]. This is an established issue, for example one study in the 1980s found that there was often no correlation between the drug concentration in particular topical corticosteroid preparations and skin absorption [32]. Currently there is a proprietary 0.1% hydrocortisone cream available in the UK which is clinically equivalent to 1% hydrocortisone cream BP [33] again providing evidence that skin penetration does not always correlate with drug concentration and that formulation is critical with regard to delivery of the drug across the skin. However, formulation factors other than the thermodynamic activity of the drug in the formulation are also important.

Formulations for application to the skin may include excipients that will improve drug absorption, known as penetration enhancers. Many different excipients can have this effect, and common examples of such molecules include ethanol, propylene glycol and isopropyl myristate. For example, the propylene glycol content of a formulation has been linked with the formulation's capability to transport aciclovir across the skin [34]. Ethanol has been shown to increase the concentrations of terbinafine in the stratum corneum [35] and is used in a terbinafine-containing product designed to treat athlete's foot with a single application. Other examples of excipients that can increase drug absorption include fatty acid derivatives such as isopropyl myristate, surfactants, fatty acids, terpenes and dimethylsulphoxide [36]. When more than one penetration-enhancing excipient is used in a formulation, synergy can occur between them, further increasing drug absorption [37, 38]. Penetration-enhancing excipients may be included in the formulations for reasons other than improving drug absorption; for example, they may be solubilisers, enabling the target drug concentration to be incorporated in the formulation, or surfactants that are being used to stabilise an emulsion. Nonetheless, their presence may increase drug absorption. Selection of penetration enhancers to include in a formulation depends upon the formulation type and the nature of the drug; however, as with all other excipients, consideration also needs to be given to the effect of the enhancer on the formulation's cosmetic acceptability, compatibility with other formulation ingredients as well as its potential to cause irritancy or toxicity.

Formulations that are occlusive may also increase skin drug absorption [39]. Through slowing the evaporation of water across the (reducing transepidermal water loss), occlusive formulations cause the stratum corneum to hydrate and swell, increasing drug permeability [40]. Ointment bases typically are occlusive, and for example topical corticosteroid ointments are often more potent than the corresponding cream formulation [41, 42]. Additionally, in infants, the nappy may act as an occlusive dressing and increase drug absorption in the treatment of nappy rash.

12.7.1 Formulation Type

A wide range of formulation types are available for application to the skin. Most commonly semisolid and liquid formulations are used for topical applications with transdermal patches being the most common formulation when a systemic effect is required. The three most commonly encountered semisolid topical formulations are ointments, creams and gels with a range of liquids also being available. Vehicle aesthetics are particularly important when the product is to be applied frequently over long periods of time, as poor aesthetic properties are believed to be linked to poor patient compliance and treatment outcomes. To a large extent these aesthetics will relate not only to patient preference but also to the condition and the specific site. For example, greasier formulations such as ointments may be preferred for dry scaly lesions with liquids or foams being preferred for hairy areas such as the scalp as they are much easier to apply. However, the recent Best Emollient for Eczema (BEE) trial found no difference between the effectiveness of ointment, cream, gel and lotion emollients, highlighting that the best choice would be the one that the patient is most likely to use [43]. It has been estimated that patient adherence to topical treatment regimes may be as low between 32% and 61% for a variety of skin conditions [44]. Commonly reported reasons for poor patient compliance include perceived lack of efficacy, staining of clothes, time required for application, interference with daily activities and fear of side-effects as well as patient preference for a particular product type [45]. Formulations typically go through significant physical changes once they are applied to the surface of the skin, affecting their aesthetic properties. For example rubbing a thixotropic formulation will reduce its viscosity, affecting its 'feel' and may also affect drug release from the formulation and skin absorption. Evaporation of any volatile solvent present will alter the nature of what resides on the skin surface and may reduce the solubility of the drug and result in drug precipitation, or potentially induce supersaturation, where the thermodynamic activity of the drug in increased above that of a saturated drug solution, improving drug delivery. However, the residue left on the skin surface may seem quite different to what was applied thus affecting patient acceptability.

Ointments principally consist of a mixture of wax/fat and oil into which a drug is incorporated. They can be very simple dosage forms, for example consisting of just the drug in a wax phase, and may be preferred for chronic dermatological disorders where dry, scaly skin presents, such as in atopic dermatitis and psoriasis. This is because they are occlusive which is beneficial for the treatment of dry skin and may also improve the delivery of the drug. Commonly hydrocarbons such as white soft paraffin or liquid paraffin are used to form the ointment base, though triglyceride derivatives and, more recently, silicones may also be used. Through altering the proportions of wax and oil the consistency of the product can be modified with an increased oil content making it looser and a higher wax content making it more viscous. This approach may be used to try to improve the aesthetic properties and ease with which the product can be applied. However, ointments are usually considered tacky and greasy upon application, regardless of how they are formulated. This may reduce cosmetic acceptability, reducing their use by patients and ultimately any therapeutic benefit. Although ointments may be desirable as a formulation type for a range of conditions, given their emollient properties, the development of an ointment formulation may be difficult, given the poor solubility of many drugs in ointment bases. For such situations drug solubility can be improved by including suitable solvents in which the drug has good solubility and are miscible with the base. Examples that are used include isopropyl myristate and propylene glycol. Additionally, surfactants have occasionally been added to ointments; this helps improve their water miscibility, making them easier to wash off. Whilst the vast majority of ointments are hydrophobic, hydrophilic ointment bases are also available which are formed from polyethylene glycols (PEGs). These are not occlusive but are useful for application to wounds; hydrophobic ointments tend not to be suitable for this task.

Creams are semisolid emulsions consisting of either an oil dispersed in water (known as an o/w emulsion) or water dispersed in oil (known as a w/o emulsion). Similar materials used for an ointment base can be utilised for the oil phase in the cream. Emulsions are thermodynamically unstable and require emulsifying agents to ensure that the product remains in a suitable physical form over its shelf life. Emulsifying agents are usually anionic or non-ionic surfactants or a mixture of surfactants, although polymeric emulsifiers may also be used, and they are the main determinant affecting the type of emulsion formed (o/w or w/o). The eventual use for the product is an important consideration when deciding on an emulsion type for a particular product. From a cosmetic acceptability point of view, o/w emulsions are generally considered less 'greasy' and thus have greater acceptability. However, w/o creams may be deemed preferable for drier skin conditions. In general o/w emulsions are much more common compared to w/o, which may be related to the greater availability of suitable emulsifying agents stabilising o/w emulsions in comparison to w/o emulsions as much as the improved aesthetic properties of such formulations. Similar to ointments, creams which may contain a high proportion of lipid provide an emollient effect hydrating the skin by restricting transepidermal water loss content and are often used for dry and inflammatory skin conditions such as in patients with dermatitis, psoriasis and eczema. Indeed, emollient products commonly used in the management of dry skin conditions are sometimes simply cream or ointment bases with the base of some proprietary topical corticosteroids also being commercially available as emollients.

Gels consist of a liquid phase to which a thickening (gelling) agent, usually a polysaccharide or acrylate polymer, is added. The liquid phase is often an aqueous- or alcohol-based cosolvent system, and the consistency of the overall product is modified through choice of the particular gelling agent and the concentration used. They are generally considered aesthetically pleasing and are often translucent or transparent. Humectants such as glycerol or propylene glycol may be included in the formulation to ensure that the polymer residue formed following application when the gel dries retains water and exists as a thin flexible film rather than as solid lumps of polymer. These humectants may also improve drug solubility in the vehicle, and additional solvents such as ethanol, isopropyl alcohol and PEG may also be included in the gel for this reason. Gels formed from an aqueous (or alcohol-based) solvent system typically lack the occlusive capabilities of ointments and creams and are usually used for treatments where this effect is not advantageous. They may also be suitable for application to wounds. In contrast emulsified gels are essentially biphasic systems containing an aqueous gel dispersed with a lipid phase and may be similar to creams. These are more occlusive than conventional gels, and the lipid phase facilitates the incorporation of hydrophobic drugs. Gels may also be formed from hydrophobic liquids, an example being Dovobet gel, which is used for the treatment of psoriasis.

In addition to semisolid topical formulations, there are also a number of liquid formulations. These may be solutions, suspensions, emulsions or liquids thickened with a gelling agent. They are, however, all characterised by a decreased viscosity, which allows them to flow freely. Much of the discussion above for topical semisolids also applies to liquids. The main advantage of their use is their ease of application to large areas of skin and to the scalp, where the presence of hair makes application of semisolids difficult. Another relatively recent formulation type which is used particularly for application to the scalp is medicated foam. Pharmaceutical foams are essentially a dispersion of gas in a liquid and are commonly generated by using a pressurised aerosol. Foaming agents, commonly surfactants, are needed to ensure the foam forms [46]. Foams typically leave a low quantity of residue on the skin following application and have been shown to increase drug bioavailability and patient acceptability over conventional formulations [47].

In addition to the drug and the main ingredients used to produce a particular formulation type, a number of other excipients may also be included in a formulation to perform a variety of functional roles. These roles may include improvement of drug solubility in the formulation to allow incorporation of the drug at a particular concentration, improvement of product aesthetics to increase patient compliance, improvement of drug and/or formulation stability and prevention of microbial growth and contamination. For many formulations, such as o/w emulsions and aqueous gels, water is the main drug solvent, and various water miscible solvents such as PEG, propylene glycol and alcohols (e.g. ethanol) can be included to improve drug solubility. Drugs in aqueous solutions may be susceptible to oxidative degradation, and therefore such products may require the addition of an antioxidant. The stability of drugs and the behaviour of certain excipients can be affected by pH requiring the addition of buffers to the formulation. For example, the stability of betamethasone valerate is pH dependent, and formulation at an inappropriate pH can cause loss of efficacy. Additionally, the control of pH may be important for particular excipients; carbomers which are common gelling agents have a particular pH range in which they can function effectively. Antimicrobial preservatives are usually included in formulations containing water to prevent contamination by microorganisms and spoilage of the product. In contrast, for non-aqueous systems such as ointments, preservatives are rarely required. The selected preservative(s) need to exhibit activity against a wide spectrum of microorganisms, and in addition, for multiphase formulations such as emulsions, care must be taken to ensure that there is a suitable concentration of the preservative in the appropriate phase. This is because the preservative will partition between the oil and the aqueous phases of the emulsion. Bacterial contamination typically occurs in the aqueous phase and at the oil/water interface. Thus, appropriate preservative concentrations need to be present in these regions. Other considerations that influence the actual concentration of preservative required include the presence of other excipients within the formulation that have antimicrobial activity. For example, ethanol which may be included in a formulation as a penetration enhancer is also a preservative. Examples of some preservatives commonly used in topical dosage forms include hydroxybenzoates, benzyl alcohol and phenoxyethanol.

12.7.2 Transdermal Systems

Transdermal treatments which provide a systemic effect are used much less frequently in the paediatric age group than topical treatments. Only a few transdermal products are licensed for children, but examples include scopolamine, fentanyl and methylphenidate patches with fentanyl being licensed for the treatment of pain for children 2 years and over, methylphenidate for the treatment of attention deficit hyperactivity disorder from 6 years and scopolamine for motion sickness for those 10 years and over. Recently a new transdermal patch for ADHD containing dextramphetamine, a drug which has been available for many years in oral form, was launched highlighting the usefulness of transdermal administration [48, 49]. Although other forms of transdermal formulations are available, such as ointments that are applied and secured in place with surgical tape, adhesive transdermal patches which are clearly defined single-unit systems are the most commonly used. The transdermal route offers particular advantages for the systemic administration of drugs. In particular, these include providing controlled, prolonged release of the drug, potentially over several days, so that the dosing frequency can be reduced, improving patient compliance. As a route of administration, it is easily accessible and is generally considered to have good patient acceptability. Moreover, for drugs such as fentanyl the avoidance of first-pass metabolism is useful therapeutically. The ability to control the release rate of a medication, prolonging the dosing interval, is a particular advantage for transdermal patches in children, as generally to achieve this with oral medications, either a relatively large tablet formulation or controlled release granules which are sprinkled on soft food such as yoghurt or apple sauce or in liquid such as water are used. Large tablets are typically difficult for children to swallow, and the controlled release granules must not be chewed before ingestion, something that may be difficult to ensure during administration. Transdermal patch removal can enable termination of drug delivery and enable dose duration modification for individual patients. A study that compared parent preferences for different methylphenidate formulations found that approximately one-third of parents would prefer a transdermal patch over oral treatments [50]. Similarly parents and clinicians are often satisfied with the performance and use of transdermal fentanyl patches [51]. However, for small children the ease of accessibility of transdermal patches can be problematic as they may be able to remove them, and they may be recommended to be applied to the upper back to prevent this. The ability of children to remove transdermal patches is a wider issue than simply the loss of therapeutic effect that would be expected with patch removal. As a result of the skin's very good barrier to drug permeation, drug absorption across the skin is slow, and patches often contain much more drug than they actually deliver. Should a child remove a patch and then ingest or suck the patch, they may then be exposed to a toxic concentration of the drug. Infant fatalities following ingestion of discarded, used fentanyl patches have been reported [52]. Other challenges with transdermal patches include sensitivity to adhesives, adhesives failing to adhere to the skin and delayed onset of action. For example, ADHD treatment patches typically need to be applied approximately 2 hours before the effects are needed, and scopolamine patches should be applied 5–6 hours before their effects are needed [48, 53].

There are two main transdermal patch designs, matrix and reservoir systems. Matrix systems can be further divided into two categories, the simplest drug in adhesive design consisting of a backing membrane, an adhesive layer into which the drug is incorporated and a release liner. To apply the patch the release liner is removed and the patch pressed onto the skin. The development of such matrix patches is not trivial; however, as the adhesive layer needs to perform two functions appropriately, attach the patch to the skin surface and deliver the drug into the body, and it may take a significant amount of development time to ensure that the adhesive formulation performs these two functions appropriately. However, once the formulation has been developed, manufacturing of these types of transdermal patches is relatively straightforward and inexpensive. A more complex matrix patch design may consist of two separate drugs in adhesive layers separated by a rate-limiting membrane. In contrast reservoir patches have the drug in a liquid or semisolid reservoir separated from the adhesive by a semipermeable membrane which controls the rate of drug delivery. This typically makes development of the particular formulation more straightforward in the first instance but increases the complexity of the manufacturing process, increasing costs. Additionally, the compartmentalisation of the drug increases the risk of problems occurring with the product, for example if the semi-permeable membrane separating the drug-containing reservoir from the adhesive fails, too much drug may be released from the patch causing toxicity.

There are some notable differences in the excipients used for transdermal patches in comparison to semisolid preparations, the main one being the pressure-sensitive adhesive which maintains the patch in contact with the skin. The main adhesives used are acrylic copolymers, silicone polymers or rubber. Acrylic co-polymers have the advantage of being relatively inexpensive, the properties of the adhesive can be easily customised by selection of the monomers used in the polymer and the degree of cross-linking of the polymer chains, and they tend to have good compatibility with drugs and other excipients. However, a level of residual monomer may remain in the adhesive, which may cause skin irritation. Silicone adhesives in contrast have low irritancy potential and the ability to deliver a relatively large proportion of the incorporated drug; however, they are expensive, and many drugs and excipients have a low solubility in them which may make formulation development more difficult. Rubber adhesives can be made from either synthetic or natural rubber with synthetic derivatives being more common, given the potential for natural rubber to cause allergic reactions. They are relatively inexpensive; however, similar to silicone adhesives they tend to have a low capability to dissolve drugs which can increase the required time for formulation development. In addition to the adhesive and drug, adhesive modifiers may be needed to adjust the properties of the adhesive such that it performs suitably. For example, ingredients such as insoluble polymers or metal oxides may be used to make the adhesive more viscous. Other excipients may be used to act as penetration enhancers to increase drug permeation across the skin as discussed for topical treatments previously, or to form the reservoir for reservoir patches, ethanol being a commonly used example.

Other important constituent parts of the transdermal patch include the backing film and release liner which may affect product performance. For example Daytrana, the methylphenidate patch, has had problems with the adhesive sticking to the release liner affecting patient use [54]. Backing films can affect drug delivery depending on the level of occlusion they provides with those that are too occlusive potentially causing skin maceration and irritation.

12.8 Skin Irritation

An important consideration in the design of a topical or transdermal dosage form is the potential for the drug and in particular, formulation excipients to produce skin irritation. This is perhaps of even greater importance when formulating for children as there is some evidence that they are at greater risk of suffering irritation [21]. A wide number of different types of molecules may induce irritation, which can occur when the irritant traverses the stratum corneum and interacts with keratinocytes. This can induce the keratinocytes to release inflammatory cytokines, inducing the infiltration of inflammatory cells and keratinocyte proliferation. Keratinocytes also play a pivotal role in allergic contact dermatitis by responding to allergens producing a range of pro-inflammatory cytokines which eventually induce the specific immune response. One common source of irritants in topical formulations is the surfactants used to stabilise emulsions. In particular ionic surfactants are capable of causing irritation [55]. Aqueous Cream BP, an emollient commonly available in the UK, was found to induce irritation in 56% of children with atopic dermatitis [56], and negative issues associated with the use of this cream have been ascribed to the sodium lauryl sulphate (an anionic surfactant) content of the formulation [57]. As a result of the irritancy potential of such surfactants, many formulations are now developed with non-ionic surfactants which have lower irritation potential, and polymeric emulsifiers such as carbomers, celluloses and polyacrylates are also being used to stabilise emulsions. In particular, the large molecular weight of polymeric emulsifiers restricts their ability to be absorbed across the stratum corneum, thus reducing their potential to cause irritation. However, surfactants are not the only excipients that can cause irritation, and many other excipients including penetration enhancers, preservatives, fragrances and adhesives also have the potential to do so. Topical products that are specifically marketed for children include sunscreens and emollients which typically are designed to reduce the chance of irritation. Sunscreens marketed for children are likely to have high levels of sun protection and are more likely to contain mineral rather than chemical sunscreens, which are less likely to cause irritation. Similarly, some emollient products for children lack the fragrance that is included in the equivalent adult product. However, in other cases topical products marketed separately for children and their adult equivalent have the same product licence number, indicating that the only difference between them relates to marketing/branding.

12.9 Conclusions

Topical treatments for localised skin conditions are commonly used in the paediatric group. They offer the advantage of providing high, localised skin concentrations which can optimise therapeutic treatment and minimise systemic side effects. Though the skin barrier to the absorption of chemicals in full-term infants is similar to that of adults, because of the differences in the surface area to body mass ratio, children are at a greater risk of systemic toxicity than adults. Transdermal products are less commonly used in children, but a few products are licensed for use. They offer controlled drug delivery, and some parents seem to prefer them over oral treatments. Overall selection of topical or transdermal formulations depends on the specific indication, the suitability of a particular formulation for a particular condition and its overall aesthetic properties. With the considerable skin barrier properties much effort has been made to investigate in trying to maximise skin absorption, but this must be balanced against considerations such as the avoidance of skin irritation to ensure optimal patient treatment.

References

1. Karimkhani C, Dellavalle RP, Coffeng LE, Flohr C, Hay RJ, Langan SM, et al. Global skin disease morbidity and mortality: an update from the global burden of disease study 2013. JAMA Dermatol. 2017;153(5):406–12.
2. Cork MJ, Danby SG, Ogg GS. Atopic dermatitis epidemiology and unmet need in the United Kingdom. J Dermatol Treat. 2020;31(8):801–9.
3. Ghodsi SZ, Orawa H, Zouboulis CC. Prevalence, severity, and severity risk factors of acne in high school pupils: a community-based study. J Invest Dermatol. 2009;129(9):2136–41.
4. Wolkenstein P, Machovcová A, Szepietowski JC, Tennstedt D, Veraldi S, Delarue A. Acne prevalence and associations with lifestyle: a cross-sectional online survey of adolescents/young adults in 7 European countries. J Eur Acad Dermatol Venereol. 2018;32(2):298–306.
5. Goulden V, Stables GI, Cunliffe WJ. Prevalence of facial acne in adults. J Am Acad Dermatol. 1999;41(4):577–80.
6. Virgili A, Corazza M, Califano A. Diaper dermatitis in an adult - a case of erythema papuloerosive of Sevestre and Jacquet. J Reprod Med. 1998;43(11):949–51.
7. Jordan WE, Lawson KD, Berg RW, Franxman JJ, Marrer AM. Diaper dermatitis: frequency and severity among a general infant population. Pediatr Dermatol. 1986;3(3):198–207.
8. Philipp R, Hughes A, Golding J. Getting to the bottom of nappy rash. Br J Gen Pract. 1997;47(421):493–7.
9. Augustin M, Glaeske G, Radtke MA, Christophers E, Reich K, Schaefer I. Epidemiology and comorbidity of psoriasis in children. Br J Dermatol. 2010;162(3):633–6.
10. Incyte corporation. Study of Ruxolitinib Cream in Children With Atopic Dermatitis 2023 [Available from: https://clinicaltrials.gov/ct2/show/NCT05034822?term=ruxolitinib&draw=2&rank=1.
11. Journey Medical Corporation. Amzeeq prescribing information 2022 [Available from: https://amzeeqhcp.com/assets/pdf/AMZEEQ-Prescribing-Information.pdf.
12. Wiedersberg S, Naik A, Leopold CS, Guy RH. Pharmacodynamics and dermatopharmacokinetics of betamethasone 17-valerate: assessment of topical bioavailability. Br J Dermatol. 2009;160(3):676–86.
13. O'Regan GM, Sandilands A, McLean WHI, Irvine AD. Filaggrin in atopic dermatitis (Reprinted from J Allergy Clin Immunol vol 122, pg 689–93, 2008). J Allergy Clin Immunol. 2009;124(3):R2–6.
14. Janssens M, van Smeden J, Gooris GS, Bras W, Portale G, Caspers PJ, et al. Increase in short-chain ceramides correlates with an altered lipid organization and decreased barrier function in atopic eczema patients. J Lipid Res. 2012;53(12):2755–66.
15. Hardman M, Byrne C. Skin structural development. In: Elias P, Feingold K, editors. Skin Barrier. New York: Taylor & Francis; 2006.
16. Cartwright RG, Cartlidge PHT, Rutter N, Melia CD, Davis SS. Transdermal delivery of theophylline to premature-infants using a hydrogel disk system. Br J Clin Pharmacol. 1990;29(5):533–9.
17. Kalia YN, Nonato LB, Lund CH, Guy RH. Development of skin barrier function in premature infants. J Invest Dermatol. 1998;111(2):320–6.
18. Delgado-Charro MB, Guy RH. Effective use of transdermal drug delivery in children. Adv Drug Deliv Rev. 2014;73:63–82.
19. Mack MC, Chu MR, Tierney NK, Ruvolo E Jr, Stamatas GN, Kollias N, et al. Water-holding and transport properties of skin stratum corneum of infants and toddlers are different from those of adults: studies in three geographical regions and four ethnic groups. Pediatr Dermatol. 2016;33(3):275–82.
20. Stamatas GN, Nikolovski J, Luedtke MA, Kollias N, Wiegand BC. Infant skin microstructure assessed in vivo differs from adult skin in organization and at the cellular level. Pediatr Dermatol. 2010;27(2):125–31.

21. Zhai HB, Meier-Davis SR, Cayme B, Shudo J, Maibach H. Irritant contact dermatitis: effect of age. Cutan Ocul Toxicol. 2012;31(2):138–43.
22. Coureau B, Bussieres J-F, Tremblay S. Cushing's syndrome induced by misuse of moderate- to high-potency topical corticosteroids. Ann Pharmacother. 2008;42(12):1903–7.
23. Singal A, Thami GP. Lindane neurotoxicity in childhood. Am J Ther. 2006;13(3):277–80.
24. Wester R, Maibach H. Regional variation in percutaneous absorption: principles and applications to human risk assessment. In: Bronaugh RL, Maibach HI, editors. Percutaneous absorption Drugs - Cosmetics - Mechanisms - Methodology. 4th ed. Boca Raton: Taylor & Francis Group; 2005. p. 85–93.
25. Ogiso T, Shiraki T, Okajima K, Tanino T, Iwaki M, Wada T. Transfollicular drug delivery: penetration of drugs through human scalp skin and comparison of penetration between scalp and abdominal skins in vitro. J Drug Target. 2002;10(5):369–78.
26. Ortiz PG, Hansen SH, Shah VP, Menne T, Benfeldt E. Impact of adult atopic dermatitis on topical drug penetration: assessment by cutaneous microdialysis and tape stripping. Acta Derm Venereol. 2009;89(1):33–8.
27. Ozon A, Cetinkaya S, Alikasifoglu A, Gonc EN, Sen Y, Kandemir N. Inappropriate use of potent topical glucocorticoids in infants. J Pediatr Endocrinol Metab. 2007;20(2):219–25.
28. Gould AR, Sharp PJ, Smith DR, Stegink AJ, Chase CJ, Kovacs JC, et al. Increased permeability of psoriatic skin to the protein, plasminogen activator inhibitor 2. Arch Dermatol Res. 2003;295(6):249–54.
29. Singh P, Roberts MS. Skin permeability and local tissue concentrations of nonsteroidal antiinflammatory drugs after topical application. J Pharmacol Exp Ther. 1994;268(1):144–51.
30. Magnusson BM, Pugh WJ, Roberts MS. Simple rules defining the potential of compounds for transdermal delivery or toxicity. Pharm Res. 2004;21(6):1047–54.
31. Reid ML, Jones SA, Brown MB. Transient drug supersaturation kinetics of beclomethasone dipropionate in rapidly drying films. Int J Pharm. 2009;371(1–2):114–9.
32. Stoughton RB, Wullich K. The same glucocorticoid in brand-name products - does increasing the concentration result in greater topical biologic activity. Arch Dermatol. 1989;125(11):1509–11.
33. Marlborough Pharmaceuticals Ltd. Dermacort hydrocortisone cream summary of product characteristics 2021 [Available from: https://www.medicines.org.uk/emc/product/5849/smpc.
34. Trottet L, Owen H, Holme P, Heylings J, Collin IP, Breen AP, et al. Are all aciclovir cream formulations bioequivalent? Int J Pharm. 2005;304(1–2):63–71.
35. Alberti I, Kalia YN, Naik A, Bonny JD, Guy RH. Effect of ethanol and isopropyl myristate on the availability of topical terbinafine in human stratum corneum, in vivo. Int J Pharm. 2001;219(1–2):11–9.
36. Williams AC, Barry BW. Penetration enhancers. Adv Drug Deliv Rev. 2004;56(5):603–18.
37. Lee PJ, Ahmad N, Langer R, Mitragotri S, Shastri VP. Evaluation of chemical enhancers in the transdermal delivery of lidocaine. Int J Pharm. 2006;308(1–2):33–9.
38. Sato K, Sugibayashi K, Morimoto Y. Effect and mode of action of aliphatic esters on the invitro skin permeation of nicorandil. Int J Pharm. 1988;43(1–2):31–40.
39. Hollingsbee DA, Fairbrother JE, Martin GP, Marriott C, Monger L. The effect of a hydrocolloid dermatological patch (actiderm) in potentiating the skin blanching activity of triamcinolone acetonide. Int J Pharm. 1991;77(2–3):199–209.
40. Pershing LK, Silver BS, Krueger GG, Shah VP, Skelley JP. Feasibility of measuring the bioavailability of topical betamethasone dipropionate in commercial formulations using drug content in skin and a skin blanching bioassay. Pharm Res. 1992;9(1):45–51.
41. Curdy C, Naik A, Kalia YN, Alberti I, Guy RH. Non-invasive assessment of the effect of formulation excipients on stratum corneum barrier function in vivo. Int J Pharm. 2004;271(1–2):251–6.
42. Topical treatments for psoriasis, including steroids: National Psoriasis Foundation; 2009 [Available from: http://www.psoriasis.org/document.doc?id=164.

43. Ridd MJ, Santer M, MacNeill SJ, Sanderson E, Wells S, Webb D, et al. Effectiveness and safety of lotion, cream, gel, and ointment emollients for childhood eczema: a pragmatic, randomised, phase 4, superiority trial. Lancet Child Adolesc Health. 2022;6(8):522–32.
44. Gupta G, Mallefet P, Kress DW, Sergeant A. Adherence to topical dermatological therapy: lessons from oral drug treatment. Br J Dermatol. 2009;161(2):221–7.
45. Ellis RM, Koch LH, McGuire E, Williams JV. Potential barriers to adherence in pediatric dermatology. Pediatr Dermatol. 2011;28(3):242–4.
46. Arzhavitina A, Steckel H. Foams for pharmaceutical and cosmetic application. Int J Pharm. 2010;394(1–2):1–17.
47. Andreassi L, Giannetti A, Milani M. Efficacy of betamethasone valerate mousse in comparison with standard therapies on scalp psoriasis: an open, multicentre, randomized, controlled, crossover study on 241 patients. Br J Dermatol. 2003;148(1):134–8.
48. Cutler AJ, Suzuki K, Starling B, Balakrishnan K, Komaroff M, Castelli M, et al. Efficacy and safety of dextroamphetamine transdermal system for the treatment of attention-deficit/hyperactivity disorder in children and adolescents: results from a pivotal phase 2 study. J Child Adolesc Psychopharmacol. 2022;32(2):89–97.
49. Noven Pharmaceuticals. Xelstrym prescribing information 2022 [Available from: https://www.xelstrym.com/wp-content/themes/xelstrym/pdfs/prescribing-information.pdf.
50. Lloyd A, Hodgkins P, Dewilde S, Sasane R, Falconer S, Barke ES. Methylphenidate delivery mechanisms for the treatment of children with attention deficit hyperactivity disorder: heterogeneity in parent preferences. Int J Technol Assess Health Care. 2011;27(3):215–23.
51. Zernikow B, Michel E, Anderson B. Transdermal fentanyl in childhood and adolescence: a comprehensive literature review. J Pain. 2007;8(3):187–207.
52. Teske J, Weller J-P, Larsch K, Troeger HD, Karst M. Fatal outcome in a child after ingestion of a transdermal fentanyl patch. Int J Legal Med. 2007;121(2):147–51.
53. Haleon UK Trading Ltd. Scopoderm summary of product characteristics 2022 [Available from: https://www.medicines.org.uk/emc/product/3276/smpc.
54. Wokovich AM, Shen M, Doub WH, Machado SG, Buhse LF. Evaluating elevated release liner adhesion of a transdermal drug delivery system (TDDS): a study of Daytrana (TM) methylphenidate transdermal system. Drug Dev Ind Pharm. 2011;37(10):1217–24.
55. Effendy I, Maibach HI. Detergent and skin irritation. Clin Dermatol. 1996;14(1):15–21.
56. Cork M, Timmins J, Holden C, Carr J, Berry V, Tazi-Ahnini R, et al. An audit of adverse drug reactions to aqueous cream in children with atopic eczema. Pharm J. 2003;271:747–8.
57. Tsang M, Guy RH. Effect of Aqueous Cream BP on human stratum corneum in vivo. Br J Dermatol. 2010;163(5):954–8.

Chapter 13
Parenteral

Fiona O'Brien and Kamelia Krysiak

Abstract Paediatric parenteral formulation development is a complex and multifactorial process. It requires careful consideration of the route of administration, choice of excipient and age-appropriate formulation alongside an understanding of the challenges associated with paediatric drug administration. These considerations and challenges vary depending on the physiological development of the child, access to the route, pharmacokinetics (absorption, distribution, metabolism, excretion) and toxicity characteristics of the drug and formulation. Formulation considerations in relation to parenteral administration to neonates are similar to those for paediatric patients but with additional concerns and considerations. Acceptability and pain felt by the patient, especially for chronic diseases where co-operation and adherence may become problematic, should also be taken into account. It should be remembered that the youngest children and neonates often cannot verbally communicate the pain experienced. The formulation development strategy also needs to take into consideration the global locations of the target population, often rural, resource-limited settings in developing countries. These present issues such as lack of cold storage, shortage of sterile diluents and administration equipment and increased risk of infections. This chapter addresses the major considerations which should be taken into account with respect to paediatric parenteral formulation development.

Keywords Parenteral · Paediatric · Neonate · Formulation development

F. O'Brien (✉) · K. Krysiak
School of Pharmacy and Biomolecular Sciences, Royal College of Surgeons in Ireland, Dublin, Ireland
e-mail: fionaobrien@rcsi.ie

H. Batchelor, K. Rose (eds.), *Pediatric Formulations*, AAPS Advances in the Pharmaceutical Sciences Series 58, https://doi.org/10.1007/978-3-031-77240-5_13

13.1 Introduction

The oral route of drug delivery is the most commonly used; however, the use of this route is not always practical. Circumstances such as dysphagia, inability to tolerate oral or enteral feeds, gastrointestinal disease, low or unpredictable enteral absorption, high first-pass metabolism, instability or degradation in the gastrointestinal tract may preclude or limit the effectiveness of drug delivery via the oral route. In such cases, medicines can be administered using the parenteral and more specifically the injectable route of administration [1, 2].

While the oral route of administration is most commonly used, the injectable route, including injections and infusions, is one of the main routes of administration in hospitals and emergency care. It is also the most widely used method of drug administration in very ill preterm and term neonates and in the critical care of children. It is used in patients who may be unconsciousness, have suffered trauma, where there is a requirement for rapid drug action or where no oral formulation exists. The injectable route of administration can also be used in the intermittent administration of drugs, for the continuous infusion of drugs, in the maintenance of blood levels via depot formulation and for the administration of maintenance nutrition. Despite its advantages, parenteral and injectable delivery is associated with several limitations, including higher formulation costs, specialised staff training and equipment and the requirement for asepsis [1, 3–5]. Many injectable drugs are included in the high-risk medications list by the Institute for Safe Medication Practices, and most of the drug categories listed in A-PINCH are used in the neonatal intensive care unit (NICU) and the paediatric intensive care unit (PICU) [6, 7].

The choice of which route to use, i.e. intravenous (IV), subcutaneous (SC) or intramuscular (IM), is dependent on clinical effect, circumstances and formulation considerations. In many clinical cases, parenteral drug delivery may be necessary; however, the challenges associated with this route for the paediatric population should be borne in mind and addressed during product development. Challenges associated with the administration of injectable products include that they usually have to be administered by professionally trained staff and can cause pain, and the anxiety associated with needle phobia in the paediatric population can be significant. There are also psychological and social conflicts experienced by children in integrating chronic parenteral medication regimes into their daily routines, which also have effects on parents and carers. Furthermore, in consideration of the development of a parenteral product for paediatric use, the effect of the choice of the route (e.g. SC or IM may not be suitable for premature neonates with low subcutaneous fat and fragile skin), formulation, volume, presentation and administration on several clinical and practice parameters must also be considered. These are summarised in Fig. 13.1 and discussed within this chapter.

Formulation considerations in relation to parenteral administration to neonates are similar to those for paediatric patients but with additional concerns. Injectable preparations are the most commonly used products in the treatment of clinically unstable term and preterm neonates.

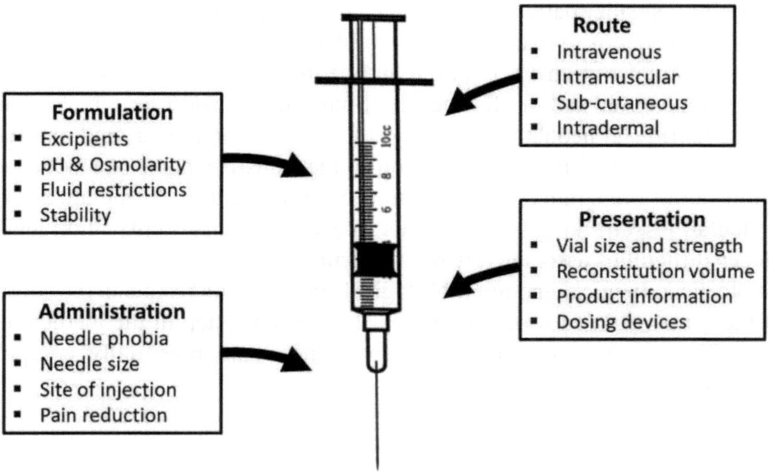

Fig. 13.1 Summary of considerations in developing paediatric parenteral formulations

13.2 Route

When selecting a parenteral route, it is important to consider the physiological development of the child, access to the route, pharmacokinetics (absorption, distribution, metabolism, excretion) and toxicity characteristics of the drug and formulation. The acceptability and pain felt by the patient, especially for chronic diseases where co-operation and adherence may become problematic, should also be taken into account. It should be remembered that the youngest children and neonates often cannot verbally communicate the pain experienced. Published literature shows that neonates in intensive care undergo many painful or stressful procedures, with an average reported number of 14- 16 procedures a day, including the administration of parenteral medications [8, 9].

Depending on the clinical condition of the patients, the target population may already have venous access for drug administration via an indwelling peripheral or central venous cannula. This can serve to reduce the pain and fear associated with multiple injections or short infusions (e.g. antibiotics are often administered via short infusions of up to 30 min). Thus, the intravenous route is more commonly used than the intramuscular or subcutaneous route in hospitalised patients. Other routes, e.g. intra-osseous, may be used as an alternative in emergency situations where venous access cannot be established. Depending on the therapeutic category and indication of the drug in question, the suitability of the use of such routes should be determined and information provided in the Summary of Product Characteristics (SPC).

An added complication in this patient cohort is the fact that many drugs used in paediatric and neonatal populations are used off-label [10, 11]. This issue arises from the exclusion of children from clinical trials because of perceived ethical considerations, smaller patient cohorts and the complexity of obtaining consent [12, 13]. The modifications of existing drug formulations to provide an appropriate dose or route of administration for children and neonates pose formulation challenges and lead to unlicensed and/or off-label use. Some authors addressed the formulation gap in paediatric and neonatal medications [14]. For example, around one-sixth of medications prescribed in the NICU require the transformation of a solid dosage form into liquid preparation [15]. Dilution of an adult drug formulation requires dose calculations, checking for excipients'safety and compatibility of diluents, and introduces difficulties associated with measuring low drug volumes. Moreover, the lack of physico-chemical and microbiological stability data results in the short shelf life of these manipulated formulations. Each additional step in the medication use process offers extra opportunities for error and could potentially compromise patient's outcomes.

13.2.1 Intravenous (IV)

Venous access is gained by insertion of small cannulae into peripheral veins or by larger catheters into central veins or by semi-permanent central venous catheters with subcutaneous reservoirs.

Administration via a peripheral vein cannula is simpler, less invasive and easier to manage compared to administration via a central vein, but cannula blockages are more common. Peripheral vein administration is used for a short period of time, and only single lumen lines are available. The slower blood flow in peripheral veins means formulation characteristics such as hyperosmolarity, extremes of pH, and irritant drugs and excipients can lead to phlebitis and extravasation, which may lead to loss of the vein access, tissue damage and the development of phlebitis. Also, catheter-related infection and sepsis may further complicate the clinical picture. The growth of opportunistic bacteria, e.g. *Escherichia Coli* and *Pseudomonas aeruginosa*, at the catheter site could lead to loss of access through that line. Insertion of a catheter into a central vein requires skilled staff, general anaesthesia, requires an X-ray confirmation of the catheter position, has associated morbidity, has a higher infection risk and is expensive to insert. However, a central line can be kept in place on a long-term basis, and the faster blood flow allows for the administration of hyperosmolar and irritant formulations (e.g. some standard concentration infusions, concentrated electrolytes and cytotoxics) as they are rapidly diluted in the circulation. This route is also valuable for fluid-restricted patients where concentrated formulations, usually diluted for peripheral vein delivery, can be given in low volumes as a more concentrated solution. Catheters with multiple lumens allow for the simultaneous administration of more than one drug where drug-drug incompatibility may preclude co-administration.

The IV route usually involves frequent infusions requiring preparation under sterile conditions, which is resource intensive and can delay treatment. The infection risk is also higher with the IV route compared to other routes. In relation to paediatrics, intravenous access in young children may be challenging. For example, peripheral venous access can be very difficult due to smaller veins in children. Neonates and infants may be especially prone to tissue necrosis related to extravasation because of their inability to communicate pain and because of the limited extravascular tissue space on the scalp and the dorsum of the hand and foot [16].

In the neonatal patient, limited IV access can make the delivery of multiple IV infusions particularly problematic. This problem is often acute in the neonatal patient where a lack of compatibility information necessitates the use of dedicated IV lines. Those dedicated lines are often not feasible due to limited vascular access in the smallest patients. Hence, add-on devices and multi-lumen extensions, like stopcocks or octopus, are used. However, these add-on devices introduce additional dead volume to the infusion setup and may compromise timely drug delivery into patient's circulation.

Chemical incompatibilities can lead to drug degradation, drug-drug interactions, or drug precipitation, potentially resulting in the loss of therapeutic drug effect or fatal or life-threatening embolism. Loss of IV access due to precipitation of incompatible IV infusions can lead to blockage of small vessels and has the potential for serious adverse outcomes in the neonatal population. A well-known example in neonatal clinical practice is a life-threatening interaction between ceftriaxone and calcium products [17]. Studies examining the compatibility of the medicinal product when co-administered with other products, particularly parenteral nutrition, and diluents used commonly in paediatric and neonatal patients are lacking [18].

Some hospital units have implemented standard concentration infusions (SCI) into their treatment regimens for critically ill in-patients. A standard concentration infusion is an IV infusion at a standardised concentration for a selected group of patients, which remains unchanged for all patients in that weight category. In contrast, the infusion rate is altered according to the patient's body weight to produce an appropriate dose of medication. The Institute for Safe Medication Practices and Vermont Oxford Network have collaboratively published a list of 'Standard Concentrations of Neonatal Drug Infusions', which gives an overview of recommended neonatal SCIs [19]. The aim is to reduce the burden of individual patient-based drug preparation and decrease the medication error rate [20–22]. However, preparation of SCI poses similar challenges as preparation of regular infusions, e.g. they may require reconstitution and/or dilution of a very low volume of an adult formulation to produce an appropriate drug concentration. Consequently, the administration of potent, highly concentrated IV infusions may cause significant adverse reactions even with relatively small administration inaccuracy. The use of SCI poses a requirement for the use of modern infusion pumps, often with built-in libraries.

The IV route can also result in metabolic acidosis, a direct effect of hydrogen ions generated from TPN. Normally, the excess hydrogens are cleared in patients with normal organ function but can be fatal in those with compromised renal or respiratory systems [23].

13.2.2 Intramuscular (IM)

Intramuscular (IM) injections are administered into muscles around the body, usually the deltoid (upper arm), quadriceps (thigh) or gluteal (buttock) muscles, although the gluteal region is not recommended until the age of 2 years [24]. Drugs administered intramuscularly enter the systemic circulation either by diffusion through the interstitial fluids and fenestrations of the vascular capillaries or by convective flow of the interstitial fluids through the lymphatic system [23]. There may be inter-individual and age-related differences in drug absorption in the paediatric population following intramuscular delivery due to a higher variation in muscle mass, depth of muscle and fat layers and muscle blood flow. The limited muscle mass in preterm and term neonates limits the use of IM injections [25]. Absorption of drugs following IM injection in children is affected by the muscle volume, muscle blood-perfusion, and the contraction of the muscle. Neonates and infants typically have decreased blood flow to muscle, smaller muscle volume, larger proportion of water and feeble muscle contraction compared to those of older children and adults resulting in reduced absorption [26]. Muscle mass increases with age: a premature neonate has <20% muscle mass (of the body composition); this increases to 30% in term neonates; 40% by 6 months of age; 50% by 12 months; and ~55% in adults [24].

Intramuscular injections are painful, and the selection of the right needle size is important. The volume administered will affect the pain felt and is usually restricted to 1–3 mL in neonates and paediatric patients and as low as 0.5 mL in premature neonates [27]. Adverse effects of the IM route commonly include persistent pain which may affect mobility, erythema and haematoma and rarely include muscle contracture, nerve damage, abscess formation, bleeding, tissue necrosis, cellulitis and gangrene [28, 29], and thus this route is avoided in children wherever possible. The complications of muscle contracture and nerve damage are also influenced by the appropriateness of the injection site, needle size and angle of injection for the age and size of the child and the nature of the preparation.

Another consideration, when there is oral and parenteral drug formulation available, is the compliance of a child's carer. One of the commonly administered medications in postnatal wards is phytomenadione (synthetic vitamin K_1) which is indicated for the prophylaxis and treatment of vitamin K deficiency bleeding in neonates. There is a particular formulation available that can be given both orally and IM, where 'in the mixed micelle solution, vitamin K_1 is solubilised by means of a physiological colloidal system consisting of lecithin and a bile acid' [30]. The oral dose has the availability of around 50% with a wide range of inter-individual variability, and two to three repeated doses are needed at day 7 and week 4 of life. However, only one IM dose is sufficient. The systemic availability following oral dosing is approximately 50%, with a wide range of inter-individual variability. It may be beneficial to a child to receive one IM dose given by a healthcare professional; however, this is subject to parents' consent.

IM formulations usually differ from IV formulations and are not interchangeable. Extra vigilance is required to avoid medication error with SALAD (sound-alike, look-alike) medications for IM and IV use; otherwise the consequences can be detrimental. The known example is fatal IV injection of benzathine benzylpenicillin in neonates [31]. Benzathine benzylpenicillin is to be given only IM for treatment of congenital syphilis in infants. Benzylpenicillin is a broad-spectrum antibiotic used as IV infusion or injection for empiric treatment of suspected neonatal sepsis and treatment of infection with susceptible organisms. Also, the correct needle size needs to be used for administration of benzathine benzylpenicillin IM injection to reduce the risk of patient not receiving the full dose due to needle blockage with a high concentration of suspended drug [32].

13.2.3 Subcutaneous (SC)

Subcutaneous (SC) injections are administered into the adipose layer of the skin. The needle length determines whether the contents are deposited into the subcutaneous tissue; too short and this results in deposition in the dermis, and too long and this will enter the intramuscular layer. In children the thickness of the skin increases significantly with age and by body site (from arms (with the lowest values) to thighs to abdomen to buttocks (with the highest values)) [33]. Measurements of skin plus subcutaneous layers showed a similar trend in terms of increasing thickness with age, although there was greater variability in the measurements [33]. Key factors that affect the bioavailability from an SC injection include the composition of interstitial fluid (e.g. ionic content, pH and charge), the architecture of the SC tissue (e.g. collagen and elastin content), and/or proteins and cell types/density present—and these components may well vary between ages and within regions over time. There is little information on the differences in these parameters between children and adults.

Within the subcutaneous layer the major barrier to absorption is connective tissue as drugs need to travel through this tissue, following the path of least resistance before reaching the systemic circulation. The subcutaneous matrix of connective tissue is bathed in interstitial fluid (ISF); limited information exists on differences in ISF between adults and children. The colloid osmotic pressure of ISF has been shown to be similar in children compared to adults [34]. Subcutaneous blood flow (SBF) will affect subcutaneous absorption where an increase in blood flow will increase absorption. The site of administration can affect SBF with greater flow in the abdomen and arm compared to the thigh and buttocks and also in a supine compared to a sitting position [35].

Repeated injections into the same area can induce lipodystrophic changes in the subcutaneous tissue, resulting in more fibrous tissue with a poorer blood supply; the consequence of which is reduced and more variable absorption. It has been acknowledged that despite the anatomical similarities between porcine models and human skin, there is no reliable preclinical model to predict pharmacokinetics of SC administered protein therapeutics in humans [36]. A key difference that is relevant for extrapolation into paediatric populations includes the volume administered; the volume needs to consider the pressure within the SC so that it does not cause mechanical damage; thus, the relative size needs to be carefully considered.

The SC route is limited to formulations that are non-irritating to the tissue and do not cause necrosis and sloughing at the injection site. The volume administered needs to be small to avoid pain. Whilst the volume in adults should be ≤2 mL [37], for children it is usually restricted to ≤1 mL. Side effects associated with the subcutaneous route include large local oedema, erythema, pain in the genital area and ecchymosis. There also may be a smaller risk of blood vessel puncture, necrosis and infection [37].

13.2.4 Intradermal (ID)

The intradermal route is usually used to elicit an immune response, for example to test for previous tuberculosis infection through ID injection of Tuberculin Purified Protein Derivative. Depot corticosteroid preparations are sometimes given by this route. The distinctive immunological properties of the skin make the epidermis and dermis attractive sites for prophylactic vaccination, and ID vaccines have been considered safe, immunogenic and feasible alternatives to IM and SC routes that warrant further investigation [38]. There has also been a renewed interest in dose-sparing strategies to mitigate future vaccine shortages following the recent H1N1 influenza virus pandemic and previous shortages. The ID route may offer similar or better immunogenicity compared to SC and IM routes, thus enabling the use of reduced doses. In terms of safety, sparse influenza vaccine studies in children have reported that the ID route resulted in a higher incidence of febrile reactions compared to the SC route and a greater frequency of local reactions (erythema) compared to the IM route. Nevertheless, ID vaccination might be more immunogenic in children than in adults, and recent developments of new injection devices mean that the future of ID vaccination is promising [38]. Accurate ID injection in terms of depth and volume injected using conventional needles is particularly challenging and painful in neonates, infants and toddlers, and the recent advances in delivery devices that alleviate these issues should be utilised if considering this route.

Reactions reported specific to the ID route include hyperpigmentation, erythema, swelling, induration and pain when compared to the IM and SC routes. While a wheal formation is required to confirm positive ID injection in adults, skin thickness is observed in the neonatal setting. A shorter needle is recommended for the ID route, preferably less than 1.2 mm for infants and children [37].

13.2.5 Intraspinal

Intraspinal injections of local anaesthetics via lumbar puncture are used effectively in neonates through to adolescents to provide anaesthesia, analgesia and sympathetic and motor block, mainly for surgery to the lower part of the body and also for cardiac surgery due to the effectiveness of this route in suppressing cardiovascular and stress responses to surgery. The flexibility of the spine, the short distance between skin and spinal space and a wide lumbar spinal canal make dural puncture easier in children than adults [39]. However, for babies weighing <5 kg epidural administration via an indwelling catheter is preferred. The advantages over general anaesthesia are that respiratory support is not usually required unless the patient is sedated, spinal block can be given on a fed stomach, and there is a lower incidence of emesis. It is also a useful route in resource-limited settings as it is safe, simple and relatively inexpensive compared to general anaesthesia. The drawback of single-injection spinal anaesthesia is the short duration of action and related inter-individual variability, limiting its use to procedures no longer than 75 min. When an extended action is required, adjuvant drugs are combined with spinal anaesthetics to modify the onset, intensity, and duration of spinal block. Adverse effects of using this route are similar to those in adults and include a position-dependent headache that worsens in the upright position and improves on lying down and transient neurological symptoms such as pain in the gluteal region and tingling in the feet [39]. In contrast to adults, cardiovascular complications are uncommon in children even at higher doses [39].

13.3 Formulation

13.3.1 Excipients

Some excipients used in parenteral formulations may be unsuitable for younger children as their metabolic pathways are still developing, particularly in neonates and infants. The safety profile of excipients in target paediatric populations should be considered during formulation development and selection, recognising that in clinical practice, off-label use is prevalent in children and that a formulation strategy that moves towards minimising risks in unlicensed age groups as well is preferred by the clinical community. The lowest feasible quantities of excipients should be used. There is a general lack of paediatric safety data for excipients since such testing is usually conducted in adults and therefore the maximum safe exposure levels in younger children for many excipients still need to be determined. The daily intake of some excipients may exceed the accepted safe limits per kilogram of body weight in adults. Consideration should be given to the daily and cumulative exposure and accumulation. Juvenile animal studies can be useful to identify mechanisms and adverse outcomes that can be monitored in clinical studies and post-marketing surveillance. It may be feasible to assess the acute toxicity of excipients currently in paediatric use through marketed products (licensed and unlicensed uses), by monitoring effects on biochemical parameters from existing exposures.

Examples that highlight differences in paediatric handling of excipients include benzyl alcohol, used as preservative in injections, which is detoxified by a saturable conjugation pathway in the liver. The detoxification pathway is immature in neonates and can lead to the potentially fatal accumulation of benzoic acid. Current recommendations are to exclude benzyl alcohol in products for neonates and to avoid its use in children up to 3 years of age unless carefully evaluated. Benzyl alcohol may also cause pain on injection and hypersensitivity. Benzoic acid, sodium benzoate and potassium benzoate when used in parenteral dosage forms may increase the risk of jaundice in neonates [40]. Benzyl alcohol may oxidise and form benzaldehyde dibenzyl acetal, an unwanted precipitant affecting the stability and solubility profiles of the product [41]. Isomerisation and polymerisation can occur, resulting in reactions with a primary amino group forming a new chemical [40].

Propylene glycol (PG) is used in injectable products of poorly water-soluble drugs and multivitamin concentrates. Children below 4 years of age have a lower alcohol dehydrogenase activity to metabolise PG, leading to accumulation, with a half-life of 16.9 h in neonates compared to 5 h in adults. Therefore, high levels of PG should not be administered to children below 4 years of age. Adverse effects include central nervous system depression, seizures, lactic acidosis, hyperosmolality (causing laxative effects), renal-, hepatic- and ototoxicity and cardiovascular effects. A reduction of the dose by 50% should be considered in neonates with underlying risk factors [42]. Recent work has shown that a median exposure of 34 mg/kg/day for 48 h does not affect maturational changes in renal, hepatic and metabolic function in preterm neonates [43].

The antioxidant sodium metabisulphite contributes to the formation of neurotoxic cysteine in neonates and can have hypersensitivity reactions such as bronchospasm and anaphylaxis.

Polysorbate 80, used as an emulsifier in pharmaceutical preparations, has been linked to toxicities and deaths in premature infants administered an intravenous Vitamin E preparation in the United States in 1984 [44, 45]. PS20 and PS80 have been associated with syndrome characterised by thrombocytopenia, pulmonary deterioration, ascites, and liver and renal failure, secondary to vasculopathic hepatotoxicity [46]. A Polysorbate 80 dose >72 mg/kg/day was associated with adverse effects in these patients.

It has been reported that the neonates cared for in a NICU are exposed to aluminium [47–50], bisphenol A [51–53], parabens [51, 52, 54, 55] and phthalates, including di(2-Ethylhexyl) phthalate (DEHP) and its metabolites [55–61]. These xenobiotics may originate from a manufacturing process or leach from a medical device into infusion solution. Release rate depends and may be greater under certain conditions, such as higher temperature, exposure to light, elevated pH or longer contact time, an infusion rate and a lipophilic character of an infusion solution, storage time and percentage of DEHP in plastic material, among others [55, 56, 62, 63]. IV infusion in a NICU will likely happen under these conditions. Loff et al. (2002)

showed that an increase in temperature by 6 °C (from 27 to 33 °C) elevated the extraction rate of DEHP by almost 30% [56]. However, the dynamic of leaching is not fully understood and may depend on the material of a device and the nature of the chemical compound [47, 56, 64].

Daily exposure in a NICU can easily exceed the levels defined as safe by national and international committees. It is concerning that manufacturers do not always provide sufficient information on the content of raw materials in medical disposables and devices [51]. Moreover, there is an indication that excipients can add up the total amount of xenobiotics available to the patient [47, 65, 66] and that the manipulation of a drug formulation can increase the total level of contamination [47]. Premature neonates are at particularly significant risk of toxicity due to the immaturity of their body organs and higher exposure to medicines and medical devices in early life. Prematurity of the liver, kidneys and enzymatic pathways disables effective metabolism and excretion of xenobiotics. Moreover, when xenobiotics are introduced through an IV infusion, they directly reach the bloodstream facing no physiological barriers. Human exposure to some xenobiotics was associated with neurotoxicity [48], hepatic toxicity, metabolism and hormonal disruption [67–69].

Additionally, some infusion solutions may promote leaching from medical devices and infusion lines they are in contact with, e.g. higher DEHP extraction rate with lipid components of TPN or calcium gluconate increasing aluminium extraction. The extraction rate depends on contact time and temperature, among other factors.

The suitability of excipients for the different parenteral routes in terms of irritancy and local tissue damage must be assessed. For the IM route, mixed cosolvent systems containing propylene glycol and ethanol have an additive effect, whilst those containing polyethylene glycol 400 may have a protective effect on the myotoxicity generated by IM injections [70]. Despite propylene glycol being mainly active, with high concentrations, cases of lactic acidosis, convulsions and bradycardia have been reported.

Liposomes prepared from phospholipids are highly biocompatible with skeletal muscle and are able to reduce the interaction between the encapsulated drug and the surrounding tissue, thus reducing the degree of myotoxicity following IM injection [71].

The choice of diluent may reduce both short-term and long-term discomfort associated with IM injection, and the use of lidocaine instead of sterile water as a diluent may be of benefit [72]. Furthermore, buffering to a higher pH (approximately 7.4) increases the amount of pharmacologically active lidocaine and may result in more effective anaesthesia [73].

For the intraspinal route, the close proximity of subarachnoidally injected drugs with neural tissue requires consideration of the potential neurotoxicity of drugs and formulations to be injected into the cerebrospinal fluid. Formulations should not contain antioxidants and preservatives that have a potential for neurotoxicity.

13.3.2 pH and Osmolarity

Irritant intravenous formulations can cause phlebitis, constriction and occlusion of smaller fragile peripheral veins, especially in neonates and infants. The resultant high backpressure can lead to extravasation (infiltration) of the medication into the subcutaneous tissue and leakage at the point of cannula insertion into the vein. Extremes of pH and non-iso-osmolar formulations contribute to irritancy, may cause more tissue damage on extravasation and cause pain on injection for most parenteral routes. Therefore, iso-osmolar formulations at physiological pH are preferred where stability allows. It is advised to use newer solutions of TPN, because they can function as a potential base and lower the incidence of metabolic acidosis [74].

Buffer characteristics and pH also influence muscle damage. In vitro studies have indicated that myotoxicity of buffers containing carboxylic acid groups (acetate, succinate and citrate) is directly affected by the pH of the solution and can be minimised by formulating at low buffer capacity and near physiological pH [75]. Furthermore, a significant reduction in pain upon IM administration of MMR vaccines has been associated with moving from an acidic to a more physiological pH (7.2–7.6) of the formulation [72].

For SC injections, citrate buffers are associated with local pain [76], and it has been shown that when non-physiological pH must be used for stability reasons, the lowest possible buffer concentration should be used to minimise pain and may enable more rapid normalisation of the pH at the site of injection [77]. Citrate supplementation also lowers the incidence of nephrocalcinosis in neonates and is recommended as preventative therapy [78].

Hyper-osmolar formulations are those with an osmolarity higher than that of plasma (>290 mosmol/L), which leads to a reversible dehydration and crenation of blood cells. Hypo-osmolar injections can cause blood cells to swell and burst, and this haemolysis can be dangerous if a large number of cells are affected. Therefore, excipient concentrations are important. Sodium chloride and glucose are appropriate tonicity-adjusting agents that can be used to attain an iso-osmolar formulation. Alternatively, dilution should be carried out prior to administration, but fluid restrictions need to be considered. Hyper-osmolar injections may be suitable for administration without further dilution via a faster flowing central vein as previously mentioned. Infusions should be of neutral pH and iso-osmolar unless intended to be co-infused with other solutions. Clear warnings must be given about the need for dilution or co-infusion.

13.3.3 Fluid Restrictions

Intravenous drug therapy has a major impact on the daily fluid allowance in children, which depends on the child's age and weight. For example, neonates are usually restricted to 150 mL per day (including nutritional intake). The need for numerous medications that require large dilution volumes and flushing into the circulation can significantly deplete the daily fluid and electrolyte allowance, such that both drug and nutritional therapy are compromised. Approximate daily fluid and electrolyte requirements for paediatric patients based on body weight can be calculated from Table 13.1.

Manufacturers should take into account fluid allowances during development of intravenous injections, and recommendations for dilution and flushing of intravenous drugs should be aimed towards the lowest manageable volumes whilst addressing issues of osmolarity, pH and chemical irritancy.

Daily fluid allowance in neonates restricts the volume of fluid, which can be administered. This allowance must also take into account the administration of essential fluid nutrition. Management of allowed volume of fluid must take also take into account birth weight, gestational age and corrected age and may also be influenced by disease pathophysiology or treatment (e.g. phototherapy) which may affect fluid and electrolyte requirements in the unwell infant. Fluid requirements/allowances for newborns of less than 3 kg start at about 40–60 mL/kg per 24 h, rising as the baby's body weight increases [79]. Residual volumes in giving sets and intravenous lines may be significant for neonates, and thus special low volume medical devices should be considered for this age group. The flexibility to be able to administer more concentrated preparations provides significant advantages and can be clinically crucial when fluid intake is restricted.

13.3.4 Stability

For products used in adults, additional in-use chemical and physical stability data of reconstituted and/or further diluted products to be used in paediatric settings should be generated, since preparation and administration may differ. For example,

Table 13.1 Fluid requirements per 24 h [79]

Body weight < 3 kg	150 mL/kg (start at 40–60 mL/kg if newborn)
3–10 kg For each kg between 10 and 20 kg For each kg over 20 kg	100 mL/kg Add 50 mL/kg Add 20 mL/kg to maximum of 2000 mL in adult female and 2500 mL in adult male
Sodium requirement	3 mmol/kg
Potassium requirement	2 mmol/kg
Glucose requirement	2.4–4.8 g/kg

preferred dilution fluids may differ, and for products aimed at neonates, the environmental conditions within the neonatal unit that affect drug stability should be considered, e.g. temperature, humidity, and UV light [80]. Compatibility of IV formulations with typical diluents, syringes, tubing and infusion bags also needs to be considered. The preferred infusion fluids for paediatric patients are glucose 5% and 10%, sodium chloride 0.45% and 0.9%, and combinations of glucose and saline.

13.4 Presentation

Where parenteral products are marketed in inappropriate strengths or dose-volumes for use in children, the requirement for dose calculation, measurement of very small volumes, part-usage of vials and multiple dilutions all increase the risk of medication errors [81]. Incorrect rate of intravenous administration, incorrect dose and incorrect administration technique have all been reported with parenteral delivery to paediatric patients [82]. Simple, unambiguous instructions on dose preparation and administration are required to reduce the risk of errors [83]. Special care needs to be taken when administering TPN with hepatic impairment, renal impairment, pregnancy, breastfeeding, and with cases of COVID-19 [84], and it is important to balance the need for dilution for osmolarity and dilution for dose measurement with daily fluid allowances.

13.4.1 Reconstitution and Displacement Volumes

Injectable drugs presented as lyophilised powders need to be reconstituted with an appropriate diluent. Whilst for adults, displacement volumes are usually regarded as negligible, they need to be taken into account in paediatric patients. Thus, the displacement volume should be clearly stated, and the volume of diluent to be added should result in a final drug concentration that allows simple calculation of proportional doses where some patients will require a dose volume less than the total volume after reconstitution.

Marketing a range of vial sizes (volumes) and concentrations to cater for the different age and weight ranges to be treated alleviates the issues of dose measurability, dose errors and wastage. The need to perform serial dilutions to obtain a measurable dose volume should be avoided. However, it may not be economically viable to produce several variations of a drug product, and it may in fact contribute to an increased risk of medication error through confusion during prescribing and selection of the wrong strength during dispensing or administration. Hence, a rational limited number of vial sizes and concentrations should be made available. In certain circumstances where a small volume vial is made available for neonates or infants, the use of multiple vials to draw a dose for single injection for elder age groups may be deemed acceptable. It is thought that a dull needle, caused by penetration by the stopper, may cause pain. In a study, it was found that no bruising or systemic reaction occurred when using the one or two needle technique when administering the DPT vaccine in children [85].

Where small volumes need to be measured, the size of the syringe that allows accurate measurement should be specified. Volumes below 0.1 mL are difficult to measure, may increase the risk of error and should be avoided [86]. In these instances, a paediatric cartridge rather than a traditional vial could be considered for use with automatic dosing devices similar to insulin injector pens to deliver accurate low doses (see Sect. 13.5). Compatibility with infusion fluids should be stated. Paediatric-specific administration information should be given about dilution volumes and rates of infusion or injection.

13.5 Administration

Anecdotal evidence suggests infants as young as 5 months will react to the sight of an injection if they have had one before. Negative early experiences may lead to persistent challenges of engagement with healthcare. If children struggle, there is a risk of injury to themselves and/or their carers. In addition, the impact of hidden parental distress should be taken into consideration as needle procedures are stressful events for parents during their child's treatment. In severe cases of non-compliance, there may be a need for play specialists or restraints. Other safety concerns include the risk of needle-stick injuries during preparation, cross-contamination and safe disposal of sharps. Evidence suggests breastfeeding and sucking on sucrose solution during the injection reduces pain and discomfort in newborns. Additional interventions include skin-to-skin care, application of pressure, and lidocaine and prilocaine topical agents [72, 87–89]. It has been reported that being held upright or held is more comforting than a supine position [90].

The appropriateness of the chosen needle dimensions for the intended route of administration should be assessed based on paediatric physiology, as they are a factor in the incidence of pain and local reactions. Increasing the needle length for IM injections may reduce pain and adverse effects since the needle is more likely to penetrate the muscle mass [72].

A better immune response for IM compared to SC has been seen with several vaccines, and both the injection technique and the needle length are crucial for ensuring the proper IM delivery and thus are directly related to vaccine safety and immunogenicity [91]. It is generally agreed that the anterior thigh (vastus lateralis) should be used for intramuscular immunisation injections until the age of 18–36 months due to the relatively large muscle mass and lack of vital structures, followed by the upper arm (deltoid) area for older children [72].

To reduce the risk of postdural puncture headache following intraspinal injection, a small diameter atraumatic needle with a stylet should be used. 27-gauge spinal needles with lengths of 25–38 mm for infants, 50–90 mm for small children and school-aged children are used [38]. Post puncture complications can be difficult to assess in younger children due to the inability of infants to verbally communicate their complaints, leading to misinterpretation of physical and behavioural changes by clinicians.

Several pharmacological and non-pharmacological techniques have been proposed for the reduction of pain associated with IM injections in children, including topical anaesthetic creams and patches, vapocoolant spray, oral sucrose solutions, oral tactile stimulation and parental holding.

The use of paediatric syringe pumps and microbore tubing allows the delivery of precise and regulated infusion rates of medications that can be prepared in 1–60 mL syringes and removes the need for multiple injections. However, this hinders patient mobility, and drug adsorption to syringes and tubes should be determined which may require administration sets to be primed before use [83].

Various indwelling catheters have been developed to aid compliance in patient groups that have needle phobia or experience painful injections. They are placed subcutaneously and remain in place for an average of 3–5 days. Such systems have been used for administering doses as low as 0.5 IU of insulin via syringes or pen devices [92]. Where spinal anaesthetic delivery is difficult in neonates <5 kg body weight, epidural delivery via an indwelling catheter technique allows the titration of the optimal dose, and can be used for postoperative pain management in this vulnerable group [39].

Compared to the vial and syringe method of drug delivery, pen devices such as those used for insulin therapy have improved patient acceptability and adherence to chronic injection regimes. Although more expensive, these devices have also reduced the pain felt and allow more accurate dosing, especially of low volume doses. Despite its safety, it was observed that the monthly expenditure was higher in the insulin pen users compared to the insulin syringe users group [93]. Whilst advances in pen design and function have been made recently, the choice of pens targeted at children is still limited and further development is required [94].

Needle-free liquid and powder jet injectors have been developed to address needle phobia and pain. They also remove the hazards associated with handling and disposing of needles. These devices operate by using compressed gas or a spring mechanism which is used to eject a jet of liquid or powder under pressure from the device onto the skin; the formulation then penetrates to the subcutaneous or muscle layer.

However, liquid jet technology has shown variable adverse reactions (soreness, redness and swelling) and patient acceptability, which may be due to the limited flexibility in their settings [95]. Recently, pulsed microjets that limit the penetration depth of the liquid jets into skin and thus potentially minimise these effects have shown effective delivery of insulin to rats [96], and development of such devices may improve acceptability for children. Liquid jet injectors have been used to deliver a range of vaccines, proteins such as insulin, growth hormone, erythropoietin and interferon, ampicillin, lidocaine, midazolam, steroids and bleomycin. These jet injectors are claimed to be amenable to parenteral formulations intended for needle-based injection. However, efficacy and safety criteria need to be met. Factors affecting drug penetration that require further investigation are mechanical properties of the skin, injection volume and the distance between the injector orifice and skin when the device is actuated (stand-off distance). Determination of the size and shape of the jet-induced hole in skin, development of predictive models that require an understanding of fluid dynamics of the skin, skin failure mechanisms and fluid dispersion into tissue is also needed. Importantly, these factors need to be investigated in relation to age. The stability of drugs in jets needs to be established as shear forces are higher compared to needle-based injections. A clear understanding of the pain caused by this administration method and local reactions is needed and whether they are drug, formulation or device specific [97].

Powder jet injectors provide the advantage of ease of storage and improved stability compared to liquid formulations. These injectors deliver drugs in dry powder form into the superficial layers of skin. As some particles are retained in the stratum corneum, impact velocity, particle size and particle density become important design parameters in determining the depth of penetration into the skin layers [95, 98]. In addition, increasing relative humidity and temperature has been shown to increase penetration depth [99]. However, the final particle location can be affected by inter-individual differences in skin layer thickness [98]. Reports have suggested that pain-free delivery can be achieved, but mild erythema, hyper-pigmentation, flaking and discolouration at the injection site following administration of dry powder DNA vaccines to adults have been recorded, although most reactions resolved within 1 month [95]. It is unknown whether repeated administration would result in persistent formulation or device-related adverse effects. If injection site reactions are related to the excipients within a liquid formulation and not the drug itself, they may be reduced through reformulating to a powder for jet injection. The authors are not aware of any products currently in advanced development using this technology. The disadvantages associated with the development of jet injectors are the cost of the technology and the noise on activation of the devices. Furthermore, strict specifications for the gas pressure and nozzle geometry of the device and for the particle size, shape, morphology and density may pose technical challenges.

13.6 Conclusions

Parenteral product development for children requires a matrix of overlapping considerations involving the rationalisation of choice of route, use of non-toxic excipients, appropriate formulation and presentation coupled to the practicalities and challenges associated with drug administration in paediatric practice. Administration devices and methods therefore play a significant role in the acceptability of parenteral products. The development strategy also needs to take into consideration the global locations of the target population, often rural, resource-limited settings in developing countries. These present issues such as lack of cold storage, shortage of sterile diluents and administration equipment and increased risk of infections.

References

1. Gulati N, Gupta H. Parenteral drug delivery: a review. Recent Pat Drug Deliv Formul. 2011;5(2):133–45.
2. Dornblaser E, Way S, Brazeau G. Dosage forms: parenterals. In: Swarbrick J, editor. Encyclopedia of pharmaceutical science and technology. CRC Press; 2013. p. 884–97.
3. Stoner K, Harder H, Fallowfield L, Jenkins V. Intravenous versus subcutaneous drug administration. Which do patients prefer? A systematic review. Patient. 2014;8(2):145–53.
4. American Cancer Society. Tubes, lines, ports, and catheters used in cancer treatment. 2023. Available from: https://www.cancer.org/treatment/treatments-and-side-effects/planning-managing/tubes-lines-ports-catheters.html.
5. Neonatal and Paediatric Pharmacists Group. A survey of current opinion on the suitability of the Medusa injectable medicines guide for paediatrics and alternative resources used by NPPG members. Medusa. Injectable medicines guide.
6. Krzyżaniak N, Pawłowska I, Bajorek B. Review of drug utilization patterns in NICUs worldwide. J Clin Pharm Ther. 2016;41(6):612–20.
7. Kaushal R, Bates D, Landrigan C, McKenna K, Clapp M, Federico F, et al. Medication errors and adverse drug events in pediatric inpatients. JAMA. 2001;285(16):2114–20.
8. Simons SH, van Dijk M, Anand KS, Roofthooft D, van Lingen RA, Tibboel D. Do we still hurt newborn babies? A prospective study of procedural pain and analgesia in neonates. Arch Pediatr Adolesc Med. 2003;157(11):1058–64.
9. Carbajal R, Rousset A, Danan C, Coquery S, Nolent P, Ducrocq S, et al. Epidemiology and treatment of painful procedures in neonates in intensive care units. JAMA. 2008;300(1):60–70.
10. Moulis F, Durrieu G, Lapeyre-Mestre M. Off-label and unlicensed drug use in children population. Therapie. 2018;73(2):135–49.
11. Conroy S, McIntyre J. The use of unlicensed and off-label medicines in the neonate. Semin Fetal Neonatal Med. 2005;10(2):115–22.
12. Ku L, Smith P. Dosing in neonates: special considerations in physiology and trial design. Pediatr Res. 2015;77(1–1):2–9.
13. Shirkey H. Therapeutic orphans. J Pediatr. 1968;72(1):119–20.
14. O'Brien F, Clapham D, Krysiak K, Batchelor H, Field P, Caivano G, et al. Making medicines baby size: the challenges in bridging the formulation gap in neonatal medicine. Int J Mol Sci. 2019;20(11)
15. Costa H, Costa T, Martins R, Oliveira A. Use of off-label and unlicensed medicines in neonatal intensive care. PLoS One. 2018;13(9):e0204427.
16. Phelps S, Helms R. Risk factors affecting infiltration of peripheral venous lines in infants. J Pediatr. 1987;111(3):384–9.
17. Bradley J, Wassel R, Lee L, Nambiar S. Intravenous ceftriaxone and calcium in the neonate: assessing the risk for cardiopulmonary adverse events. Pediatrics. 2009;123(4):e609–13.
18. Farhan M, McCallion N, Bennett J, Cram A, O'Brien F. Stability and compatibility of parenteral nutrition solutions; a review of influencing factors. Eur J Pharm Biopharm. 2023;187:87–95.
19. Institute for Safe Medication Practices, Vermont Oxford Network. Neonatal Drug Concentrations. 2022 [December 2023]. Available from: https://public.vtoxford.org/neonatal-drug-concentrations/.
20. The Joint Commission. Sentinel Event Alert Preventing pediatric medication errors. 2021(39).
21. Christie-Taylor S, Tait P. Implementation of standard concentration medication infusions for preterm infants. Infant. 2012;8(5):155–9.
22. Arenas-López S, Stanley IM, Tunstell P, Aguado-Lorenzo V, Philip J, Perkins J, et al. Safe implementation of standard concentration infusions in paediatric intensive care. J Pharm Pharmacol. 2017;69(5):529–36.
23. Kushner R. Total parenteral nutrition-associated metabolic acidosis. J Parenter Enteral Nutr. 1986;10(3):306–10.

24. Tom-Revzon C. Erratic absorption of intramuscular antimicrobial delivery in infants and children. Expert Opin Drug Metab Toxicol. 2007;3(5):733–40.
25. Kristensen HG. WHO guideline development of paediatric medicines: points to consider in pharmaceutical development. Int J Pharm. 2012;435(2):134–5.
26. Lim SY, Pettit RS. Pharmacokinetic considerations in pediatric pharmacotherapy. Am J Health Syst Pharm. 2019;76(19):1472–80.
27. Department of Pediatric Newborn Medicine, Brigham and Women's Hospital. Intramuscular and Subcutaneous Injections in Neonates and Infants. Available from: https://www.brighamandwomens.org/assets/BWH/pediatric-newborn-medicine/pdfs/intramuscular-subcutanteous-injection.pdf.
28. Bergeson P, Singer S, Kaplan A. Intramuscular injections in children. Pediatrics. 1982;70(6):944–8.
29. deWit C. Fundamental concepts and skills for nursing. 1st ed. Saunders; 2001.
30. CHEPLAPHARM Arzneimittel GmbH. Summary of product characteristics, Konakion MM Paediatric Ampoules 2 mg/0.2 ml oral solution or solution for injection. Health Products Regulatory Authority.
31. Institute for Safe Medication Practices. Lesson from Denver: look beyond blaming individuals for errors. 1998.
32. The Society of Hospital Pharmacists of Australia. Australian injectable drugs handbook. Benzathine benzylpenicillin. 9th ed. The Society of Hospital Pharmacists of Australia; 2023.
33. Lo Presti D, Ingegnosi C, Strauss K. Skin and subcutaneous thickness at injecting sites in children with diabetes: ultrasound findings and recommendations for giving injection. Pediatr Diabetes. 2012;13(7):525–33.
34. Guthe HJ, Indrebø M, Nedrebø T, Norgård G, Wiig H, Berg A. Interstitial fluid colloid osmotic pressure in healthy children. PLoS One. 2015;10(4):e0122779.
35. Gradel AKJ, Porsgaard T, Lykkesfeldt J, Seested T, Gram-Nielsen S, Kristensen NR, et al. Factors affecting the absorption of subcutaneously administered insulin: effect on variability. J Diabetes Res. 2018;2018:1205121.
36. McDonald TA, Zepeda ML, Tomlinson MJ, Bee WH, Ivens IA. Subcutaneous administration of biotherapeutics: current experience in animal models. Curr Opin Mol Ther. 2010;12(4):461–70.
37. Ansel H, Allen L, Popovich N. Pharmaceutical dosage forms and drug delivery systems. 12th ed. Lippincott Williams & Wilkins; 2021.
38. Vankerckhoven V, Van Damme P. Clinical studies assessing immunogenicity and safety of intradermally administered influenza vaccines. Expert Opin Drug Deliv. 2010;7(9):1109–25.
39. Kokki H. Spinal blocks. Paediatr Anaesth. 2012;22(1):56–64.
40. European Medicines Agency, Committee for Medicinal Products for Human Use (CHMP). Reflection paper: formulations of choice for the paediatric population. 2006.
41. Rayaprolu B, Strawser J, Anyarambhatla G. Excipients in parenteral formulations: selection considerations and effective utilization with small molecules and biologics. Drug Dev Ind Pharm. 2018;44(10):1565–71.
42. Lim TY, Poole RL, Pageler NM. Propylene glycol toxicity in children. J Pediatr Pharmacol Ther. 2014;19(4):277–82. https://doi.org/10.5863/1551-6776-19.4.277.
43. Allegaert K, Vanhaesebrouck S, Kulo A, Cosaert K, Verbesselt R, Debeer A, et al. Prospective assessment of short-term propylene glycol tolerance in neonates. Arch Dis Child. 2010;95(12):1054–8.
44. Alade S, Brown R, Paquet A Jr. Polysorbate 80 and E-Ferol toxicity. Pediatrics. 1986;77(4):593–7.
45. Martone W, Williams W, Mortensen M, Gaynes R, White J, Lorch V, et al. Illness with fatalities in premature infants: association with an intravenous vitamin E preparation, E-Ferol. Pediatrics. 1986;78(4):591–600.
46. Kriegel C, Festag M, Kishore RSK, Roethlisberger D, Schmitt G. Pediatric safety of polysorbates in drug formulations. Children (Basel). 2019;7(1):1. https://doi.org/10.3390/children7010001. PMID: 31877624; PMCID: PMC7022221.

47. de Oliveira S, Bohrer D, Garcia S, do Nascimento P, Noremberg S. Aluminum content in intravenous solutions for administration to neonates: role of product preparation and administration methods. J Parenter Enter Nutr. 2010;34(3):322–8.
48. Bishop N, Morley R, Day J, Lucas A. Aluminum neurotoxicity in preterm infants receiving intravenous-feeding solutions. N Engl J Med. 1997;336(22):1557–61.
49. Lima-Rogel V, Romano-Moreno S, de Jesús López-López E, de Jesús Escalante-Padrón F, Hurtado-Torres GF. Aluminum contamination in parenteral nutrition admixtures for low-birth-weight preterm infants in Mexico. J Parenter Enter Nutr. 2016;40(7):1014–20.
50. Bohrer D, Oliveira S, Garcia S, Nascimento P, Carvalho L. Aluminum loading in preterm neonates revisited. J Pediatr Gastroenterol Nutr. 2010;51(2):237–41.
51. Iribarne-Durán L, Artacho-Cordón F, Peña-Caballero M, Molina-Molina JM, Jiménez-Díaz I, Vela-Soria F, et al. Presence of Bisphenol A and parabens in a neonatal intensive care unit: an exploratory study of potential sources of exposure. Environ Health Perspect. 2019;127(11):117004.
52. Calafat A, Weuve J, Ye X, Jia L, Hu H, Ringer S, et al. Exposure to bisphenol A and other phenols in neonatal intensive care unit premature infants. Environ Health Perspect. 2009;117(4):639–44.
53. Duty S, Mendonca K, Hauser R, Calafat A, Ye X, Meeker J, et al. Potential sources of bisphenol A in the neonatal intensive care unit. Pediatrics. 2013;131(3):483–9.
54. Mulla H, Yakkundi S, McElnay J, Lutsar I, Metsvaht T, Varendi H, et al. An observational study of blood concentrations and kinetics of methyl- and propyl-parabens in neonates. Pharm Res. 2015;32(3):1084–93.
55. Calafat A, Needham L, Silva M, Lambert G. Exposure to di-(2-ethylhexyl) phthalate among premature neonates in a neonatal intensive care unit. Pediatrics. 2004;113(5):e429–34.
56. Loff S, Kabs F, Subotic U, Schaible T, Reinecke F, Langbein M. Kinetics of diethylhexyl-phthalate extraction from polyvinylchloride-infusion lines. J Parenter Enter Nutr. 2002;26(5):305–9.
57. Stroustrup A, Bragg J, Busgang S, Andra S, Curtin P, Spear E, et al. Sources of clinically significant neonatal intensive care unit phthalate exposure. J Expo Sci Environ Epidemiol. 2020;30(1):137–48.
58. Weuve J, Sánchez B, Calafat A, Schettler T, Green R, Hu H, et al. Exposure to phthalates in neonatal intensive care unit infants: urinary concentrations of monoesters and oxidative metabolites. Environ Health Perspect. 2006;114(9):1424–31.
59. Su P, Chang Y, Chang H, Wang S, Haung H, Huang P, et al. Exposure to di(2-ethylhexyl) phthalate in premature neonates in a neonatal intensive care unit in Taiwan. Pediatr Crit Care Med. 2012;13(6):671–7.
60. Green R, Hauser R, Calafat A, Weuve J, Schettler T, Ringer S, et al. Use of di(2-ethylhexyl) phthalate-containing medical products and urinary levels of mono(2-ethylhexyl) phthalate in neonatal intensive care unit infants. Environ Health Perspect. 2005;113(9):1222–5.
61. von Rettberg H, Hannman T, Subotic U, Brade J, Schaible T, Waag K, et al. Use of di(2-ethylhexyl)phthalate-containing infusion systems increases the risk for cholestasis. Pediatrics. 2009;124(2):710–6.
62. Geens T, Goeyens L, Covaci A. Are potential sources for human exposure to bisphenol-A overlooked? Int J Hyg Environ Health. 2011;214(5):339–47.
63. Mercea P. Physicochemical processes involved in migration of Bisphenol A from polycarbonate. J Appl Polym Sci. 2009;112(2):579–93.
64. Marcel Y. Determination of di-2-ethylhexyl phthalate levels in human blood plasma and cryoprecipitates. Environ Health Perspect. 1973;3:119–21.
65. Bohrer D, Bertagnolli D, de Oliveira S, do Nascimento P, de Carvalho L, Pomblum S. Drugs as a hidden source of aluminium for chronic renal patients. Nephrol Dial Transplant. 2007;22(2):605–11.
66. Bohrer D, do Nascimento PC, Binotto R, Becker E, Pomblum S. Contribution of the raw material to the aluminum contamination in parenterals. J Parenter Enter Nutr. 2002;26(6):382–8.

67. Darbre P, Harvey P. Paraben esters: review of recent studies of endocrine toxicity, absorption, esterase and human exposure, and discussion of potential human health risks. J Appl Toxicol. 2008;28(5):561–78.
68. Giulivo M, Lopez de Alda M, Capri E, Barceló D. Human exposure to endocrine disrupting compounds: their role in reproductive systems, metabolic syndrome and breast cancer. A review. Environ Res. 2016;151:251–64.
69. Boberg J, Taxvig C, Christiansen S, Hass U. Possible endocrine disrupting effects of parabens and their metabolites. Reprod Toxicol. 2010;30(2):301–12.
70. Brazeau G, Fung H. Use of an in vitro model for the assessment of muscle damage from intramuscular injections: in vitro-in vivo correlation and predictability with mixed solvent systems. Pharm Res. 1989;6(9):766–71.
71. Lima E, Oliveira A. Tissue tolerance of diclofenac sodium encapsulated in liposomes after intramuscular administration. Drug Dev Ind Pharm. 2002;28(6):673–80.
72. Schechter N, Zempsky W, Cohen L, McGrath P, McMurtry C, Bright N. Pain reduction during pediatric immunizations: evidence-based review and recommendations. Pediatrics. 2007;119(5):e1184–98.
73. Hayward C, Nafziger A, Kohlhepp S, Bertino JS Jr. Investigation of bioequivalence and tolerability of intramuscular ceftriaxone injections by using 1% lidocaine, buffered lidocaine, and sterile water diluents. Antimicrob Agents Chemother. 1996;40(2):485–7.
74. Dounousi E, Zikou X, Koulouras V, Katopodis K. Metabolic acidosis during parenteral nutrition: pathophysiological mechanisms. Indian J Crit Care Med. 2015;19(5):270–4.
75. Napaporn J, Thomas M, Svetic K, Shahrokh Z, Brazeau G. Assessment of the myotoxicity of pharmaceutical buffers using an in vitro muscle model: effect of pH, capacity, tonicity, and buffer type. Pharm Dev Technol. 2000;5(1):123–30.
76. Frenken L, van Lier H, Jordans J, Leunissen K, van Leusen R, Verstappen V, et al. Identification of the component part in an epoetin alfa preparation that causes pain after subcutaneous injection. Am J Kidney Dis. 1993;22(4):553–6.
77. Fransson J, Espander-Jansson A. Local tolerance of subcutaneous injections. J Pharm Pharmacol. 1996;48(10):1012–5.
78. Schell-Feith EA, Moerdijk A, van Zwieten PH, Zonderland HM, Holscher HC, Kist-van Holthe J, van der Heijden BJ. Does citrate prevent nephrocalcinosis in preterm neonates? Pediatr Nephrol. 2006;21(12):1830–6. https://doi.org/10.1007/s00467-006-0274-4. Epub 2006 Oct 13
79. https://www.rcpch.ac.uk/resources/standardising-intravenous-infusion-concentrations-children-uk. Last accessed 28 Feb 24.
80. European Medicines Agency, Committee for medicinal producs for human use and paediatric committee. Guideline on the investigation of medicinal products in the term and preterm neonate. 2007.
81. Beaney AM. Preparation of parenteral medicines in clinical areas: how can the risks be managed – a UK perspective? J Clin Nurs. 2010;19(11–12):1569–77.
82. Ghaleb MA, Barber N, Franklin BD, Wong IC. The incidence and nature of prescribing and medication administration errors in paediatric inpatients. Arch Dis Child. 2010;95(2):113–8.
83. Walsh J, Bickmann D, Breitkreutz J, Chariot-Goulet M. Delivery devices for the administration of paediatric formulations: overview of current practice, challenges and recent developments. Int J Pharm. 2011;415(1–2):221–31.
84. Hamdan M, Puckett Y. Total parenteral nutrition. StatPearls. Treasure Island (FL): StatPearls Publishing. Copyright © 2024, StatPearls Publishing LLC 2024.
85. Lamblet L, Meira E, Torres S, Ferreira B, Martucchi S. Randomized clinical trial to assess pain and bruising in medicines administered by means of subcutaneous and intramuscular needle injection: is it necessary to have needles changed? Rev Lat Am Enfermagem. 2011;19(5):1063–71.
86. Uppal N, Yasseen B, Seto W, Parshuram CS. Drug formulations that require less than 0.1 mL of stock solution to prepare doses for infants and children. Can Med Assoc J. 2011;183(4):E246–8.

87. Dilli D, Küçük IG, Dallar Y. Interventions to reduce pain during vaccination in infancy. J Pediatr. 2009;154(3):385–90.
88. Kashaninia Z, Sajedi F, Rahgozar M, Noghabi FA. The effect of Kangaroo care on behavioral responses to pain of an intramuscular injection in neonates. J Spec Pediatr Nurs. 2008;13(4):275–80.
89. Taddio A, Ilersich A, Ipp M, Kikuta A, Shah V. Physical interventions and injection techniques for reducing injection pain during routine childhood immunizations: systematic review of randomized controlled trials and quasi-randomized controlled trials. Clin Ther. 2009;31(Suppl 2):S48–76.
90. Taddio A, Appleton M, Bortolussi R, Chambers C, Dubey V, Halperin S, et al. Reducing the pain of childhood vaccination: an evidence-based clinical practice guideline. Can Med Assoc J. 2010;182(18):E843–55.
91. Groswasser J, Kahn A, Bouche B, Hanquinet S, Perlmuter N, Hessel L. Needle length and injection technique for efficient intramuscular vaccine delivery in infants and children evaluated through an ultrasonographic determination of subcutaneous and muscle layer thickness. Pediatrics. 1997;100(3 Pt 1):400–3.
92. Hanas R. Reducing injection pain in children and adolescents with diabetes: a review of indwelling catheters. Pediatr Diabetes. 2004;5(2):102–11.
93. Singh R, Samuel C, Jacob JJ. A comparison of insulin pen devices and disposable plastic syringes – simplicity, safety, convenience and cost differences. Eur Endocrinol. 2018;14(1):47–51. https://doi.org/10.17925/EE.2018.14.1.47. Epub 2018 Apr 18. PMID: 29922352; PMCID: PMC5954595.
94. Hanas R, de Beaufort C, Hoey H, Anderson B. Insulin delivery by injection in children and adolescents with diabetes. Pediatr Diabetes. 2011;12(5):518–26.
95. Arora A, Prausnitz M, Mitragotri S. Micro-scale devices for transdermal drug delivery. Int J Pharm. 2008;364(2):227–36.
96. Arora A, Hakim I, Baxter J, Rathnasingham R, Srinivasan R, Fletcher D, et al. Needle-free delivery of macromolecules across the skin by nanoliter-volume pulsed microjets. Proc Natl Acad Sci USA. 2007;104(11):4255–60.
97. Shah UU, Roberts M, Orlu Gul M, Tuleu C, Beresford M. Needle-free and microneedle drug delivery in children: a case for disease-modifying antirheumatic drugs (DMARDs). Int J Pharm. 2011;416(1):1–11.
98. Kendall M, Mitchell T, Wrighton-Smith P. Intradermal ballistic delivery of micro-particles into excised human skin for pharmaceutical applications. J Biomech. 2004;37(11):1733–41.
99. Kendall M, Rishworth S, Carter F, Mitchell T. Effects of relative humidity and ambient temperature on the ballistic delivery of micro-particles to excised porcine skin. J Invest Dermatol. 2004;122(3):739–46.

Chapter 14
The Challenges of Paediatric Pulmonary Drug Delivery

Darragh Murnane, Laura Urbano, and Marc B. Brown

Abstract The lungs offer a (relatively) easy access route by inhalation for drug delivery, both in terms of topical therapy of lung conditions as well as for systemically acting therapies. In paediatric therapy, it is the former which has been most exploited for the delivery of agents such as bronchodilators, corticosteroids and antibiotics where localized drug delivery can minimize systemic exposure and resultant side-effects.

Keywords Corticosteroid · Lung deposition · Bronchodilator · Aerosol · Impaction · Sedimentation · Diffusion deposition

14.1 Introduction

The widespread use of gaseous anaesthetics reveals also the potential of the large lung surface area, thin epithelial barrier (typically <1 μm) and extensive vascularization of the lungs for systemic delivery of compounds, including peptides and proteins. Pulmonary drug delivery is achieved effectively through the inhalation of pharmaceutical aerosols. However, the lung has evolved anatomical and physiological mechanisms to exclude potentially harmful aerosols. Investigation of these

D. Murnane (✉)
School of Health, Medicine and Life Science, University of Hertfordshire, Hatfield, UK

Informix Pharma Ltd., Bury St. Edmunds, UK
e-mail: d.murnane@herts.ac.uk

L. Urbano
School of Health, Medicine and Life Science, University of Hertfordshire, Hatfield, UK

M. B. Brown
School of Health, Medicine and Life Science, University of Hertfordshire, Hatfield, UK

MLBT Investments Ltd., Aylesbury, UK

H. Batchelor, K. Rose (eds.), *Pediatric Formulations*, AAPS Advances in the Pharmaceutical Sciences Series 58, https://doi.org/10.1007/978-3-031-77240-5_14

barriers in pulmonary drug delivery has resulted in the development of devices and aerosol formulations to achieve effective pharmacological therapy.

Most inhalation products have been developed with the adult patient in mind and translated to paediatric populations. Children have special needs both in terms of pulmonary physiology and anatomy but also cognitive and physical ability which means that adult therapies are not always well translated into paediatric care. In this chapter, the potential uses of aerosol therapy in childhood diseases will be highlighted and the particular issues pertaining to aerosol therapy of the developing respiratory tract will be addressed. Rather than justifying pharmacological therapy of specific diseases or populations, the emphasis of the chapter will be on physicochemical properties of aerosol that influence effective drug delivery as well as the formulation strategies that are employed for their delivery. It is the aim of this chapter to provide the reader with the necessary knowledge to judge the appropriateness of inhaled therapy in clinical situations.

14.2 Cognitive, Anatomical and Respiratory Development During Childhood

In order to achieve effective therapeutic outcomes, adherence and concordance with therapy are of paramount importance. The dramatic changes in cognitive, social and emotional skills provide barriers to effective use of therapies. However, the anatomical and physiological development during childhood also provides barriers to achieving effective aerosol delivery in paediatric patients. Ultimately, children must be *able* to use their medicine accordingly, whether they are reliant on a parent for their care or not. Often, children are not able to adopt the correct inhalation manoeuvre for their inhaler [1], particularly when they are below school-going age. Conversely, cognitive barriers may represent a barrier to the use of a device in children who can control their inhalation, and many children commit errors when using their inhaler [2].

The ultimate goal of therapy is to achieve deposition of an aerosol in the lung; however, deposition depends on the anatomy of the airways as well as the inhalation process. Children's airways are smaller than adults, and the ventilation and anatomy change substantially during childhood [3]. Infants tend to breathe diaphragmatically at high frequency and low flow rates through the nose [4]. The lung volume increases by alveolar enlargement (thereby increasing the surface area), and breathing becomes mixed nasal–oral between 2 and 5 years of age. Between 5 and 11 years, parallel to skeletal growth, airways increase in calibre, thereby lowering resistance and breathing becomes slower or deeper. By this age, many children are also able to control their oral breathing suitable for inhalation of aerosol therapy. Infants breathe mainly through the nose with irregular patterns of high flow rate and spend longer in exhalation or crying, leading to reduced lung penetration of aerosols [5]. Thus, the success of lung dosing depends critically on the control of the process of inhalation, which many children are unable to achieve.

14.3 Can Inhaled Medicines Be Delivered for More Than Lung Therapy in Infancy and Childhood?

The majority of inhaled therapies are intended to treat respiratory diseases such as infectious diseases or chronic lung disease such as asthma. The discussion of inhaled therapies which follows in this chapter will focus on topical therapy of respiratory diseases. However, the high surface area, thin epithelial barrier and absence of mucociliary clearance make the terminal bronchioles and alveolar regions of the lung, a particularly effective deposition site for drug absorption. Even in adults, the challenges of achieving distal lung delivery of aerosols are daunting. However, the substantial increases in bioavailability which are possible for proteins ultimately culminated in the marketing (and subsequent withdrawal) of inhaled insulin as Exubera® (Nektar/Pfizer) [6]. There have also been reports of development of inhaled therapies of fentanyl and other opioids for pain treatment [7] or dyspnoea [8]. With the exception of the development of inhaled growth hormone [9], systemic therapies have not been developed with children in mind, and inhaled insulin was not licensed for paediatric use. Systemic inhaled therapy is certainly worth consideration in adolescent populations where an adult product is available. In younger children, however, the ability to achieve reproducible deep lung deposition is less assured with conventional inhalers. However, where a suitable solution formulation can be developed, nebulized drug delivery may provide a drug delivery stratagem.

14.4 Aerosol Deposition in the Lungs

Deposition is the term applied to the processes by which the particles within an aerosol cloud leave the inhaled airstream and accumulate on the airway epithelium. Accepting that childhood is a period of immense change in pulmonary anatomy and physiology, it is necessary to consider the processes which occur when an aerosol is inhaled into an "ideal" lung to provide the scientific framework for analysing the problems posed by the requirement to access the pulmonary route for drug delivery in children. The upper airways extend from the nasal cavity and the mouth to the larynx and are responsible for conditioning inhaled air. Nasal breathing is rather effective at filtering particles from the air; however, the high velocity of airflow in the upper airways imposes turbulence, which leads to particle deposition in the oropharynx even during mouth breathing (Fig. 14.1). The airways branch 23 times with a decrease in calibre but increase in number [12]. Thus, through the conducting and pulmonary airways, the surface area of the lungs increases until the air reaches the alveolar epithelium. The changes in airflow direction at each bifurcation lead to inertial impaction of particles. Additionally, the airflow velocity becomes laminar-to-stagnant by the bronchiolar–alveolar region such that aerosol particle deposition can occur by sedimentation and diffusion [13] (Fig. 14.2).

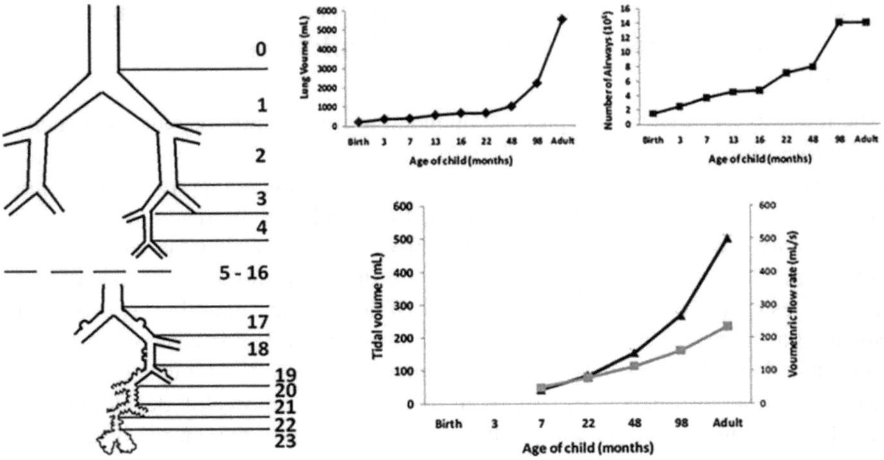

Fig. 14.1 Schematic of the pulmonary anatomy according to the Weibel model of the adult lung. The number of airways increases with each branch point, but the airflow velocity and diameter of the airways decrease. As the child ages from infancy, there is an increase in lung volume (*top left graph*) and the number of airways (*top right graph*) to adult levels, as shown by Dunnill [10]. Of relevance to paediatric drug delivery, the inhalation flow rate (gray line) and tidal volume (black line) under sedentary conditions increase with age to adulthood (*bottom graph*). In infancy, the flow rate is determined by the tidal volume [11]

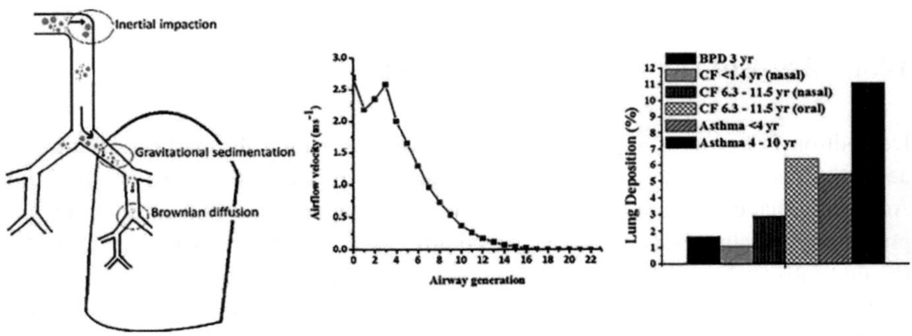

Fig. 14.2 Schematic of the deposition mechanisms for aerosols inhaled into the respiratory tract (*left*). Inertial impaction occurs for the largest particles in airstreams with high velocity. Sedimentation and diffusive deposition occur for particles with small sizes in stagnant air. A simulation showing airflow in the pulmonary airways decreasing to stagnant conditions due to the dichotomous branching of the airways (*centre*), simulated after [13] with morphometric scaling after [14]. Representation of lung deposition fractions in children of different ages suffering from lung diseases. BPD bronchopulmonary dysplasia, CF cystic fibrosis. (Data taken from [4, 15, 16])

14.4.1 Aerosol Properties Affecting Pulmonary Deposition

Achieving deposition of particles within the lung for therapeutic effect is not trivial. The strongest evidence (albeit in the case of adults) is that for lung deposition to occur, aerosol particles must possess a particle diameter below 10 μm [17]. Particles with a size of 1–10 μm deposit by a combination of impaction and sedimentation, and there is an overlap between deposition by sedimentation and diffusion for particles with a size of 0.1–1.0 μm. However, particles <0.1 μm deposit solely by diffusion [18, 19]. Inertial impaction dominates in the conducting airways where airflow is of high velocity and turbulent. Diffusion and sedimentation which occur in laminar airflow are time-dependent processes; hence, deposition fractions increase with a long aerosol residence time in the lung and a close proximity of particles to the airway wall.

The deposition of particles by impaction and sedimentation depends on the aerodynamic diameter, d_{ae}, of the aerosol, which is defined as the diameter of a unit density sphere that possesses the same settling velocity as the particle being considered:

$$d_{ae} = d_g \sqrt{\frac{\rho}{\chi}}$$

where d_g is the geometric diameter, ρ is the particle density (g cm^{-3}) and χ is the dynamic shape correction factor.

Specifically, a d_{ae} of 2–6 μm is typically regarded as being required for deposition in the conducting and tracheobronchiolar airways, whilst a size of 1–3 μm is preferable for targeting deposition to the peripheral (smaller) airways and for systemic delivery. From the above equation, it can be seen that the effective size of the particles depends on the density and shape of the particle. Hence, a large particle may penetrate the deep lung if it is either needle-shaped or possesses low density.

14.4.2 Physiological and Pathophysiological Factors Affecting Pulmonary Deposition

The major physiological factors to be considered in connection with particle deposition relate to the airflow and geometry of the airways. The interaction of a particle with the airway walls is of higher probability in smaller calibre airways. In this context, it is important to consider also the pathological situation, when airways are obstructed (e.g. by accumulation of secretions, narrowing of the airway or alteration of bifurcation angles). For example, due to central and peripheral airway obstruction in asthma, there is a marked decrease in the peripheral deposition of smaller particles, which should deposit in peripheral airways [20]. The lung ventilation characteristics (both tidal volume and frequency) are equally important as the airways anatomy in influencing drug deposition. High ventilation rates with shallow volume decrease the residence time of the aerosol in the lung, and consequentially, sedimentation and diffusional deposition are decreased [17]. This leads to the routine clinical advice for inhalation of the aerosol with a deep breath followed by a period of breath-holding. High inhalation flow rates (e.g. panting) favour inertial impaction in accordance with:

$$S = Bmv$$

where B is the mobility of the particle (i.e. the velocity per unit of force), m is the particle mass (i.e. size) and v is velocity with which the particle is travelling.

Thus, particles which are large or travelling with high velocity will fail to progress beyond a sharp directional change. With the exception of nebulizer use, patients inhale medicines with a pronounced inspiratory peak rather than sinus breathing [21]. Thus, there is a high oropharyngeal deposition in those who inhale from aerosol products such as metered dose or dry powder inhalers (DPIs). With an increased inhalation rate, the deposition profiles of aerosols are shifted towards the central and upper airways because of enhanced inertial impaction as well as increased turbulence in generations 0–10 [22]. It is also possible to alter the deposition profiles of inhaled aerosols by controlling the ventilation profile of the patient [23].

14.4.3 Deposition in the Paediatric Drug Delivery

The same principles governing respiratory tract particle deposition apply to children as adults; however, the ventilation characteristics and anatomy dictate diverse outcomes for aerosols with similar physicochemical properties. Because younger children (<4 years) breathe nasally, the impact of nasal deposition when children use facemasks must be considered for children. It may therefore be necessary to design an aerosol with different physicochemical properties for children in order to mitigate the enhanced extra-thoracic airway deposition patterns. In particular, it is necessary to avoid nasal/oropharyngeal deposition that leads to side effects and possible systemic absorption. Because of low inhaled volumes, longer relative exhalation period and high turbulence during irregular breathing, the relative central/peripheral airway deposition ratios may change in children with the reduced time for small particle deposition in the deep lung.

Young children who inhale their aerosol through a mouthpiece rather than a facemask receive lower total doses because of the contribution of their concomitant nasal breathing [24]. However, extra-thoracic (i.e. nasal) deposition is very high when a facemask is used leading to lower relative lung dosing [25]. Thus, children should be encouraged to use mouthpieces as soon as possible to maximize pulmonary deposition.

Unlike adults, the evidence regarding sites and extent of deposition for aerosols in children as a function of physicochemical properties is inconclusive [14]. The ideal particle size for deposition is suggested to be smaller than 3.6 μm for infants [26]. Particles <2.1 μm are relatively insensitive to ventilation changes in comparison to larger particles. However, unfortunately, the marketed formulations may not always be suitable. Larger particles are mainly affected by inertial impaction, and nasal and upper airway deposition is enhanced for particles in the range of 3–5 μm. However, equally particles in the range of 0.5–3 μm, which deposit by sedimentation require breath-holding for effective deposition. Thus, it is difficult to accurately assign the ideal particle size for inhalation therapy in infants. Studies have shown that effective deposition fractions in younger children can be achieved with extra-fine aerosol (<2 μm) [26, 27]. However, the clinical consequences of enhanced deposition of extra-fine aerosols in children [28] must be balanced with the risks of increased systemic exposure and potential toxicity.

14.5 Aerosol Generation Systems for Pulmonary Therapy

The main platforms for the production of aerosols on demand are techniques which produce droplets from bulk solutions or suspension (i.e. nebulization), techniques which release dry powder particle clouds (i.e. DPIs) or techniques which release condensation vapours (e.g. cigarettes). With the exception of solution-based inhalation formulations, the majority of orally inhaled medicinal products contain aerosolizable drug in particulate form. The latter particles must be produced in the low micron size range for formulation into drug delivery products. Particles are typically produced by crystallization and subsequent particle size reduction (micronization). It will become obvious that the dosage manufacture is highly specialized for inhalation. Manufacture often requires expensive equipment or complex processes, which are incompatible with extemporaneous product manipulation or small-batch operations (e.g. pressurized systems cannot readily be manipulated, nor dry powder blends readily diluted without altering the inhaled drug dose). However, the final of the three dosage forms, nebulization, may offer some scope for extemporaneous production in-house for treatment of children in the absence of other suitable dosage forms.

14.6 Pressurized Metered Dose Inhalers

The pressurized metered dose inhaler (pMDI) is a cheap, convenient and portable dosage form that contains multiple doses. All pMDI products have a roughly similar design consisting of one or more active pharmaceutical ingredients (in solution or suspension), a liquefied propellant with excipients, all filled into a canister and sealed with a metering valve. A metering valve that limits the volume dispensed upon actuation controls the dose available for the patient to inhale. The formulation

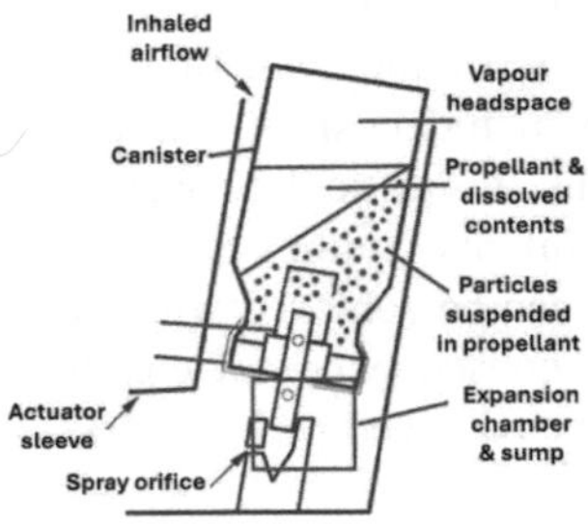

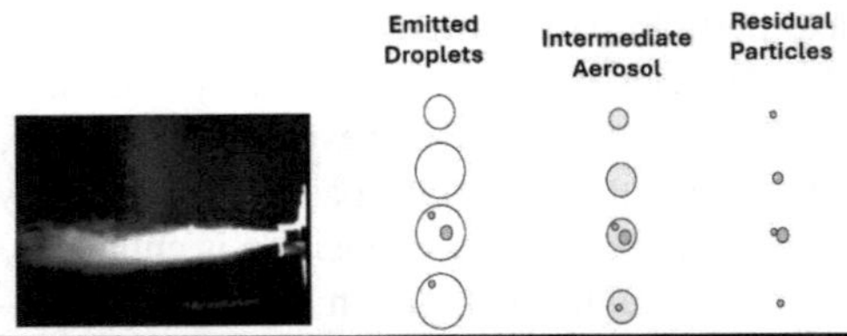

Device	Age range	Effectiveness	
pMDI + Antistatic spacer[1]	<4 yr	5.4 % lung deposition	
	4 – 10 yr	9.6 % lung deposition	
pMDI + electrostatic spacer[2]	7 – 12 yr	Plasma level 1.9 ng/mL	
pMDI + non-electrostatic spacer[2]	7 – 12 yr	Plasma level 4.3 ng/mL	
		Facemask	Mouthpiece
pMDI + spacer3	1 – 2 months	3 % respirable	-
	2 – 3 yr	21 % respirable	-
	4 – 6 yr	35 % respirable	22 % respirable
	10 – 15 yr	-	40 % respirable

Fig. 14.3 The components and principles of operation of a pressurized metered dose inhaler (pMDI, *left*). Upon actuation, a high velocity aerosol cloud of polydisperse droplets is emitted (photograph). The propellant in large emitted aerosol droplets evaporates to form intermediate droplets (which may contain suspended drug microparticles of polydisperse size distribution) and finally residual respirable particles. Volatile solution formulations result in particles of smaller diameter than suspension formulations. Spacer devices are often employed to improve the specificity of lung deposition; however, lung deposition is age-dependent and is affected by electrostatic charge on the spacer, as well as the choice of facemask or mouthpiece. ([1]Wildhaber et al. [16]; [2]Anhoj et al. [29]; [3]Dubus and Anhoj [24])

is a liquid under pressure in the canister; however, upon actuation of the valve, the high vapour pressure of the propellant causes a two-phase fluid (liquid–vapour) to exit the device at high velocity through an atomization (spray) orifice which is a component of a plastic actuator. pMDIs are the most used inhalation device worldwide (Fig. 14.3).

14.6.1 Formulation Components and Aerosol Formation

The propellant constitutes greater than 80% of a formulation, and the toxicology, vapour pressure, solvency power, flammability and density are all important properties. The currently employed propellants are hydrofluoroalkanes (HFA) with high vapour pressures, low boiling points and low densities. The low density leads to the rapid sedimentation of denser suspended drug microcrystals, resulting in unstable suspensions and dose variability unless the pMDI is shaken appropriately. HFA propellants show comparatively higher water and drug substance solubility in comparison to other propellants such as chlorofluorocarbons. Ingress of water can cause solvency changes and dissolved drug can precipitate, or Ostwald ripening of suspensions can occur. One of the greatest problems is the low solubility of many surfactants in HFAs, leading to difficulties stabilizing suspension formulations. Thus, it is important for patients to shake their inhalers effectively to disperse drug particle sediments.

Upon actuation of a pMDI, the opening of the metering chamber to the atmospheric pressure across the spray orifice promotes evaporation of the propellant. A gas–liquid mixture is thus propelled through the spray orifice, resulting in the formation of an atomized aerosol cloud. The vapour pressure and viscosity of the formulation determine the fluid flow and hence the aerosol properties. To produce small droplets, a combination of a high vapour pressure formulation and a small spray orifice diameter is necessary [30]. The primary aerosol is typically comprised of large droplets which may contain some or no drug or excipients. The droplets evaporate, resulting in aerosol particles which may be of a size suitable for inhalation. For solution pMDI formulations (e.g. QVAR™ or Modulite™), the inhaled particle size distribution is determined entirely by the size of the initial droplets and the evaporation rate, and can be carefully tuned to produce droplets of the required diameter for inhalation. The smallest possible particle size for a suspension formulation is that of the original micronized particles (or their aggregates) following evaporation of the propellant. To maximize lung delivery, it is thus important to formulate a stable suspension with minimal particle aggregation [31].

14.6.2 Patient Use of Metered Dose Inhalers

The pMDI generates an aerosol cloud under the atomizing force of the propellant independent of a patient's respiratory effort. However, the aerosol produced is polydisperse in size and contains large droplets when inhaled directly from the actuator, typically in the order of tens of micrometres. The large droplets are also emitted at high velocity, decreasing from approximately 58 to 5 ms^{-1} depending on the distance from the spray orifice. The consequences of the large particle size and the high velocity are the ballistic impaction of a large fraction of the aerosol cloud in the oropharynx [32]. Throat deposition is high in those patients who do not breathe at the same time as pMDI actuation (i.e. poor actuation coordination). This is one of the most common causes of cognitive error in pMDI use; however, even patients with good technique show lung deposition of less than 1/3 of the emitted dose [33]. Although smaller particles from solution pMDIs are advocated to maximize lung deposition in children, high oropharyngeal deposition is still observed in children using breath-actuated inhalers that were designed to overcome the need for coordination [28]. Thus, the pMDI including breath-actuated pMDIs may not be the best device for use in children unless combined with an auxiliary device.

14.6.3 Metered Dose Inhalers and Spacer Devices: The Paediatric Option

Valved-holding chambers (VHCs) are a spacer device that can reduce oropharyngeal deposition of pMDIs by temporally separating the actuation and inhalation events. Modern VHCs include face masks and encouragement aids designed to appeal to children. VHCs are perceived to reduce extra-thoracic deposition while not altering fine particle lung deposition. However, some pMDI–VHC combinations have demonstrated substantial oropharyngeal deposition [34]. Early studies showed dose retention in the VHCs of 50–80% for salbutamol [35]. Dose deposition in the VHC can occur by impaction in smaller devices; however, evaporation of the aerosol cloud in the VHC promotes formation of a small particle aerosol. The residence time in the VHC dictates the period available for not only evaporation of the aerosol cloud but also sedimentation of the particles. The residence time is a function of the inhaled volume, inhalation frequency and flow rate of the patient and paediatric patients will require multiple inhalations to inhale the dose. Upper airway deposition is physiologically enhanced in children, but the sedimentation occurring during long residence times can result in lower lung deposition in younger children due to VHC retention [36].

The problem of electrostatic interactions between the (typically) plastic VHC devices and the charged pMDI aerosol cloud is a well-known phenomenon, leading to VHC-particle retention. This is particularly the case for paediatric patients who require long inhalation times when sufficient time is provided for electrostatic deposition to occur. The consequences are that following multiple actuations dose retention in the spacer decreases [37]. Hence, bioavailability will alter during the use of the VHC, and the electrostatic effect would become important each time the device is washed with water. Washing with a surfactant solution coats the plastic with an antistatic layer, thus avoiding the problem. However, it is more usual that modern VHCs are constructed using conducting plastics or with antistatic linings in order to remove the requirement for parent/patient co-operation with washing and appropriate drying techniques. Antistatic spacers are unaffected by washing and have been reported to improve drug dose delivery for a fluticasone propionate pMDI compared to conventional VHC [38].

14.7 Dry Powder Inhalers

DPIs are also portable but do not require coordination of inspiration and actuation of the dose. DPIs are also suitable for the administration of doses as high as mg quantities, unlike pMDIs. DPIs are either single-dose devices, where the formulation is typically contained in a capsule, or multiple single-unit dose devices, where the formulation is contained in a series of unit-dose blisters on a strip. Multidose devices are also manufactured, where the formulation is contained in a reservoir and metered by the patient prior to inhalation. The majority of marketed DPIs require a patient's inhalation manoeuvre to aerosolize the drug particles from the formulation and are termed passive devices. In recent years, several active devices have also been developed, which use external energy (e.g. compressed air) to achieve aerosolization. When considering DPIs, therefore, it is important to consider the properties of the inhalable particles, the formulation containing those particles and the interactions that occur when the patient inhales through the device.

14.7.1 Formulation Components and Aerosolization of Powder Inhalers

Particles in this respirable size range (<10 μm) exhibit high cohesivity and adhesivity, leading to the particles forming powder agglomerates which are difficult to efficiently aerosolize. The micronized particles for use in DPIs naturally tend to adhere to the device or cohere as large agglomerates. In order to achieve an aerosol suitable for lung deposition (i.e. small particle size), the agglomerated particles must be effectively dispersed. It is difficult to achieve uniform filling of capsules and devices with particles of this size due to poor flow properties, and hence, micronized particles are not particularly suitable for high-dose formulations. Rather a host of particle design techniques have been developed to improve powder flow and dispersibility, including spray drying. Such products have now reached the market and include Tobramycin Inhalation Powder™ [39]. In the case of low-dose therapies (<1 mg), it is necessary to blend micronized particles with a diluent such as lactose monohydrate particles (typically with a size in the range 50–100 μm) to form an ordered mixture known as a carrier-based formulation. The Turbuhaler® products from AstraZeneca employ a granulation approach where the micronized API particles are granulated with an appropriate wetting liquid with/or without the addition of a bulking agent to aid uniform dose metering.

During aerosolization, the forces of adhesion between the particles in the granules or in the carrier-based blends must be overcome. Not every inhaled drug particle possesses the correct balance of cohesive and adhesive forces in the lactose blend for aerosolization to be effective. Strong adhesive bonds to the carrier result in poor aerosolization upon inhalation. However, although high cohesive forces lead to good re-dispersal, uniformity of dose content can be difficult to achieve. It is possible to employ so-called ternary agents in the form of micronized lactose [40] or low surface free energy materials (e.g. leucine or magnesium stearate) that modify the cohesive/adhesive force balance with the carrier [41]. For DPIs, the product per formance has been shown to depend on the properties and history of the micronized particles in the formulations, the physicochemical properties of the carrier, the blending process employed for drug and carrier and the force of the airflow which produces aerosolization. The important properties of the carrier have been extensively reviewed [42]. For example, the lactose crystal size and shape are extremely important properties to consider as is the ratio of the drug particles to the carrier component. The micronization process used for the drug particles results in solid-state damage, particularly on particle surfaces [43]. Such regions of disorder are susceptible to adsorption of water vapour, leading to increased adhesion forces to the carrier. For this reason, DPIs must be protected from exposure to excessive environmental relative humidity.

14.7.2 Patient Use of DPIs

When a patient inhales through a DPI the powder is fluidized, entrained and ultimately de-agglomerated under aerodynamic shear to form the aerosol. Impaction of entrained powder on device walls is also important, while vibrational or centrifugal forces upon spinning of the capsule in capsule-based products (e.g. Cyclohaler™ or Handihaler™) also contribute to aerosolization. In order to achieve effective de-agglomeration into a respirable aerosol, a fast, forceful inhalation is required [44]. Otherwise, a large particle size aerosol, or an aerosol where the drug particles adhere to the carrier, is produced, leading to high throat deposition. Devices are classified as low, medium or high resistance devices, and the resistance dictates the airflow which a patient can achieve through the device to achieve aerosolization. For example, many patients with respiratory disease are unable to inhale effectively through medium to high-resistance devices [45]. As patients exhibit variability in their inhalation strength, there is also a high degree of inter-patient variability in lung deposition from traditional DPIs (range: 5–28%) [32]. The suitability of a device for a particular patient can be assessed using devices such as the In-Check Dial produced by Clement Clarke International. The general assumiption is, that if a patient cannot produce a suitable peak inspiratory flow rate (PIFR) through a specific inhaler device, then aerosolization will not be effective for that device. It is clear that many patients with the cognitive abilities to use their DPI may simply not be able to inhale through their device with sufficient strength to achieve deagglomeration of the formulation suitable for pulmonary deposition.

14.7.3 The Role of DPIs in Paediatric Therapy

DPIs require the patient to possess appropriate manual dexterity, be able to sequence dose priming correctly, and to use an appropriate inhalation manoeuvre, and therefore appear on the face of it, to be the least suitable option for paediatric therapy. The evidence for the place of DPIs in therapy is conflicting because of the high rate of errors when children use DPIs [46]. Older children (>8 years) show greater competence, and detailed training can be used effectively; thus, there appears to be an age-dependence in the suitability of DPI use. Given a choice, older children and adolescents favour DPIs over pMDI + VHC combinations, due to convenience and portability.

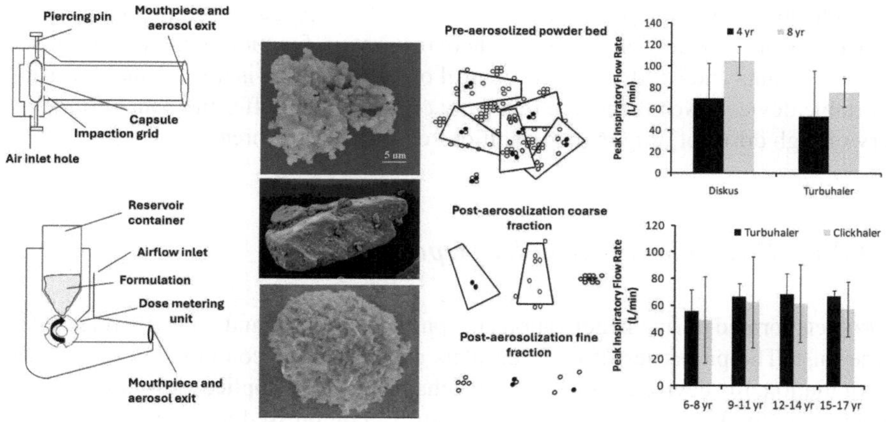

Fig. 14.4 Example designs of a single-dose capsule-based (*top left*) and reservoir multiple-dose (based on Teva's Airmax® device) dry powder inhalers (DPI). Powders in the respirable size range (e.g. salmeterol xinafoate in the *top micrograph*) agglomerate into non-respirable fractions. Micronized particles are typically blended with a large particle size diluent (e.g. lactose monohydrate, *centre micrograph*) or granulated (*bottom micrograph*). The powder blend disperses under the force of inhalation by a patient, generating a non-respirable coarse fraction of carrier, carrier-adhered drug and large agglomerates, or a fine fraction of respirable particles or small agglomerates. The magnitude and variability of the inspiratory force generated by children vary according to the device design and age of the child data are mean ± RSD, from [49, 51, 52]

Not all children have the appropriate degree of control of lung function to use DPIs correctly. It is important to inhale with a high flow rate and volume through the device, and this is a particular problem with high-resistance devices such as the HandiHaler™ and Turbuhaler™ for children with low lung capacity. In children as much as adults, a prescribing tool such as the In-Check™ Dial is useful in judging the suitability of DPI use, but such tools may also be used as training aids for children [47]. Children as young as 6 years [48, 49] or even 3 years of age [50] are able to generate a suitable inspiratory flow with low-medium resistance DPIs (Fig. 14.4). However, the evidence for high resistance devices is more ambiguous, and low-to-medium resistance devices appear to offer greater consistency in aerosolization performance in younger children. Clinical efficacy has, however, been shown for children using high-, medium-, and low-resistance devices. Therefore, it is crucial to assess the respiratory capacity and dexterity of the paediatric patient when judging the suitability of DPIs for paediatric inhalation therapy.

14.8 Nebulized Drug Delivery

Nebulizers are systems that use an external source of energy to create aerosol droplets of drug solutions or suspensions, which the patient inhales as a cloud. The most typical external source of energy is a jet of compressed air; however, nebulizers also use ultrasonic vibrations to achieve aerosol formation. Nebulizers have an important place in acute care delivery settings, where respiratory function is compromised and to treat infants and children whose control of lung function is not optimal for using portable devices. Nebulizers are relatively expensive, but offer the benefit of delivery of high doses of drug under conditions of normal tidal breathing.

14.8.1 Principles of Nebulizer Operations

Droplets formed during nebulization are spherical in shape and typically released as fine mist. The properties of these droplets depend on the combination of the feed formulation, the device geometry and mechanical energy applied. Air-jet nebulizers operate on the Bernoulli principle and require a compressed air supply. Ultrasonic nebulizers supply mechanical energy produced by vibrations of a piezoelectric crystal or by vibration of a membrane. Nebulizers produce slow-moving aerosol clouds, that are compatible with many ventilation systems used in respiratory support (e.g. in pre-term neonatal care). In jet nebulization, compressed air is passed through a small orifice entraining the formulation liquid from one or more adjacent capillaries, leading to droplet dispersal. Jet nebulizers are suitable for the administration of suspensions and high-concentration solutions [53]. The droplet size distribution of the aerosol is determined by the jet-capillary configuration, the airflow rate and the formulation (e.g. viscosity and surface tension). Vibrations in ultrasonic nebulizers generate capillary waves in the fill liquid leading to droplet generation at the surface of the fill liquid. To produce small droplet diameters, a high airflow is required for jet nebulizers. In ultrasonic nebulization, the droplet formation is determined by the frequency of vibration.

The droplets formed from nebulizers are polydisperse, and baffles must be used to retain large droplets in the chamber. The formed droplets evaporate and/or coalesce before inhalation, and this depends on the adequacy of the diluting carrier airflow. Ultrasonic nebulization may be unsuitable for thermolabile substances, including biopharmaceuticals because the fill volume tends to increase in temperature. They are also not suitable for nebulization of suspension formulations [54]. Jet nebulizers are suitable for use with suspensions but require fill volumes of at least 2–3 mL and long nebulization times. The drug output rate for both depends on the drug concentration in the formulation, and an optimum formulation exists; thus, dilution is not advisable unless the dose emission characteristics can be confirmed [55].

14.8.2 Aerosol Formulation Considerations

The majority of formulations are aqueous solutions using water-soluble salt of the drug for its high solubility (e.g. salbutamol sulphate) and to prevent precipitation upon storage. In suspension formulations, it is necessary to include surfactants (e.g.

polysorbate 80) to prevent particle aggregation, which would increase the eventual particle size of the aerosol. Aqueous formulations should be adjusted to isotonicity to avoid bronchoconstriction upon inhalation. Cosolvents may be required, including ethanol and glycerol, to solubilize poorly soluble compounds (e.g. Ventavis™ iloprost trometamol). Other typical excipients included in formulations include antimicrobial preservatives (e.g. benzalkonium chloride), chelating agents (e.g. disodium edetate) and protein peptide formulations often require additional stabilizing agents [56]. It is important to note that the aerosol cloud properties are determined by the properties and components of the formulations (e.g. volatility due to cosolvent use or concentration of non-volatiles, surface tension or viscosity). Hence, the practice of mixing formulations or dilution of nebulization liquids is best avoided in order to achieve consistency of product performance, especially given the fact that such a degree of variability exists from nebulizer to nebulizer.

14.8.3 Improved Nebulizer Designs

Vented nebulizers overcome the problem with traditional jet nebulizers where the airflow is not usually sufficient for tidal breathing. By providing extra make-up air droplet coalescence nebulization time is reduced. Breath-assisted vented nebulizers restrict aerosol output to the inspiratory phase of the inhalation cycle, thus improving the specificity of lung deposition and improving output rates [57]. Similar technologies have been developed to control lung site deposition by controlling the timing of aerosol release/generation during the breathing cycle [58]. The latter technologies are usually termed adaptive aerosol delivery systems and require microcomputer monitoring of the inhalation process for a patient.

Vibrating mesh nebulizers represent the latest development of nebulization technology offering high portability and low power requirements (AA batteries). By controlling the diameter of the mesh aperture, the droplet size can be controlled in the size range 1–6 μm, and no baffles are required since the majority of the primary aerosol is inhalable. In active devices, the vibration of the membrane achieves aerosolization, but in passive devices, the vibration is achieved with an ultrasonic horn in contact with the fill liquid. Mesh nebulizers are suitable for delivery of viscous, high-concentration and suspension formulations [59]. However, excessive viscosity and surface tension impede aerosolization rates. Many mesh devices are also suitable for use with low fill volumes [60] and produce slow-moving aerosol clouds, the generation of which can be coordinated with patient inhalation. Thus, mesh nebulizers offer a potential solution as a portable drug delivery device for patients who cannot use pMDIs or DPIs correctly.

14.8.4 Nebulizers–Patient Interactions in the Paediatric Setting

For infants and younger children, nebulization offers an excellent drug delivery options since the aerosol is created without requiring any specialized breathing manoeuvres. Marketed nebulizers differ substantially in their delivery performance for a formulation; hence, regulators recommend specific nebulizers to be used with licensed formulations. With traditional jet nebulizers, airflow of 6–8 L min^{-1} is required for effective aerosol delivery [53]. However, this can be above the tidal breathing of infants leading to waste to the environment rather than inhalation. The dose emission from jet nebulizers is also continuous regardless of the breathing cycle, leading to further waste and potential for facial deposition. For children from about 6 months to 1 year upwards, inspiratory flow rates approximately match nebulizer outputs. When using vented nebulizers that provide for make-up air, older children with higher lung volumes dilute the nebulized aerosol when compared to a younger child. Therefore, the overall inhaled dose is lower in older children [61].

Nasal inhalation results in reduced lung deposition fractions in comparison to oral inhalation regardless of delivery device. When sealed masks are used with infants and children, facial deposition can occur leading to side effects or irritation [62]. However, when the facemask is held at a distance (even as small as 2 cm) from the child, the dose available for lung deposition decreases to less than half that with a tightly fitting mask [63]. Vibrating mesh nebulizers may be more suitable than jet nebulizers for use with younger children because the small droplets produced are more likely to escape nasal and oropharyngeal deposition when using facemasks. Vibrating meshes are also useful for speeding up administration times without compromising bioavailability in children [64]. Adaptive aerosol delivery systems have also been shown to be acceptable for children under 4 years when using a face mask [65] (Fig. 14.5).

14.9 Concluding Remarks

The field of pulmonary drug delivery is one which is fraught with many uncertainties for the paediatric specialist. The benefits of local therapy of airway diseases are clear and desirable, and systemic drug delivery offers real potential for improvements in therapy, even in children. However, a relatively simple analysis demonstrates that there is a high incidence of side-effects following inhaled therapy that means the proposed benefits may not always be met. Unlike other delivery routes, the basic anatomy and physiology that constitute the barrier to drug delivery change vastly from pre-term infant to the adolescent. Thus, the patient population are far from small adults and very heterogeneous. Infants and young children, in particular, represent a difficult patient group to treat. However, with an appropriately chosen aerosol (i.e. small particle size) and device (i.e. nebulizer of pMDI with a VHC), drug delivery to these children is possible. It is clear that the earlier a switch to oral

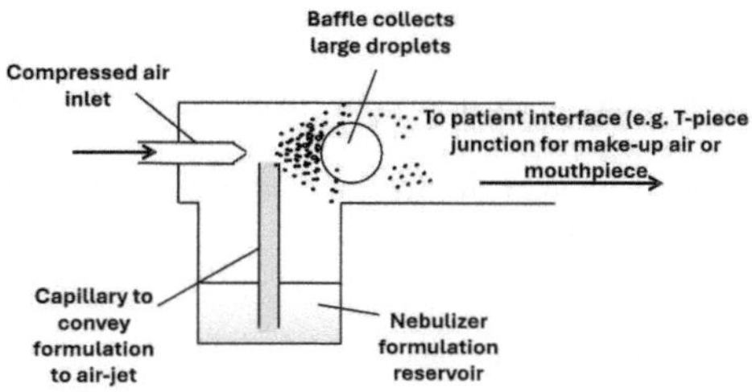

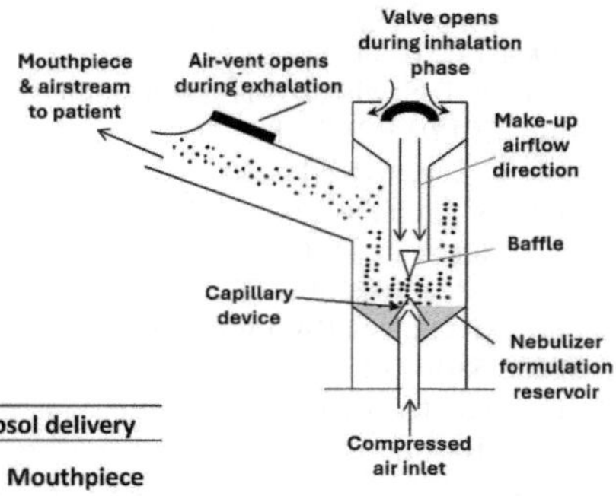

Age (yr)	Lung deposition (%) following nebulized aerosol delivery			
	Facemask			Mouthpiece
	Not crying	Crying	Poorly fitting	
<0.5	0.8 [1]			
0.5-1.4	1.3 [1]			
2-3	8.0 [2]	1.0 [2]	0.1 [2]	
<4	5.4 [3]	1.3 [3]		
6-8	2.9 [1]			6.0 [1]

[1]Chua et al. (1994), [2] Keller et al. (2004), [3] Wildhaber et al. (1999)

Fig. 14.5 Schematic of a traditional air-jet nebulizer (*left*) and a more modern breath-enhanced nebulizer (*right*). Upon inhalation, additional make-up air is inhaled through a one-way valve, thus entraining the aerosol for inhalation. Upon exhalation, the inlet valve closes and an exhalation vent opens, thereby reducing aerosol emission during the exhalation period. Because the make-up is drawn from the nebulizer, breath-enhanced systems are less sensitive to changes in environmental humidity. The data demonstrate the low specific lung deposition during crying periods as well as that achieved in young children using face masks compared to a mouthpiece. ([1]Chua et al. [15]; [2]Keller et al. [66]; [3]Wildhaber et al. [16])

inhalation using a mouthpiece can be achieved, the better pulmonary drug deposition will be. However, the choice of drug delivery platform in such competent children is less equivocal owing to the heterogeneity of the patient population. Every child should be assessed individually for his/her cognitive and physical capacity to use a range of inhalation devices, and parents and children should be trained and routinely counselled on correct use. It is important, however, to consider that the worst device is the one a child will not use. Ultimately, the decision to use inhaled therapy requires not only prescribing a device the child can use, but also being sure it is a device that they will use because concordance and adherence to therapy achieve the best outcomes for children.

References

1. Kamin W, Kreplin A. Teaching the inhalation manoeuvre to asthmatic children by means of visual feedback. Pneumologie. 2007;61(3):150–6. (Article in German)
2. Lavorini F, Magnan A, Dubus JC, Voshaar T, Corbetta L, Broeders M, Dekhuijzen R, Sanchis J, Viejo JL, Barnes P, Corrigan C, Levy M, Crompton GK. Effect of incorrect use of dry powder inhalers on management of patients with asthma and COPD. Respir Med. 2008;102(4):593–604.

3. Couriel JM, Child F. Applied physiology: lung function testing in children. Curr Paediatr. 2004;14(5):444–51.
4. Amirav I, Newhouse MT. Deposition of small particles in the developing lung. Paediatr Respir Rev. 2012;13(2):73–8.
5. Iles R, Lister P, Edmunds AT. Crying significantly reduces absorption of aerosolised drug in infants. Arch Dis Child. 1999;81(2):163–5.
6. Mastrandrea LD, Quattrin T. Clinical evaluation of inhaled insulin. Adv Drug Deliv Rev. 2006;58(9–10):1061–75.
7. Farr SJ, Otulana BA. Pulmonary delivery of opioids as pain therapeutics. Adv Drug Deliv Rev. 2006;58(9–10):1076–88.
8. Dudgeon DJ, Lertzman M. Dyspnea in the advanced cancer patient. J Pain Symptom Manag. 1998;16(4):212–9.
9. Walvoord EC, de la Pena A, Park S, Silverman B, Cuttler L, Rose SR, Cutler G, Drop S, Chipman J. Inhaled growth hormone (GH) compared with subcutaneous GH in children with GH deficiency: pharmacokinetics, pharmacodynamics, and safety. J Clin Endocrinol Metab. 2009;94(6):2052–9.
10. Dunnill MS. Postnatal growth of the lung. Thorax. 1962;17:329–33.
11. Musante CJ, Martonen TB. Computer simulations of particle deposition in the developing human lung. J Air Waste Manage Assoc. 2000;50(8):1426–32.
12. Weibel ER. Morphometry of the human lung. Berlin: Springer; 1963.
13. Martonen TB. Mathematical-model for the selective deposition of inhaled pharmaceuticals. J Pharm Sci. 1993;82(12):1191–9.
14. Isaacs KK, Martonen TB. Particle deposition in children's lungs: theory and experiment. J Aerosol Med. 2005;18(3):337–53.
15. Chua HL, Collis GG, Newbury AM, Chan K, Bower GD, Sly PD, Lesouef PN. The influence of age on aerosol deposition in children with cystic-fibrosis. Eur Respir J. 1994;7(12):2185–91.
16. Wildhaber JH, Dore ND, Wilson JM, Devadason SG, LeSouef PN. Inhalation therapy in asthma: nebulizer or pressurized metered-dose inhaler with holding chamber? In vivo comparison of lung deposition in children. J Pediatr. 1999;135(1):28–33.
17. Heyder J. Deposition of inhaled particles in the human respiratory tract and consequences for regional targeting in respiratory drug delivery. Proc Am Thorac Soc. 2004;1(4):315–20.
18. Carvalho TC, Peters JI, Williams RO. Influence of particle size on regional lung deposition: what evidence is there? Int J Pharm. 2011;406(1–2):1–10.
19. Rostami AA. Computational modeling of aerosol deposition in respiratory tract: a review. Inhal Toxicol. 2009;21(4):262–90.
20. Sbirlea-Apiou G, Lemaire M, Katz I, Conway J, Fleming J, Martonen T. Simulation of the regional manifestation of asthma. J Pharm Sci. 2004;93(5):1205–16.
21. Broeders MEAC, Molema J, Hop WCJ, Folgering HTM. Inhalation profiles in asthmatics and COPD patients: reproducibility and effect of instruction. J Aerosol Med. 2003;16(2):131–41.
22. Martonen TB, Katz IM. Deposition patterns of aerosolized drugs within human lungs: effects of ventilatory parameters. Pharm Res. 1993;10(6):871–8.
23. Brand P, Friemel I, Meyer T, Schulz H, Heyder J, Haubetainger K. Total deposition of therapeutic particles during spontaneous and controlled inhalations. J Pharm Sci. 2000;89(6):724–31.
24. Dubus JC, Anhoj J. Inhaled steroid delivery from small-volume holding chambers depends on age, holding chamber, and interface in children. J Aerosol Med. 2004;17(3):225–30.
25. Le Souef PN, Devadason SG. Lung dose of inhaled drugs in children with acute asthma. J Aerosol Med. 2002;15(3):347–9.
26. Mallol J, Rattray S, Walker G, Cook D, Robertson CF. Aerosol deposition in infants with cystic fibrosis. Pediatr Pulmonol. 1996;21(5):276–81.
27. Schuepp KG, Devadason S, Roller C, Wildhaber FH. A complementary combination of delivery device and drug formulation for inhalation therapy in preschool children. Swiss Med Wkly. 2004;134(13–14):198–200.

28. Devadason SG, Huang T, Walker S, Troedson R, Le Souef PN. Distribution of technetium-99m-labelled QVAR™ delivered using an Autohaler™ device in children. Eur Respir J. 2003;21(6):1007–11.
29. Anhoj J, Bisgaard H, Lipworth BJ. Effect of electrostatic charge in plastic spacers on the lung delivery of HFA-salbutamol in children. Br J Clin Pharmacol. 1999;47(3):333–6.
30. Brambilla G, Ganderton D, Garzia R, Lewis D, Meakin B, Ventura P. Modulation of aerosol clouds produced by pressurised inhalation aerosols. Int J Pharm. 1999;186(1):53–61.
31. Murnane D, Martin GP, Marriott C. Investigations into the formulation of metered dose inhalers of salmeterol xinafoate and fluticasone propionate microcrystals. Pharm Res. 2008;25(10):2283–91.
32. Pauwels R, Newman S, Borgstrom L. Airway deposition and airway effects of antiasthma drugs delivered from metered-dose inhalers. Eur Respir J. 1997;10(9):2127–38.
33. Newman SP, Brown J, Steed KP, Reader SJ, Kladders H. Lung deposition of fenoterol and flunisolide delivered using a novel device for inhaled medicines – comparison of RESPIMAT with conventional metered-dose inhalers with and without spacer devices. Chest. 1998;113(4):957–63.
34. Hirst PH, Bacon RE, Pitcairn GR, Silvasti M, Newman SP. A comparison of the lung deposition of budesonide from Easyhaler®, Turbuhaler® and pMDI plus spacer in asthmatic patients. Respir Med. 2001;95(9):720–7.
35. Dubus JC, Dolovich M. Emitted doses of salbutamol pressurized metered-dose inhaler from five different plastic spacer devices. Fundam Clin Pharmacol. 2000;14(3):219–24.
36. Barry PW, O'Callaghan C. A comparative analysis of the particle size output of beclomethasone di propionate, salmeterol xinafoate and fluticasone propionate metered dose inhalers used with the Babyhaler, Volumatic and Aerochamber spacer devices. Br J Clin Pharmacol. 1999;47(4):357–60.
37. Clark DJ, Lipworth BJ. Effect of multiple actuations, delayed inhalation and antistatic treatment on the lung bioavailability of salbutamol via a spacer device. Thorax. 1996;51(10):981–4.
38. Louca E, Leung K, Coates AL, Mitchell JP, Nagel MW. Comparison of three valved holding chambers for the delivery of fluticasone propionate-HFA to an infant face model. J Aerosol Med. 2006;19(2):160–7.
39. Geller DE, Weers J, Heuerding S. Development of an inhaled dry-powder formulation of tobramycin using PulmoSphere™ technology. J Aerosol Med Pulm Drug Deliv. 2011;24(4):175–82.
40. Adi H, Larson I, Chiou H, Young P, Traini D, Stewart P. Agglomerate strength and dispersion of salmeterol xinafoate from powder mixtures for inhalation. Pharm Res. 2006;23(11):2556–65.
41. Begat P, Price R, Harris H, Morton DAV, Staniforth JN. The influence of force control agents on the cohesive-adhesive balance in dry powder inhaler formulations. "KONA" Powder Sci Technol Jpn. 2005;23:109.
42. Kou X, Chan LW, Steckel H, Heng PWS. Physico-chemical aspects of lactose for inhalation. Adv Drug Deliv Rev. 2012;64(3):220–32.
43. Buckton G. Characterisation of small changes in the physical properties of powders of significance for dry powder inhaler formulations. Adv Drug Deliv Rev. 1997;26(1):17–27.
44. Srichana T, Martin GP, Marriott C. Dry powder inhalers: the influence of device resistance and powder formulation on drug and lactose deposition in vitro. Eur J Pharm Sci. 1998;7(1):73–80.
45. Cegla UHU. Pressure and inspiratory flow characteristics of dry powder inhalers. Respir Med. 2004;98(Suppl 1):S22–8.
46. Kamps AWA, van Ewijk B, Roorda RJ, Brand PLP. Poor inhalation technique, even after inhalation instructions, in children with asthma. Pediatr Pulmonol. 2000;29(1):39–42.
47. Desager KN, Geldhof J, Claes R, De Backer W. Measurement of inspiratory flow through three different dry powder inhalation devices using In-Check™ in children with asthma. Pediatr Asthma Allergy Immunol. 2006;19(1):6–13.
48. Bronsky EA, Grossman J, Henis MJ, Gallo PP, Yegen U, Della Cioppa G, Kottakis J, Mehra S. Inspiratory flow rates and volumes with the Aerolizer dry powder inhaler in asthmatic children and adults. Curr Med Res Opin. 2004;20(2):131–7.

49. Parry-Billings M, Birrell C, Oldham L, O'Callaghan C. Inspiratory flow rate through a dry powder inhaler (Clickhaler®) in children with asthma. Pediatr Pulmonol. 2003;35(3):220–6.
50. Nielsen KG, Skov M, Klug B, Ifversen M, Bisgaard H. Flow-dependent effect of formoterol dry-powder inhaled from the Aerolizer®. Eur Respir J. 1997;10(9):2105–9.
51. Bisgaard H, Klug B, Sumby BS, Burnell PKP. Fine particle mass from the Diskus inhaler and Turbuhaler inhaler in children with asthma. Eur Respir J. 1998;11(5):1111–5.
52. Wildhaber JH, Devadason SG, Wilson JM, Roller C, Lagana T, Borgstrom L, LeSouef PN. Lung deposition of budesonide from Turbuhaler in asthmatic children. Eur J Pediatr. 1998;157(12):1017–22.
53. Hess DR. Aerosol delivery devices in the treatment of asthma. Respir Care. 2008;53(6):699–723.
54. Nikander K, Turpeinen M, Wollmer P. The conventional ultrasonic nebulizer proved inefficient in nebulizing a suspension. J Aerosol Med. 1999;12(2):47–53.
55. Le Brun PPH, de Boer AH, Heijerman HGM, Frijlink HW. A review of the technical aspects of drug nebulization. Pharm World Sci. 2000;22(3):75–81.
56. Gonda I. Inhalation therapy with recombinant human deoxyribonuclease I. Adv Drug Deliv Rev. 1996;19(1):37–46.
57. Le Brun PPH, de Boer AH, Gjaltema D, Hagedoorn P, Heijerman HGM, Frijlink HW. Inhalation of tobramycin in cystic fibrosis – part 1: the choice of a nebulizer. Int J Pharm. 1999;189(2):205–14.
58. Bennett WD, Brown JS, Zeman KL, Hu SC, Scheuch G, Sommerer K. Targeting delivery of aerosols to different lung regions. J Aerosol Med. 2002;15(2):179–88.
59. Dhand R, Sohal H. Pulmonary drug delivery system for inhalation therapy in mechanically ventilated patients. Expert Rev Med Devices. 2008;5(1):9–18.
60. Newman S, Gee-Turner A. The Omron MicroAir vibrating mesh technology nebuliser, a 21st century approach to inhalation therapy. J Ther Res. 2005;5(4):29–33.
61. Collis GG, Cole CH, Lesouef PN. Dilution of nebulized aerosols by air entrainment in children. Lancet. 1990;336(8711):341–3.
62. Brodie T, Adalat S. Unilateral fixed dilated pupil in a well child. Arch Dis Child. 2006;91(12):961.
63. Everard ML, Clark AR, Milner AD. Drug delivery from jet nebulizers. Arch Dis Child. 1992;67(5):586–91.
64. Coates AL, Denk O, Leung K, Ribeiro N, Chan J, Green M, Martin S, Charron M, Edwardes M, Keller M. Higher tobramycin concentration and vibrating mesh technology can shorten antibiotic treatment time in cystic fibrosis. Pediatr Pulmonol. 2011;46(4):401–8.
65. Nikander K, Arheden L, Denyer J, Cobos N. Parents' adherence with nebulizer treatment of their children when using an adaptive aerosol delivery (AAD™) system. J Aerosol Med. 2003;16(3):273–81.
66. Keller M, Jauernig J, Schueepp K, Stangl R, Ohl S, Roller C, Devadason S, Wildhaber J. Using infant deposition models to improve inhaler system design. Respiratory Drug Delivery IX, Palm Desert; 2004.

Chapter 15
Paediatric Formulations: Nasal, Ocular and Otic Drug Delivery

Hannah Batchelor

Abstract Non-invasive routes of administration of drugs are attractive in paediatric populations. The nose, eye and ear offer opportunities for drug delivery for localised (and systemic) treatment. This chapter considers anatomical and physiological differences in children compared to adults such that delivery of medicines at these sites is safe and efficacious. In addition, the benefits and limitations of formulations used at these sites are discussed with particular relevance for the paediatric population.

Keywords Aqueous humour · Ocular surface · Cochlear implant · Diabetic macular oedema · Posterior segment

15.1 Nasal Drug Delivery

Nasal preparations are typically used to treat local diseases, including nasal allergy, rhinitis, bacterial sinusitis or nasal polyps in children. However, there has also been interest in delivery via the nasal cavity for systemic therapeutics, particularly where a rapid onset of action is required.

Nasal administrations of drugs have several significant advantages over current practices. The nose has a very rich vascular supply, and it facilitates direct absorption to the systemic blood supply and increases bioavailability of the drug, compared to oral administration [1]. The highly vascularised nasal mucosa and the olfactory tissue being located in direct contact with the central nervous system allow nasally administered drugs to be rapidly transported into the bloodstream and brain, with onsets of action approaching that of intravenous therapy, following

H. Batchelor (✉)
Strathclyde Institute of Pharmacy and Biomedical Sciences, University of Strathclyde, Glasgow, UK
e-mail: Hannah.batchelor@strath.ac.uk

H. Batchelor, K. Rose (eds.), *Pediatric Formulations*, AAPS Advances in the Pharmaceutical Sciences Series 58, https://doi.org/10.1007/978-3-031-77240-5_15

administration of the medicine the child can sit or stand up due to the rapid absorption. In addition, there is no requirement for the medicine to be sterile.

15.1.1 Structure of the Nose

In adult humans, the nasal cavity volume is 15–20 mL; the surface area is 150–180 cm^2 and is covered by a 2–4 mm thick mucosa, of which 5–10 cm^2 is olfactory, and the remaining 145–170 cm^2 is respiratory [2]. Although the surface area of the olfactory epithelium is relatively small, it can enable direct access to the central nervous system by bypassing the blood-brain barrier. The absorptive surface includes cells that are ciliated; these ciliated cells have approximately 100 cilia on their apical surface, which are used to transport mucus towards the nasopharynx [3]. Other cells within the nasal cavity are covered with microvilli, which also act to enlarge the overall surface area.

Mucociliary clearance involves the combined actions of the mucus layer and the cilia, and is an important factor in the physiological defence of the respiratory tract against inhaled hazardous particles. It is assumed that the speed of mucociliary clearance in healthy human adults is about 5 mm/min [4], although this is affected by pharmaceutical excipients [5] and diseases (including cystic fibrosis and diabetes) [6]. These differences in mucociliary clearance can affect exposure to intranasally administered drugs.

Nasal mucus, which is secreted mostly from submucosal glands, is composed of water (95%), glycoproteins (2%), other proteins (1%), inorganic salts (1%) and lipids (<1%) [7]. Hydrophilic drugs are highly soluble in mucus, which makes them susceptible to mucociliary clearance. Medicines deposited on the nasal surface are cleared via mucociliary clearance mechanisms; inhaled particles are eliminated within 15–30 min [8].

Intranasal baseline pH is approximately 6.3 in adults; this will affect the solubility of drugs whose pKa is around this value.

Drugs are absorbed across the nasal epithelium according to the drug's intrinsic physicochemical properties. For passive diffusion, the concentration gradient drives absorption; therefore, a highly concentrated solution within the nasal cavity will drive absorption.

To ensure adequate coverage of the nasal epithelium, the maximum volume to be administered in each nostril is 150 μL in adults to avoid runoff into the pharynx [9]. The unit volume administered is also important because it appears that the administration of a single volume of 100 μL leads to deposition over a greater surface area than that obtained with the administration of two 50 μL volumes [10].

15.1.2 Age-Related Differences in the Nose

It is evident that adults and children have substantially different nasal airways with respect to their size, shape and age-dependent characteristics. In general, the nasal cavity of a child is smaller than that of an adult. This means that the total surface area of the nasal epithelium available for drug absorption is smaller in children, which can affect the absorption rate and extent of drug uptake. Porter et al. (1997) demonstrated that nasal volume was shown to correlate with age, height and weight in children [11]. Xi et al. (2011) reported that the nasal airway volume and surface area of the 5-year-old child are 40.3% and 65.7% that of an adult, respectively [12]. Further work showed that the relative values are 6.4%, 18.8% and 24.2% in a 10-day, 7-month and 3-year subject, respectively. Dose adjustments made based on weight and height should match the overall surface area such that the applied dose is similar in adults and children, and absorption will be matched. The parameters of intranasal spaces have been reported to depend on age and gender, where growth is completed by the age of 16, and after this age, nasal cavities are bigger in men than in women [13].

There are no reports on the differences in nasal mucus, nasal pH, or mucociliary clearance in paediatric patients compared to adults. Therefore, it is assumed that these properties remain the same in children as in adults.

Balbani et al. (2004) reported that in Brazil between 1996 and 2000 there were 233 cases of toxicity in children caused by topical nasal medicines, the main causes of toxicity were accidental intake and error in administration [14].

15.1.3 Impact of Differences in Paediatric Nose Structure and Physiology on Drug Performance

There are no reported differences in toxicity or clinical efficacy resulting from nasal drug delivery when comparing paediatric to adult patients. The dose adjustment is typically undertaken by instillation of fewer nasal drops or fewer nasal sprays to ensure that the relevant dose is administered.

15.1.4 Local Delivery

Localised delivery to the nose is often measured in terms of administered dose, droplet or particle size distribution, spray pattern and plume geometry. Spray pattern and plume geometry define the shape of the expanding aerosol cloud, while droplet size determines the likelihood of deposition within the nasal cavity by inertial impaction. The distribution within the nasal cavity determines the efficacy of the treatment. When formulations are compared, it is the aspect of the dispersion that is used to document bioavailability and bioequivalence [10]. The nasal turbinates, which are bony structures that project into the nasal cavity, are proportionally larger in children. This can create a more turbulent airflow in the nasal cavity, which can affect the distribution of the drug and the amount that reaches the site of absorption [12]; the use of age-appropriate models of the nasal passageways is recommended.

15.1.5 Systemic Delivery

Systemic delivery of drugs via the nasal cavity is attractive due to the low barrier to permeation and the rich vascular nature. Examples of marketed nasal products include drugs for the treatment of migraine, e.g. zolmatriptan (Zomig®), sumatriptan (Imitrex®) and butorphanol tartrate (Stadol NS); treatment of severe pain, e.g. fentanyl (PecFent®; Instanyl®); for smoking cessation (Nicorette®); and for treatment of menopausal symptoms (Aerodiol®).

15.1.6 Nasal Formulations

Intranasal medications can be delivered in several methods. Drops can be applied from a syringe, and the drug can be nebulised or given through pressurised aerosol. All have been demonstrated to be effective. While it is believed that metered-dose systems provide the greatest dose accuracy and reproducibility, the ease of use of device can vary significantly.

15.1.6.1 Nasal Liquids (Drops and Sprays)

Many different vehicles can be used for delivery intranasally; polymers and gels can be used to increase viscosity and thereby increase residence time within the nasal cavity; bioadhesive microspheres may also increase residence time. The viscosity of the solution can affect the deposition area, with solutions of higher viscosity having a reduced deposition area [15].

An atomised nasal spray results in the distribution of the nasal formulation over a larger surface area compared to nasal drops; this leads to less drug loss to the pharynx and improved clinical effectiveness. The spray administration and plume angles are key determinants of optimal nasal drug delivery. The combination of an administration angle of 30° and a plume angle of 30° led to deposition primarily in the anterior region of the nose, with a deposition efficiency close to 90% [16].

Nasal solutions use the drug in a high concentration within the formulation as only a small volume can be administered (approximately 150 μL). Suspensions for nasal administration typically contain micronised drug to ensure rapid dissolution in situ.

Nasal liquid systems drain from the nose to the gastrointestinal tract; upon passage, these can result in an unpleasant or bitter taste. Although there have been no reports of flavoured commercial formulations, there are patents that detail the use of taste-masking or flavouring agents within nasal liquid formulations. The use of flavours may well enhance acceptability within the paediatric population. Xylometazoline nasal spray to treat nasal congestion contains menthol and eucalyptus odours, which may be present both to boost perceived efficacy and to aid the acceptability of the product.

The intranasal formulation of zolmitriptan was recently approved by the FDA in the USA for use in children aged 12 years and over [17].

15.1.6.2 Nasal Powders

Nasal powders are inhaled by patients. The supply of a powder will increase the stability of the drug compared to a liquid formulation; in addition, it is less likely to require taste masking. It is believed that the powder forms a viscous gel within the nasal mucus, which may assist in retention within the nasal cavity. The first powder formulations that entered the market were locally acting products against hay fever that were used in Japan. In 2016, Onzetra Xsail (sumatriptan; Currax Pharmaceuticals, Morristown, NJ, USA) was the first nasal powder formulation for systemic action (to treat migraine) approved by the Food and Drug Administration (FDA), followed by Baqsimi (glucagon, Eli Lilly, Indianapolis, IN, USA) in 2019 [18]. Baqsimi is indicated for use in children aged 4 years and older. Glucagon without promoters is not absorbed through the nose, and the commercial formulation includes beta-cyclodextrin plus dodecyl-phospho-choline as the absorption enhancer [19]. The same formulation is used in both adult and paediatric populations.

15.1.6.3 Nasal Particulate Delivery Systems

Particulate drug carrier systems may be used to aid in systemic uptake from nasal administration, where the carrier may protect the drug from enzymatic degradation, increase the drug dissolution rate, intensify the contact of the formulation with the mucosa, enhance the uptake by the epithelium and act as a controlled release system resulting in prolonged blood concentrations [20].

Nanocarriers can effectively increase the amount of drug that reaches the systemic system. For a full review, the reader is directed to [21–23].

15.1.6.4 Nasal Ointments and Emulsions

Emulsion formulations provide prolonged residence time within the nasal cavity to enhance the duration of action of the drug. Emulsions that are ionic can also form electrostatic interactions with the mucus present within the nasal cavity to further enhance retention. The presence of oil within these formulations can lead to improved absorption although the disadvantages of ointments and emulsions are associated with lack of dose accuracy and poor patient acceptability. Viferon is a nasal ointment licensed for use as a part of complex therapy for influenza and other viral infections in children over 1 year [24].

15.1.6.5 Nasal Gels

Gels can be localised within the nasal cavity to promote retention and/or absorption. They can incorporate agents to reduce the irritation that is often associated with liquids or powders. The disadvantage of gels is their distribution where viscous gels can be localised and not homogenously distributed. Specialist devices are often used to dispense gels due to their higher viscosity. Octenisan nasal gel is an antibacterial agent used to moisten and prevent decontamination by external cleaning (physical cleaning) of nasal vestibules.

15.1.6.6 Nasal Vaccines

Within the possibilities for mucosal vaccination, the nasal cavity is one of the most promising sites due to (i) its reduced enzymatic activity compared to other possible administration routes (e.g. oral route), (ii) its moderately permeable epithelium and (iii) the high availability of immuno-reactive sites. Typically, particulate formulations are used for nasal vaccination.

Fluenz™ influenza nasal vaccine (AstraZeneca) was licensed in 2011 and available in European markets for the 2012–2013 influenza season. Fluenz™ has been tested to demonstrate that it is appropriate for use in children over 2 years of age. The acceptability of Fluenz was reported to be easy to administer with a high degree of satisfaction with use from both health workers and parents [25].

The COVID-19 vaccine products used in children were identical formulations to those used in adults, demonstrating that the formulation aspects are less critical compared to the dose adjustments required.

15.1.7 In Vitro Models to Predict Exposure in Paediatric Populations

Current regulatory guidance recommends that several in vitro tests (emitted dose, droplet or particle size distribution, spray pattern, plume geometry) can be used to characterise or compare nasal drug products in support of regulatory applications for new drugs or for generic drugs. Several in vitro models have been developed that mimic the nasal anatomy to demonstrate the deposition of nasal formulations.

The Sophia Anatomical Infant Nose-Throat (Saint) Model is an anatomically accurate model of the upper airways of a 9-month-old infant [26]. The Premature Infant Nose Throat-Model (PrINT-Model) is an upper airway replica of a premature neonate (32 weeks gestational age), which has also been useful in predicting deposition from an aerosol [27]. There has been further work on the development of nasal airway models for children [28].

15.1.8 Discussion/Conclusions

Nasal drug formulations are used within paediatric populations routinely. There are no reported paediatric specific formulations that differ from adult products. However, the similarities in anatomy and physiology ensure that products are likely to perform in the same way in adults compared to children with a few reported adverse effects following nasal drug delivery. A review of prescribed intranasal products in a paediatric population revealed that most products were used to treat local conditions, and saline was the most commonly prescribed product [29].

15.2 Ocular Drug Delivery

Many ocular medications are used in children to treat common bacterial and viral infections, inflammation and allergy, uveitis and glaucoma, as well as other conditions including myopia, amblyopia and strabismus. Eye conditions are prevalent in paediatric populations; within the United Kingdom, more than 5% of children have had at least one eye condition by the age of three [30].

The ocular tissues can be reached either by local or systemic drug administration. The route of administration can be broadly divided into those targeting the anterior or posterior sections of the eye. The absorption, distribution and elimination of drugs may be altered in children compared to adults; the effects of this on ocular drug delivery are detailed.

15.2.1 Structure of the Eye

The anterior segment of the eye consists of external cornea, conjunctiva, aqueous humour, iris-ciliary body and lens. The cornea and lens obtain most of their necessary nutrients from aqueous humour, with tear fluid providing additional nutrients to the cornea. The iris-ciliary body and conjunctiva are highly vascular tissues. The aqueous humour is a dynamic watery fluid that is continuously secreted by the ciliary body and drained via the canal of Schlemm. A schematic diagram of the structure of the eye is shown in Fig. 15.1.

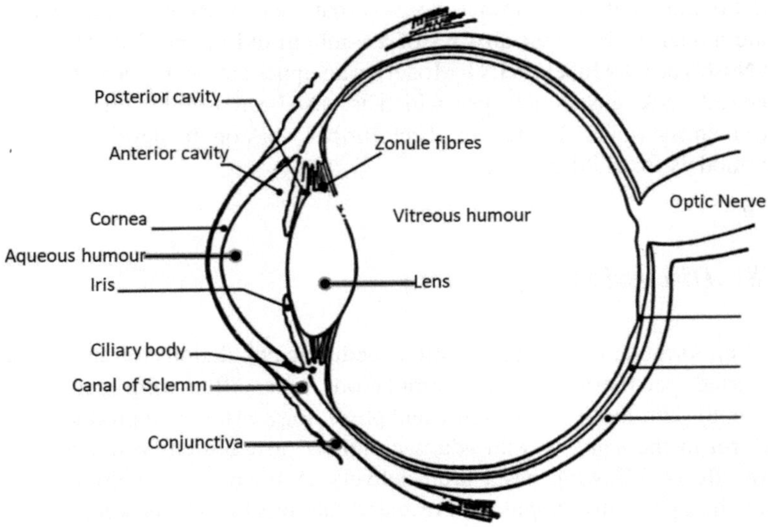

Fig. 15.1 A schematic representation of the eye

The posterior segment of the eye consists of outer sclera, choroid, retina and vitreous humour. The sclera is an avascular tissue that acts as a protective layer. Beneath the sclera is the highly vascular choroid that supplies nutrients to the outer sclera and inner retina. The retina is primarily responsible for image formation and therefore vision. The blood–retina barrier is comprised of retinal pigment epithelium and the endothelium of retinal blood vessels. Unlike the aqueous humour, the vitreous humour is a clear watery viscous fluid that is replaced at a slow rate.

15.2.2 Age-Related Differences in the Eye

The eye of the newborn is roughly two-thirds of its adult size at birth; it reaches adult size around ages 3–4 year [31]. In the eye, membranes are thin in neonates and infants; absorption and corneal permeation may be more rapid in infants and neonates. The cornea of the neonate has 70% of the absorptive surface of the adult cornea, but the total intraocular volume is barely one-third of the adult eye [32]. The area of contact between the posterior conjunctiva and the eye globe has been approximated to 4 cm^2 in adults; this surface area would be reduced in children. The ratio of surface area to internal volume could lead to drugs becoming somewhat concentrated within the eye of paediatric patients. A study that explored the use of Travopost 0.004% ophthalmic solution in children aged 2–18 years showed no clear relationship between age, body surface area and Cmax, suggesting that eye surface area is not a relevant factor for the pharmacokinetics of eye drops [33].

15.2.2.1 Tear Volume

Basal tear volume increases with age; typical volumes reported are 0.5 μL (range 0.6–2 μL) for neonates, 2.5 μL (range 1.4–7.75 μL) for infants (mean age 7 weeks) at an older age and 6 μL (range 2.73–12.75 μL) in adults [34]. This age-related reduction in tear volume can lead to topically applied medications becoming concentrated in younger patients.

Adult pharmacokinetic studies have shown that a 20 μL eye drop achieves the maximum tear film concentration as the low capacity of the precorneal area limits the overall retained volume. The average size of a conventional ophthalmic eyedrop is 42.6 μL (ranging from 33.3 to 59.0 μL) [35].

There is no literature data regarding the capacity of the precorneal area in paediatric patients. However, Lawrenson et al. reported that the dimensions of the palpebral aperture change with age, with a rapid increase in width and length occurring from birth to 1 year [36]. These changes in palpebral aperture size have an effect on the exposed ocular surface area, with a 50% increase in mean surface observed from 0–17 to 36–53 weeks [36].

Therefore, it could be assumed that the precorneal area in children is reduced compared to adult values with the greatest rate of change being in the first year of life.

The bulk of the tear volume drains through the nasolacrimal system; however, a significant proportion of tears is lost to evaporation. Tears drained via the nasolacrimal system enter the gastrointestinal tract, where any drug contained within the tear fluid is subject to gastrointestinal absorption processes.

15.2.2.2 Tear Composition

Tear fluid is a complex aqueous solution containing proteins, metabolites, electrolytes and lipids. The normal pH range of tears in adults was reported as 6.5–7.6 with a mean value of 7.0 by Abelson (1981) [37]. Carney and Hill (1976) reported that tear pH in adults ranged from 5.2 to 8.6 with a mean value of 7.45, and a reduction to 7.25 was observed in the closed eye [38]. Tear pH was measured in 173 newborn infants, and a mean pH value of 6.74 ± 0.26 was recorded [39]. The lower pH value observed in newborn infants may be a result of the fact that they have their eyes closed for longer periods than adult subjects. Slight changes in pH can affect the solubility of drugs; ciprofloxacin has a minimum solubility at pH 7; therefore, small change in pH around pH 7 can greatly affect its solubility.

The composition of tear fluid in adults has been shown to be affected by eye closure; immediately upon waking, higher levels of IgA as well as reactive complement components and activated polymorphonuclear leukocytes have been reported [40]. The equivalent studies have not been carried out in neonates. However, since neonates spend a large proportion of their time sleeping, coupled with a low turnover of tears, a tear composition closer to the adult closed eye is likely.

15.2.2.3 Blinking Rate

Upon instillation of an eyedrop, the tear film will increase until a blink occurs. Thus, any excess instilled volume is rapidly diminished by reflex blinking causing overflow to the check coupled with the extended drainage capacity through the nasolacimal system.

From a developmental perspective, blink rate increases with age and reaches adult levels by age 20 years. The blink reflex was assessed following a glabellar tap in 164 infants and children from ages 2 days to 18 years and in 18 adults aged 18–50 years [41]. The results showed that the mean number of glabellar taps required to induce a blink increased from 2.7 at 0–2 months of age to a peak of 13.3 at age 3–4, remained at more than 10 until age 6, after which a rapid decline occurred, reaching the adult level of 2–5 taps at age 12 years [41]. Spontaneous blink rates were measured in 269 children and 179 adults, with blink rate increasing with age as shown in the table below [41].

Age	Mean spontaneous blink rate (per minute)
0–2 months	0.7
1–4 years	3.4
5–10 years	6.1
11–15 years	10.3
15–20 years	11.3
>20 years	16.0

The rate of blinking will affect the clearance of the excess solution from the eye with more rapid clearance expected in adults compared to paediatric patients.

15.2.3 *Impact of Differences in Paediatric Eye Structure and Physiology on Drug Performance*

Topical application of ocular drugs may cause serious adverse ocular or systemic side effects. Children are at greater risk of systemic side effects because ocular dosing is not weight-adjusted, and infants are especially vulnerable, particularly in cases where drug metabolism is reduced in the young and/or an immature blood–brain barrier.

There are examples where topical administration of ophthalmic medicines in children has led to elevated systemic drug concentrations or systemic side effects; examples include:

Elevated plasma levels of brimonidine 1459 pg/mL and 700 pg/mL following instillation by eye (compared to reported adult studies that show a maximum concentration of 60 pg/mL), leading to somnolence or coma [42].

Blood levels of timolol in 5 small children ranged from 3.5 to 34 ng/mL, in contrast to 2.45 ng/mL in adults following topical ocular administration [43].

Systemic exposure of latanoprost ophthalmic solution 0.005% once daily was higher in a <3-year age group (166 pg/mL) versus other groups (49, 16 and 26 pg/mL for the 3–<12-year, 12–<18-year and ≥18-year age groups, respectively [44].

The increased systemic exposure observed in paediatric patients has been attributed to absorption of eyedrops into the systemic circulation, where the reduced size of the patient results in higher plasma levels of circulating drug.

Calculating dosages for paediatric patients is complex; body weight, surface area, development, metabolism, other medications taken and physiologic function can all affect the dosage. Pharmacokinetic models that adjust dosage based on aqueous humour volume ratio have previously been proposed for pilocarpine, a drug delivered topically [32]. It is estimated that a newborn requires only one-half of the adult dosage of eyedrops to obtain an equivalent ocular concentration, two-thirds of the adult dosage is required at age 3, and 90% of this dosage at age 6 [45].

15.2.4 Systemic Ocular Drug Delivery

Systemic delivery of therapeutic agents to the eye is poor. Systemic delivery to the eye requires permeation of the blood–aqueous barrier to reach the anterior chamber or the blood–retinal barrier to reach the posterior section of the eye. The blood–aqueous barrier, composed of the uveal capillary endothelia and ciliary epithelia, limits the access of compounds from the systemic circulation to the anterior chamber, whereas the blood–retina barrier limits the drug diffusion from the systemic blood to the retina and vice versa.

Access to the aqueous humour of the anterior and posterior chambers is restricted by the blood–aqueous barrier. Compounds administered systemically can penetrate the vessels of the ciliary body and diffuse through the iris into the aqueous humour in the posterior chamber. Movement into the posterior chamber from the anterior chamber is restricted by the diaphragm-like action of the iris on the lens. The permeability of the blood–aqueous barrier is reported to be unaffected by age between 13 and 72 years [46].

15.2.5 Local Targeting of the Anterior Segment

Local delivery of therapeutic agents ensures that the drug is present at higher concentrations than may be achieved following systemic administration with reduced incidence of side effects.

Drug absorption following topical delivery can occur by two routes: the corneal and non-corneal (conjunctival/scleral) pathways. In the precorneal space, solution drainage, lacrimation and tear dilution, tear turnover, conjunctival absorption and the corneal epithelium limit transcorneal penetration. The low fraction of applied drug reaching the anterior chamber further undergoes rapid elimination from the intraocular tissues and fluids by distribution into non-target tissues and as a consequence of aqueous humour flow.

Corneal permeability has been shown to decrease with age in a rabbit model, where the permeability of pilocarpine was higher in the younger cornea (representing a neonate) compared to that representing a 3-year-old child [47]. Additional studies [48–50] support this data and indicate that corneal permeability is well correlated with epithelial maturation. The permeability differences due to age may be attributed to structural variations; for example, the overall corneal thickness or may in fact reflect more subtle differences in one or more of the corneal layers [47].

The non-corneal or conjunctival/scleral route of absorption involves penetration across the conjunctiva and sclera and then into the intraocular tissues. This mechanism of absorption was once thought to be non-productive. However, many studies have shown that the conjunctival/scleral route is significant for compounds with poor corneal permeability, such as inulin, timolol maleate, gentamicin, bimatoprost and prostaglandin PGF2-ALPHA [51–54]. Indeed, rabbit conjunctival tissue is relatively more permeable than corneal tissue in terms of paracellular permeability [48].

Typically, the anterior segment of the eye (cornea, conjunctiva, sclera, anterior uvea) is treated using formulations including solutions, suspensions, ointments and gels which contain excipients to increase the contact time of the formulation with the eye to maximise the bioavailability of the drug from the formulation. Eye drops are the most commonly used formulation, yet they are rapidly drained from the ocular surface, and, therefore, the time for drug absorption is only a few minutes, and bioavailability is very low, typically less than 5% [55, 56]. The high rate of clearance of formulations from the eye is rapid due to tear turnover, although differences in tear production and blinking rate may lead to slower clearance in neonates compared to adults.

15.2.6 Targeting the Posterior Segment

The anatomic and physiologic barriers of the eye render drug delivery to the posterior segment (vitreous humour, retinal pigmented epithelium, retina and choroid) tissues a major challenge. A fundamental mechanistic understanding of the absorption, distribution and elimination pathways for delivery of drugs to the posterior segment is required for rational treatment paradigms.

15.2.6.1 Via Systemic Routes

Penetration of drugs into the posterior segment of the eye from the systemic circulation is restricted by the blood–retinal barrier (BRB). For a drug to cross the BRB, it should exhibit either optimum membrane partition characteristics or should target one of the membrane transporters present on the retinal pigmented epithelial (RPE) cells [57].

The blood-aqueous barrier (formed by the nonpigmented layer of the epithelium of the ciliary body and the endothelium of the blood vessels of the iris) is dual-facing as it is exposed to both the anterior and posterior eye; typically, hydrophilic substances penetrate the anterior chamber, whereas lipophilic substances (e.g. chloramphenicol and tetracycline) penetrate the posterior segment of the eye [57].

Following drug diffusion into the posterior segment of the eye, the drug is often localised at the site of appearance rather than distributed homogenously throughout the vitreous humour, and negligible drug concentrations are present at the centre of the vitreous body. For a drug to have high concentration in the posterior segment following systemic administration, the drug must be able to traverse the RPE, have some depot effect and have a prolonged concentration gradient to drive the diffusion process.

Most of the drugs used in the treatment of posterior segment diseases are not able to penetrate the membranes to reach the desired site of action. Therefore, in practice, large doses of intravenously administered drugs (e.g. steroids, antibiotics, antivirals) are required to create the necessary concentration gradients to drive transport into the vitreous humour. Typically, drugs are delivered via intravitreal and periocular routes to target the posterior segment of the eye. However, there are very few paediatric medicines licensed that target the posterior segment of the eye.

15.2.6.2 Via Topical Routes

Penetration and distribution of a drug into the posterior tissues of the eye after topical administration are rare. However, memantine HCl and brimonidine tartrate have been shown to reach the vitreous and retina after topical ocular delivery [57].

To maximise penetration of topically administered eyedrops to the posterior segment, the patient should be supine to allow the formulation to flood the conjunctival cul de sac, and the diffusion into the posterior sclera is maximised [56]. Lipophilic drugs are more likely to be able to traverse the membranes and target the posterior segment following topical application.

15.2.6.3 Via Intra-Ocular Injections

Direct intra-ocular administration circumvents the many barriers that prevent drugs from penetrating the eye when delivered either topically or systemically. However, this route has low patient acceptability and requires a health professional to administer the medicine. More recent research has looked at solid metal microneedles measuring 500–750 μm in length as a means of delivering drugs into the eye via intrascleral and intracorneal routes in a minimally invasive manner. This theory was assessed by Jiang (2007) who demonstrated that microneedles were mechanically strong enough to penetrate into human cadaveric sclera and that the drug coating rapidly dissolved off the needles within the scleral tissue within 30 s after insertion [58].

15.2.7 Ocular Formulations

15.2.7.1 Solutions

The precorneal residence of an ophthalmic solution can be increased by the inclusion of viscosity enhancing or bioadhesive polymers as excipients. Various polymers have been used including carbopol gels, cellulose derivatives, dextran, gelatin glycerin, polyethylene glycol, poloxamer 407, polysorbate 80, propylene glycol, polyvinyl alcohol and polyvinyl pyrrolidone to prolong the ocular residence time. The use of such excipients, however, remains applicable to only hydrophilic drugs, and the advantage of increasing the viscosity must be balanced against the potential disadvantage of inducing ocular disturbances due to the blurring of vision as a result of a change in the refractive index on the ocular surface.

Certain polymers have additional properties that increase the residence time, including those able to interact with the mucus layer covering the corneal and scleral surfaces. Furthermore, polymers that undergo a phase transition from a solution to a gel within the eye can have benefits (e.g. pH, temperature or mono/divalent ion concentration). For a full report on polymers and their function in ocular drug delivery, the reader is referred to [59]. The use of microspheres and liposomes in targeting the anterior segment of the eye has largely focused on improving corneal adhesion and permeation by incorporating various bioadhesive and penetration-enhancing polymers within the outer surface of the microparticles [60].

A significant issue with solutions is that the volume of liquid dispensed from each drop is difficult to control which leads to wastage. Using an appropriate drop volume would reduce wastage of product and may assist in reducing systemic side effects in younger children [61]. Reducing drop size cannot be achieved by just changing the size and opening of the tip of the dropper. Formulations vary in their surface tension, which results in different-sized drops when the volume of fluid is the same. Investigations have focused on altering surface tension by changing the inactive ingredients; alternatively, a spray system may offer benefits in terms of smaller droplets.

Due to the mechanism by which eye drop solutions are cleared, taste of the formulation may need to be considered in certain cases. An unpleasant bitter taste following eye drops has been reported for many drug products (e.g. in adults using topical therapy for lowering intraocular pressure [62]). There are currently no flavoured eye drops available for paediatric patients that may improve acceptability of the formulation although there are several patents in this area, including orange and lemon flavoured Opcon-A® eye drops [63]. The regulatory burden associated with inclusion of a flavouring agent within an ocular solution is significant as full ocular toxicity testing is likely to be required for any flavouring agent.

15.2.7.2 Suspensions

The use of suspensions in ophthalmology is limited as suspensions present problems with regard to physical stability and method of sterilisation, having to be prepared aseptically due to the unsuitability of other techniques. Sieg and Robinson (1975) claimed that for a suspension formulation to show any advantage over a saturated solution in terms of ocular bioavailability, the dissolution rate of the particles must be equal to or greater than their rate of clearance from the precorneal area [64]. As described previously, tear fluid may be retained for longer in very young children; therefore, there may be greater drug concentrations within the tear fluid as more may be dissolved. This may lead to a greater effect of a drug administered as a solution to the very young compared to an adult patient.

15.2.7.3 Ointments

Ophthalmic ointments typically have greater retention in the precorneal region compared to aqueous formulations [65] meaning that the frequency of application can be reduced. However, their major limitation is the associated blurring of vision caused by administration of the ointment; this factor is less of an issue for night-time use and may be less problematic in younger children. Ointments are administered following instructions such as "Apply 1.5 cm of ointment every 3 hours"; this type of instruction allows for dosage adjustments to be made to account for younger patients if required. This is an advantage compared to solutions and suspensions, where the drop size is determined by the formulation and dropper supplied.

15.2.7.4 Gels

Eye gels are used as a means to deliver longer-acting agents with the most used example being timolol maleate. The product can be formulated to be applied as a solution that gels upon contact with the ocular surface. The systemic absorption can also be reduced by the use of a gel as the drug is retained at the ocular surface rather than cleared via the nasolacrimal route. There are no particular differences with this formulation for use in children as compared to adults. Timolol-LA is an example of a gel-forming ocular product although it is not licensed for use in paediatric populations; however, there have been clinical trials that evaluate its use in a paediatric population (Clinical trial.gov identifier: NCT01652664).

15.2.7.5 Inserts

Ocular inserts are solid or semi-solid devices, usually made of polymeric materials, that are placed in the conjunctival sac to deliver drugs to the ocular surface. Biodegradable silicone-based ocular implants have been developed to continuously release cyclosporine A for several years. These devices have higher bioavailability for deep ocular tissues compared to topical delivery [66, 67]. Insertion and replacement of these CyA implants require surgery, and thus, these appear to be more suitable for severe ocular diseases that need long-term treatment. The potential advantages offered by inserts include accurate dosing for adults, increased ocular residence time, reduction in systemic side effects, improved patient compliance due to reduced frequency of administration and potential to control the release of drugs over time as well as increased shelf-life. Despite the advantages of ocular inserts, their main disadvantage is the foreign body sensation associated with their initial administration. However, this disadvantage did not prevent the adoption of this technology in several successfully marketed ocular inserts (Ocusert®, Ocufit® SR, and Minidisc®). The accuracy of dose, although measured for adults, is not adapted for paediatric patients therefore these may only be suitable for older children.

15.2.7.6 Punctal Plugs with Active Ingredient

Occlusion of puncta through punctal plugs is a common non-medical treatment of dry eyes as punctal plugs block the flow of tears through the canaliculi, which connect the eyes to the nose. Insertion of punctal plugs has been reported to improve tear film stability and tear osmolarity and also to improve vision for dry eye patients [68]. Commercial punctal plugs range in length from 1.1 to about 2 mm and in diameter from 0.4 to 1.1 mm. Typical drug-eluding punctal plugs consist of cylindrical cores coated with an impermeable shell where the drug essentially diffuses out from the circular cross-section in contact with the tears. The rate of drug diffusion can be controlled to ensure delivery of therapeutic doses over a prolonged period, as demonstrated by Gupta et al. (2011) using cyclosporine A as an example drug [69].

Although dry eye severe enough to warrant treatment is rare in children the use of punctual plugs in children with a mean age of 7 years was reported by Mataftsi et al. [70]; however, there are limited reports of punctual plugs incorporating active ingredients in children. Recently, the Dextenza dexamethasone insert was trialled in the treatment of postoperative ocular inflammation in children undergoing cataract surgery with positive results suggesting that the dexamethasone insert reduced or eliminated the need for postoperative steroid drops [71].

15.2.7.7 Contact Lens

Ophthalmic drug delivery through contact lenses significantly increases drug residence time and bioavailability compared to drug delivery via eye drops.

Nanoparticle-laden lenses for extended delivery of several drugs, including cyclosporine A, have been developed [72]. Additionally, biomimetic and "imprinted" contact lenses to increase the drug loading and release durations have been developed with most work undertaken on timolol [73–75]. Also, recently, Ciolino et al.

(2009) developed a lens containing a layer of drug-loaded PLGA (Poly[lactic-co-glycolic acid]) film sandwiched between layers of pHEMA (poly[hydroxyethyl methacrylate]) for extended delivery of ciprofloxacin [76].

To date, there are no commercially available drug-releasing contact lenses although Vistakon Pharmaceuticals has completed a clinical trial of a contact lens that would release ketotifen-4 to treat allergic conjunctivitis; the eligibility for this study was from aged 8 years so there is potential for this device to be used in older paediatric patients.

15.2.7.7.1 Visulex® Noninvasive Iontophoretic Ocular Drug Delivery Device

The Visulex® (Aciont Inc., Salt Lake City, UT) is a noninvasive, iontophoretic ocular drug delivery scleral-lens shaped application device, which maintains a drug reservoir in the same location on the eye's conjunctival surface to minimise solution leakage, exposure of the drug to the cornea and surface clearance effects. Iontophoresis is a method of drug delivery that utilise electric current to deliver ionised molecules to intraocular tissues, where the ionised molecules are driven by specially designed ocular surface electrodes. Iontophoresis offers the benefits of targeting drugs to specific regions within the eye. There is no literature evidence that details the use of iontophoresis in paediatric populations.

15.2.7.8 Subconjunctival Injection

Subconjunctival injections are routinely used for the administration of anti-infective drugs, mydriatics or corticosteroids for conditions not responding to topical therapy. Following injection, the drug will diffuse through the sclera to reach the anterior and posterior sections and the vitreous humour avoiding the cornea, which has low permeability compared to the sclera. As the subconjunctival space is limited, this type of formulation is only suitable for drugs that are readily soluble. In addition, to avoid frequent administration of subconjunctival injections, a slow-release formulation is preferred. Typically, a concentration of 1 microgram per millilitre may be achieved in the vitreous following this type of administration [77].

15.2.7.9 Intravitreal Injection

Direct intravitreal injection of drugs into the vitreous cavity is employed to achieve higher drug concentrations in the vitreous and the retina. However, repeated injections are needed to maintain drug concentrations at an effective therapeutic level over a certain period of time since the half-life of drugs in the vitreous is relatively short. Repeated intravitreal injections result in extreme patient discomfort and may lead to complications such as vitreous haemorrhage, infection, and lens or retinal injury. Formulation strategies focus on improving the retention site at the site of injection or the use of drugs with a long half-life to avoid frequent administration of intravitreal injections.

15.2.7.10 Microparticles for Sustained Release Injection Formulations

Drugs can be formulated to be encapsulated within microparticles, such as microspheres or liposomes, which can serve as the basis for intravitreal injections to provide sustained drug release. Microspheres of biodegradable polymers such as poly(lactic acid) (PLA) or poly(lactide-co-glycolide) (PLGA) have been shown to effectively deliver drugs to the vitreous and retina and can be tolerated by the ocular tissues [78].

15.2.7.11 Nanomedicines for Ocular Delivery

The use of nanoparticles in ocular drug delivery provides an opportunity to enhance the bioavailability of topically administered drugs. Nanomedicines are able to target ocular tissue and exert sustained efficacy via controlled release. The use of nanomedicines has proven promising for adults, but there are yet to be studies conducted in paediatric populations [79].

15.2.7.12 Intravitreal Inserts

Solid biocompatible implantable devices for sustained or controlled intravitreal drug delivery to the posterior segment of the eye have been developed employing diverse approaches and include the use of implantable devices such as osmotic mini-pumps, nonbioerodible and bioerodible drug-loaded pellets, configured capillary fibres, biodegradable scleral plugs, scleral discs, polymeric matrices and scaffolds of various geometries providing unique mechanisms of drug release for the delivery of drugs to the posterior segment of the eye [80].

Nonbioerodible devices are able to offer the advantages of sustained release and reduced host response. However, bioerodible intravitreal drug delivery devices have gained much popularity over nonbioerodible devices due to the fact that they are eventually absorbed or excreted by the body, eliminating the need for surgical removal of the device after the drug load has been depleted, thereby increasing patient acceptance and compliance [80].

The rate of polymer erosion from bioerodible ocular inserts can be affected by numerous factors, including pH or temperature, which may cause a transient increase or decrease in the erosion rate of the device. The rate of erosion may differ in paediatric patients if the pH within the eye is different; currently, this is unknown, although as stated previously, the pH of tears in neonates is lower than in adults. The surface area of the drug delivery device also plays a significant role in its erosion;

geometrical shapes with surface areas that remain relatively constant over time are typically preferred. During erosion, the dimensions of the device compared to the overall eye shape may influence the rate of surface area decrease and again this may be different in paediatric patients compared to adults. Most implantable devices have fixed dimensions that control the dose administered; therefore, these are of limited use in very young patients.

15.2.7.13 OphthaCoil®

Pijls et al. (2005) developed an intraocular drug delivery device called "OphthaCoil". This device consists of a drug-loaded mucoadhesive hydrogel on a thin coiled metallic wire [81]. On contact with tear fluid, the hydrogel coating swells, and the drug is released into the tear film. A disadvantage of the OphthaCoil device is the fact that removal of the device is required after the drug load is depleted. Despite this device being described many years ago, there is no commercialisation activity currently.

15.2.7.14 Vitrasert®

Vitrasert (Chiron Vision Inc., Irvine, CA) is a nonbiorerodible device that requires a 4–5 mm sclerotomy at the pars plana for implantation. It was indicated for the local treatment of cytomegalovirus (CMV) retinitis in patients with acquired immunodeficiency syndrome (AIDS). Safety and effectiveness in paediatric patients below 12 years of age were not established.

15.2.7.15 Retisert® and Medidur® Devices

Retisert (Bausch & Lomb, Rochester, NY) is a nonbioerodible device that was designed to release 0.59 μg/day of fluocinolone acetonide into the vitreous cavity to treat diabetic macular edema. Safety and effectiveness in paediatric patients below 12 years of age have not been established.

Medidur® device (Alimera Sciences Inc., Atlanta, GA and pSivida Inc., Watertown, MA) also contains fluocinolone, but it is a much smaller device, and the insertion procedure is similar to an intravitreal injection. This device is authorised as Iluvien® within the UK to treat diabetic macular oedema. The EMA granted a waiver for the use of fluocinolone in paediatric patients as diabetic macular oedema does not occur in paediatric patients under the age of 12, and the medicinal product does not represent a significant therapeutic benefit in children aged 12–18 as clinical studies are not feasible [82].

15.2.7.16 I-vation®

The I-vation® technology (SurModics (Pty) Ltd., Eden Prairie, MN) consists of a helical coil with an eluting polymer containing triamcinolone. The device is self-anchoring within the sclera. This product is not currently available clinically. The EMA granted a waiver for triamcinolone for paediatric populations in groups where clinical studies cannot address the therapeutic needs of these patients [83].

15.2.8 Discussion/Conclusions

Ocular drug delivery systems are prescribed for use in paediatric populations despite there being limited information about their rational use. Anatomical and physiological differences in the eye of neonates and infants leave them vulnerable to systemic effects of topically administered ocular drugs. Further studies are required to understand how formulations behave in a paediatric population. In addition, there may be a need for a paediatric delivery device to provide a smaller dose of topically applied medicines.

15.3 Otic Drug Delivery

Many inner ear disorders cannot be adequately treated by systemic drug delivery. A blood–cochlear barrier exists, similar physiologically to the blood–brain barrier, which limits the concentration and size of molecules able to leave the circulation and gain access to the cells of the inner ear.

15.3.1 Introduction

Drugs that act on the ear in paediatric populations include therapies for otitis externa, otitis media and the removal of earwax. These medicines are most frequently applied as ear drops and sprays. A small volume is generally used as excess will be lost out of the ear passage.

15.3.2 Structure of the Ear

The ear can be divided into three parts: external ear, middle ear and inner ear. The external auditory canal (EAC) is a tube that leads from the external to the middle ear. Salvinelli et al. (1991) [84] measured the length of the EAC in adult cadavers and reported the length as 23.5 ± 2.5 mm and the greatest and least diameters to be 9.3 ± 0.9 mm and 4.8 ± 0.5 mm, respectively.

The tympanic membrane separates the external ear from the middle ear; within the middle ear is the tympanic cavity which is a small epithelial lined cavity hollowed out of the temporal bone that contains three auditory ossicles. The distal part of the middle ear is continuous with numerous mastoid air spaces in the temporal bone. The middle ear is the site at which infections can spread, for example, head colds can lead to mastoid infections.

The inner ear (also called the labyrinth due to its complex structure), which contains both the organ of hearing, the cochlea, and the organ of balance, the vestibular system, is embedded deep within the skull near the brainstem in the petrous bone. The extreme inaccessibility of the cochlea, coupled with its very small size, renders cochlear drug delivery difficult. The complexity of the cochlear structures and their extreme sensitivity to changes in fluid volume also must be considered in the design of formulations and delivery systems as the sensory cells of the cochlea must be protected from noise and surgical trauma.

The pH of the external auditory canal varies between 5.0 and 5.7 and is therefore slightly acidic; these conditions inhibit bacterial growth [85].

15.3.3 Age-Related Differences in the Ear

The outer ear in humans is not completely mature at birth, and various anatomical and physiological changes occur with age. The external auditory canal (EAC) of an infant is straighter, narrower and much shorter than the adult's EAC.

Previous techniques used to estimate ear canal length include optical and acoustic measurements. Measurements of ear canal volume use alternative techniques which are less invasive. Noh and Lee (2012) measured EAC volume in 194 children with a mean age of 58 months, reporting a mean value of 0.56 mL [86]. Previously, the volume was reported to increase from 0.42 to 0.97 mL from 2.8 to 5.8 years of age [87]. The EAC volume in adults is reported to be approximately 0.696–0.979 mL [88].

Dosing devices allow smaller doses to be administered in paediatric patients, and as there is no significant systemic uptake from medicines administered aurally, there are little anticipated differences in treatments in paediatric patients compared to adults.

15.3.4 Topical Otic Delivery

Topical formulations are used to provide high drug concentrations at the disease site; for example, anti-inflammatory drugs are used to treat acute otitis externa.

Formulations applied to the ear are administered directly into the ear canal; this structure should be manipulated to ensure that the medicine enters the ear rather than being lost externally. Typically, a fixed number of drops is used, or the ear canal can be filled with solution. The ear can be plugged post administration using cotton wool.

15.3.5 Inner Ear Drug Delivery

Inner ear drug delivery methods can be divided into two main categories based on the location of entry of the drug. Intratympanic delivery involves depositing the therapeutic agent in the middle ear, relying primarily on diffusion through the round window membrane (RWM) for access to the scala tympani. The second method, intracochlear, depends on a cochleostomy with direct delivery into the inner ear space, completely bypassing the middle ear.

Intratympanic drug delivery relies on high drug concentrations driving diffusion into the scala tympani from the middle ear. However, the RWM is known to be variable in terms of permeability, which limits dosing accuracy [89]. In addition, drug delivered to the middle ear can be lost into the pharynx via the Eustachian tube. Typical formulations used include biodegradable polymers, hydrogel-based systems, nanoparticles, microcatheters and osmotic pumps. Typically, delivery systems used in intratympanic drug delivery use local triggers to stimulate drug release, including temperature or pH.

The thickness of RWM has been reported to decrease in ageing mice [90], which is likely to increase permeability across this membrane, although similar studies in humans showed no change in the mean thickness with ageing [91]. Typically, membranes are also thinner in the very young; therefore, there may be greater permeability in the youngest patients compared to adult populations although there is currently no data to support this theory.

15.3.6 Intracochlear Drug Delivery

Direct intracochlear drug delivery involves placement of drugs within cochlear perilymphatic spaces. Molecules perfused into a perilymph compartment (ideally, the scala tympani) have direct access to the cells of the inner ear [92]. Methods of delivery include direct perfusion using micropumps and osmotic pumps. Research is being conducted on modifying the electrodes of cochlear prostheses to integrate drug delivery components, while others are developing novel implantable delivery devices [93].

Direct injections and infusions delivered via syringe have been administered both in research and in a clinical setting; however, these do not allow for prolonged delivery. Numerous other devices which can be used for sustained release or multi-dose delivery to the inner ear have been developed in attempts to overcome the shortcomings of direct transtympanic injection. These include the Silverstein MicroWick (Micromedics, Eaton, MN), the Round Window Microcatheter or µCath™ (Durect Corp. Cupertino, CA) in conjunction with an electronically controlled pump, the Alzet osmotic pump (Durect Corp. Cupertino, CA) and other devices still in earlier stages of development.

15.3.7 Formulations

15.3.7.1 Otic Solutions

The pH of otic solutions can be as low as the reduced pH can inhibit bacterial growth. Historically, otic formulations may have been sterile and isotonic; this was mainly due to ocular formulations being used within the ear, rather than a clinical requirement. There is no need for otic formulations to be either isotonic or sterile.

Typically, otic solutions are simple solutions of drugs in water or other solvents (e.g. glycerol, propylene glycol, alcohol/water, mineral oil). The choice of vehicles is determined by the solubility of the drug within the product as the concentration needs to be high to account for the low volume typically administered. Excipients to increase the viscosity may be included to assist with administration or retention within the ear canal. Otic solutions are usually administered as drops, sprays or washes.

Lipophilic vehicles are used in treatments designed to assist in the removal of earwax, as their oily nature assists in solubilisation to remove the wax.

15.3.7.2 Otic Gel

The use of gels as drug delivery systems to prolong trans-tympanic delivery to the middle ear has been described by Khoo et al. (2013) [94]. Previously, this route of administration was limited by the impermeability of the intact tympanic membrane to small hydrophilic drugs; however, research demonstrated that incorporation of chemical permeation enhancers into the formulation provided enhanced permeability. This formulation is yet to be commercialised but offers promise in the treatment of acute otitis media in paediatric patientsPaediatric patients.

15.3.7.3 Otic Foam

Otic Pharma developed a foam formulation technology (FoamOtic™) that aims to overcome the disadvantages of eardrops, including: ineffective delivery to the infection site within the ear canal; retention of the drug within the ear; the administration of the product to the patient; and the need for the patient to lie down following administration. A foam formulation, a ciprofloxacin foam, has been developed that was shown to be as efficacious as the solution treatment for acute otitis externa [95].

15.3.7.4 Microwick Technology

The Silverstein MicroWick™ consists of an absorbent wick that is passed through a vent hole in the tympanic membrane and inserted to contact the round window within the middle ear. The wick absorbs medication that is administered in the external ear and transports it to the RWM for diffusion into the inner ear [96].

15.3.7.5 Osmotic Pump

Osmotic pumps have been employed to test the effectiveness of steroid treatment to prevent hearing loss from noise trauma. The small volume pump is subcutaneously implanted, and a cannula is routed to the round window niche or directly into the cochlea. Various models allow for reservoir volumes of 0.1–2 mL and flow rates from 0.1 to 10 μL/h, and the pumps provide continuous infusion for 1 day to 6 weeks [97]. The pump operates by osmotic pressure in the outer section of the pump, forcing the drug in an inner, impermeable chamber out through the cannula. Delivery cannot be started or stopped, nor can the flow rate be changed in situ. The benefits of osmotic pumps are that they offer a controlled delivery profile and a proven surgical technique.

15.3.7.6 Cochlear Implants

Cochlear implants stimulate the auditory neurons in the cochlea through an electrode inserted into the scala tympani. Med-El and Cochlear Ltd. are examples of companies investigating the combination of drug delivery with an implant. Combining a drug delivery platform with the implant electrode, drugs such as neurotrophic factors may prevent further auditory degradation and maintain the number of surviving nerve fibres, thus improving the effectiveness of cochlear implants for long-term use [98].

15.3.8 Discussion/Conclusions

Otic drug formulations are used within paediatric populations routinely. There are no reported paediatric specific formulations that differ from adult products. However, the similarities in anatomy and physiology ensure that products are likely to perform in the same way in adults compared to children with few reported adverse effects following otic drug delivery. There is a need for some caution in drug delivery to the inner ear as the membranes may be more permeable in the youngest within the paediatric population, which may lead to increased drug absorption and a greater likelihood of adverse events.

References

1. Illum L. Nasal drug delivery—recent developments and future prospects. J Control Release. 2012;161(2):254–63.
2. Sarkar MA. Drug metabolism in the nasal mucosa. Pharm Res. 1992;9(1):1–9.
3. Petruson B, Hansson HA, Karlsson G. Structural and functional aspects of cells in the nasal mucociliary system. Arch Otolaryngol (Chicago, Ill: 1960). 1984;110(9):576–81.
4. Proctor DF, Lundqvist G. Clearance of inhaled particles from the human nose. Arch Intern Med. 1973;131(1):132–9.
5. Morgan K, Patterson D, Gross E. Responses of the nasal mucociliary apparatus to airborne irritants. In: Barrow C, editor. Toxicology of the nasal passages. London: Hemisphere Publishing Corporation; 1968. p. 123–33.
6. Grassin-Delyle S, Buenestado A, Naline E, Faisy C, Blouquit-Laye S, Couderc L-J, et al. Intranasal drug delivery: an efficient and non-invasive route for systemic administration: focus on opioids. Pharmacol Ther. 2012;134(3):366–79.
7. Kaliner M, Marom Z, Patow C, Shelhamer J. Human respiratory mucus. J Allergy Clin Immunol. 1984;73(3):318–23.
8. Marttin E, Schipper NGM, Verhoef JC, Merkus FWHM. Nasal mucociliary clearance as a factor in nasal drug delivery. 1998 1–2:13–38. Available from: http://www.sciencedirect.com/science?_ob=GatewayURL&_origin=ScienceSearch&_method=citationSearch&_piikey=S0169409X97000598&_version=1&_returnURL=http%3A%2F%2Fwww.scirus.com%2Fsrsapp%2F&md5=048655bbe65c873a9cd96d837edd818d.
9. Dale O, Hjortkjær R, Kharasch ED. Nasal administration of opioids for pain management in adults. Acta Anaesthesiol Scand. 2002;46(7):759–70.
10. Newman SP, Pitcairn GR, Dalby RN. Drug delivery to the nasal cavity: in vitro and in vivo assessment. Crit Rev Ther Drug Carrier Syst. 2004;21(1):21–66.
11. Porter MJ, Maw AR, Kerridge DH, Williamson IM. Manometric rhinometry: a new method of measuring the volume of the air in the nasal cavity. Acta Otolaryngol. 1997;117(2):298–301.
12. Xi J, Si X, Longest W. Electrostatic charge effects on pharmaceutical aerosol deposition in human nasal–laryngeal airways. Pharmaceutics. 2014;6(1):26–35.
13. Samoliński B, Rapiejko P, Krzych-Fałta E, Gotlib T, Emeryk A, Buczylko K, et al. Standards of nasal provocation tests. Postępy Dermatologii i Alergologii/Adv Dermatol Allergol. 2010;27:149–61.
14. Balbani APS, Duarte JG, Montovani JC. Retrospective analysis of toxicity of eardrops, topical nasal and oropharyngeal medicines, documented in São Paulo, Brazil. Rev Assoc Med Bras (1992). 2004;50(4):433–8.
15. Kundoor V, Dalby RN. Effect of formulation- and administration-related variables on deposition pattern of nasal spray pumps evaluated using a nasal cast. Pharm Res. 2011;28(8):1895–904.
16. Foo MY, Cheng YS, Su WC, Donovan MD. The influence of spray properties on intranasal deposition. J Aerosol Med. 2007;20(4):495–508.
17. McKeage K. Zolmitriptan nasal spray: a review in acute migraine in pediatric patients 12 years of age or older. Pediatr Drugs. 2016;18:75–81.
18. Trenkel M, Scherließ R. Nasal powder formulations: in-vitro characterisation of the impact of powders on nasal residence time and sensory effects. Pharmaceutics. 2021;13(3)

19. La Sala L, Pontiroli AE. New fast acting glucagon for recovery from hypoglycemia, a life-threatening situation: nasal powder and injected stable solutions. Int J Mol Sci. 2021;22(19):10643.
20. Jain AK, Khar RK, Ahmed FJ, Diwan PV. Effective insulin delivery using starch nanoparticles as a potential trans-nasal mucoadhesive carrier. Eur J Pharm Biopharm. 2008;69(2):426–35.
21. Tanna V, Sawarkar SP, Ravikumar P. Exploring nose to brain nano delivery for effective management of migraine. Curr Drug Deliv. 2023;20(2):144–57.
22. Mardikasari SA, Sipos B, Csóka I, Katona G. Nasal route for antibiotics delivery: advances, challenges and future opportunities applying the quality by design concepts. J Drug Deliv Sci Technol. 2022;77
23. Jeong SH, Jang JH, Lee YB. Drug delivery to the brain via the nasal route of administration: exploration of key targets and major consideration factors. J Pharma Invest. 2023;53(1):119–52.
24. Vavilova VP, Vavilov AM, Cherkaeva AK. Experience of using viferon® in children during the period of adaptation to preschool educational institutions. Pediatriya. 2018;97(6):50–6.
25. Gasparini C, Acunzo M, Biuso A, Roncaglia S, Migliavacca F, Borriello CR, et al. Nasal spray live attenuated influenza vaccine: the first experience in Italy in children and adolescents during the 2020–21 season. Ital J Pediatr. 2021;47(1)
26. Hettie M, Janssens JCJ, Fokkens WJ, Robben SGF, Wouters K, Tiddens HAWM. The Sophia anatomical infant nose-throat (saint) model: a valuable tool to study aerosol deposition in infants. J Aerosol Med. 2001;14(4):433–41.
27. Minocchieri S, Burren JM, Bachmann MA, Stern G, Wildhaber J, Buob S, et al. Development of the premature infant nose throat-model (PrINT-model)—an upper airway replica of a premature neonate for the study of aerosol delivery. Pediatr Res. 2008;64(2):141–6.
28. Esmaeili AR, Wilkins JV, Hosseini S, Alfaifi A, Hejazi M, Hindle M, et al. In vitro evaluation of intersubject variability in pediatric intranasal drug delivery using nasal spray suspension products. J Aerosol Sci. 2024;179:106387.
29. Bhadury A, Roy UK, Ghosh T, Barman D, Mandal P. Assessment of prescribing pattern and safety profile of drugs used in intranasal route in paediatric age group of patients in a tertiary care hospital. Kathmandu Univ Med J. 2021;19(1):62–8.
30. Cumberland PM, Pathai S, Rahi JS, Millennium Cohort Study Child Health G. Prevalence of eye disease in early childhood and associated factors: findings from the millennium cohort study. Ophthalmology. 2010;117(11):2184–90.e3.
31. Isenberg SAL, McCarty J, Cooper LL, Lim L, Del Signore M. Development of tearing in preterm and term neonates. Arch Ophthalmol. 1998;116(6):773–6.
32. Patton TF. Pediatric dosing considerations in ophthalmology–dosage adjustments based on aqueous humor volume ratio. J Pediatr Ophthalmol. 1977;14(4):254–6.
33. Dixon ER, Landry T, Venkataraman S, Gustafson N, Salem C, Bradfield Y, et al. A 3-month safety and efficacy study of travoprost 0.004% ophthalmic solution compared with timolol in pediatric patients with glaucoma or ocular hypertension. J Am Assoc Pediatric Ophthalmol Strabismus. 2017;21(5):370–4.e1.
34. Esmaeelpour M, Watts PO, Boulton ME, Cai J, Murphy PJ. Tear film volume and protein analysis in full-term newborn infants. Cornea. 2011;30(4):400–4.
35. Apt L, Gaffney WL. Adverse effects of topical eye Medicationin INfants and children. In: Tasman W, Jaeger EA, editors. Duane's ophthalmology 2009 edition. Lippincott Williams & Wilkins; 2009.
36. Lawrenson JG, Murphy PJ, Esmaeelpour M. The neonatal tear film. Cont Lens Anterior Eye. 2003;26(4):197–202.
37. Abelson MB, Udell IJ, Weston JH. Normal human tear ph by direct measurement. Arch Ophthalmol. 1981;99(2):301.
38. Carney LG, Hill RM. Human tear pH. Diurnal variations. Arch Ophthalmol. 1976;94(5):821–4.
39. Dahl H, Dahl C. Hydrogen ion concentration of tear fluid in newborn infants. Acta Ophthalmol. 1985;63(6):692–4.

40. Sack RA, Beaton A, Sathe S, Morris C, Willcox M, Bogart B. Towards a closed eye model of the pre-ocular tear layer. Prog Retin Eye Res. 2000;19(6):649–68.
41. Zametkin AJ, Stevens JR, Pittman R. Ontogeny of spontaneous blinking and of habituation of the blink reflex. Ann Neurol. 1979;5(5):453–7.
42. Berlin RJ, Lee UT, Samples JR, Rich LF, Tang-Liu DDS, Sing KA, et al. Ophthalmic drops causing coma in an infant. J Pediatr. 2001;138(3):441–3.
43. Passo MS, Palmer EA, Van Buskirk EM. Plasma timolol in glaucoma patients. Ophthalmology. 1984;91(11):1361–3.
44. Raber S, Courtney R, Maeda-Chubachi T, Simons BD, Freedman SF, Wirostko B. Latanoprost systemic exposure in pediatric and adult patients with glaucoma: a phase 1, open-label study. Ophthalmology. 2011;118(10):2022–7.
45. Patton TF, Robinson JR. Pediatric dosing considerations in ophthalmology. J Pediatr Ophthalmol. 1976;13(3):171–8.
46. van Best JA, Kappelhof JP, Laterveer L, Oosterhuis JA. Blood aqueous barrier permeability versus age by fluorophotometry. Curr Eye Res. 1987;6(7):855–63.
47. Francoeur M, Ahmed I, Sitek S, Patton TF. Age-related differences in ophthalmic drug disposition III. Corneal permeability of pilocarpine in rabbits. Int J Pharm. 1983;16(2):203–13.
48. Huang AJ, Tseng SC, Kenyon KR. Paracellular permeability of corneal and conjunctival epithelia. Invest Ophthalmol Vis Sci. 1989;30(4):684–9.
49. Mishima S, Hedbys B. The permeability of the corneal epithelium and endothelium to water. Exp Eye Res. 1967;6:10.
50. Klyce S, Crosson C. Transport processes across the rabbit corneal epithelium: A review. Curr Eye Res. 1985;4:323.
51. Schoenwald RD, Deshpande GS, Rethwisch DG, Barfknecht CF. Penetration into the anterior chamber via the conjunctival/scleral pathway. J Ocul Pharmacol Ther. 1997;13(1):41–59.
52. Bito LZ, Baroody RA. The penetration of exogenous prostaglandin and arachidonic acid into, and their distribution within, the mammalian eye. Curr Eye Res. 1981;1(11):659–69.
53. Ahmed I, Patton TF. Importance of the noncorneal absorption route in topical ophthalmic drug delivery. Investig Ophthalmol Vis Sci. 1985;26(4):584–7.
54. Bloomfield SE, Miyata T, Dunn MW. Soluble gentamicin ophthalmic inserts as a drug delivery system. Arch Ophthalmol. 1978;96(5):885–7.
55. Maurice DM, Mishima S. Ocular pharmacokinetics. ML S, editor. Berlin/Heidelberg: Springer Verlag; 1984.
56. Maurice DM. Drug delivery to the posterior segment from drops. Surv Ophthalmol. 2002;47(4 SUPPL. 1):S41–52.
57. Hughes PM, Olejnik O, Chang-Lin J-E, Wilson CG. Topical and systemic drug delivery to the posterior segments. Adv Drug Deliv Rev. 2005;57(14):2010–32.
58. Jiang J, Gill HS, Ghate D, McCarey BE, Patel SR, Edelhauser HF, et al. Coated microneedles for drug delivery to the eye. Invest Ophthalmol Vis Sci. 2007;48(9):4038–43.
59. Kuno N, Fujii S. Recent advances in ocular drug delivery systems. Polymers. 2011;3:193–221.
60. Mishra GP, Bagui M, Tamboli V, Mitra AK. Recent applications of liposomes in ophthalmic drug delivery. J Drug Deliv. 2011;2011:1–14.
61. Kumar S, Karki R, Meena M, Prakash T, Rajeswari T, Goli D. Reduction in drop size of ophthalmic topical drop preparations and the impact of treatment. J Adv Pharm Technol Res. 2011;2(3):192–4.
62. Jampel HD, Schwartz GF, Robin AL, Abrams DA, Johnson E, Miller RB. Patient preferences for eye drop characteristics: a willingness-to-pay analysis. Arch Ophthalmol. 2003;121(4):540–6.
63. Salamore J C XE, inventorOphthalmic solution with a flavoring agent. US patent US 2006/0292188. 2006.
64. Sieg JW, Robinson JR. Vehicle effects on ocular drug bioavailability I: evaluation of fluorometholone. J Pharm Sci. 1975;64(6):931–6.
65. Takaoka M, Yokoi N, Ishibashi T, Kinoshita S. Assessment of retention and drug-release of ophthalmic ointment on the ocular surface. Nihon Ganka Gakkai zasshi. 2004;108(5):307–11.

66. Robinson MR, Lee SS, Rubin BI, Wayne AS, Pavletic SZ, Bishop MR, et al. Topical corticosteroid therapy for cicatricial conjunctivitis associated with chronic graft-versus-host disease. Bone Marrow Transplant. 2004;33(10):1031–5.
67. BenEzra D, Maftzir G. Ocular penetration of cyclosporine A in the rat eye. Arch Ophthalmol. 1990;108(4):584–7.
68. Burgess PI, Koay P, Clark P. SmartPlug versus silicone punctal plug therapy for dry eye: A prospective randomized trial. Cornea. 2008;27(4):391–4.
69. Gupta C, Chauhan A. Ophthalmic delivery of cyclosporine A by punctal plugs. J Control Release. 2011;150(1):70–6.
70. Mataftsi A, Subbu RG, Jones S, Nischal KK. The use of punctal plugs in children. Br J Ophthalmol. 2012;96(1):90–2.
71. Trivedi RH, Wilson ME. A sustained-release intracanalicular dexamethasone insert (Dextenza) for pediatric cataract surgery. J AAPOS. 2021;25(1):43–5.
72. Kapoor Y, Chauhan A. Ophthalmic delivery of Cyclosporine A from Brij-97 microemulsion and surfactant-laden p-HEMA hydrogels. Int J Pharm. 2008;361(1–2):222–9.
73. Ali M, Horikawa S, Venkatesh S, Saha J, Hong JW, Byrne ME. Zero-order therapeutic release from imprinted hydrogel contact lenses within in vitro physiological ocular tear flow. J Control Release. 2007;124(3):154–62.
74. Venkatesh S, Sizemore SP, Byrne ME. Biomimetic hydrogels for enhanced loading and extended release of ocular therapeutics. Biomaterials. 2007;28(4):717–24.
75. Hiratani H, Fujiwara A, Tamiya Y, Mizutani Y, Alvarez-Lorenzo C. Ocular release of timolol from molecularly imprinted soft contact lenses. Biomaterials. 2005;26(11):1293–8.
76. Ciolino JB, Hoare TR, Iwata NG, Behlau I, Dohlman CH, Langer R, et al. A drug-eluting contact lens. Investig Ophthalmol Vis Sci. 2009;50(7):3346–52.
77. Lee TW-Y, Robinson JR. Drug delivery to the posterior segment of the eye IV: theoretical formulation of a drug delivery system for subconjunctival injection. J Ocul Pharmacol Ther. 2009;25(1):29–37.
78. Kimura H, Ogura Y, Hashizoe M, Nishiwaki H, Honda Y, Ikada Y. A new vitreal drug-delivery system using an implantable biodegradable polymeric device. Invest Ophthalmol Vis Sci. 1994;35(6):2815–9.
79. Sheybani ND, Yang H. Pediatric ocular nanomedicines: challenges and opportunities. Chin Chem Lett. 2017;28(9):1817–21.
80. Choonara YE, Pillay V, Danckwerts MP, Carmichael TR, du Toit LC. A review of implantable intravitreal drug delivery technologies for the treatment of posterior segment eye diseases. J Pharm Sci. 2010;99(5):2219–39.
81. Pijls RT, Sonderkamp T, Daube GW, Krebber R, Hanssen HHL, Nuijts RMMA, et al. Studies on a new device for drug delivery to the eye. Eur J Pharm Biopharm. 2005;59(2):283–8.
82. European Medicines Agency. European Medicines Agency decision. P/91/2010. 2010.
83. European Medicines Agency. European Medicines Agency decision of 15 June 2009 on the granting of a product specific waiver for triamcinolone acetonide (EMEA-000465-PIP01–08) in accordance with Regulation (EC) No 1901/2006 of the European Parliament and of the Council as amended. 2009.
84. Salvinelli F, Maurizi M, Calamita S, D'Alatri L, Capelli A, Carbone A. The external ear and the tympanic membrane. Scand Audiol. 1991;20(4):253–6.
85. Neher A, Nagl M, Scholtz AW. Otitis externa: etiology, diagnostic and therapy. HNO. 2008;56(10):1067–79.
86. Noh H, Lee D-H. Direct measurement of ear canal volume in a pediatric population: can we explain its individual variation in terms of age and body weight? Int J Pediatr Otorhinolaryngol. 2012;76(5):658–62.
87. Margolis RH, Heller JW. Screening tympanometry: criteria for medical referral. Audiology. 1987;26(4):197–208.
88. Shanks JE, Lilly DJ. An evaluation of tympanometric estimates of ear canal volume. J Speech Hear Res. 1981;24(4):557–66.

89. Banerjee A, Parnes LS. The biology of intratympanic drug administration and pharmacodynamics of round window drug absorption. Otolaryngol Clin N Am. 2004;37(5):1035–51.
90. Kitamura Y, Teranishi M-A, Sone M, Nakashima T. Round window membrane in young and aged C57BL/6 mice. Hear Res. 2002;174(1–2):142–8.
91. Sahni RS, Paparella MM, Schachern PA, Goycoolea MV, Le CT. Thickness of the human round window membrane in different forms of otitis media. Arch Otolaryngol Head Neck Surg. 1987;113(6):630–4.
92. Swan EEL, Mescher MJ, Sewell WF, Tao SL, Borenstein JT. Inner ear drug delivery for auditory applications. Adv Drug Deliv Rev. 2008;60(15):1583–99.
93. Swan EEL. A fully implantable intracochlear drug delivery device: development and characterization. 2009.
94. Khoo X, Simons EJ, Chiang HH, Hickey JM, Sabharwal V, Pelton SI, et al. Formulations for trans-tympanic antibiotic delivery. Biomaterials. 2013;34(4):1281–8.
95. Marom T, Yelin R, Goldfarb A, Rakover Y, Shlizerman L, Eilat E, et al. Comparison of safety and efficacy of foam-based versus solution-based ciprofloxacin for acute otitis externa. Otolaryngol Head Neck Surg. 2010;143(4):492–9.
96. Silverstein H. Use of a new device, the MicroWick(TM), to deliver medication to the inner ear. Ear Nose Throat J. 1999;78(8):595–600.
97. Pararas EEL, Borkholder DA, Borenstein JT. Microsystems technologies for drug delivery to the inner ear. Adv Drug Deliv Rev. 2012;64(0):1650.
98. Duan ML, Ulfendahl M, Laurell G, Counter AS, Pyykkö I, Borg E, et al. Protection and treatment of sensorineural hearing disorders caused by exogenous factors: experimental findings and potential clinical application. Hear Res. 2002;169(1–2):169–78.

88. Rene [illegible] A. [illegible] of drug [illegible] for [illegible] based [illegible] ... (2018) [illegible]
89. [illegible] and [illegible]
90. Sahil [illegible] The [illegible] of the human [illegible] Med Sci [illegible] ... [illegible] Surg [illegible]
92. [illegible] ... [illegible] Rigatelli [illegible]
93. [illegible] ... [illegible] tion [illegible] and [illegible]
94. [illegible] ... [illegible] medical [illegible]
95. [illegible] ... [illegible] Shaver [illegible] and [illegible] ... Oncology [illegible] ... [illegible] (2021) [illegible]
96. [illegible] ... [illegible] a cervical [illegible] Force [illegible]
97. [illegible] ... [illegible]
98. [illegible] ... [illegible] India [illegible] ... (2021) [illegible]

Chapter 16
Rectal Drug Delivery

Sara Hanning, Trusha Purohit, and Catherine Tuleu

Abstract The rectal route of administration is particularly useful for infants and children who have difficulty swallowing oral medicine. This route is also used in cases of nausea and vomiting, or where upper intestinal disorders present may affect the absorption of a drug. There is no need for taste masking of drugs delivered rectally.

Keywords Rectal administration · Suppositories · Peadiatric rectal drug delivery · Absorption variability · Acceptability

16.1 Introduction

The rectal route of administration has been in use for centuries for both localised and systemic drug delivery. Rectal delivery is appropriate for infants and children who may have difficulty swallowing oral medicine. It is also useful in cases of nausea and vomiting, or in those with upper intestinal disorders that may affect the absorption of a drug. There is no need for taste masking of drugs delivered rectally, and they do not require trained personnel for administration. Rectal preparations include suppositories, enemas, foams, creams, ointments and sprays.

Important drawbacks of the rectal route of drug administration include the patient and carer's lack of acceptance (cultural reluctance in some countries) and the interruption of drug absorption (removal of the delivery system) by defaecation.

S. Hanning · T. Purohit
School of Pharmacy, Faculty of Medical and Health Sciences, The University of Auckland, Auckland, New Zealand

C. Tuleu (✉)
UCL School of Pharmacy, UCL, London, UK
e-mail: c.tuleu@ucl.ac.uk

H. Batchelor, K. Rose (eds.), *Pediatric Formulations*, AAPS Advances in the Pharmaceutical Sciences Series 58, https://doi.org/10.1007/978-3-031-77240-5_16

16.2 Structure of the Rectum

The rectum is a cylindrical organ that forms the distal portion of the large intestine, deviating at three (upper, middle and lower) lateral curves. The rectal surface comprises epithelial mucosa with no surface villi and is mainly associated with water absorption and re-absorption from the gastrointestinal contents. The literature shows some disagreement regarding the anatomy of the rectum, which has led to variations in reported rectal lengths—ranging from 10–15 cm in certain studies to 18–20 cm in others [1]. When empty, an adult rectum contains approximately 3 mL of mucus in a layer about 100 μm thick.

The blood supply to the rectum changes with distance from the anus; the inferior rectal vein is closest to the anus, and the superior rectal artery is further inside the rectal cavity. Drug absorbed at the inferior portion enters the rectal veins, which drain to the internal pudendal veins, bypassing the liver. However, drug absorbed further inside the rectum enters the superior haemorrhoidal veins, which connect to the portal vein where drugs are transported to the liver. This difference is important for systemic absorption of drugs that are extensively metabolised because the plasma levels reached will depend upon where absorption occurred. The main mechanism of absorption from the rectum is via passive diffusion across the mucous membrane.

16.3 Age-Related Differences in the Rectum and Their Impact on Medicines Performance

There is limited knowledge about rectal physiology, including colonic contents, motility and transit time in the paediatric population [2], despite these factors being important when considering rectal drug delivery. However, it is agreed that rectal length and subsequent surface area increase with age, with adult dimensions being reached from around 10 years of age [3] (Table 16.1). The diameter of the rectum in children aged 7 has been approximated as 21 mm [4]. The surface of the rectum is

Table 16.1 Approximated rectal surface area based on rectal length, assuming a linear relationship between rectum length and surface area by age

Age	Rectal length (cm)	Approximate rectal surface area (cm^2)
Neonate	4	114
1 year	7	171
5 years	8	200
10 years	9	257
15 years	10	286
Adult	10–15	300

Rectal length data sourced from Valentin 2002 [14]

epithelial mucosa that does not have villi at the surface; thus, the surface area is estimated to be 200–400 cm^2 in adults, based on the dimension of a cylinder [5]. In adults, the circumference varies from 15 cm at the rectosigmoid junction to 35 cm or more at its widest portion [5].

The pH within the rectal cavity in adults is between 7.2 and 7.4, although rectal fluids have little buffering capacity [6]. The mean pH within the rectal cavity of children has been reported to be 9.6 (range 7.2–12.1); the pH value was not affected by age and gender [7]. The elevated pH in children compared to adult values is of importance in drugs which are ionised and whose pKa is significant at this value (e.g. codeine base has a pKa of 8.1). In a previous study, rectal absorption of morphine was increased with increased pH of the delivery vehicle [8]. Mixing of drug with rectal fluids may elevate the pH and therefore have an impact on exposure. This pH difference may account for some of the variability noted in rectal drug absorption in paediatric patients [9].

As with all formulations, efficacy and safety need to be established in the target population(s). Variable absorption both within and between patients, particularly between developmental age groups, remains a challenge [10]. Historically, rectal dosage forms are avoided in neonates, especially pre-term neonates, due to concerns about erratic absorption and the risk of damage to the rectal lining, which could lead to infection [11]. However, evidence is emerging that suggests in vivo safety and efficacy of rectally administered ibuprofen [12] and paracetamol [13] in neonates.

Dose adjustments in paediatric patients are typically made based on weight, height or body surface area. In terms of rectal drug delivery for systemic effects, the rectal absorptive surface area is an important factor to consider. The relationship between rectal surface area, body weight and body surface area is also shown in Fig. 16.1.

The data in Fig. 16.1 show that a paediatric patient's height most closely correlates to their rectal surface area compared to other typically used measures. Although dose adjustments are relatively simple for solutions and suspensions, the use of suppositories restricts simple dose adjustment in many cases. Using paracetamol as an example drug, the rectal dosages recommended within the British National Formulary (BNF) for children are quoted as mg/kg; therefore, the required dose is adjusted depending on bodyweight [15]. Other drugs may be dosed based on age of patient, independent of bodyweight or height. For example, the recommended rectal dose of diazepam is 1.25–2.5 mg for neonates and 5 mg for 1-month to 2-year-old infants [15].

16.4 Drug Delivery Via the Rectal Route

Rectal drug delivery may be a good option when the oral route is not available, for example, due to swallowing difficulties or patient refusal. It is also useful for drugs that are not suitable for oral administration due to poor stability, solubility or permeability. Depending on how they are administered, hepatic metabolism can be

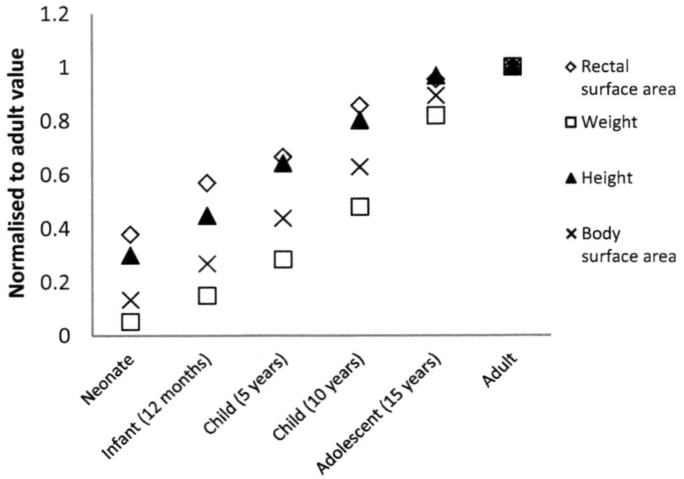

Fig. 16.1 Comparison of rectal surface area to typical dose adjustment parameters (weight, height and body surface area) by age of paediatric patient. (Data sourced from Valentin 2002 [14])

partially or completely bypassed, so rectal delivery can be useful for drugs that undergo high hepatic first-pass metabolism [5, 11, 16]. Rectal preparations can be formulated for local or systemic drug delivery.

16.4.1 Local Delivery

Agents used locally within the rectum include therapies for constipation, haemorrhoids, local anaesthetics and soothing agents. Glycerin suppositories are formulated for the treatment of constipation in infants right through to older children and adults. They are also given to premature neonates to facilitate meconium evacuation and transition to enteral feeding, although evidence to support this treatment is inconclusive [17, 18].

Rectal preparations for local delivery are formulated such that a thin coating of drug-containing formulation is retained within the rectum. Typically, suppositories, ointments, creams and enemas are used for this application.

16.4.2 Systemic Delivery

The rectum is used as a site of administration for a range of systemically acting agents. Due to the absence of active transport in the rectum, drugs with a smaller molecular weight and higher Log P values are best suited for systemic delivery [19]. Systemic agents need to be absorbed from the rectum, and the formulation selected can determine the rate of absorption. Administration of formulations in solution provides a rapid onset of action as the drug is readily available at the site of absorption, whereas drug within solid or semi-solid formulations needs to diffuse and dissolve prior to presentation at the absorptive surface. In children, as in adults, the rapid onset of action provided by a solution formulation is desirable in certain therapeutic areas. For example, one study showed that a diazepam rectal solution provided rapid systemic concentrations and improved clinical action compared to a diazepam suppository in children [20].

Examples of commercially available rectal formulations for children under the age of two include agents for pain relief (e.g. paracetamol, diclofenac sodium, morphine); antibacterials (e.g. metronidazole); seizures (e.g. sodium valproate); and for nausea and vomiting (e.g. ondansetron). Due to advantages including low manufacturing cost, ease of administration by untrained personnel and absence of manipulation required prior to administration, the rectal route has been investigated for systemic delivery of drugs to manage diseases in low- and middle-income countries [21]. In particular, amoxicillin [19, 22], azithromycin [23] and artesunate [24, 25]. Rectal artesunate capsules for pre-referral treatment of severe malaria have been shown to improve survival in children under 5 [25], but unfortunately, this was not shown in a "real-world" multi-country observational study [26]. This is likely due to underlying health system issues, such as quality of care and not following guidelines fully, and reviews are underway to establish conditions that ensure the safe and effective implementation of this therapy for children under 6 years old [27].

16.5 Rectal Formulations

Rectal dosage forms can be formulated as solids (suppositories, tablets), liquids (suspensions, solutions) or semi-solids (foams, ointments) [28]. Suppositories are traditionally the most common rectal dosage form. Studies of both suppositories and solutions have shown migration of the administered drug, but this appears to be limited to the rectum area [29]. Migration of enemas with as little volume as 80–100 ml may result in the drug found in the colon and even in the ileum [30].

16.5.1 Enemas

Enema formulations are delivered as rectal injections and are typically non-viscous solutions or suspensions. The vehicles can be aqueous or oily, and in some cases, the vehicle can be the therapeutic agent, for example, bowel evacuation resulting from arachis oil retention enema.

Historically, oral liquid preparations or injectable solutions have been administered rectally when oral administration was not appropriate. The rectal delivery of oral liquid preparations of antiepileptic agents was investigated by Graves and Kriel [31]; they found that most were well tolerated and demonstrated clinical efficacy.

Rectal solutions typically have a neutral pH and are near isotonic to minimise irritation to the local tissue. However, phosphate enemas have a lower pH (5–5.8) and are highly concentrated to promote fluid movement from the intestines into the intestinal tract, resulting in peristalsis to aid in colonic evacuation.

In adults, enema volume can vary from 1 to 20 mL for a micro-enema to volumes greater than 50 mL for a macro-enema (also known as a retention enema) [32]. Evacuation enemas can have significantly greater volumes (> 500 mL). In paediatric patients, examples of micro-enemas include sodium citrate (5–10 mL) for children aged 3–18 years. Based on a phosphate enema formula, typical volumes of enemas for children are 45–64 mL for children aged 3–7 years; 65–100 mL for children aged 7–10 years and 100–128 mL for children from 12 to 18 years [15].

Enema formulations can be used across a range of ages as dose adjustments can be made by alteration of the volume administered. The excipients used in adult formulations are appropriate for use in children.

16.5.2 Suppositories

Suppositories can be manufactured in a range of shapes and sizes although a torpedo shape is most commonly used. The typical dimensions of suppositories are listed in Table 16.2.

Suppositories are usually solid suspensions or emulsions, formulated by incorporating the drug into a lipophilic or hydrophilic base. Lipophilic suppositories require melting or liquification of the base following administration to allow drug release. Following liquification, the drug must diffuse through the rectal mucus layer prior to absorption through the membrane, if required. For solid hydrophilic suppositories, the hydrophilic base dissolves in rectal fluid to release the drug. Gelatin capsules that contain liquids can also be used as rectal suppository formulations (e.g. artesunate rectal capsules).

Paediatric suppositories are typically manufactured at an appropriate size for children (1 g nominal size). In certain cases, portions of adult suppositories may be used in paediatric patients as it is assumed that there is a uniform distribution of the active substance in the suppository matrix. However, there is unlikely to be any accuracy or stability data to support such a practice, and the resulting shape may not be optimal for rectal insertion.

Table 16.2 Typical dimensions of suppositories with a nominal size of 1 g, 2 g and 4 g

Nominal size	Dimensions	
	Length (mm)	Diameter (mm)
1 g	20–25	8–10
2 g	22–28	8–15
4 g	30–35	10–20

16.5.3 Semi-Solid Preparations

Semi-solid preparations include creams, ointments and gels. They have dosage flexibility advantages yet some products require manual application which can be messy and awkward for self-administration, coupled with concerns from carers regarding discomfort for administration to children. Administration devices such as applicators used in conjunction with creams, ointments and gels are important in improving ease of administration.

Rectal creams and ointments are readily available for paediatric use, usually for local drug delivery. In addition, rectal gels may offer the potential to increase retention time within the rectal cavity. A two-compartment, mucoadhesive gel formulation of artesunate was developed for use in children; this formulation was prepared as a solid for reconstitution to provide the required stability for use in tropical climates [33].

16.5.4 Rectal Foam

A foam is a dispersion of gas within a liquid. Rectal foams combine a drug formulation with gas that is administered rectally via an applicator extension to an aerosol. The rectal foam expands within the rectal cavity, and the drug formulation makes contact with the rectal mucosa enabling even application. The advantages of a foam include the use of an applicator to administer the formulation, which avoids contact with hands and the greasiness of an ointment. In addition, the retention of the drug within the rectal cavity is increased via the use of a foam as the reduced density of the product reduces the risk of seepage. Examples of rectal foams used in children include prednisolone, mesalazine, hydrocortisone and budesonide.

16.5.5 Rectal Spray

A rectal spray is a solution formulation that is dispensed from a spray pump to ensure even distribution within the rectal cavity. A spray minimises the excessive volume that may be used with a simple solution as the dispenser ensures that the product makes contact with the mucosal surface. This results in less drug wastage and is also preferred by patients as there is less seepage of the drug.

16.6 Acceptability of the Rectal Route

The term "acceptability" refers to the ability of the patient and caregiver, or the "user", to use a medicinal product as intended or authorised. This factor can greatly influence the patient's adherence to the treatment regimen, which in turn can affect the safety and effectiveness of the product [34]. For rectal use, aspects to consider include volume, size and shape (solids) of the dosage form as well as ease of retention (effect caused after application) [35] In addition, acceptability of the route of administration from the perspective of the child (the "user"), their caregiver and the prescriber all need to be considered. If the child and/or caregiver do not perceive rectal administration to be acceptable, then adherence may be an issue, whereas if the prescriber does not favour this route of administration, then this could be a barrier to having medicine prescribed for rectal administration. The acceptability of the rectal route for drug administration is generally considered to be poor, particularly in the United Kingdom [11]. The reasons for this are unclear, but feedback from a young person advisory group indicates that not all children in the United Kingdom are against the rectal route of drug administration [36]. Appropriate education targeted at both children and their caregivers is required to ensure awareness and improve acceptability for this route of delivery. Enabling self-administration in children as early as possible is an important factor to consider [36].

16.7 The Future of Rectal Drug Delivery Systems

Recent advances in rectal drug delivery have focused on modified drug release by controlled spreading and improved retention. Some examples of these rectal drug delivery systems include hollow-type, liquid, thermo-gelling and mucoadhesive liquid suppositories as well as solid lipid nanoparticles. Moreover, nano- and microparticulate or vesicular systems suspended in the suppository matrix have also been investigated recently.

Hollow-type suppositories allow co-administration of multiple drugs as well as the potential to deliver solid, semi-solid or liquid formulations within the suppository core [11, 37, 38]. Moreover, liquid suppositories and thermos-gelling systems have enabled localised as well as systemic drug absorption by improving the residence time of the formulation by modulating intra-rectal spreading hence bioavailability [39, 40].

In terms of the manufacture of rectal dosage forms, there has been some recent development of three-dimensional (3D) printing techniques using various excipients [41–43], including for children [44]. 3D printing has an emerging role in pharmaceutical development of all formulations, with the ability to create rapid prototypes. If 3D printing can be applied to develop dosage forms with improved stability and bioavailability compared with those manufactured using conventional methods, then this could be a promising technique in the future.

16.8 Conclusions

The formulation of paediatric dosage forms for rectal administration follows the same principles as for adult products. It is a viable alternative route of administration, especially in cases where oral administration is not an option. Dosage adjustments in paediatric populations need to be considered carefully for systemically absorbed rectal products. The dimensions and volumes of paediatric rectal dosage forms should be considered to maximise patient and carer acceptability. Pros and cons of rectal drug delivery should be discussed between healthcare professionals, patients and caregivers so that end-users are adequately informed and comfortable with this route of administration.

References

1. Kenig J, Richter P. Definition of the rectum and level of the peritoneal reflection—still a matter of debate? Videosurg Other Miniinvas Tech. 2013;3:183–6.
2. Wollmer E, Ungell AL, Nicolas JM, Klein S. Review of paediatric gastrointestinal physiology relevant to the absorption of orally administered medicines. Adv Drug Deliv Rev. 2022;181:114084.
3. Woody RC, Golladay ES, Fiedorek SC. Rectal anticonvulsants in seizure patients undergoing gastrointestinal surgery. J Pediatr Surg. 1989;24(5):474–7.
4. Joensson IM, Siggaard C, Rittig S, Hagstroem S, Djurhuus JC. Transabdominal ultrasound of rectum as a diagnostic tool in childhood constipation. J Urol. 2008;179(5):1997–2002.
5. De Boer AG, Moolenaar F, De Leede LGJ, Breimer DD. Rectal drug administration: clinical pharmacokinetic considerations. Clin Pharmacokinet. 1982;7(4):285–311.
6. Lukhvidova ZM, Persits BP, Novoselets SA, Matofanova LF. Determination of pH of the rectal mucosa with the aid of a new recorder. Eksp Khir Anesteziol. 1974;3:38–9.
7. Jantzen JP, Tzanova I, Witton PK, Klein AM. Rectal pH in children. Can J Anaesth. 1989;36(6):665–7.
8. Moolenaar F, Yska JP, Visser J, Meijer DK. Drastic improvement in the rectal absorption profile of morphine in man. Eur J Clin Pharmacol. 1985;29(1):119–21.
9. Jantzen JP, Erdmann K, Witton PK, Klein AM. The effect of rectal pH values on the absorption of methohexital. Anaesthesist. 1986;35(8):496–9.
10. Kaneria NS, Tuleu C, Ernest T. Opportunities for enteral drug delivery for neonates, infants, and toddlers: a critical exploration. Expert Opin Drug Deliv. 2022;19(5):475–519.
11. Jannin V, Lemagnen G, Gueroult P, Larrouture D, Tuleu C. Rectal route in the 21st century to treat children. Adv Drug Deliv Rev. 2014;73:34–49.
12. Demir N, Peker E, Ece İ, Balahoroğlu R, Tuncer O. Efficacy and safety of rectal ibuprofen for patent ductus arteriosus closure in very low birth weight preterm infants. J Matern Fetal Neonatal Med. 2017;30(17):2119–25.
13. Chen L, Zhang M, Yung J, Chen J, McNair C, Lee KS. Safety of rectal administration of acetaminophen in neonates. Can J Hosp Pharm. 2018;71(6):364–9.
14. Valentin J. Basic anatomical and physiological data for use in radiological protection: reference values: ICRP publication 89. Ann ICRP. 2002;32(3–4):1–277.
15. British national forrmulary for children. London: BMJ Group and Pharmaceutical Press; 2012-2013.
16. Hua S. Physiological and pharmaceutical considerations for rectal drug formulations. Front Pharmacol. 2019;10:1196.

17. Livingston MH, Elliott T, Williams C, Jones SA, Rosenbaum PL, Walton JM. Glycerin suppositories used prophylactically in premature infants (supp): a pilot randomized controlled trial. J Neonatal-Perinatal Med. 2020;13(4):495–505.
18. Livingston MH, Shawyer AC, Rosenbaum PL, Williams C, Jones SA, Walton JM. Glycerin enemas and suppositories in premature infants: a meta-analysis. Pediatrics. 2015;135(6):1093–106.
19. Purohit TJ, Hanning SM, Amirapu S, Wu Z. Rectal bioavailability of amoxicillin sodium in rabbits: effects of suppository base and drug dose. J Control Release. 2021;338:858–69.
20. Dhillon S, Ngwane E, Richens A. Rectal absorption of diazepam in epileptic children. Arch Dis Child. 1982;57(4):264–7.
21. Gerrard SE, Walsh J, Bowers N, Salunke S, Hershenson S. Innovations in pediatric drug formulations and administration technologies for low resource settings. Pharmaceutics. 2019;11(10):518.
22. Hanning SM, Matiz S, Krasser K, Orlu M, Dodoo C, Gaisford S, et al. Characterisation of rectal amoxicillin (RAMOX) for the treatment of pneumonia in children. Drug Deliv Transl Res. 2020;11(3):944–55.
23. Kauss T, Gaubert A, Boyer C, Ba BB, Manse M, Massip S, et al. Pharmaceutical development and optimization of azithromycin suppository for paediatric use. Int J Pharm. 2013;441(1):218–26.
24. de Carvalho LP, Kreidenweiss A, Held J. The preclinical discovery and development of rectal artesunate for the treatment of malaria in young children: a review of the evidence. Expert Opin Drug Discov. 2020:1–10.
25. Gomes MF, Faiz MA, Gyapong JO, Warsame M, Agbenyega T, Babiker A, et al. Pre-referral rectal artesunate to prevent death and disability in severe malaria: a placebo-controlled trial. Lancet. 2009;373(9663):557–66.
26. Hetzel MW, Okitawutshu J, Tshefu A, Omoluabi E, Awor P, Signorell A, et al. Effectiveness of rectal artesunate as pre-referral treatment for severe malaria in children under 5 years of age: a multi-country observational study. BMC Med. 2022;20(1):343.
27. Global Malaria Programme. WHO Marlia Policy Advisory Group (MPAG) meeting report 2021 [9 Dec 2022]. Available from: https://apps.who.int/iris/handle/10665/348220.
28. Kaneria NS, Tuleu C, Ernest T. Opportunities for enteral drug delivery for neonates, infants, and toddlers: a critical exploration. Expert Opin Drug Deliv. 2022:1–45.
29. Jay M, Beihn RM, Digenis GA. Disposition of radiolabelled suppositories in humans. J Pharm Pharmacol. 1985;37(4):266–8.
30. Van Hoogdalem EJ, De Boer AG, Breimer DD. Pharmacokinetics of rectal drug administration, part II. Clinical applications of peripherally acting drugs, and conclusions. Clin Pharmacokinet. 1991;21(2):110–28.
31. Graves NM, Kriel RL. Rectal administration of antiepileptic drugs in children. Pediatr Neurol. 1987;3(6):321–6.
32. Washington N, Wilson CG, Washington C. Physiological pharmaceutics: barriers to drug absorption. 2nd ed. London: Taylor and Francis; 2001.
33. Gaudin K, Barbaud A, Boyer C, Langlois M-H, Lagueny A-M, Dubost J-P, et al. In vitro release and stability of an artesunate rectal gel suitable for pediatric use. Int J Pharm. 2008;353(1–2):1–7.
34. Committee for Medicinal Products for Human Use (CHMP) Paediatric Committee (PDCO). Guideline on pharmaceutical development of medicines for paediatric use (EMA/CHMP/QWP/805880/2012 Rev. 2) [Internet]. London: European Medicines Agency (EMA); 2014. Available from: https://www.ema.europa.eu/en/documents/scientific-guideline/guideline-pharmaceutical-development-medicines-paediatric-use_en.pdf.
35. Kozarewicz P. Regulatory perspectives on acceptability testing of dosage forms in children. Int J Pharm. 2014;469(2):245–8.
36. Hanning SM, Walker E, Sutcliffe E, Tuleu C. The rectal route of medicine administration for children: let's get to the bottom of it! Eur J Pharm Biopharm. 2020;157:25–7.

37. Purohit TJ, Hanning SM, Wu Z. Advances in rectal drug delivery systems. Pharm Dev Technol. 2018;23(10):942–52.
38. Ameer Z, Laith HS, Nawal A. Preparation and in-vitro evaluation of metoclopramide Hcl hollow-type suppository. Int J Pharm Pharm Sci. 2013;5(2):660–6.
39. Lin HR, Tseng CC, Lin YJ, Ling MH. A novel in-situ-gelling liquid suppository for site-targeting delivery of anti-colorectal cancer drugs. J Biomater Sci Polym Ed. 2012;23(6):807–22.
40. Lo YL, Lin Y, Lin HR. Evaluation of epirubicin in thermogelling and bioadhesive liquid and solid suppository formulations for rectal administration. Int J Mol Sci. 2013;15(1):342–60.
41. Chatzitaki A-T, Tsongas K, Tzimtzimis EK, Tzetzis D, Bouropoulos N, Barmpalexis P, et al. 3D printing of patient-tailored SNEDDS-based suppositories of lidocaine. J Drug Delivery Sci Technol. 2021;61:102292.
42. Seoane-Viano I, Ong JJ, Luzardo-Alvarez A, Gonzalez-Barcia M, Basit AW, Otero-Espinar FJ, et al. 3D printed tacrolimus suppositories for the treatment of ulcerative colitis. Asian J Pharma Sci. 2021;16(1):110–9.
43. Tagami T, Hayashi N, Sakai N, Ozeki T. 3D printing of unique water-soluble polymer-based suppository shell for controlled drug release. Int J Pharm. 2019;568:118494.
44. Persaud S, Eid S, Swiderski N, Serris I, Cho H. Preparations of rectal suppositories containing artesunate. Pharmaceutics. 2020;12(3)

Chapter 17
Excipients and Active Pharmaceutical Ingredients

Eleanor Dakin, Smita Salunke, Hudson Polonini, Peter Kuehl, and Kevin Hughes

Abstract Excipients play a critical role in the formulation of pediatric medications, impacting both safety and efficacy. The challenge lies in recognizing that not all excipients are innocuous bystanders; some have raised concerns about their potential impact on the growth and development of young patients. This chapter provides a comprehensive overview of excipients, focusing on their unique considerations for the pediatric population. Key sections cover the physiological characteristics of children that influence excipient selection, including their developing organ systems, metabolism, and absorption pathways. The chapter also addresses the potential for excipient toxicity, particularly in neonates and young children, and explores the regulatory frameworks that guide excipient use in pediatric formulations. Excipient interactions within dosage forms, especially with active pharmaceutical ingredients, are discussed to highlight stability and performance considerations. Finally, an overview of the most commonly used excipients in pediatric products is provided, emphasizing their roles in improving palatability, dosage accuracy, and overall drug acceptability for young patients. This chapter aims to guide formulators and regulatory professionals in making informed decisions when developing safe and effective pediatric medications.

E. Dakin
Department of Pharmaceutics, School of Pharmacy, University College London, London, UK

S. Salunke (✉)
Department of Pharmaceutics, School of Pharmacy, University College London, London, UK

European Paediatric Formulations Initiative (EuPFI), London, UK
e-mail: s.salunke@ucl.ac.uk

H. Polonini
Fagron BV, Rotterdam, The Netherlands

P. Kuehl
Pharmaceutical Research and Development, F. Hoffman-La Roche, Basel, Switzerland

K. Hughes
IPEC (International Pharmaceutical Excipients Council) and Colorcon Ltd, Dartford, UK

H. Batchelor, K. Rose (eds.), *Pediatric Formulations*, AAPS Advances in the Pharmaceutical Sciences Series 58, https://doi.org/10.1007/978-3-031-77240-5_17

Keywords Pediatric · Formulation · Excipient · Inactive ingredients · Safety · Children · Toxicity · Interactions · Regulatory considerations

17.1 Introduction

In the intricate landscape of pediatric pharmaceutical formulations, the spotlight often falls on active pharmaceutical ingredients (APIs) and their therapeutic activity. Excipients are often overlooked and have only recently gained prominent attention from regulatory authorities worldwide. Excipients, food ingredients, and APIs intersect in a nuanced interplay of safety and efficacy. The challenge lies in recognizing that not all excipients are innocuous bystanders; some have raised concerns about their potential impact on the growth and development of young patients [1]. While the Federal Food, Drug, and Cosmetic (FD&C) Act of 1938 mandates rigorous safety evaluations for raw materials in drug products, very few excipients have undergone randomized clinical trials specific to the pediatric subpopulation. Excipients are defined in monographs in various pharmacopoeias, and batches are released on certificates of analysis against monograph specifications, using monograph test methods. However, such monographs usually cover use in adults only, and some excipients are known to be tolerated less by children, especially young children whose physiological systems are still developing [2]. The vast variability in absorption, distribution, metabolism, and excretion (ADME) profiles across the pediatric subpopulation, influenced by growth and developmental changes, implies that neonates and infants may exhibit heightened sensitivity to certain excipients in comparison to toddlers [3]. Changes in metabolic capacity, distribution sites, and organ function all affect the way in which excipients are handled in the very young [4]. Due to the vast differences in metabolism and clearance of medications in various stages of development, it is difficult to ascertain the precise harmful levels of pharmaceutical excipients and the extent to which they will impact patients. Consequently, there are only a few excipients that can be confidently deemed "safe" within the pediatric subpopulation. This adds an additional layer of complexity to the selection of safe excipients. The current guidelines proposed by the Food and Drug Administration (FDA) and the European Medicines Agency (EMA) on the use of excipients in children are limited and conflicting [5]. Important research is being conducted into identifying the safe level of use for excipients in different age categories of pediatric patients. This is essential as excipients that are safely used in adult populations may be toxic, carcinogenic, mutagenic, teratogenic, or affect reproduction in specific subpopulations (e.g., ethanol in neonates). This chapter provides an overview of selected excipients frequently found in pediatric products, examining their chemical properties, safety, toxicity, and acceptable daily intake, which will ease the risk-benefit analysis for excipients.

17.2 Excipients

Excipients are any component, other than the active pharmaceutical ingredient (API), included in a dosage form to convert the API into a pharmaceutical product that can be administered to, or taken by, the patient and that is acceptable to them. The FDA describes an excipient as an inactive ingredient intentionally added to therapeutic and diagnostic products, which is not intended to exert therapeutic effects at the intended dosage, but which may act to improve product delivery (e.g., enhance absorption or control release of the drug substance) [6, 7]. Advancements in the functionality of excipients have now rendered the traditional view of excipients as "simple inert pharmaceutical fillers" obsolete. They are highly diverse and play multifaceted roles, from ensuring the proper delivery and absorption of the active compound to preserving the drug's quality, stability, appearance, and taste. In traditional oral solid dosage forms, many excipients are used as bulking agents, fillers, and binders to form a pill or improve direct compression processes. However, for complex drug products, excipients often play a functional role in the dosage form, improving or modifying the delivery of an API. Some excipients, such as polyethylene glycol, sodium pyrophosphate, and mannitol, have the potential to expedite the transit of orally administered active substances through the gastrointestinal tract. This can, in some cases, detrimentally impact the absorption of the active principle in the gastrointestinal system [8]. There is growing evidence that they can change the pharmacokinetics of APIs through various mechanisms, such as P-glycoprotein (P-gp) inhibition and CYP450 inhibition [9]. Additionally, they may contribute to reactions that lead to degradation or interactions between the drug and the excipient [10, 11]. Adding to the complexity, these effects can vary depending on the specific drug, dosage, formulation, and even the individual subject. For example, when it comes to parenteral products, excipients can alter absorption, potentially leading to changes in exposure patterns and subsequently impacting both safety and efficacy outcomes. Thus, while excipients are essential components of the formulation, they could also affect the safety profile of an API and may be the cause of certain adverse events in patients. The effect on API bioavailability with excipients is even more pronounced across the pediatric population, as physiological differences in systems like the gastrointestinal tract can further enhance or suppress API absorption and activity.

17.3 Excipients in Pediatric Formulations

17.3.1 Key Considerations for Excipients in the Pediatric Population: Developing Physiological Characteristics

The pediatric population is physiologically distinct from the adult population and is in itself extremely heterogenous. Children's body composition's swift evolution and progression lead to a continuous spectrum of growth and developmental stages.

From birth to adulthood, a typical child's body size and weight can increase as much as 20 times, and dosage variations during childhood may vary by up to 100-fold [12, 13]. These differences can alter the pharmacodynamic (PD) and pharmacokinetic (PK) profiles of excipients. PK processes of API liberation, absorption, distribution, metabolism, and elimination (LADME) are influenced by the varied developmental stages within the pediatric population, therefore affecting the extent of excipient and API exposure in different (sub)populations [14]. Release, or liberation, of the API from the dosage form is determined by the excipients included, but also by physiological conditions, such as pH, which can differ across age groups [15]. Significant developmental disparities in absorptive surfaces, such as the gastrointestinal tract, skin, and pulmonary tract, can affect the rate and extent of API and excipient bioavailability [16]. For example, cutaneous absorption is elevated in neonates as their skin barrier is thin and permeable [17]; topical formulations containing the excipients ethanol and propylene glycol are known to readily permeate and cause systemic toxicity in neonates [18, 19]. Distribution is dependent on pH, permeability across organ systems, and body composition [20]. The high body water composition at birth (80%) readily decreases over one's lifetime, with body fat increasing in the first 9 months, then similarly decreasing with age; these effects lead to different volumes of distribution for lipophilic and water-soluble drugs or excipients [20]. The immaturity of the blood–brain barrier (BBB) in neonates allows for the passage of potentially toxic compounds into the central nervous system (CNS) [21]. Notably, propylene glycol is readily able to pass the BBB in neonates, exerting dose-dependent CNS effects [21]. Clearance of excipients is stunted, particularly in the first few months of life, given the immaturity of the renal and hepatic systems. The activity of metabolizing enzymes is generally limited in neonates, leading to rapid saturation, with different rates of maturation occurring across individual enzymatic pathways [17]. The potential of accumulation then increases for excipients that undergo extensive metabolism, like benzyl alcohol [21]. In neonates, renal elimination is limited by the reduced glomerular filtration rate (GFR) and blood flow; adult levels of renal clearance are not reached until toddler age [21]. Therefore, reduced hepatic and renal clearance can cause accumulation of excipients and/or their toxic metabolites; renal function is often affected in neonates and infants exposed to high concentrations of excipients [18]. The extent of accumulation is dependent on the size and timing of the dose, duration of exposure to the excipient, and concomitant exposure to the same excipient from multiple medicines. The non-linear growth of the pediatric population and intra-individual variability in development complicates the ability to discern at which age or weight an excipient may become toxic [17, 20]. The condition of the child can also influence the PK behavior of excipients, as seen with critically ill neonates with substantial renal or hepatic failure who therefore have depleted capacities to metabolize and eliminate [22].

17.4 Excipient Toxicity in Children

Certain excipients deemed acceptable in adult formulations, such as benzyl alcohol, ethanol, propylene glycol, and parabens, are associated with heightened toxicological risks and safety concerns when employed in pediatric medications, even when used in proportionally reduced concentrations. Although excipients are typically not the primary focus of drug safety concerns, they have been associated with adverse reactions and even fatalities in some cases [23, 24]. For instance, various excipients may lead to adverse reactions in specific patient populations, including preterm infants, newborns, children under six years of age, and hypersensitive patients, underscoring the importance of considering excipient selection for patient safety. Common issues arising from specific excipients are presented in Table 17.1.

Indeed, in 1998, the World Health Organization (WHO) brought a sobering statistic to light: over 500 patient deaths have occurred due to intoxication from contaminated medicines since 1937 [34]. One of the most well-known tragedies occurred in Haiti between 1995 and 1996, where at least 86 children, aged three

Table 17.1 Excipients with known safety concerns

Excipient	Route of administration	Effect	Reference
Preterm infants, newborns, children (< 6 years old)			
Benzyl alcohol	Oral, parenteral	Neurotoxicity, metabolic acidosis, gasping syndrome	[25]
Ethanol	Oral, parenteral	Neurotoxicity	[26]
Propylene glycol	Parenteral	Metabolic acidosis	[27]
Polysorbate 20 and 80	Parenteral	E-ferol syndrome (low-birth-weight newborns): Thrombocytopenia, renal dysfunction, hepatomegaly, cholestasis, ascites, hypotension, and metabolic acidosis.	[28]
Preterm infants and newborns with jaundice, kernicterus, or hyperbilirubinemia			
Benzoic acid and sodium benzoate	Oral	Hyperbilirubinemia, metabolic acidosis, respiratory disorders	[8, 29]
Parabens	Oral, parenteral	Hyperbilirubinemia	[26]
Patients with reduced renal function			
Aluminum salts	Oral, parenteral	Encephalopathy, microcytic anemia, osteodystrophy	[30, 31]
Patients with metabolic disorders			
Fructose, sorbitol, and sucrose	Oral, parenteral	Hereditary Fructose Intolerance: Hypoglycemia, renal dysfunction, and liver failure.	[32]
Lactose	Oral	Lactose intolerance: Diarrhea	[33]
Saccharin	Oral	Hypersensitivity reactions, allergies to sulfonamides	[1]

months to 13 years, tragically lost their lives after consuming syrup contaminated with diethylene glycol (DEG) [35].

Tragically, the list of fatal incidents caused by contaminated pharmaceutical products has continued to grow. In 2007, Panama faced DEG contamination, while 2008 saw heparin adulterated with over-sulphated chondroitin sulphate (OSCS), impacting the USA and Germany. In 2009, Nigeria experienced DEG contamination. In 2012, Italy confronted an alarming situation when sodium nitrite was incorrectly marked and used as sorbitol [36].

Following a momentary lull in WHO Medical Product Alerts from 2012 to 2021, a surge in Alerts in 2022 and 2023 has surfaced. In 2022, Gambia and Indonesia grappled again with DEG contaminations [37, 38], while a devastating case of microbiological contamination in methotrexate affected Yemen and Lebanon [39]. The year 2023 brought further grim news, with DEG contaminations in Uzbekistan, Cambodia, Cameroon, Iraq, Marshal Islands, and Micronesia—those last two also witnessed ethylene glycol (EG) contaminations [40–43].

The resumption of such alerts coincides with recent news and reports. In 2023, India's regulatory authorities discovered toxic syrups, leading to serious concerns about their impact on public health. Multiple cough syrups—many of which were marketed toward children or indicated for use in children—were found to contain unacceptable levels of DEG and EG, and their use was linked to fatal outcomes, raising alarms within the healthcare community. The WHO is actively investigating these events, as these products have also been exported to other regions [44].

Analyzing these alarming trends shows that the issue of contaminated pharmaceutical products is not isolated to any one region, and the root cause often varies from deliberate criminal adulteration for economic gain to a failure of Good Distribution Practice (GDP) or Good Manufacturing Practice (GMP). The WHO has been diligently tracking these alerts. From 2019 to 2023, there have been seven deadly alerts related to contaminated medicines [37, 38, 40]. These incidents, while undeniably tragic, serve as reminders of the need for comprehensive knowledge of excipient supply chains and strict quality control of pharmaceutical excipients. It is essential to recognize the excipients' significant role in drug safety, particularly when considering patient health and well-being.

17.5 Regulatory Frameworks

Pharmaceutical excipients are, in the vast majority, well-established ingredients that are used in medicinal products for many years; however, typically not to the same extent in children as in adults. These well-established excipients are described in monographs of the official national or regional compendia, among those the United States Pharmacopeia—National Formulary (USP/NF), the European Pharmacopeia (Ph.Eur.), and the Japanese Pharmacopeia (JP) together with the Japanese Pharmaceutical Excipients Compendium (JPE) set precedence also in other regions and are commonly referenced in new drug applications. When an excipient is described in the applicable pharmacopeia of a country, the conformance with the defined quality and purity is regulatory binding and provides the prerequisite for the approvable use in a medicinal product. Health authorities typically accept references to excipient monographs from third countries in the approval process in case the monograph does not exist in the concerned local or regional pharmacopeia. The Pharmacopoeial Discussion Group[1] (PDG) coordinates the international harmonization of excipient monographs between the members' pharmacopeia in order to promote the global alignment of quality standards. In cases where no pharmaceutical compendia monograph is available, reference to a monograph of registered and approved food additives can also be acceptable. Food compendia, such as Food Chemicals Codex (FCC)[2] or the "union list" by the EU on EC legislation, include in fact many excipients that are also used in pharmaceutical products [45, 46]. Pharmaceutical compendia adopt a new excipient into a monograph if at least one commercial product is approved and not protected by patents anymore.

In an approval process for a new product, the use of an excipient is justified without further provision of toxicity and safety data if precedencies of comparable use are existing with marketed products. Health authorities acknowledge the use in market products as established use with accumulated human safety data [47]. The comparability of use depends on the quantitative levels in approved drug products, the route of administration, the length of treatment, and the indicated patient populations. For use in children, history of existing market product often does not provide sufficient evidence of established use for pediatric patients.

Whenever the targeted use of a known excipient departs from the established use in terms of route of administration, patient population and quantitative levels with proven safety based on accumulated safety data, health authorities expect additional preclinical and clinical safety data to support the application. Since there is a much more limited legacy of pediatric use of drug products in different age groups, the use of excipients based on established use of marketed products is often less comprehensive and additional data and assessments are required to support the intended use in a pediatric product. If the excipient is also established as a food additive (which is in fact, for the majority of oral excipients, the case), the applicant can rely on a greater, more consolidated evidence of safety based on wider use of the excipient. Worldwide food additive status is established by JECFA, a committee, which is formed in 1956 by the UN, bodies FAO[3] responsible for specifications and test methods, and WHO[4] responsible for non-clinical tox assessments. In the USA, under FDA[5] regulation, food ingredients are exempted from the regulation for food

---

[1] Consortium of European Pharmacopoeia (Ph. Eur.), the Japanese Pharmacopoeia (JP) and the United States Pharmacopeia (USP), Indian Pharmacopoeia Commission (IPC), with the World Health Organization (WHO)

[2] Food Chemicals Codex, https://www.foodchemicalscodex.org/

[3] Food and Agriculture Organization, United Nations

[4] World Health Organization, United Nations

[5] U.S. Food & Drug Administration, https://www.fda.gov/

additives[6] if they are notified by the FDA as ingredients that are "generally recognized as safe" (GRAS[7]) at the assessed levels of daily intake with food. Food legislation bodies responsible for the safe use of food additives, such as the EFSA,[8] FDA, and the JECFA, are regularly evaluating the available toxicity and safety data, and updating the safety thresholds of food additives if necessary. The EFSA further issues regularly report on scientific opinions and advises on existing and emerging food risks. Acceptable daily intake levels are determined where it deemed necessary for their safe use based on the estimated overall daily intake that is derived from food consumption data. This includes all age populations that consume the concerned food additive but does not consider age specific sensitivity or toxicity for children.

Therefore, applications for new pediatric products need to provide age-specific assessment of potential safety risks of all excipients including those excipients that are registered as food additives or used only in approved products for adults. The EMA[9] guideline for pediatric development describes in more in detail how to evaluate pediatric use of excipients in pediatric medicines [48]. An important element to waive an extensive assessment of the safety in the targeted age population is to identify approved pediatric products with similar quantitative intake levels, similar ages, and similar length of treatment. Since quantitative information about excipients are typically not available publicly, often assumption of comparable intake levels needs to be made. The inactive ingredients database (IID[10]) of the FDA provides an important source of information about the approved use of a certain excipient for a certain route of administration in different dosage forms. Although no information of pediatric use is indicated, certain dosage forms may be recognized as particular pediatric dosage forms. A maximum daily exposure (MDE) is provided for each product registered, which enables the identification of an upper limit to which safe use in humans in assumed.

The use of compendial grade excipients generally provides a high level of confidence in an appropriate quality, purity and its control. When a new excipient is adopted to a pharmacopeia, preclinical safety assessments are available and reviewed. Compendial excipients are considered generally to be non-toxic or have low toxicity for intended quantitative level and administration route in scope of the excipient use but not automatically safe for children.

The use of a new excipient, which is used first time in a drug product of a certain route of administration, is classified as "novel excipients" in the regulatory terminology [49]. Such a novel excipient requires a full dossier with data package for the

[6] Approval of a food additive petition as set forth in 21 CFR 171

[7] (GRAS) pursuant to Title 21, U.S. Code of Federal Regulations, Parts 182, 184 or 186 (21 CFR 182, 184 & 186)

[8] European Food Safety Authority, https://www.efsa.europa.eu/

[9] European Medicines Agency, https://www.ema.europa.eu/

[10] Inactive Ingredient Search for Approved Drug Products, FDA/Center for Drug Evaluation and Research, https://www.accessdata.fda.gov/scripts/cder/iig/index.cfm

evaluation of the quality and safety and full details of manufacture and control in similar manner as for active ingredients.

In the USA, there are no standalone approval procedures for pharmaceutical excipients, and the necessary safety and quality information has to be submitted either in conjunction with the drug product dossier or via determination of a regulatory status as food additive or GRAS ingredient. The FDA has published a guidance for "nonclinical studies for the safety evaluation of pharmaceutical excipients," which proposes the conduction of toxicological studies in accordance with ICH S7 [7]. However, the specific aspects of preclinical assessment of pediatric use are covered in the FDA guidance related to pediatric products and addressing the potential benefit of juvenile toxicity studies [7]. Furthermore, the USP defines in the general chapter <1074> the necessary toxicity testing dependent on route of administration and treatment duration [50]. Juvenile toxicity studies to support pediatric use of excipients is not included in the chapter. In the EU, a novel excipient (new chemical entity) has to be approved on the basis of the same level of chemical, pharmaceutical, and biological information as submitted in the marketing authorization application for a new drug substance. The evaluation of safety of a novel excipient in humans is firstly approached with drug products in adults before the concerned excipient will be introduced in medicines for children. Hence, novel excipients will not play a relevant role for pediatric use.

Flavors are products containing complex mixtures of flavoring agents that are only authorized by food regulations. The quantitative composition of a flavor is the intellectual property of the flavor manufacturer and therefore not disclosed to the applicants. Via a "letter of access" procedure, health authorities get access to the excipient dossier, e.g., the drug master file, during the drug product review. Therefore, the applicant is only required to provide a qualitative composition of the flavor. EMA guidance for excipients in the dossier is explicitly referring to standards of flavoring agents as food additives [6]. International acceptance of a flavoring ingredient is provided when eligible institution such as the FAO, FCC, WHO, or FEMA[11] have established a monograph including purity specifications. In the EU an approved list of flavoring ingredients is established as part of a dedicated regulation "on flavorings and certain food ingredients with flavoring properties for use in and on foods" [51].

Another group of excipients that is frequently used in pediatric products is coloring agents. In the EU, the regulation of coloring agents for use in pharmaceuticals is deferred per directive to food additives regulation, and no pharmaceutical monographs are accordingly established for this excipient group [52]. Safety evaluation of coloring agents in medicinal products and their approved use depend on the evaluation as food additives. However, in some cases, the EMA has also published scientific evaluation for the use in pharmaceutical products. In the USA, color agents are certified by the FDA under the legislation of the Food, Drug and Cosmetic Act

[11] Flavor and Extract Manufacturers Association, United States, https://www.femaflavor.org/

and maintain an official list of color additives in which approved color additives are labelled with the term "FD&C" in the name [53].

The appropriate, regulatory acceptable selection of excipients in a drug product is an integral part of the pharmaceutical development, including the formulation design. In principle, the regulatory guidance ICH Q8 for pharmaceutical development provides all the elements that are also suitable to ensure that pharmaceutical companies meet appropriately specific aspects of pediatric requirements and objectives [54]. The guideline, which has introduced elements such as the quality target product profile as a starting point for development of a new product, the approach of quality by design, and the application of risk assessments to guide development, also addresses the aspects of appropriate selection of excipients. As part of the quality risk assessments, the impact of excipients on the drug product performance and manufacturability should be discussed, and the intended functionality of an excipient has to be demonstrated. With this principle, the inclusion of an excipient at a certain level is linked to a justification of use and thereby enables risk-benefit evaluations in combination with safety information of the concerned excipient.

These key elements are further explicated for excipients in the EMA guideline "on pharmaceutical development of medicines for paediatric use" with a stronger emphasis on safety information since excipients used in children have to be evaluated more cautiously due to the limited availability of relevant pediatric safety data [48]. The guideline highlights the particular importance to justify the use of an excipient under consideration of all safety risks, benefits, advantages, and disadvantages of its use for the targeted conditions of use and pediatric patient populations and age groups. For the assessment of the safety profile of excipients in a pediatric formulation, a comprehensive checklist in hierarchical order is proposed to evaluate all available safety information. However, the exact methodology on how to balance safety risks against the actual excipient levels used and the benefit of its use is not defined further. Certain functional groups of excipients, such as coloring agents, preservatives, flavors, and sweeteners, are subjected to dedicated paragraphs since excipients from these groups are frequently used in pediatric products, and some of them present known potential for adverse effects and safety risks. Particularly, oral liquid dosage forms need to make use of antimicrobial preservatives and antioxidants. Independent of pediatric use, these two excipient groups are subject to stricter requirements regarding justification of the level used. The EMA guideline "on Excipients in the Dossier for Application for Marketing Authorization of a medicinal product" explicitly referring to pediatric use claims the general requirement to avoid or at least to minimize as far as possible and to justify the use of preservatives and antioxidants [6]. The evidence has to be provided that the "lowest feasible level" has been selected for a product. Use of synthetic coloring agents should not be used in pediatric products. The use of any specific coloring agent in a pediatric preparation should be discussed and justified in terms of allergenic potential, minimal toxicological implications in the target age group(s), patient acceptability, and the need to avoid accidental dosing errors.

In general, it is the responsibility of the applicants for a new drug product to collect and compile the relevant preclinical and clinical safety information for excipients. A unique regulatory compilation of safety information represents the Annex of the EMA guideline on "excipients in the labelling and package leaflet of medicinal products from human use" [5]. The list of excipients contains binding information about excipients with known effects that have to be labelled in the EU. For pediatric use, this document is of particular relevance since it represents a comprehensive list of excipients with safety concerns in pediatric populations and defines thresholds of daily intake levels for each of outlined adverse effects. For the future development of pediatric products, such publicly shared information of known excipient effects in children is very useful, and the associated determination of thresholds of daily intake should for this reason be further elaborated based on data-based, comprehensible approaches.

Since the regulations for pharmaceutical excipients do not consider an approved status of use independently of the approval of a drug product, a significant uncertainty exists for an appropriate safe use level in children. The regulations for pharmaceutical use of excipients should, in the interest of a safe use in children, elaborate on ensuring the availability of clear safety limits and quantitative information on pediatric excipient intake. For the benefit of global approvability of pediatric products, the international harmonization of regulations for pediatric use of excipient should be further fostered [55].

17.6 Excipient Interactions in a Dosage Form

Excipients may contain chemical groups that initiate, propagate, or interact with active ingredients and/or other excipients [56]. They can also introduce the risk of microbial contamination and give rise to unwanted effects like irritation, sensitization, or alterations in product aesthetics [57].

Physical interactions can have both favorable and unfavorable effects on pharmaceutical formulations. For instance, some materials can adsorb drugs, enhancing surface activity and improving dissolution rates, while strong attraction forces can hinder drug release and absorption. An example is the use of lipophilic lubricants (e.g., magnesium stearate), which can reduce dissolution and bioavailability when finely dispersed within drug particles [57]. Adsorption can also trigger chemical decomposition, such as colloidal silica catalyzing nitrazepam degradation in tablet form [58]. On a positive note, these physical interactions can be harnessed to optimize formulations, such as the complexation of APIs with cyclodextrins [59, 60].

Conversely, chemical interactions can lead to API degradation and the formation of degradation impurities. In liquid formulations, hydrolysis is the most common reaction due to water being the primary solvent—but also in solid dosage forms with hygroscopic components; moisture control is vital to prevent hydrolysis-related degradation [57]. Other reactions are also found in the literature:

(i) *Ionic Charge Interactions:* Soluble ionizable excipients can interact with ionizable active ingredients, forming insoluble compounds.

For example, sodium carboxymethyl cellulose with polymyxin, leading to precipitation.
(ii) *Hydrogen Donation Interactions:* Polyvinylpyrrolidone (PVP) can interact with compounds containing hydrogen-donating functional groups (such as lansoprazole), causing degradation reactions [56].
(iii) *Lactose Reactions:* Lactose can react with compounds containing primary (e.g., amino acids) or secondary amines (e.g., fluoxetine), leading to product grouping and color development (the Maillard reaction) [61].
(iv) *Silicon Dioxide Reactions:* Silicon dioxide can act as a Lewis acid under anhydrous conditions, causing reactions such as dehydration, hydrolysis, epimerization, cyclization, and transesterification (with, for example, nitrazepam) [58].

These interactions can significantly impact drug stability and should be considered when selecting excipients and designing stability studies.

Additionally, excipients often contain manufacturing residues and impurities. These can arise from using catalysts, reagents, and solvents during the extraction and purification of various materials. These residual components can interact with the API and other excipients, possibly leading to API degradation and posing safety concerns, which is why it is important to select the correct excipient grade and have a robust supplier qualification program in place when choosing excipients. Examples of some excipients and their associated residues are benzyl alcohol (benzaldehyde), croscarmellose sodium (sodium monochloroacetate, nitrite, nitrate), hydroxypropylmethyl/ethyl cellulose (glyoxal), lubricants such as stearates (alkaline residues), microcrystalline cellulose (lignin, hemicelluloses, aldehydes, reducing sugar), polyethylene glycol (aldehydes, peroxides, organic acids such as formic acid), starch (formaldehyde, aldose), and talc (heavy metals) [56].

17.7 Excipients Commonly Used in Pediatric Products

Pediatric patients present a unique challenge when it comes to the administration of medications, as they span a wide age range, from newborns to adolescents, each with distinct physiological and developmental characteristics. To cater to this diversity, pharmaceutical formulations for children are available in a variety of dosage forms, each designed to ensure safe and effective drug delivery while accommodating the preferences and needs of young patients.

- *Oral Liquid Formulations:* These are often the preferred choice for infants and young children due to their ease of administration. Oral liquid medications may include solutions, suspensions, or syrups. In these formulations, excipients such as sweetening agents (e.g., aspartame), flavoring agents, and viscosity modifiers are commonly used to improve taste, mask bitterness, and enhance palatability.
- *Oral Solid Dosage Forms:* Tablets, chewable tablets, granules, and powders are frequently used for older children who can swallow solid forms. In these formu-

lations, excipients like disintegrants, binders, and lubricants play a crucial role in ensuring the tablet's integrity and disintegration properties.

- *Orally Disintegrating Tablets (ODTs):* ODTs are designed to rapidly disintegrate in the mouth, eliminating the need for water for administration. Excipients such as superdisintegrants and taste-masking agents are essential to create these user-friendly dosage forms, which are particularly suitable for children who may have difficulty swallowing traditional tablets.
- *Topical Formulations:* Creams, ointments, and gels are applied to the skin for various skin conditions. Excipients in these formulations often include emollients, stabilizers, and penetration enhancers.
- *Rectal Suppositories:* Suppositories are designed for rectal administration and are especially useful for pediatric patients who have difficulty swallowing or retaining oral medications. Excipients in suppositories include base materials and additives for consistency and release.
- *Injectable Formulations:* For certain medical conditions, injectable drugs are required. These formulations may include solutions or suspensions for intramuscular, subcutaneous, or intravenous administration. Excipients in injectables help maintain stability and adjust the release profile of the drug.
- *Inhalation Formulations:* Inhalers, nebulizers, and dry powder inhalers are used for respiratory conditions. Excipients are chosen to facilitate aerosolization, stabilize the drug, and enhance patient compliance, especially in the case of children who may find inhalation therapy challenging.

The meticulous screening and selection of excipients in pediatric medicinal products has emerged as one of the pivotal facets of pharmaceutical development [62]. The selection of excipients in these pediatric formulations must consider not only the pharmaceutical and technical aspects but also the unique requirements of the age group being treated. The excipients chosen, their concentration, and their relevance to the drug product's function must be substantiated with respect to safety for the targeted age group, treatment, route of administration, treatment duration, allergies, and the severity of the disease. The need to be carefully chosen to ensure safety, efficacy, and palatability while addressing the specific needs of children, making them a critical component in the development of medications tailored for pediatric patients. A major challenge for formulators of pediatric medicines is the limited choice of excipients that meet the acceptance criteria.

A risk-benefit analysis is commonly initiated as the benefits of treatment may outweigh the risk of excipient-related adverse effects. Decreasing excipient content or substituting excipients may be necessary, although that may decrease the efficacy of the final formulation. Guidelines for pediatric development emphasize the need to individually select excipients, justifying their inclusion based on a careful consideration of the benefit-risk approach. However, obtaining safety information on excipients for pediatric patients can be challenging. Efforts are being made, exemplified by the EMA publishing various excipient Q&A documents and reflection papers. Valuable sources for information include food safety reviews, published literature encompassing human safety and animal toxicity studies, and insights from excipient suppliers. The European Paediatric Formulation Initiative (EuPFI) has introduced the Safety and Toxicity of Excipients in Paediatrics (STEP) database to enhance access to safety information on over 78 excipients [63]. Additionally, the EuPFI excipients workstream in collaboration with the Innovation and Quality Consortium (IQ) pediatric workgroup members have designed a Paediatric Excipient Risk Assessment (PERA) framework and a tool that provides a structured method for the selection and overall risk assessment of excipients for pediatric formulations [55]. It provides a structured, systematic decision-making framework via a customizable tool and processes that can help improve transparency and communications on the selection and justification of the use of excipients in a pediatric formulation. In instances where pediatric safety data is lacking, the suggestion is to conduct juvenile animal toxicology studies to facilitate the benefit-risk evaluation process. The concept of a Progressive Pediatric Safety Factor has been proposed, allowing the calculation of maximum acceptable doses of an excipient in pediatric patients when reliable safety data for neonates, infants, and children are unavailable [64].

The properties of excipients, categorized by function, is presented in the following section with Tables (17.2, 17.3, 17.4, 17.5, 17.6, and 17.7) with corresponding physical characteristics, regulatory status, and pediatric toxicity profiles.

17.7.1 Diluents/Fillers/Binders (Table 17.2)

Diluents, fillers, and binders make up a considerable portion of solid dosage forms like tablets. These excipients are used to bulk up the tablet to a suitable size for swallowing for low-dose drugs, ensure desired tablet integrity (i.e., reduce friability), and allow adequate distribution of the active throughout the tablet for adequate dissolution and solubility [65]. Multifunctional binders include the saccharides mannitol, sucrose, and sorbitol, which also impart a sweet flavor to the tablet.

17.7.2 Sweeteners and Flavors (Table 17.3)

Younger children have a greater propensity toward sweet tastes and are found to be more sensitive toward, and therefore less likely to consume, bitter tastes [73]. Many APIs are naturally bitter, which requires the use of "taste-masking" agents to improve children's acceptability and therefore compliance to orally administered medicines. Some sugars, such as mannitol and sucrose, are multifunctional, acting as binders, vehicles, and diluents. A combination of sweeteners may be necessary as nutritive sweeteners may not provide adequate sweetness, while non-nutritive, artificial sweeteners may not provide adequate robustness to the formulation [74]. Flavor preference is also dependent on the age group, gender, region, and individual [75].

Table 17.2 Diluents, fillers, binders, and other bulk excipients used in pediatric formulations

Excipient	Information	Regulatory status	Adverse effects in pediatrics	References
Lactose 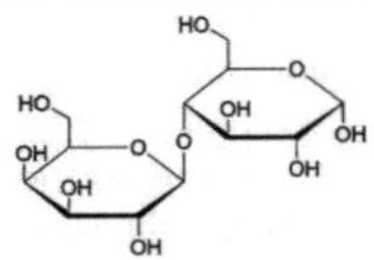 *Formula:* $C_{12}H_{22}O_{11}$ *Molecular weight:* 342.3 g/mol *Solubility:* Soluble in water *Appearance:* White powder *Synonyms:* Lactin; D-glucose; galactinum; lactobiose; milk sugar; D-lactose	Lactose is a disaccharide of galactose and glucose. As a filler and diluent, lactose is used in one of three forms: Alpha-lactose (monohydrate), beta-lactose (crystalline, anhydrous), and lactose (amorphous). Amorphous and anhydrous lactose may undergo a Maillard reaction with primary amines in APIs to form potentially harmful electrophiles	*GRAS:* Yes *ADI:* Not specified for lactose-intolerant and galactosemia patients	*Oral* In children with lactase deficiency (lactose intolerance): Diarrhea, bloating, vomiting, malabsorption, and metabolic acidosis. In children with galactosemia: Vomiting, diarrhea, cataracts, and delayed symptoms, including cognitive defects and stunted growth	[8, 66, 67]
Polyvinylpyrrolidone 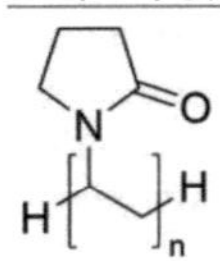 *Formula:* $(C_6H_9NO)_n$ *Molecular weight:* 360,000 g/mol (average *n*) *Solubility:* Freely soluble in water *Appearance:* White hygroscopic powder *Synonyms:* Povidone; polyvidone; cross-linked povidone; PVP; PVPP	Polyvinylpyrrolidone is found in tablets and eye drops, acting as a binder, disintegrant, and dissolution enhancer	*GRAS:* Not listed *ADI:* Not specified	*Topical* Hypersensitivity reactions (contact dermatitis, anaphylaxis)	[66, 68, 69]

(continued)

Table 17.2 (continued)

Excipient	Information	Regulatory status	Adverse effects in pediatrics	References
Hydroxypropyl methylcellulose (HPMC) 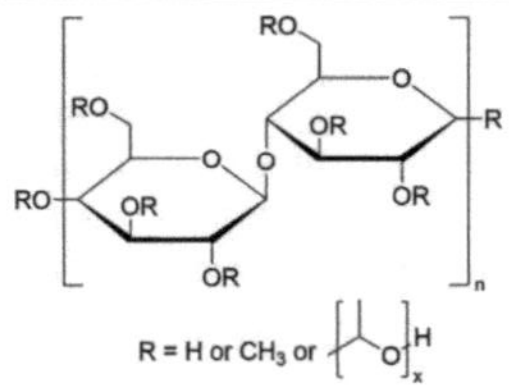 *Solubility:* Soluble in water *Appearance:* Off-white powder or granules *Synonyms:* Hypromellose; methylcellulose propylene glycol ether; methyl hydroxypropyl cellulose; HPMC	HPMC has multiple functions (emulsifier, film-former, mucoadhesive, suspending agent, binder, and coating agent) and is present in topical formulations (gels, ointments), capsules, and tablets	*GRAS:* Yes *ADI:* Not specified. Suggested tolerable intake of 5 mg/kg/day	*Oral* May react with APIs (duloxetine hydrochloride) to form toxic byproducts glyoxal or dialdehyde	[66, 70, 71]
(maize, potato, rice, corn, wheat) starch 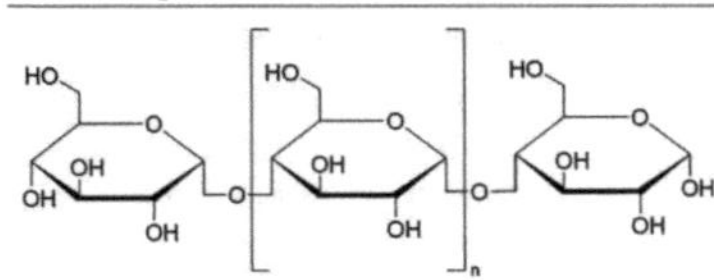 *Formula:* $(C_6H_{10}O_5)_n$ *Molecular weight:* - *Solubility:* Insoluble in ethanol and water. *Appearance:* White powder *Synonyms:* Amido; amylum	Starch is made from linear amylose and branched amylopectin. Starch is used as a diluent, binder, and tablet disintegrant	*GRAS:* Yes *ADI:* Not specified	*Oral* Allergic reactions (rare). Impurities (lignin, maltose, acetylformoin) can react with APIs with primary or secondary amines	[66, 72]

Excipient	Information	Regulatory status	Adverse effects in pediatrics	References
Microcrystalline cellulose (MCC) 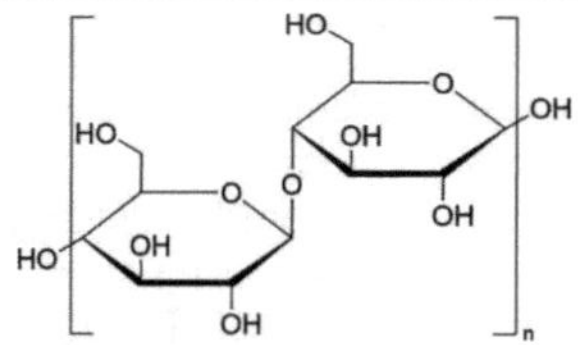 *Formula:* $(C_6H_{10}O_5)_n$ *Molecular weight:* - *Solubility:* Practically insoluble in water *Appearance:* White crystalline powder *Synonyms:* Cellulose gel; crystalline cellulose	MCC is available in different particle sizes with different properties. Depending on the grade, MMC can be used as a lubricant, binder, or diluent	*GRAS:* Yes *ADI:* Not specified	*Oral* Limited toxic potential as it is not absorbed systemically. Laxative effects at large amounts	[66]
Liquid paraffin *Formula:* C_nH_{2n+2} *Molecular weight:* - *Solubility:* Insoluble in water *Appearance:* Translucent, colorless, or white greasy solid *Synonyms:* Wax; paraffin wax	Paraffin is a mixture of saturated hydrocarbons, acting as an occlusive agent in ointments or creams. Paraffin may also be used as a lubricant or coating for tablets and capsules	*GRAS:* Not listed *ADI:* 12 mg/kg/day	*Parenteral* Granuloma formation *Inhalation* Lipoid pneumonia	[66]

Table 17.3 Sweeteners and flavors in pediatric formulations

Excipient	Information	Regulatory status	Adverse effects in pediatrics	References
Non-artificial				
Sucrose (fructose) 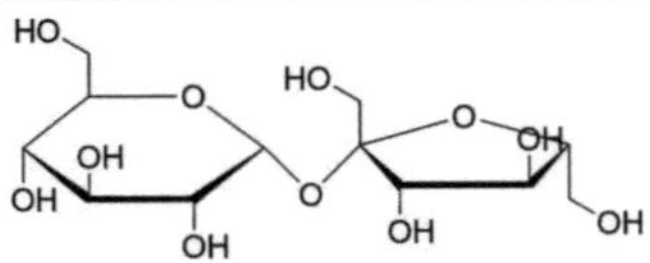 *Formula:* $C_{12}H_{22}O_{11}$ ($C_6H_{12}O_6$) *Molecular weight:* 342.3 g/mol (180.2 g/mol) *Appearance:* White crystalline solid *Synonyms:* Table sugar; D-sucrose; Saccharose; Saccharum (D-fructose; D-Fructopyranose; Levulose)	Sucrose is a disaccharide of fructose and glucose used as a binding agent and sweetener. Fructose is more soluble than sucrose and is used as a diluent and sweetener	*GRAS:* Yes *ADI:* Not specified *HFI tolerable dose:* 5 mg/kg/day	*Oral, parenteral* Various dental effects (cavities, decayed tooth enamel) can occur in children High doses of sucrose may be carcinogenic. In patients with diabetes mellitus: Hyperglycemia (> 5 g) In children with hereditary fructose intolerance (HFI): Vomiting, bloating, diarrhea, and severe adverse effects (liver failure, death) from accumulation of the toxic fructose 1-phosphate metabolite	[1, 76, 77]
Sorbitol 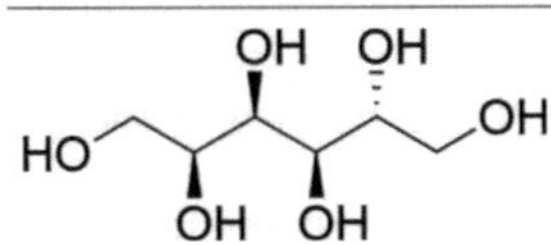 *Formula:* $C_6H_{14}O_6$ *Molecular weight:* 182.2 g/mol *Appearance:* White hygroscopic powder *Synonyms:* D-Glucitol; Diakarmon; Sorbit; Sorbol	Sorbitol is a polyol (sugar alcohol) that is poorly absorbed in the GIT. Sorbitol is metabolized into fructose and glucose by sorbitol dehydrogenase. Sorbitol acts as a diluent in tablets and a vehicle for liquid formulations	*GRAS:* Yes *ADI:* Not specified *HFI tolerable dose*: 5 mg/kg/day	*Oral, parenteral* Laxative effect (> 140 mg/kg/day), abdominal pain, bloating, osmotic diarrhea, and accumulation in neonates/infants can lead to diabetic-like complications (e.g., cataracts). In children with hereditary fructose intolerance (HFI): Vomiting, bloating, diarrhea, and severe adverse effects (liver failure, death) from accumulation of the toxic fructose 1-phosphate metabolite	[1, 66, 77]

Excipient	Information	Regulatory status	Adverse effects in pediatrics	References
Mannitol 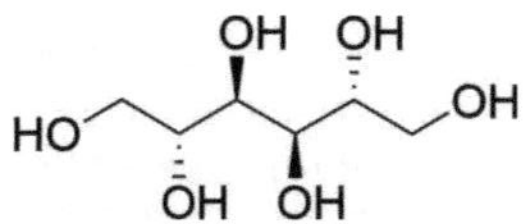 *Formula:* $C_6H_{14}O_6$ *Molecular weight:* 182.2 g/mol *Appearance:* White crystalline, non-hygroscopic powder *Synonyms:* Mannite; Osmitrol	Mannitol is a polyol derived from fructose. It has limited absorption in the GIT and is mostly excreted, unchanged, by the kidneys. Aside from being a sweetener, mannitol is also used as a diluent for tablets and capsules, and carrier for lyophilized preparations	*GRAS:* Yes *ADI:* Not specified	*Oral* Laxative effect (> 20 g) and risk of allergic reaction in pediatrics	[1, 66]
Xylitol 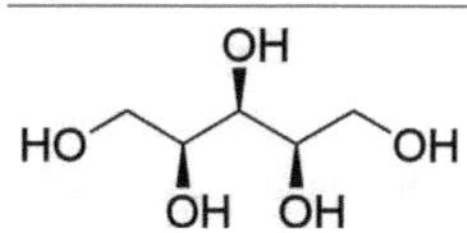 *Formula:* $C_5H_{12}O_5$ *Molecular weight:* 152.2 g/mol *Appearance:* White crystalline powder *Synonyms:* Birch sugar; Xylite; Xylitolum	Xylitol is a polyol that is poorly absorbed in the GIT. Xylitol is used in lozenges, cough syrups, and ODTs as a coating agent, diluent, filler, and sweetening agent	*GRAS:* Yes *ADI:* Not specified	*Oral* Laxative effect at high doses (> 0.5–1.0 g/kg BW/day)	[66]
Artificial				

(continued)

Table 17.3 (continued)

Excipient	Information	Regulatory status	Adverse effects in pediatrics	References
Aspartame 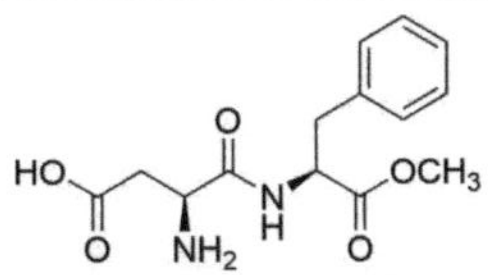 *Formula:* $C_{14}H_{18}N_2O_5$ *Molecular weight:* 294.3 g/mol *Appearance:* White crystalline powder *Synonyms:* Nutrasweet™; equal™; aspartyl phenylalanine methyl Ester	Aspartame metabolism results in the formation of L-phenylalanine and L-aspartic acid	*GRAS:* Yes *ADI:* 40 mg/kg/day (does not apply to children <12 weeks old)	*Oral* In children with phenylketonuria (PKU), L-phenylalanine may accumulate and cause neurological effects (seizures, neurotoxicity, microcephaly) Hyperactivity, hyperreactivity, and headaches, grand mal seizures, with long-term, high-dose exposure	[24, 78]
Saccharin 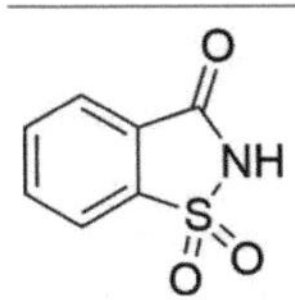 *Formula:* $C_7H_5NO_3S$ *Molecular weight:* 183.2 g/mol *Appearance:* White crystals *Synonyms:* o-benzoic Sulfimide; o-Sulfobenzimide; Saccharimide; saccharine	Saccharin is 300–600 times sweeter than sucrose; its sodium salt is typically used	*GRAS:* Yes. *ADI:* 5 mg/kg/day. Previously banned in Canada, but now permitted	*Oral* Potentially carcinogenic (bladder cancer) in pediatric populations. Hypersensitivity reactions (urticaria, dermatitis, photosensitivity, hives) and nausea. Hypertonia, insomnia, irritability, and opisthotonus also occur in children. Cross-reactivity with sulfonamides	[8, 24, 66]

Excipient	Information	Regulatory status	Adverse effects in pediatrics	References
Sodium cyclamate 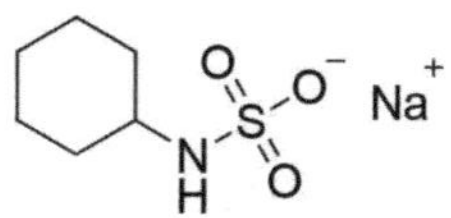 *Formula:* $C_6H_{12}NNaO_3S$ *Molecular weight:* 201.2 g/mol *Appearance:* White crystals or crystalline powder *Synonyms:* Sodium N-cyclohexyl sulfamate; Sucaryl sodium	Sodium cyclamate is usually used alongside saccharin as a sweetening agent.	*GRAS:* Not listed. *ADI:* 7 mg/kg/day. Cyclamate (and its sodium salt) was banned in the USA in 1970, but is still permitted in Canada and Europe	*Oral* Previously thought to be carcinogenic. Photosensitive dermatitis	[66]
Flavors				
Peppermint oil (menthol) 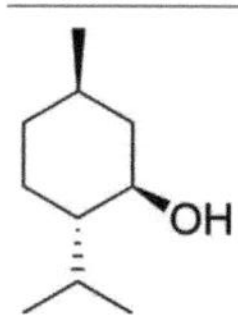 *Formula:* $C_{10}H_{20}O$ *Molecular weight:* 156.2 g/mol *Appearance:* Liquid or white solid with odor *Synonyms:* N/A	Menthol is derived from mint plants. It is a pharmacologically active molecule, able to activate "cooling" TRPM8 receptors. Menthol desensitizes against bitter flavors and imparts a desirable mint flavor	*GRAS:* Yes *ADI:* 4 mg/kg/day	*Topical* Hypersensitivity reactions (contact dermatitis) *Oral, inhalation* Severe abdominal pain and coma. Instant collapse and apnea in infants Atrial fibrillation, burning, and muscle pain in pediatric populations	[8, 66, 79]

Table 17.4 Azo dyes in pediatric formulations

Azo dyes		
Information	*Adverse effects in pediatrics*	
Azo dyes are synthetic dyes containing the R-N=N-R' functional group. The regulations involving the use of azo dyes in medicines change per legislative region. All azo dyes are approved for use in the EU	*Oral* Azo dyes have been implicated in hypersensitivity reactions (anaphylaxis, angioedema, asthma, and urticaria) and hyperactivity in children. Children who are hypersensitive to azo dyes can experience cross-sensitivity to aspirin, sodium benzoate, and indomethacin [65]	
Excipient	*Physical properties*	*Regulatory status*
Tartrazine (E 102)	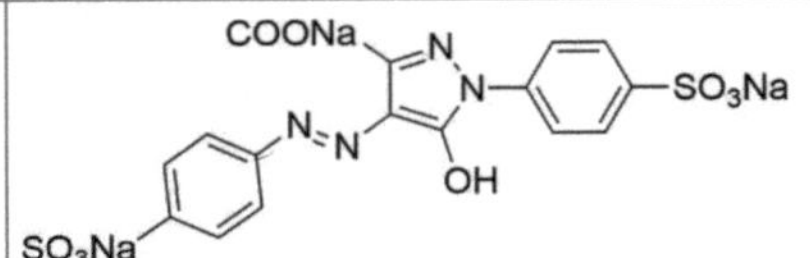 *Formula:* $C_{16}H_9N_4Na_3O_9S_2$ *Molecular weight:* 534.4 g/mol *Solubility:* Highly soluble in water *Appearance:* Yellow/Orange-yellow powder *Synonyms:* FD&C Yellow No. 5; Hydrazine yellow; Cl 19140	*ADI:* 10 mg/kg/day
Sunset yellow FCF (E 110)	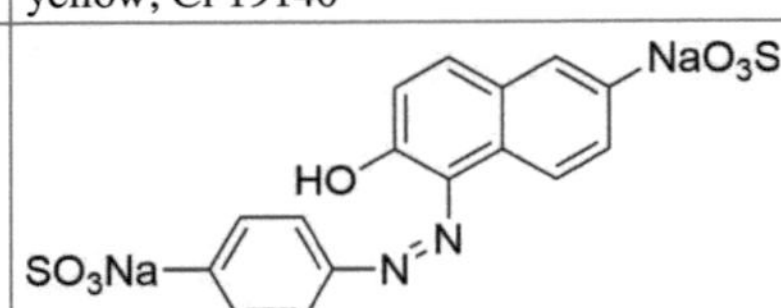 *Formula:* $C_{16}H_{10}N_2Na_2O_7S_2$ *Molecular weight:* 452.4 g/mol *Solubility:* Highly soluble in water *Appearance:* Red-yellow powder *Synonyms:* FD&C Yellow No. 6; Orange yellow S; Cl 14985	*ADI:* 4 mg/kg/day
Ponceau 4R (E 124)	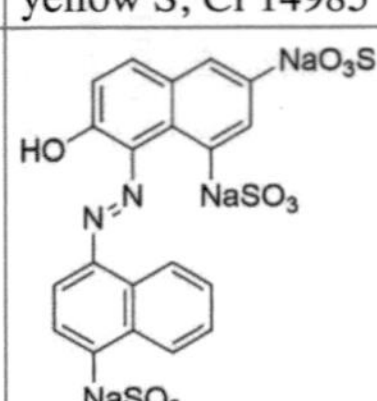 *Formula:* $C_{20}H_{11}N_2Na_3O_{10}S_3$ *Molecular weight:* 604.5 g/mol *Solubility:* Soluble in water *Appearance:* Red powder *Synonyms:* Cl 16,255; Cl Acid Red 18; New Coccine	*ADI:* 4 mg/kg/day Not permitted for use in the USA

(continued)

Table 17.4 (continued)

Azo dyes		
Azorubine (E 122)	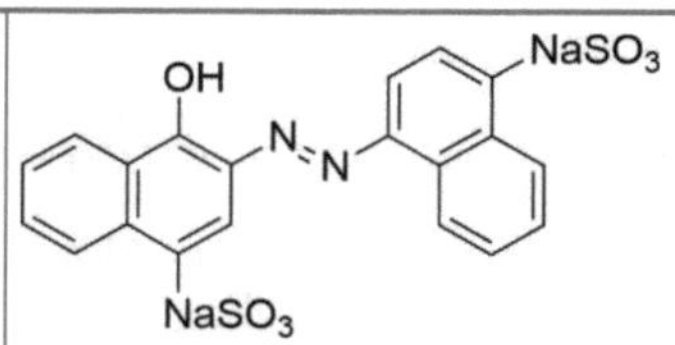 *Formula:* $C_{20}H_{12}N_2Na_2O_7S_2$ *Molecular weight*: 502.4 g/mol *Solubility:* Highly soluble in water *Appearance:* Dark maroon/reddish-brown powder or granules *Synonyms:* Carmoisine; FD&C Red No. 10; Cl 14,720	*ADI:* 4 mg/kg/day Not permitted for use in the USA
Allura red (E 129)	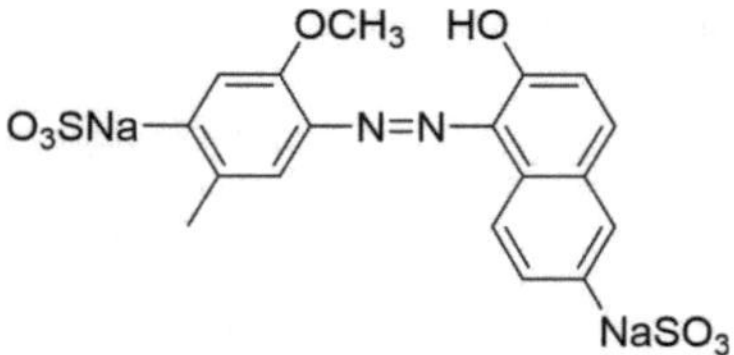 *Formula:* $C_{18}H_{14}N_2Na_2O_8S_2$ *Molecular weight:* 496.4 g/mol *Solubility:* Soluble in water *Appearance:* Dark red powder *Synonyms:* FD&C Red No. 40; Cl 16035, Food Red 17; Allura Red AC	*ADI:* 7 mg/kg/day

17.7.3 Coloring Agents (Tables 17.4 and 17.5)

According to the USFDA, a coloring agent is a pigment, dye, or other substance that imparts color to a drug, food, or cosmetic, or to the human body [80]. The use of color for use in drugs may only be permitted for application on specific surfaces. Hence, some colorants may only be used for external application as ingestion may be associated with adverse reactions. Color has an influence on patient compliance in the pediatric population. Brightly colored tablets are reminiscent of fruit-flavored, sweet candy, and therefore affect taste perception, regardless of ingredients [62, 75]. As with the food industry, a recent push toward using natural colorants (e.g., curcumin, beta-carotenes, and chlorophyll) has been made to replace artificial colors, which have been restricted for use given their link to hypersensitivity and ADHD in children [76]. Most colors used in drugs include azo dyes, triphenylmethane dyes, quinoline dyes, and xanthine dyes. Colors are not designated GRAS status by the FDA.

Table 17.5 Triphenylmethane, quinoline, and xanthine dyes in pediatric formulations

Excipient	Regulatory status	Adverse effects in pediatrics	References
Triphenylmethane dyes			
Brilliant blue FCF (E 133) 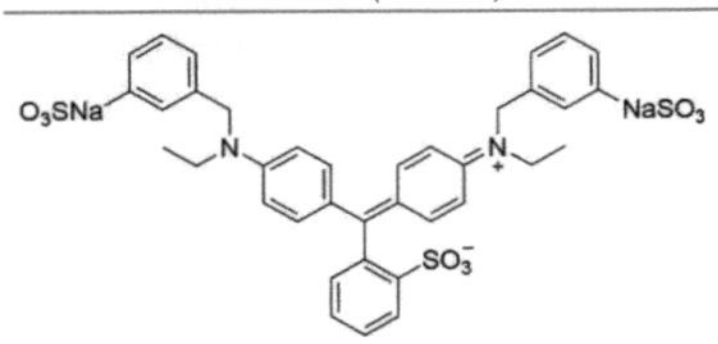 *Formula:* $C_{37}H_{34}N_2Na_2O_9S_3$ *Molecular weight:* 792.9 g/mol *Solubility:* Highly soluble in water *Appearance:* Reddish-blue/violet powder *Synonyms:* FD&C Blue No. 1; Acid Blue 9; Cl 42090	*ADI:* 6 mg/kg/day. Brilliant Blue FCF is approved for use in ingested drugs in the EU and USA	*Oral* Bronchoconstriction	[65]
Fast green FCF 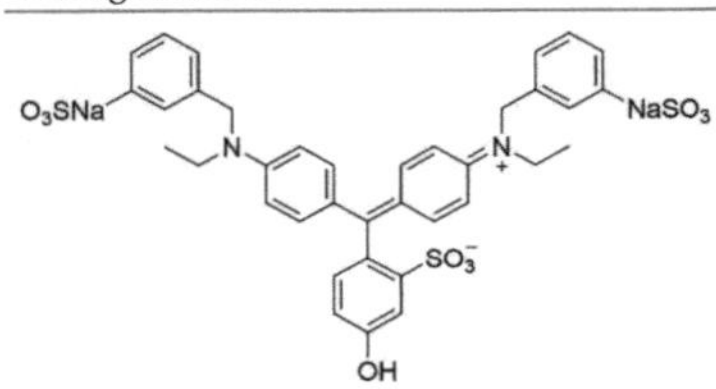 *Formula:* $C_{37}H_{34}N_2Na_2O_{10}S_3$ *Molecular weight:* 808.9 g/mol *Solubility:* Soluble in water *Appearance:* Dark green powder *Synonyms:* FD&C Green No. 3; Cl 42053; Food Green 3	*ADI:* 25 mg/kg/day. Fast Green FCF is prohibited in the EU	*Oral* Contact dermatitis	[65]
Quinoline dyes			
Quinoline yellow (E 104) 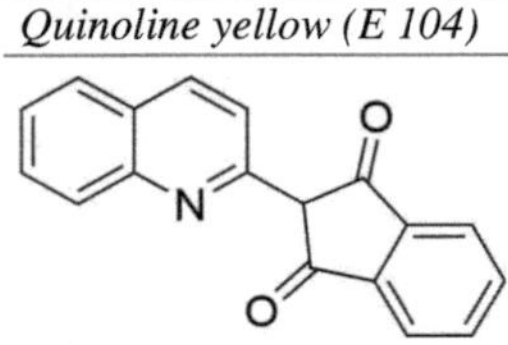 *Formula:* $C_{18}H_{11}NO_2$ *Molecular weight:* 273.3 g/mol *Solubility:* Variable *Appearance:* Greenish-yellow solid *Synonyms:* D&C Yellow No. 10; Cl 47005; Cl Acid Yellow No. 3	*ADI:* 3 mg/kg/day. Approved for use in drugs in the EU and USA	*Oral* Contact dermatitis, hyperactivity	[65, 66]
Xanthine dyes			

(continued)

Table 17.5 (continued)

Excipient	Regulatory status	Adverse effects in pediatrics	References
Erythrosine (E 127) 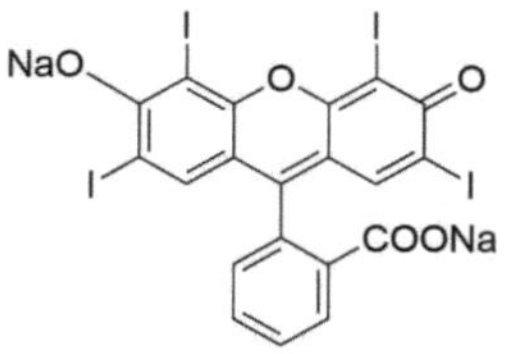 *Formula:* $C_{20}H_6I_4Na_2O_5$ *Molecular weight:* 879.9 g/mol *Solubility:* Soluble in water *Appearance:* Red powder *Synonyms:* FD&C Red No. 3; Cl 45430; Cl Acid Red 51	*ADI:* 0.1 mg/kg/day Approved for use in drugs in the EU and USA	*Oral* Potentially carcinogenic	[66]
Eosin Y 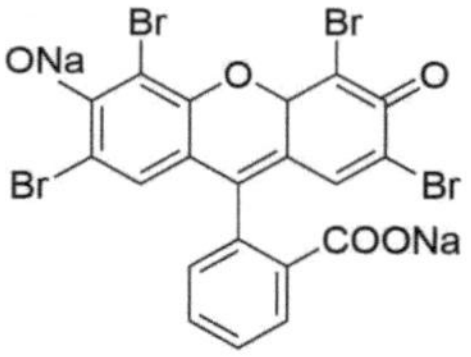 *Formula:* $C_{20}H_6Br_4Na_2O_5$ *Molecular weight:* 691.9 g/mol *Solubility:* Soluble in water *Appearance:* Red with blue undertone powder *Synonyms:* D&C Red No. 22; Cl 45380; Bromeosin	*ADI:* - Not approved for use in the EU, but approved for use in all drugs, generally, in the USA	*Oral* Photosensitizer	[65]

17.7.4 Antimicrobials, Antibacterials, Antioxidants (Table 17.6)

Antimicrobials and antioxidants are necessary in formulations to prolong shelf life and ensure stability and safety. Antioxidants prevent oxidative degradation of the formulation and API. The use of preservatives is especially important in multi-dose, liquid formulations, such as children's cough syrups, as the presence of water and repeated exposure to the environment can elevate the risk of microbial contamination. However, preservative use in medicines for children less than 3 years old is not recommended by the American Academy of Paediatrics (AAP), given evidence of increased adverse effects in this age group as a result of immature clearance capacity [1]. In situations in which preservative use cannot be avoided, it is recommended to limit the amount to the minimum effective amount to therefore minimize risk in pediatrics [76].

Table 17.6 Antibacterials, antioxidants, and antimicrobials in pediatric formulations

Excipient	Information	Regulatory status	Adverse effects in pediatrics	References
Benzalkonium chloride 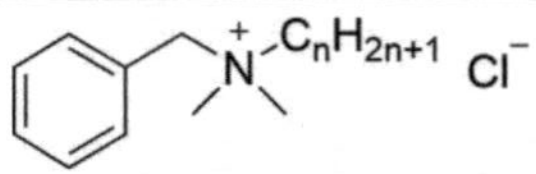 *Formula:* $[C_6H_5CH_2N(CH_3)_2R]Cl$ ($R = C_{12}$—C_{16} alkyls) *Molecular weight:* 360.0 g/mol (average) *Solubility:* Highly soluble in water *Appearance:* White/yellow-white amorphous powder, gel, or flakes. *Synonyms:* Alkylbenzyldimethylammonium chloride; BKC; BTC	Benzalkonium chloride is a potent antiseptic, disinfectant, and antimicrobial preservative. It is used (up to) in liquid nasal (0.02%), otic (0.02%), ophthalmic (0.02%), and parenteral (0.01%) formulations. Multi-dose vials contain low enough concentrations that risk is considered minimal	*GRAS:* Not listed *ADI:* -	*Topical* Irritation, hypersensitivity with prolonged contact *Inhalation* Bronchoconstriction, pruritus, and facial flushing *Oral* Vomiting, collapse, and coma, with toxic doses causing respiratory paralysis in newborns. Local irritation of the GIT	[8, 24, 66]
Sodium benzoate 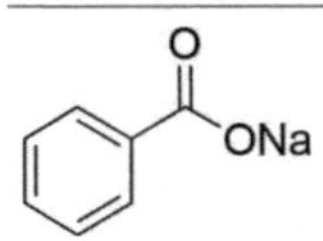 *Formula:* $C_7H_5O_2Na$ *Molecular weight:* 144.1 g/mol *Solubility:* Slightly soluble in water *Appearance:* White crystalline powder *Synonyms:* Benzoic acid sodium salt; sobenate	Sodium benzoate is the sodium salt of benzoic acid, used as a preservative from 0.01% to 0.5% in oral and parenteral formulations. Sodium benzoate is extensively metabolized by glycine conjugation into hippuric acid for renal excretion	*GRAS:* Yes *ADI:* 20 mg/kg/day	*Oral, parenteral* Urticaria, contact reactions, and anaphylaxis in infants and children. Accumulation can elevate the risk of jaundice, metabolic acidosis, and neurotoxicity (kernicterus) in newborns. Possible link to hyperactivity in 3 year olds	[1, 66]

Excipient	Information	Regulatory status	Adverse effects in pediatrics	References
Benzyl alcohol 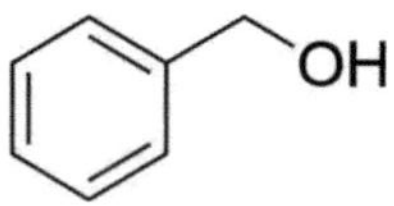 *Formula:* C_7H_8O *Molecular weight:* 108.1 g/mol *Solubility:* Moderately soluble in water *Appearance:* Transparent oily liquid *Synonyms:* Benzenemethanol; phenylmethanol; alpha-hydroxytoluene	Benzyl alcohol is an antimicrobial preservative with local anesthetic properties. It is found in oral and parenteral formulations up to 2.0%, and up to 10% as an anesthetic agent. Benzyl alcohol is metabolized into benzoic acid	*GRAS:* Not listed *ADI:* 4 mg/kg/day. Given its associated toxicity, its use has declined, and it is not recommended to be used in newborns and is limited use in children <3 years old	*Oral, parenteral, and inhalation* Fatal gasping syndrome in neonates given intravascular and intramuscular solutions containing 99–405 mg/kg. Metabolic acidosis, respiratory depression, intraventricular hemorrhage, and neurotoxicity occur Vomiting, coma, and diarrhea	[1, 66, 81]
Boric acid *Formula:* H_3BO_3 (trihydrate) *Molecular weight:* 61.8 g/mol *Solubility:* Soluble in water *Appearance:* White crystalline powder *Synonyms:* Orthoboric acid; boracic acid	Boric acid and salts are used as antimicrobial and antibacterial preservatives with limited approval for use in ophthalmic formulations	*GRAS:* Not listed *ADI:* 0.16 mg boron/kg/day (0.91 mg boric acid/kg/day) Boric acid is contraindicated in medicines for children up to 3 years old	*Oral, parenteral, and inhalation* Abdominal pain, rash, vomiting, and stimulation or depression of the CNS. Reproductive toxicity. Lethal dose is 5 g in children	[66, 82]

(continued)

Table 17.6 (continued)

Excipient	Information	Regulatory status	Adverse effects in pediatrics	References
Propylparaben 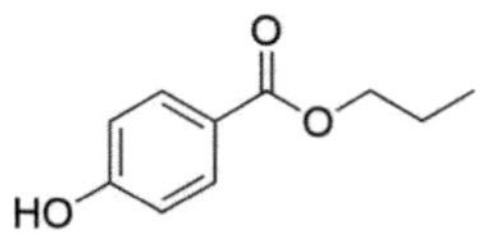 *Formula:* $C_{10}H_{12}O_3$ *Molecular weight:* 180.2 g/mol *Solubility:* Low solubility in water *Appearance:* White crystalline powder *Synonyms:* Propyl 4-hydroxybenzoate; propyl butex	Propylparaben is an antimicrobial preservative that can be combined with methylparaben to preserve oral and parenteral formulations. Levels used in oral products of methyl- (0.2%) and propyl- (0.06%) paraben are expected not to induce adverse effects	*GRAS:* Yes (up to 0.1%) *ADI:* 10 mg/kg/day (methylparaben). PDE of 2 mg/kg/day Propylparaben is not permitted for use as a food additive in the EU	*Oral, parenteral, and topical* Hypersensitivity with cross-sensitization across all parabens. Bilirubin-displacement in neonates	[8, 66, 83]
Potassium metabisulfite *Formula:* $K_2S_2O_5$ *Molecular weight:* 222.3 g/mol *Solubility:* Soluble in water *Appearance:* White granular solid *Synonyms:* Potassium metabisulfite; potassium disulfite; disulfurous acid; dipotassium pyrosulfite	Potassium metabisulfite belongs to the group of sulfite preservatives acting as an antioxidant and antimicrobial	*GRAS:* Yes. *ADI:* 0.7 mg/kg/day	*Oral, parenteral, and topical* Dermatitis, bronchospasms, hypotension, urticaria, and anaphylaxis, especially in asthmatic children	[1, 8, 66]

Excipient	Information	Regulatory status	Adverse effects in pediatrics	References
Thymol 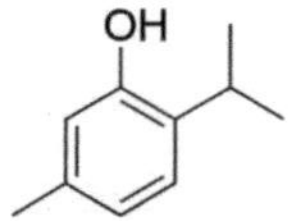 *Formula:* $C_{10}H_{14}O$ *Molecular weight:* 150.2 g/mol *Solubility:* Slightly soluble in water *Appearance:* Colorless/white crystals or powder. *Synonyms:* Acido trimico; 2-isporopyl-5-methylphenol; thymic acid; thyme camphor	Thymol is used as an antioxidant and antibacterial, with limited use as a preservative given its low solubility. Thymol is also used as an antiseptic, cooling agent, and flavoring agent	*GRAS:* Yes. *ADI:* Not specified Not recommended for children <5 years old	*Topical* Skin and eye irritant *Inhalation* Respiratory arrest due to nasal congestion and edema in a 3-week-old when thymol was used as an inhalation decongestant	[66]

Table 17.7 Surfactants and solvents in pediatric formulations

Excipient	Information	Regulatory status	Adverse effects in pediatrics	References
Surfactants				
Polysorbates (20, 40, 60, 80) *Solubility:* Non-ionic surfactant *Appearance:* Yellow, oily liquid *Synonyms:* Polyoxyethylene sorbitan fatty acid esters, tween (20, 40, 60, 80)	Polysorbates are hydrophilic non-ionic surfactants to emulsify or solubilize APIs, improving oral bioavailability. Various sizes of polysorbates are given by different amounts of ethylene oxide copolymerized	*GRAS:* Yes (60 and 80) *ADI:* 25 mg/kg/day	*Topical* Hypersensitivity *Parenteral* E-ferol syndrome: Thrombocytopenia, cholestasis, metabolic acidosis, and hepatotoxicity leading to mortality in 38 low-birth-weight infants (LBWI) given intravenous polysorbate 20 (1%) and 80 (9%)	[66, 88, 89]
Solvents				
Ethanol 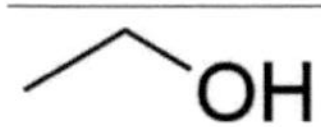 *Formula:* C_2H_6O *Molecular weight:* 46.1 g/mol *Solubility:* Soluble in water *Appearance:* Clear, colorless, liquid *Synonyms:* Ethyl alcohol, alcohol, EtOH	Ethanol is commonly used in oral medicines as a solvent or preservative. Ethanol readily crosses the blood–brain barrier (BBB) and is absorbed in the GIT. Certain countries may not allow alcohol content in medicines	*GRAS:* Yes. *ADI:* OTC content limits: 0–6 years old: 0.5% 6–12 years old: 5% >12 years old: < 10%	*Oral, parenteral* Central nervous system effects (depression, confusion, seizures, coma), abdominal pain, vomiting, hypoglycemia, and acidosis in children. Drug-interactions. Accumulation of acetaldehyde due to immature alcohol dehydrogenase and livers in neonates can cause vasodilation, headache, nausea, and dizziness	[1, 66, 90, 91]

(continued)

Excipient	Information	Regulatory status	Adverse effects in pediatrics	References
Polyethylene glycol (PEG) 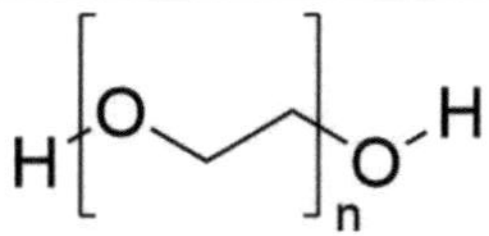 *Formula:* $(C_2H_4O)_nH_2O$ (*n* = ethylene oxide units) *Molecular weight:* *Solubility:* Soluble in water. *Appearance:* Colorless/off-white liquid (PEG 200–600) or solid (PEG 1000 <) *Synonyms:* Macrogol, polyoxyethylene glycol, Carbowax™	Depending on the grade, PEG can be found in topical, parenteral, ophthalmic, and oral formulations. PEG can stabilize emulsions, act as a suspending agent, enhance binders, and improve aqueous solubility. PEG is not well absorbed in the GIT	*GRAS:* Not listed *ADI:* 10 mg/kg/day (PEG 200–10,000)	*Oral* Laxative effects at high doses *Topical, parenteral* Systemic exposure from administration to damaged skin or parenteral administration can cause hyperosmolarity, metabolic acidosis, and renal failure	[66]
Glycerol 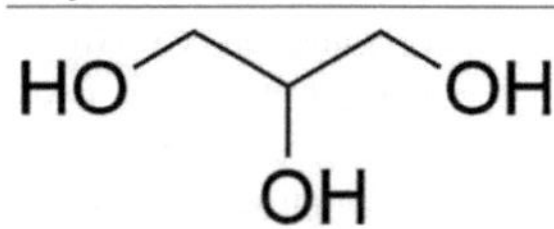 *Formula:* $C_3H_8O_3$ *Molecular weight:* 92.1 g/mol *Solubility:* Soluble in water *Appearance:* Colorless, clear, liquid *Synonyms:* Glycerine, glycerin, glycerolum, 1,2,3-propanetriol, trihydroxypropane glycerol	Glycerol acts as a solvent and cosolvent in parenteral and oral solutions, in addition to a sweetening agent in the latter	*GRAS:* Yes *ADI:* Not specified	*Oral, parenteral* At high concentrations (> 40%) mucositis and diarrhea can occur. Central nervous system toxicity in pediatrics	[65]

(continued)

Table 17.7 (continued)

Excipient	Information	Regulatory status	Adverse effects in pediatrics	References
Propylene glycol 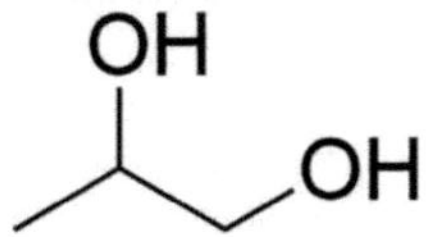 *Formula:* $C_3H_8O_2$ *Molecular weight:* 79.1 g/mol *Solubility:* Soluble in water *Appearance:* Colorless, clear, liquid *Synonyms:* 1,2-propanediol, 2-hydroxypropanol, methyl glycol	Propylene glycol acts as a solvent for poorly soluble APIs in parenteral and non-parenteral formulations. Propylene glycol is readily absorbed and metabolized by alcohol dehydrogenase (ADH) into lactic acid and pyruvic acid. ADH and renal immaturity in children less than 4 years old can prolong half-life and cause accumulation. Concomitant exposure to ethanol, a substrate of ADH, can exacerbate propylene glycol accumulation	*GRAS:* Yes. *ADI:* 25 mg/kg/day.	*Oral, parenteral* Ethanol-like CNS effects (coma, apnea, lethargy) in neonates and children. Lactic acidosis, serum hyperosmolality, cardiorespiratory arrest, acute renal failure, and seizures *Topical* Extensive cutaneous absorption to damaged skin leads to dermatitis, hyperosmolality, and cardiorespiratory arrest	[24, 66, 92]

17.7.5 Surfactants and Solvents (Table 17.7)

Surfactants and solvents are utilized to improve the dissolution and solubility of poorly soluble APIs. Surfactants work by reducing interfacial surface energy between the solvent and API, allowing enhanced dispersion and dissolution of the API. In topical preparations, surfactants can act as dispersing agents, solubilizers, and moisturizers. While in tablets and capsules, surfactants can aid in dissolution, permeation, and disintegration [84]. Solvents and cosolvents are chosen based on their ability to dissolve the API in liquid formulations. These solvents can compose a significant portion of a formulation; therefore, their safety in large amounts must be carefully considered. The total volume of an oral liquid formulation for pediatrics should not exceed 10 mL, but smaller volumes may be necessary given the patient's age and ability to swallow large volumes [85]. Incidents of fatalities in young children have been reported as a result of liquid formulations containing diethylene glycol (DEG) and ethylene glycol (EG); DEG and EG are not safe for use but may be inappropriately utilized to replace expensive solvents such as propylene glycol and glycerin [86]. Over 300 children died due to renal failure in 2023 as a result of consumption of DEG/EG-adulterated medicinal syrups [87].

17.8 Conclusion

Excipients are essential in the formulation of pediatric medications. However, their selection and application require caution due to children's distinctive physiological and metabolic traits at various stages of development. Safety profiles established for adults cannot always be directly applied to pediatric patients, who may exhibit increased sensitivities and decreased tolerance to certain substances. Notable examples include propylene glycol, ethanol, and various parabens, which have been associated with adverse reactions such as toxicity and metabolic acidosis in children. This underscores the need for pediatric-specific safety data and more comprehensive regulatory guidelines.

The current regulatory framework provides guidelines for using excipients in pediatric formulations but reveals significant gaps, particularly in the lack of pediatric-specific considerations in pharmacopeial monographs. This gap highlights the necessity for accelerated and expanded efforts by regulatory bodies to include a broader array of excipients and detailed evaluations tailored to pediatric needs.

To improve the safety and effectiveness of pediatric medications, an integrated approach involving robust preclinical and clinical research is essential to fully understand how excipients affect children. Enhanced collaboration among industry, academia, and regulatory agencies is crucial to expanding the excipient database with detailed safety and toxicity profiles reflective of pediatric requirements. This focus will improve existing medication safety and foster the development of new excipients that meet the rigorous standards of pediatric medicine, ensuring that therapies are both effective and safe for the youngest patients.

References

1. Rouaz K, Chiclana-Rodriguez B, Nardi-Ricart A, Sune-Pou M, Mercade-Frutos D, Sune-Negre JM, Perez-Lozano R, Garcia-Montoya E. Excipients in the paediatric population: a review. Pharmaceutics. 2021;13(3):387.
2. Fabiano V, Mameli C, Zuccotti GV. Paediatric pharmacology: remember the excipients. Pharamcol Res. 2011;63(5):362–5.
3. De Cock RFW, Knibbe CAJ, Kulo A, de Hoon J, Verbesselt R, Danhof M, Allegaert K. Developmental pharmacokinetics of propylene glycol in preterm and term neonates. Br J Clin Pharamcol. 2013;75(1):162–71.
4. Allegaert K, Vanhaesebrouck S, Kulo A, Cosaert K, Verbesselt R, Debeer A, de Hoon J. Prospective assessment of short-term propylene glycol tolerance in neonates. Arch Dis Child. 2010;95(12):1054–8.
5. EMA. (2017) Annex to the European Commission guideline on 'Excipients in the labelling and package leaflet of medicinal products for human use' (SANTE-2017-11668). https://www.ema.europa.eu/en/annex-european-commission-guideline-excipients-labelling-package-leaflet-medicinal-products-human-use
6. EMA. (2007) Guideline on excipients in the dossier for application for marketing authorisation of a medicinal product. https://www.tga.gov.au/resources/resource/international-scientific-guidelines/international-scientific-guideline-guideline-excipients-dossier-application-marketing-authorisation-medicinal-product
7. FDA (2005) Guidance for Industry: Nonclinical Studies for the Safety Evaluation of Pharmaceutical Excipients. https://www.fda.gov/media/72260/download
8. Belayneh A, Tadese E, Molla F. Safety and biopharmaceutical challenges of excipients in off-label pediatric formulations. Int J Gen Med. 2020;9(13):1051–66.
9. Gurjar R, Chan CYS, Curley P, Sharp J, Chiong J, Rannard S, Siccardi M, Owen A. Inhibitory effects of commonly used excipients on P-glycoprotein in vitro. Mol Pharm. 2018;15(5):4835–42.
10. Crowley, P., Martini, L. G. Excipients in pharmaceutical products in encyclopedia of pharmaceutical technology. New York: Marcel Dekker Inc. (2002).
11. Fathima N, Mamatha T, Qureshi HK, Anitha N, Rao JV. Drug-excipient interaction and its importance in dosage form development. J Appl Pharm Sci. 2011;1(6):66–71.
12. Ivanovska V, Rademaker CM, van Dijk L, Mantel-Teeuwisse AK. Pediatric drug formulations: a review of challenges and progress. Pediatrics. 2014;134(2):361–72.
13. EMA (2006) Reflection paper: formulations of choice for the paediatric population. https://www.ema.europa.eu/en/documents/scientific-guideline/reflection-paper-formulations-choice-paediatric-population_en.pdf
14. Valeur KS, Holst H, Allegaert K. Excipients in neonatal medicinal products: never prescribed, commonly administered. Pharmaceut Med. 2018;32(4):251–8.
15. Batchelor HK, Fotaki N, Klein S. Paediatric oral biopharmaceutics: key considerations and current challenges. Adv Drug Deliv Rev. 2014;73:102–26.
16. Kearns GL, Abdel-Rahman SM, Alander SW, Blowey DL, Leeder JS, Kauffman RE. Developmental pharmacology: drug disposition, action, and therapy in infants and children. N Engl J Med. 2003;349(12):1157–67.
17. Lu H, Rosenbaum S. Developmental pharmacokinetics in pediatric populations. J Pediatr Pharmacol Ther. 2014;19(4):262–76.
18. EMA (2017) Report published in support of the 'Questions and answers on propylene glycol used as an excipient in medicinal products for human use' (EMA/CHMP/704195/2013). https://www.ema.europa.eu/en/documents/report/propylene-glycol-used-excipient-report-published-support-questions-answers-propylene-glycol-used_en.pdf.
19. EMA (2018) Information for the package leaflet regarding ethanol used as an excipient in medicinal products for human use. https://www.ema.europa.eu/en/documents/scientific-

guideline/information-package-leaflet-regarding-ethanol-used-excipient-medicinal-products-human-use_en.pdf.
20. Van den Anker JN, Reed MD, Allegaert K, Kearns GL. Developmental changes in pharmacokinetics and pharmacodynamics. J Clin Pharmacol. 2018;58(Suppl 10):S10–25.
21. Khan D, Kirby D, Bryson S, Shah M, Rahman MA. Paediatric specific dosage forms: patient and formulation considerations. Int J Pharm. 2022;616:121501.
22. Allegaert K, Simons SHP, Tibboel D, Krekels EH, Knibbe CA, van den Anker JN. Non-maturational covariates for dynamic systems pharmacology models in neonates, infants, and children: filling the gaps beyond developmental pharmacology. Eur J Pharm Sci. 2017;109S:S27–31.
23. Rafidison P, Ulman K. Critical good manufacturing practice aspects to consider for pharmaceutical excipients. Business Briefing: Pharmatech. 2003:118–23.
24. American Academy of Pediatrics. "Inactive" ingredients in pharmaceutical products: update (subject review). Pediatrics. 1997;99(2):268–78.
25. Shehab N, Lewis CL, Streetman DD, Donn SM. Exposure to the pharmaceutical excipients benzyl alcohol and propylene glycol among critically ill neonates. Pediatr Crit Care Med. 2009;10(2):256–9.
26. Cuzzolin L. Neonates exposed to excipients: concern about safety. J Pediatr Neonat Individual Med. 2017;7(1):e070112.
27. Zar T, Graeber C, Perazella MA. Reviews: recognition, treatment, and prevention of propylene glycol toxicity. Semin Dial. 2007;20(3):217–9.
28. Singh SK, Koulov AV. Surfactants in biopharmaceutical development. 1st ed. Academic Press; 2023.
29. del Olmo A, Calzada J, Nuñez M. Benzoic acid and its derivatives as naturally occurring compounds in foods and as additives: uses, exposure, and controversy. Crit Rev Food Sci Nutr. 2017;57(14):3084–103.
30. American Academy of Pediatrics, Committee on Nutrition. Aluminum toxicity in infants and children. Pediatrics. 1986;78(6):1150–4.
31. Sedman A. Aluminum toxicity in childhood. Pediatr Nephrol. 1992;6(4):383–93.
32. Singh SK, Sarma MS. Hereditary fructose intolerance: A comprehensive review. World J Clin Pediatr. 2022;11(4):321–9.
33. Hebbink, G. A., Dickhoff, B. H. J. (2019) Application of lactose in the pharmaceutical industry in lactose: evolutionary role, health effects, and applications. Academic Press.
34. WHO. Report on Starting materials for pharmaceutical products: Control and Safe Trade Geneva: May 25–27. WHO/Pharm/98.605; 1998.
35. CDC. (1996) Fatalities Associated with Ingestion of Diethylene Glycol-Contaminated Glycerin Used to Manufacture Acetaminophen Syrup—Haiti, November 1995–June 1996. https://www.cdc.gov/mmwr/preview/mmwrhtml/00043194.htm.
36. WHO. https://www.who.int/teams/regulation-prequalification/incidents-and-SF/full-list-of-who-medical-product-alerts.
37. WHO. (2022) Medical product Alert N°6/2022: Substandard (contaminated) paediatric medicines. https://www.who.int/news/item/05-10-2022-medical-product-alert-n-6-2022-substandard-(contaminated)-paediatric-medicines.
38. WHO. (2022) Medical Product Alert N°7/2022: Substandard (contaminated) paediatric liquid dosage medicines. https://www.who.int/news/item/02-11-2022-medical-product-alert-n-7-2022-substandard-(contaminated)-paediatric-liquid-dosage-medicines.
39. WHO. (2022) Medical Product Alert N°8/2022: Substandard (contaminated) METHOTREX 50mg. https://www.who.int/news/item/27-12-2022-medical-product-alert-n-8-2022-substandard-(contaminated)-methotrex.
40. WHO. (2023) Medical Product Alert N°1/2023: Substandard (contaminated) liquid dosage medicines.

41. WHO. (2023) Medical Product Alert N°4/2023: Substandard (contaminated) syrup medicines. https://www.who.int/news/item/25-04-2023-medical-product-alert-n-4-2023%2D%2Dsubstandard-(contaminated)-syrup-medicines.
42. WHO. (2023) Medical Product Alert N°5/2023: Substandard (contaminated) syrup medicines. https://www.who.int/news/item/19-07-2023-medical-product-alert-n-5-2023%2D%2Dsubstandard-(contaminated)-syrup-medicines.
43. WHO. (2023) Medical Product Alert N°6/2023: Substandard (contaminated) syrup medicines. https://www.who.int/news/item/07-08-2023-medical-product-alert-n-6-2023%2D%2Dsubstandard-(contaminated)-syrup-medicines.
44. Davies J (2023) Cough Cough: Two more children's cough syrups found to be toxic—as 141 reported dead in The Sun. https://www.thesun.co.uk/health/24281603/children-cough-syrups-toxic-dead-india/#:~:text=TWO%20more%20cough%20syrups%20have,Syrup%20made%20by%20Norris%20Medicines.&text=The%20World%20Health%20Organization%20is,products%20have%20been%20exported%20elsewhere
45. EU Commission. Regulation No. 231/2012, Annex I: Amended list of specifications.
46. EU Commission. Regulation No. 1333/2008.
47. CHMP (2007). Scientific Article 5(3) Opinion On: The Potential Risks of Carcinogens, Mutagens, and Substances Toxic to Reproduction when these Substances are Used as Excipients of Medicinal Products for Human Use. https://www.ema.europa.eu/system/files/documents/other/wc500004013_en.pdf.
48. EMA (2013) Guideline on pharmaceutical development of medicines for paediatric use. https://www.ema.europa.eu/en/documents/scientific-guideline/guideline-pharmaceutical-development-medicines-paediatric-use_en.pdf.
49. ICH. (2003) Common technical document for the registration of pharmaceuticals for human use. https://www.ema.europa.eu/en/documents/scientific-guideline/ich-m-4-q-common-technical-document-registration-pharmaceuticals-human-use-quality-step-5_en.pdf.
50. 1074: Excipients Biological Safety Evaluations, USP-NF.
51. Regulation (EC) No. 1334/2008 Annex 1.
52. Regulation (EC) No. 1333/2008 Directive 2009/35/EC.
53. Code of Federal Regulations. Title 21, Parts: 74, 81, 82.
54. ICH. Q8 Note for Guidance on Pharmaceutical Development (CHMP/ICH/167068/04).
55. Saito J, Agrawal A, Patravale V, Pandya A, Orubu S, Zhao M, Andrews GP, Petit-Turcotte C, Landry H, Croker A, Nakamura H, Yamatani A, Salunke S. The current states, challenges, ongoing efforts, and future perspectives of pharmaceutical excipients in pediatric patients in each country and region. Children. 2022;9(4):453.
56. Crowley P, Martini LG. Drug-excipient interactions. Pharm Technol Eur. 2001;13:26–34.
57. Pifferi G, Restani P. The safety of pharmaceutical excipients. Farmaco. 2003;58(8):541–50.
58. Czaja J, Mielck J. Solid-state degradation kinetics of nitrazepam in the presence of colloidal silica. Pharm Acta Helv. 1982;57:144–53.
59. Uekama K. Design and evaluation of cyclodextrin-based drug formulation. Chem Pharm Bull (Tokyo). 2004;52(8):900–15.
60. Saokham P, Muankaew C, Jansook P, Loftsson T. Solubility of cyclodextrins and drug/Cyclodextrin complexes. Molecules. 2018;23(5):1161.
61. Wirth DD, Baertschi SW, Johnson RA, Maple SR, Miller MS, Hallenbeck DK, et al. Maillard reaction of lactose and fluoxetine hydrochloride, a secondary amine. J Pharm Sci. 1998;87(1):31–9.
62. Orlu M, Salunke S, Tuleu C. Paediatric pharmaceutics—the science of formulating medicines for children in pharmaceutical formulation: the science and technology of dosage forms. In: Tovey GD, editor.. Cambridge: Royal Society of Chemistry; 2018. p. 258–87.
63. EUPFI (2022) STEP Database. http://www.eupfi.org/step-database-info/.
64. Hossain AF. A systematic review on the effectiveness of toxicity reference values and tools used in determining safety limits of excipients and their application to paediatric drug development. Master's Dissertation, UCL; 2022.

65. Pawar S, Kumar A. Issues in the formulation of drugs for oral use in children: role of excipients. Paediatr Drugs. 2002;4(6):371–9.
66. Rowe RC, Sheskey PJ, Quinn ME. Handbook of pharmaceutical excipients. 6th ed. London: Pharmaceutical Press and American Pharmacists Association; 2009.
67. EMA. (2018) Draft: Information for the package leaflet regarding lactose used as an excipient in medicinal products for human use. https://www.ema.europa.eu/en/documents/scientific-guideline/draft-information-package-leaflet-regarding-lactose-used-excipient-medicinal-products-human-use_en.pdf.
68. Caballero ML, Quirce S. Immediate hypersensitivity reactions caused by drug excipients: A literature review. J Investig Allergol Clin Immunol. 2020;30(2):86–100.
69. Yoshida K, Sakurai Y, Kawahara S, Takeda T, Ishikawa T, Murakami T, et al. Anaphylaxis to polyvinylpyrrolidone in povidone-iodine for impetigo contagiosum in a boy with atopic dermatitis. Int Arch Allergy Immunol. 2008;146(2):169–73.
70. Wu Y, Levons J, Narang AS, Raghavan K, Rao VM. Reactive impurities in excipients: profiling, identification and mitigation of drug-excipient incompatibility. AAPS PharmSciTech. 2011;12(4):1248–63.
71. Burdock GA. Safety assessment of hydroxypropyl methylcellulose as a food ingredient. Food Chem Toxicol. 2007;45(12):2341–51.
72. Yu X, Buevich AV, Li M, Wang X, Rustum AM. A compatibility study of a secondary amine active pharmaceutical ingredient with starch: identification of a novel degradant formed between desloratadine and a starch impurity using LC-MS(n) and NMR spectroscopy. J Pharm Sci. 2013;102(2):717–31.
73. Mennella JA, Bobowski NK. The sweetness and bitterness of childhood: insights from basic research on taste preferences. Physiol Behav. 2015;152:502–7.
74. Chattopadhyay S, Raychaudhuri U, Chakraborty R. Artificial sweeteners—a review. J Food Sci Technol. 2014;51(4):611–21.
75. Alyami H, Dahmash E, Alyami F, Dahmash D, Huynh C, Terry D, et al. Dosage form preference consultation study in children and young adults: paving the way for patient-centred and patient-informed dosage form development. Eur J Hosp Pharm. 2017;24(6):332–7.
76. Malkawi WA, Al-Rafayah E, Al-Hazabreh M, Abu-Laita S, Al-Ghananeem AM. Formulation challenges and strategies to develop pediatric dosage forms. Children (Basel). 2022;9(4):488.
77. EMA. (2017) Information for the package leaflet regarding fructose and sorbitol used as excipients in medicinal products for human use. https://www.ema.europa.eu/en/documents/scientific-guideline/information-package-leaflet-regarding-fructose-sorbitol-used-excipients-medicinal-products-human-use_en.pdf
78. EMA. (2017) Information for the package leaflet regarding aspartame and phenylalanine used as excipients in medicinal products for human use. https://www.ema.europa.eu/en/documents/scientific-guideline/information-package-leaflet-regarding-aspartame-phenylalanine-used-excipients-medicinal-products_en.pdf
79. Farco JA, Grundmann O. Menthol--pharmacology of an important naturally medicinal "cool". Mini Rev Med Chem. 2013;13:124–31.
80. USFDA. (2022) Summary of Color Additives for Use in the United States in Foods, Drugs, Cosmetics, and Medical Devices. https://www.fda.gov/industry/color-additive-inventories/summary-color-additives-use-united-states-foods-drugs-cosmetics-and-medical-devices.
81. EMA. (2017) Report published in support of the 'Questions and answers on benzyl alcohol used as an excipient in medicinal products for human use' (EMA/CHMP/508188/2013) and the 'Questions and answers on benzoic acid and benzoates used as excipients in medicinal products for human use' (EMA/CHMP/508189/2013). https://www.ema.europa.eu/en/documents/report/benzyl-alcohol-benzoic-acid-group-used-excipients-report-published-support-questions-answers-benzyl/chmp/508188/2013-t_en.pdf.
82. EMA. (2021) Report published in support of the 'Questions and answers on boric acid and borates used as excipients in medicinal products for human use' (EMA/CHMP/619104/2013

Rev. 1) https://www.ema.europa.eu/en/documents/report/boric-acid-borate-used-excipients-report-published-support-questions-answers-boric-acid-borates-used_en-0.pdf.
83. EMA. (2015) Reflection paper on the use of methyl- and propylparaben as excipients in human medicinal products for oral use. https://www.ema.europa.eu/en/documents/scientific-guideline/reflection-paper-use-methyl-propylparaben-excipients-human-medicinal-products-oral-use_en.pdf.
84. de Freitas Araújo Reis MY, de Araújo Rêgo RI, Rocha BP, Guedes GG, de Medeiros Ramalho IM, de Medeiros Cavalcanti AL, et al. A general approach on surfactants use and properties in drug delivery systems. Curr Pharm Des. 2021;27(42):4300–14.
85. NPPG. (2020) Position statement 2020–01: Choosing an Oral Liquid Medicine for Children. https://nppg.org.uk/wp-content/uploads/2020/12/Position-Statement-Liquid-Choice-V1-November-2020.pdf.
86. Schier JG, Rubin CS, Miller D, Barr D, McGeehin MA. Medication-associated diethylene glycol mass poisoning: a review and discussion on the origin of contamination. J Public Health Policy. 2009;30(2):127–43.
87. WHO. (2023) WHO urges action to protect children from contaminated medicines https://www.who.int/news/item/23-01-2023-who-urges-action-to-protect-children-from-contaminated-medicines.
88. EMA. (2018) Draft: Information for the package leaflet regarding polysorbates used as excipients in medicinal products for human use. https://www.ema.europa.eu/en/documents/scientific-guideline/draft-information-package-leaflet-regarding-polysorbates-used-excipients-medicinal-products-human_en.pdf.
89. Bove KE, Kosmetatos N, Wedig KE, Frank DJ, Whitlatch S, Saldivar V, et al. Vasculopathic hepatotoxicity associated with E-Ferol syndrome in low-birth-weight infants. JAMA. 1985;254(17):2422–30.
90. EMA. (2018) Information for the package leaflet regarding ethanol used as an excipient in medicinal products for human use. https://www.ema.europa.eu/en/documents/scientific-guideline/information-package-leaflet-regarding-ethanol-used-excipient-medicinal-products-human-use_en.pdf.
91. Soremekun R, Ogbuefi I, Aderemi-Williams R. Prevalence of ethanol and other potentially harmful excipients in pediatric oral medicines: survey of community pharmacies in a Nigerian City. BMC Res Notes. 2019;12:460.
92. EMA. (2017) Report published in support of the 'Questions and answers on propylene glycol used as an excipient in medicinal products for human use' (EMA/CHMP/704195/2013). https://www.ema.europa.eu/en/documents/report/propylene-glycol-used-excipient-report-published-support-questions-answers-propylene-glycol-used_en.pdf

Chapter 18
Clinical Testing in Children

Klaus Rose

Abstract Clinical trials compare therapeutic interventions. Since the 1960s, proof of efficacy and safety (E&S) is required for drug approval. Regulatory trials became a cornerstone of drug development. In parallel to industry-sponsored regulatory trials, academic trials aim at improving interventions without regulatory concerns. Initially, drugs were approved in humans, and pediatricians used them as well. When toxicities of antibiotics were reported in babies, the American Academy of Pediatrics and pediatric researchers adopted the concept of children as "therapeutic orphans" and demanded separate pediatric studies, resulting eventually in the first US pediatric legislation of 1997 and the EU pediatric legislation, in force since 2007. But pediatric legislation often equalizes administrative age limits with alleged, nonexisting physiological changes at a birthday. Better understanding of the patient's developing body and consequences for drug treatment encountered an expanding regulatory framework, stronger academic and professional institutions, and the pharmaceutical industry. Pediatric legislation resulted in "pediatric" studies in adolescents that are legally still children, but physiologically already mature. Also separate proof of E&S in younger minors may be exaggerated as they are not another species. All populations require the avoidance of toxicities and adequate dose determination. Demands for age-appropriate pediatric formulations are relevant only in young children, not in the entire "pediatric population."

Keywords Pediatric drug development (PDD) · Pediatric legislation · On-label/off-label

K. Rose (✉)
klausrose Consulting, Pediatric Drug Development and More, Riehen, Switzerland
e-mail: klaus.rose@klausrose.net; https://www.klausrose.net

H. Batchelor, K. Rose (eds.), *Pediatric Formulations*, AAPS Advances in the Pharmaceutical Sciences Series 58, https://doi.org/10.1007/978-3-031-77240-5_18

18.1 Clinical Trials in Man

Clinical trials assess and compare healthcare interventions, mostly drugs, vaccines, diagnostics, and medical devices. Ancient surgeons learned from individual masters, with improvement by trial and error. In clinical trials, all questions to be answered must be written down in a protocol with pre-defined endpoints [1].

A very early study was described in 1753 by James Lind in his book, "A Treatise of the Scurvy." He compared twelve sailors suffering from scurvy, of whom, respectively, two received daily, on top of a standard diet, either cider, sulfuric acid, vinegar, seawater, citrus fruit or a spicy paste with barley water. The citrus fruit treatment stopped early when they ran out of fruit, but then one sailor was fit for duty and the other had almost recovered. Apart from citrus fruits, only cider showed some treatment effect. By today's standards, the trial had several flaws, including non-adherence to the protocol due to logistical deficiencies. But the systematic comparison of different treatments was in line with modern testing. The results of this trial, however, did not translate immediately into action in the Royal Navy [2–4].

Systemic testing was and is part of the scientific and technical revolution. Testing in man has become more frequent with the availability of effective drugs, devices, diagnostics, and scientific publications. During and after World War II, medical research expanded [5]. After 1945, the world was outraged by the murders and atrocities in adults and children by Nazi and Japanese physicians [6, 7]. The judges in the trial against Nazi medical doctors in Nuremberg, Germany, published in 1947 a list of principles that became the "Nuremberg Code." It promulgated key issues of human experimentation [8]. The outrage after World War II did not lead to the application of the Nuremberg Code to experimentation in man in the USA [9].

In 1964, the Declaration of Helsinki (DoH) was adopted by the World Medical Association (WMA) in 1964 in Helsinki as a set of ethical principles for the medical community regarding clinical research in man; children are not mentioned in the DoH but are part of the mentioned research subjects that are "legally incompetent" [10, 11].

In 1966, the US American Harvard anesthesiologist Henry K Beecher summarized 22 selected academic research projects, published in academic journals, that were unethical by contemporary ethical standards. One example was the infection of mentally retarded children with hepatitis [5]. Beecher's explanation of the reason for the massive increase in clinical research was the increased availability of government funds. Since then, the funds available for clinical research have continued to grow, and the basic challenge of hidden interests of the sponsor vs those of the study participant has continued. Large-scale clinical trials in post–World War II North America were often performed in young male adult prisoners, including the testing of drugs, toothpaste, deodorants, shampoo, and more [5, 12].

Two major types of clinical trials are frequently differentiated: academic ("investigator-initiated") clinical trials compare different interventions or a new concept with standard treatment; they have two major aims: to clinically modify and improve existing treatment; and to create scientific publications, a key factor in career progress. The other trial type is for drug development, aiming at approval, so the study design must either have the pre-trial blessings of the regulatory authority or must follow official regulatory guidance [13, 14]. There is still a broad conviction that research initiated by academia is more noble than research with commercial aims, a belief that is stronger in Europe than in North America [15–18].

For clinical research, a framework has been developed to balance sponsors' and society's interests vs. the need to protect participants. Key features are voluntary participation, the right to terminate study participation whenever, full information about potential benefits and risks, written documentation of the informed consent, and an acceptable benefit-risk ratio. These features are codified in the rules of good clinical practice (GCP) [1].

18.2 Modern Drug Labels, Pediatric Disclaimers, and Pediatric Clinical Pharmacology

With the Industrial Revolution began the chemical production of drugs on a large scale in the nineteenth and twentieth centuries [19]. Modern medicines have a two-sided face: their therapeutic potential is often enormous, see the immediate life-saving effect of antibiotics, industrially produced hormones, or even cytostatic agents; and they have the potential for disaster. In 1936, a liquid formulation of sulfanilamide, an early antibiotic, was brought to the market in the USA. There had been no safety testing for the used solvent, as this was not required then. Within days after introduction, deaths were reported. The FDA seized the entire production lot. More than 100 patients died. The public outcry led to a serious revision of the FDA legislation, mandating a first-time safety experiments in animals before a drug could be marketed [20]. In 1961/62, a second major catastrophe occurred, this time already global, when the tranquillizer thalidomide caused deformation in the growing fetus when taken by pregnant women. Thousands of children were born with shortened and deformed arms and legs. With a few exceptions, these children were born outside of the USA, as thalidomide had not been licensed there. "Only" a few children were born with such defects in the USA due to the uncontrolled distribution of thalidomide tablets handed out to doctors by sales representatives and further distributed by medical doctors to patients within so-called "clinical trials" without any documentation [14, 19, 21, 22]. Today, GCP requires documentation of each tablet and a plan for emergency call-back of medication in case of a safety issue. The thalidomide catastrophe led to the US Kefauver-Harris amendments in 1962, mandating manufacturers to prove efficacy and safety (E&S) of drugs [14, 19, 22].

The power of regulatory authorities has increased considerably. Most clinical trials organized for commercial purposes are regulatory trials to back a marketing authorization application in the EU, North America, or Japan. The development budget of a medium or large drug company exceeds the research budget of most academic institutions or networks. The costs of development of a new drug today are estimated at around US$ three billion [23, 24]. The costs of research, preclinical safety, and formulation development are comparatively low compared to phase 3 trials [23–26].

Until the 1990s, most pharmaceutical companies performed their clinical trials in-house. Since then, clinical trials are increasingly outsourced to clinical research organizations (CROs). CROs offer services ranging from advice for strategic development to protocol design and study execution, including selection of trial centers, organizing investigator meetings, coordinating patient recruitment, and supporting electronic data capturing. Trial execution has become a business in its own right; in view of the enormous costs, it can also be described as a whole industry of its own [27, 28].

Regulations on drug development and modern labels were initiated nationally. The Kefauver-Harris amendments were US national [22], followed eventually by most other states. This led to the "International Conference on Harmonization of Technical Requirements for Registration of Pharmaceuticals for Human Use" (ICH), an international framework founded in 1990 to respond to the increasingly global drug development and the need for international harmonization [29]. The name alone shows how difficult it is to get different partners on one table and to find a common name. The partners are the trade unions of research-based pharmaceutical industry in the USA (PhRMA) [30], Europe (EFPIA) [31], Japan (JPMA) [32], worldwide (IFPMA) [33]; and the regulatory authorities of USA (FDA) [34], EU (EMA) [35] and Japan (MHLW) [36], with other regulatory authorities as observers. Both sides use academic specialists for specific issues.

The framework for clinical trials is documented in the ICH guideline ICH E 6 on GCP [1]. It summarizes how clinical trials should be done so that they can be used for registration. It lists the responsibilities of the institutional review board (IRB)/ ethics committee (EC), the responsibilities of the clinical investigator and his institution, the responsibilities of the sponsor, including the trial logistics, the requirements for the study protocol, the investigator's brochure, and more. All these principles must also be adhered to in clinical trials in minors.

18.3 Pediatric Medicine and Clinical Trials in Children

Pediatrics is a rather young academic discipline compared to the history of other medical sub-specialities [37]. The focus of public attention shifted continuously with the mainstream of innovation and advances in learning. "Safe Milk Campaigns" to pasteurize milk or the application of silver nitrate in newborns' eyes to prevent blindness are today almost forgotten, as are the iron lungs that kept children with polio alive in the 1950s. Today's children enjoy the best medical care ever in history. Children have fully participated in medical progress, see the advances in vaccination, in surgery of inborn heart failures, in child transplant medicine, and more. "Pediatric drug development" promised further improvement [38, 39]. In its 2001 report to Congress, the FDA predicted further significant advances in pediatric medicine. "Superior drug treatment information is expected to permit quicker recoveries from childhood illnesses, with fewer attendant hospital stays, physician visits and parental work days lost" [40]. Reacting to the Kefauver-Harris amendments and toxicities of antibiotics in newborns reported earlier, manufacturers had introduced disclaimers into drug labels stating that the respective drug had not been tested in children. They did this to prevent being sued for alleged damages in the litigious US legal culture. The first chairman of the American Academy of Pediatrics committee on drugs, Dr. Shirkey, however, interpreted these warnings medically, claimed that children were "therapeutic orphans," and claimed that this left medical doctors in the dilemma of either prescribing drugs off-label to a child or not prescribing the drug and potentially withholding effective treatment. In his view, this conferred responsibility to the prescribing doctor and away from the manufacturer [41, 42].

In parallel to the increasing availability of modern drugs, clinical pharmacology evolved, investigating absorption, distribution, metabolism, and excretion (ADME) of drugs [43]. As a sub-specialty, pediatric clinical pharmacology initially evolved as an academic movement [44]. A baby's body is, in many aspects, not just a small adult body. Pediatric clinical pharmacology measures ADME in minors. In younger children, the organs are not yet mature. Key lessons were summarized in a publication of Kearns 2003 [45]. Its key message is that due to the different organ systems, dosing in young children cannot be deduced mechanistically from the weight or body surface of children. The consequence of the mechanistic formulas and tables (almost every European country had one or even several different pediatric formula[s]) was that in neonates and more so in preterm neonates, there was the danger of either underdose or overdose. "Children are not small adults" was a mantra that has been taught for generations to students of pharmacology, medicine, and other healthcare professions.

There are also occasional observations that adult drugs can work in completely different disease areas in children. Famous examples are indomethacin and ibuprofen, non-steroidal anti-inflammatory drugs that show efficacy in closing the arterial duct, a vessel that in the unborn child connects the pulmonary artery with the aorta. Normally, this vessel closes at birth. If it remains open, it can lead to pulmonary hypertension and impair the child's development [46]. Such exceptions, however, are rather rare.

18.4 The Catch

The term "child" has different legal/administrative and physiological meanings. The 18th birthday is in most countries the day when somebody comes of age, but it does not correspond to a physiological transformation of the body overnight. The transformation from childhood to adulthood occurs during the process of puberty, which takes years and has accelerated during the last century [47–49]. The catch in the justification of "pediatric drug development" is that it pretends a physiological transformation of the body at an administrative/legal age limit. This is in itself only a semantic blur. The detection of rapid development processes in the newborn was a fundamental scientific advance. But this detection encountered the new societal framework of new institutions with considerable self-interests, such as the regulatory authorities, the professional representations of developmental pharmacology, academic research institutions, and the life science industry. In this framework, the mentioned semantic blur became a justification of "pediatric" studies not only in those babies and infants that differ fundamentally from adults, but also in adolescents that *legally* are still children but *bodily* are already mature. The claim that off-label prescription in a child represents a doctor's dilemma of either prescribing a drug of which he cannot know if it works, or not to prescribe it at all [41, 42], is untrue. Children are not another species. They mature with age. The older they grow, the more their bodies mature. The fundamental flaw of "pediatric drug development" is that pediatric researchers and regulatory authorities have so far refused to apply the findings of developmental pharmacology and developmental physiology. The reason is banal: look where the money flows. Pediatric researchers had complained that industry funds went only into adult regulatory studies. This changed with pediatric legislation [50, 51]. A paper by FDA officers describes that in "pediatric" drug studies in adolescents 94.5% the resulting recommended doses were identical to adult doses. When we discuss testing in children, we need to be clear of what we are talking: of minors that are legally still children, but physiologically already mature, or of minors that physiologically are still children. Terms like "pediatric drug development," called by the EU also "better medicines for children" [52], blur this difference.

A strong tension between the official view and reality has occurred often in history. Life science companies have to live with this challenge. In the long term, this will change. We see already such signs on the horizon. Both FDA and EMA acknowledge that antiseizure medications (ASMs) work before and after the 18th birthday. The FDA accepts this down to the age of 2 years, the EMA down to 4 years. To safe face, both claim that for minors "extrapolation" of efficacy from adults is required. This is ridiculous. Minors are not another species, and the body does not change on the night of the 18th birthday [50, 51, 53–55].

18.5 US and EU Pediatric Pharmaceutical Legislation

The debate about "pediatric drug development" started with the described pediatric disclaimers and the fact that many drugs were not approved for minors. In 1997, the FDA Modernization Act (FDAMA) offered, the first time, a voluntary reward to the pharmaceutical industry of 6 months patent extension for the generation of "pediatric" data. This patent extension is called "pediatric exclusivity." In 2002, FDAMA was replaced by the Best Pharmaceuticals for Children Act (BPCA), later complemented by the Pediatric Research Equity Act (PREA), which authorized the FDA to mandate pediatric trials and more without reward. BPCA and FDAMA were re-authorized repeatedly and became permanent in 2012 [56].

The EU pediatric regulation, in force since 2007, was inspired by the USA but is more ambitious. Marketing Authorization Applications for new drugs must be submitted with a Pediatric Investigation Plan (PIP) already approved by the European Medicines Agency's (EMA) Pediatric Committee (PDCO), unless the EMA confirms in writing the applicability of a class waiver. Generics are exempt; orphan drugs are not. Although another EMA committee, the Committee for Medicinal Products for Human Use (CHMP), later decides on approval, a missing PIP will block a submission. The PDCO is composed of representatives of the EU member states, the CHMP, pediatric healthcare professionals, and patient advocacy groups [57]. The majority of PDCO members are employees of national regulatory authorities.

The PIP must cover all age groups as defined by ICH E 11 (R1): preterm newborns (< 36 weeks gestational age), newborns (0–27 days), infants and toddlers (28 days–23 months), children (2–11 years), and adolescents (12–17 years). The PIP should be submitted at the end of phase 1, long before proof of concept. PIPs must address preclinical testing, juvenile animal studies, formulation(s), clinical pharmacology, and clinical trials. For developers of pediatric formulations, the key message is that US and EU legislation has increased the demand for formulations beyond pills and tablets.

The EU pediatric regulation has the same flaws as the US precedence. Adolescents are *legally* still children, but no longer *physiologically*. Also, younger minors are not another species.

18.6 ICH E 11

The ICH guideline E 11 "Clinical Investigation of Medicinal Products in the Pediatric Population" was finalized in 2000. In 2017, it was updated by the version E11 (R1). It addresses the need for "pediatric drug development" but uses unreflectively the legal definition of the "pediatric population" from birth to the 16/18th birthday, depending on the region, as if the body would change overnight. But drugs treat the body, not the legal status. ICHE11 (R1) further addresses when pediatric considerations should start, types of drugs, age groups (preterm newborn infants, term newborn infants, infants and toddlers, children, adolescents), pediatric formulations, and technical issues such as withdrawal of blood [58]. It is an official international document with fundamental flaws.

The draft version of ICH Guideline E11A, released for public consultation in April 2022, addresses pediatric extrapolation [59]. It is, in principle, based on the mantra that in children everything is different, admits that the body matures with age, and explains the rules of extrapolation from "adults" (≥18y) to "children" (<18y). It represents a compromise between a dogma and the reality that has gradually been accepted by the regulatory authorities, more so by the FDA than by the EMA.

18.7 Barriers Against Clinical Trials in Children

Were there no clinical trials in children in the past? There were many. Many of the studies listed in 1966 by Beecher were performed in children [5]. The demand to include children in pharmaceutical progress is a *regulatory* justification of "pediatric" studies that are pointless in adolescents and mostly exaggerated in younger minors. Drug development has been economically successful, but a continuous barrage of criticism has contributed to a public image of the pharmaceutical industry comparable to the tobacco industry or even organized crime [15, 16, 60].

In many therapeutic areas, extensive pediatric research was done without pediatric legislation. Vaccines have been systematically developed for more than a century, with studies mostly done in children. Growth hormone and lung surfactant were developed without pediatric legislation. Where children represent a market, they attract business, also in children's clothes [61], toys, push chairs, wheelchairs, or education.

In the public opinion, the perception of children in clinical trials was ambivalent. Children should not be abused as "guinea pigs." But who would object to saving children with cancer? Until 2000, the advances of pediatric oncology resulted in about 90% of children with acute lymphatic leukemia (ALL) surviving more than adults. Pediatric oncology had started with systematic experimental use of cytotoxics that had been approved for adult cancer in the 1950/1960s. Initially, homeopathic low doses were prescribed. Experience increased survival by 10% per decade since the 1970s, achieved by higher dosing and new drug combinations [62–64]. The pediatric oncology community had hoped for decades for new compounds to further improve treatment. Most used cytotoxics had been approved for adult cancer [65]. Since the first edition of this book, young patients with ALL after failed standard chemotherapy have been successfully treated by involving the body's immune system. Blood is withdrawn, and the T-cells are programmed to attack the ALL cells and are re-infused. The youngest patient in the pivotal trial was 6-year-old Emily Whitehead, today a cancer-free young woman. She was not saved by "pediatric drug development," but despite it [66–69].

18.8 Pediatric Formulations

The present legislation ensures early formulations beyond pills and tablets. Most children ≤7y cannot swallow tablets. For preterm newborns, often special intravenous formulations are required.

18.9 Ethical Challenges of Clinical Trials with Children

Legally, a child cannot act on its own [70–72]. Today's world is full of well-intended international declarations of the rights of children, with many university-offered postgraduate studies on this. In the past, in the prevailing opinion, children should not be abused as guinea pigs. Today's view has shifted toward a position that it is equally unethical to expose children to unsafe drugs. The legitimacy of pediatric research is today much less disputed than decades ago. Children cannot give informed consent. The parents or caregivers must do this. The debate over whether one or both parents need to give consent, if this is practical, and what to do when the mother is a minor herself fills entire libraries [70–72]. Today, minors should be asked to give their assent, which requires age-appropriate oral and written explanation of the potential benefits and risks. A more elaborate explanation should be given to adolescents, corresponding to their ability to understand the study, and a simpler one for children 7–11 years.

Mainstream teaching assumes that FDA/EMA-triggered pediatric studies contribute to further advancing child healthcare. But "pediatric" studies in adolescents have limited value where disease progression is not altered in children versus adults, and safety and efficacy data is already in place from adult populations. They are massively exaggerated in school age children. Insulin, antibiotics and other drugs work before the 12th or 18th birthday and do not need separate proof of efficacy. The recommended doses for adolescents were identical to adult doses in 94.5% of all drugs for which "pediatric" studies had been performed [73].

18.10 Operational Challenges of Clinical Trials with Children

Children cannot survive alone. They are mostly part of a family, and so a clinical trial must take into consideration the entire circle of care. No carer will adhere to a rigid visit scheme that does not allow flexibility to encompass other caring responsibilities. No carer will return to a shabby hospital with unfriendly nurses.

A 10 kg child has less blood than an adult, and a 500 g preterm newborn has even less. The normal laboratory values in children are often different. The maturity of the organs is different specifically in preterm newborns, with different drug-drug interactions. Operational details are discussed in entire textbooks to account for protocol differences between pediatric and adult study designs [74, 75].

Children's attention span is also different from adults. A child will not listen for an hour. The physician has maybe 5 minutes and has to prioritize their messages. A child's world differs from the adult world in many more aspects. The emotions are stronger, the understanding of institutions is less systemic, and the understanding of time and geographic dimensions is different.

18.11 Conclusions

Clinical trials with children require a solid awareness of GCP plus special knowledge and training. Employees of pharmaceutical companies, CROs charged with pediatric studies, members of IRBs/ECs, and all other involved people should be aware of the difference between exaggerated regulatory studies and reasonable studies. The term "pediatric extrapolation" [59] is a face-saving compromise between the dogma that in "children" everything is different, and the reality that the body matures over time. It is hoped that sooner or later "pediatric drug development" will be replaced by a rational concept that instead of forcing studies of limited value, exaggerated, or even harmful studies promotes studies that allow rational drug treatment of minors.

The pediatric legislation demands special formulations for young children. Probably, this is the most reasonable part of "pediatric drug development." Although most professionals in life science companies that work in formulation development are not directly involved in clinical studies, they should be aware of their basic flaws.

References

1. ICH Guidelines for Good Clinical Practice E6, 1996. http://www.ich.org/fileadmin/Public_Web_Site/ICH_Products/Guidelines/Efficacy/E6_R1/Step4/E6_R1__Guideline.pdf.
2. Lind J. A treatise on the scurvy. 3rd ed. London: S Crowder et al; 1772.
3. Dunn PM. James Lind (1716–94) of Edinburgh and the treatment of scurvy. Arch Dis Childhood. 1997;76:F64–5.
4. Bartholomew M. James Lind's Treatise of the Scurvy (1753). Postgrad Med J. 2002;78:695–6. http://www.ncbi.nlm.nih.gov/pmc/articles/PMC1742547/pdf/v078p00695.pdf.
5. Beecher H. Ethics and clinical research. NEJM. 1966;274(24):1354–60.
6. Weyers W. The Abuse of Man. An illustrated history of dubious medical experimentation. New York: Ardor Scribendi; 2003.
7. Mozes-Kor E. The Mengele Twins and human experimentation. In: Annas G, Grodin M, editors. The Nazi Doctors And The Nuremberg Code.
8. Weindling P. The origins of informed consent: the International Scientific Commission on Medical War Crimes, and the Nuremberg Code. Bull Hist Med. 2001;75(1):37–71.
9. Childress JF, et al., editors. Belmont Revisited. Washington D.C.: Georgetown University Press; 2005.
10. WMA Declaration of Helsinki, original version 1964. In: Williams JR: The Declaration of Helsinki and public health. Bulletin of the World Health Organization August 2008, 86(8):650-652. http://www.who.int/bulletin/volumes/86/8/08-050955.pdf.
11. WMA Declaration of Helsinki amended version 2008 http://www.wma.net/en/30publications/10policies/b3/17c.pdf
12. Hornblum AM. Acres of SKIN. New York: Routledge; 1998.
13. Meinert CL, Tonascia S. Clinical trials: design, conduct, and analysis. USA: Oxford University Press; 1986. ISBN 978-0-19-503568-1.
14. Rägo L, Santo B. Drug regulation: history, present and future. In: van Boxtel CJ, Santo B, Edwards IR, editors. Drug benefits and risks: International Textbook of Clinical Pharmacology, revised 2nd edition. Uppsala: IOS Press & Uppsala Monitoring Centre; 2008, chapter 6,

pp. 65–77. https://www.who.int/medicines/technical_briefing/tbs/Drug_Regulation_History_Present_Future.pdf.
15. Goldacre B. Bad Pharma: how drug companies mislead doctors and harm patients. London, UK: HarperCollinsPublishers; 2012.
16. Angell M. The truth about the drug companies: how they deceive us and what to do about It. New York, USA: Random House Publishers; 2004.
17. Angell M. I academic medicine for sale? New Engl J Med. 2000;342(20):1516–8.
18. Angell M. Industry-sponsored clinical research. A broken system. JAMA. 2008;300(9):1069–71.
19. Hilts PJ. Protecting America's Health. New York: Alfred A. Knopf; 2003, ISBN 0-375-40466-X.
20. Wax P. Elixirs, Diluents, and the Passage of the 1938 Federal Food, drug and Cosmetic Act. Ann Intern Med. 1995;122:456–61.
21. Taussig HB. A study of the German outbreak of phocomelia. JAMA. 1962;180:1106–14.
22. Meadows M. Promoting safe and effective drugs for 100 years. FDA Consumer Magazin. The Centennial Edition/January-February 2006. http://www.fda.gov/AboutFDA/WhatWeDo/History/ProductRegulation/PromotingSafeandEffectiveDrugsfor100Years/default.htm
23. DiMasi JA, Grabowski HG, Hansen RW. Innovation in the pharmaceutical industry: New estimates of R&D costs. J Health Econ. 2016;47:20–33.
24. DiMasi JA. Research and development COSTS of new drugs. JAMA. 2020;324(5):517.
25. Rawlins MD. Cutting the cost of drug development? Nature. 2004;3:360–4.
26. Adams CP, Brantner VV. Estimating the cost of new drug development: is it really $802 million? Health Affairs. 2006;25(2):420–8.
27. Associations of Clinical Research Organisations ACRO. http://www.acrohealth.org/.
28. European CRO Federation http://www.eucrof.eu/.
29. International Conference on Harmonisation of Technical Requirements for Registration of Pharmaceuticals for Human Use (ICH). www.ich.org.
30. Pharmaceutical Research and Manufacturers of America (PhRMA). www.phrma.org.
31. European Federation of Pharmaceutical Industries and Associations (EFPIA). www.efpia.eu.
32. Japan Pharmaceutical Manufacturers Association (JPMA). www.mhlw.go.jp.
33. International Federation of Pharmaceutical Manufacturers and Associations (IFPMA). www.ifpma.org.
34. Food and Drug Administration. www.fda.gov.
35. European Medicines Agency. www.emea.europa.eu.
36. Ministry of Health, Labor and Welfare. www.mhlw.go.jp.
37. Colon AR. Nurturing children: a history of pediatrics. Greenwood Pub Group: Publisher; 1999. ISBN-10: 0313310807, SBN-13: 978-0313310805.
38. Ward RM, Benjamin DK Jr, Davis JM, Gorman RL, Kauffman R, Kearns GL, et al. The need for pediatric drug development. J Pediatr. 2018;192:13–21.
39. Rocchi F et al. The european paediatric legislation: benefits and perspectives. 2011. https://www.ncbi.nlm.nih.gov/pmc/articles/PMC2933611/pdf/1824-7288-36-56.pdf
40. FDA 2001. The Pediatric Exclusivity Provision. January 2001. Status Report to Congress. https://www.fda.gov/downloads/Drugs/DevelopmentApprovalProcess/DevelopmentResources/UCM049915.pdf
41. Shirkey H. Therapeutic Orphans. Pediatrics. 1968;72:119–20.
42. Wilson JT. An Update on the Therapeutic Orphan. Pediatrics. 1999;104(3):585–90. http://pediatrics.aappublications.org/cgi/content/full/104/3/S1/585
43. Orme M. Clinical pharmacology in research, teaching and health care. Considerations by IUPHAR, the International Union of Basic and Clinical Pharmacology. Basic Clin Pharmacol Toxicol. 2010;107:531–59. http://onlinelibrary.wiley.com/doi/10.1111/j.1742-7843.2010.00602.x/pdf.
44. Seyberth HW, Rane A, Schwab M, editors. Pediatric Clinical Pharmaocology. Handbook of Experimental Pharmacology 205. Berlin: Springer Publishers; 2011.
45. Kearns GL, et al. Developmental pharmacology - drug disposition, action, and therapy in infants and children. N Engl J Med. 2003;349:1157–67.

46. Van Overmeire B, et al. A comparison of ibuprofen and indomethacin for closure of patent ductus arteriosus. N Engl J Med. 2000;343:674–81. http://www.neonatologie.ugent.be/ibu-indo.pdf.
47. Komlos J, Lauderdale BE. The mysterious trend in American heights in the 20th century. Ann Hum Biol. 2007;34(2):206–15.
48. Cone TE Jr. Secular acceleration of height. J Pediatr. 1961;59:736–40.
49. Beunen GP, Rogol AD, Malina RM. Indicators of biological maturation and secular changes in biological maturation. Food Nutr Bull. 2006;27(4 Suppl Growth Standard):S244–56. https://journals.sagepub.com/doi/pdf/10.1177/15648265060274S508.
50. Rose K. Considering the Patient in Pediatric Drug Development. How good intentions turned into harm. London, UK: Elsevier; 2020.
51. Rose K. Blind Trust - How parents with a sick child can escape the labyrinth of lies, hypocrisy and false promises of researchers and regulatory authorities. London, UK: Hammersmith; 2022.
52. European Medicines Agency (EMA) 2015. Better Medicines for Children https://www.ema.europa.eu/en/documents/leaflet/better-medicines-children_en.pdf
53. Rose K, Ettienne EB, Grant-Kels JM, Striano P, Neubauer D, Oishi T. Neurology's vital role in preventing unnecessary and potentially harmful pediatric studies. Expert Rev Neurother. 2022; https://doi.org/10.1080/14737175.2022.2045953.
54. FDA. 2019. Drugs for treatment of partial onset seizures: full extrapolation of efficacy from adults to pediatric patients 2 years of age and olderGuidance for Industry. https://www.fda.gov/media/ 130449/download
55. Sun H, Temeck JW, Chambers W, et al. Extrapolation of efficacy in pediatric drug development and evidence-based medicine: progress and lessons learned. Ther Innov Regul Sci. 2017;2017:1–7.
56. FDA. 2018. Pediatric Product Development. https://www.fda.gov/drugs/development-resources/pediatric-product-development.
57. Regulation (EC) No 1901/2006 of the European Parliament and of the Council of 12 December 2006 on medicinal products for paediatric use and amending Regulation (EEC) No 1768/92, Directive 2001/20/EC, Directive 2001/83/EC and Regulation (EC) No 726/2004. http://eur-lex.europa.eu/LexUriServ/LexUriServ.do?uri=OJ:L:2006:378:0001:0019:en:PDF.
58. ICH Harmonized Guideline. Addendum to ICHE11: Clinical Investigation Of Medicinal Products In The Pediatric Population. E11 (R1). Final version. Adopted on 18 August 2017. https://database.ich.org/sites/default/files/E11_R1_Addendum.pdf.
59. ICH Harmonized Guideline. Pediatric Extrapolation. Draft version, April 2022 https://database.ich.org/sites/default/files/ICH_E11A_Document_Step2_Guideline_2022_0404_0.pdf.
60. Gøtzsche P. Deadly Medicines and Organised Crime: How Big Pharma Has Corrupted Healthcare. London, UK: Routledge; 2013.
61. Cook DT. The commodification of childhood: the children's clothing industry and the rise of the child consumer. Durham, NC: Duke University Press; 2004.
62. Sullivan R, Kowalczyk JR, Agarwal B, et al. Improving cancer care for children and young people 4 - New policies to address the global burden of childhood cancers. Lancet Oncol. 2013;14(3):e125–35. http://press.thelancet.com/childhoodcancer4.pdf
63. Vassal G, Zwaan CM, Ashley D. Improving cancer care for children and young people 3 - New drugs for children and adolescents with cancer: the need for novel development pathways. Lancet Oncol. 2013;14(3):e117–24. http://press.thelancet.com/childhoodcancer3.pdf.
64. Stiller CA, Kroll ME, Pritchard-Jones K. Population survival from childhood cancer in Britain during 1978–2005 by eras of entry to clinical trials. Annals of Oncology. 2012;23:2464–9.
65. Adamson PC. Improving the outcome for children with cancer: development of targeted new agents. CA Cancer J Clin. 2015;65(3):212–20. https://acsjournals.onlinelibrary.wiley.com/doi/epdf/10.3322/caac.21273.
66. Maude SL, Laetsch TW, Buechner J, et al. Tisagenlecleucel in children and young adults with B-cell lymphoblastic leukemia. N Engl J Med. 2018;378:439–48.

67. First Child To Receive CAR T-Cell Therapy Is 10 Years Cancer-Free. https://www.youtube.com/watch?v=0ei1yRxbdfU.
68. 7 Year Old Girl Beats Acute Lympoblastic Leukemia: Emily Whitehead's Immunotherapy Story. https://www.youtube.com/watch?v=Sz11e0r2L_s.
69. New Cancer Cure: World's First Child Treated with Immunotherapy Is Cancer-free 10 Years Later. https://www.youtube.com/watch?v=DP7fFoKPWcM.
70. Neubauer D, Laitinen-Parkkonen P, Matthys D. Ethical challenges of clinical research in children. In: Rose K, Van Den Anker J, editors. Guide to paediatric clinical research. Basel: Karger Publishers; 2007.
71. Grodin MA, Glanth LH, editors. Children as research subjects. New York: Oxford University Press; 1994.
72. Ross LF. Children in medical research: access versus protection. Oxford University Press; 2006.
73. Momper JD, Mulugeta Y, Green DJ, et al. Adolescent dosing and labeling since the Food and Drug Administration Amendments Act of 2007. JAMA Pediatr. 2013;167(10):926–32.
74. Mulberg AE, Silber SA, van den Anker JN, editors. Pediatric drug development – concepts and applications. Hoboken, NJ: Wiley-Blackwell; 2009.
75. Mulberg AE, Murphy MD, Dunne J, Mathis LL, editors. Pediatric drug development: concepts and applications. 2nd ed. Chichester, UK: John Wiley & Sons; 2013.

Chapter 19
Bridging of Formulations Used During Clinical Development

Norbert Pöllinger

Abstract Any change in formulation within a clinical development program requires bridging to ensure that the clinical data generated is relevant for a new formulation. As pediatric formulation development typically follows development of the adult product, this can require bridging of safety and efficacy data from the adult product to the pediatric formulation. This chapter covers the principles of bridging formulations in clinical development and uses case study examples to highlight the pathways that can be taken when using multiparticulate formulations.

Keywords Integrated development concept with micropellets · Relative bioavailability studies · Bioequivalence studies · Pediatric formulation bridging · Enabling formulations

19.1 Introduction

Pharmaceutical bridging in clinical development involves the use of relative bioavailability studies and bioequivalence assessments to expedite the approval process for new drug products. Their purpose is to demonstrate equivalence between formulations used in the development of a new drug, which may involve a difference from a liquid to a solid or even a change in the route of administration.

Relative bioavailability studies compare the rate and extent of absorption of a drug from different formulations or routes of administration. Bioequivalence studies focus on demonstrating the equivalence of two drug products in terms of their bioavailability. These studies are typically conducted using healthy adult volunteers and compare the pharmacokinetic parameters of the new formulation and the reference formulation. Parameters such as maximum concentration (Cmax), area under

N. Pöllinger (✉)
Glatt Pharmaceutical Services GmbH & Co. KG, Binzen, Germany
e-mail: Norbert.Poellinger@glatt.com

H. Batchelor, K. Rose (eds.), *Pediatric Formulations*, AAPS Advances in the Pharmaceutical Sciences Series 58, https://doi.org/10.1007/978-3-031-77240-5_19

the curve (AUC), and elimination half-life are assessed to establish if the two drugs are bioequivalent.

To ensure the validity and reliability of bioequivalence studies, strict guidelines and regulations are in place. These guidelines outline the acceptable limits for pharmacokinetic parameters and statistical methodologies to assess bioequivalence. The studies must adhere to Good Clinical Practice (GCP) guidelines and be conducted by qualified investigators.

19.2 Pediatric Clinical Development and Bridging Concepts

Pediatric clinical studies are usually initiated when sufficient clinical data have been collected in adult volunteers and patients (e.g., in phase 1, 2a and 2b clinical studies), which provide the opportunity to learn from knowledge in adults (Fig. 19.1). Ideally, both safety and efficacy have already been established in adults in the same indication.

Development of pediatric formulations may be technically more complex and challenging than the development of a common tablet for adults. In order not to delay the pediatric clinical studies, a simple *enabling formulation* may play an important role in the pediatric development strategy, as it may allow rapid initiation of early, exploratory, non-pivotal pediatric studies [1]. Examples of enabling formulations may be powder mixes, granulates, or pellets in bottles. As enabling formulations are preliminary formulations, often requiring some degree of ad hoc preparation, they should be optimized to more complex and elegant market-viable final pediatric formulations. A pediatric program may therefore include *enabling*

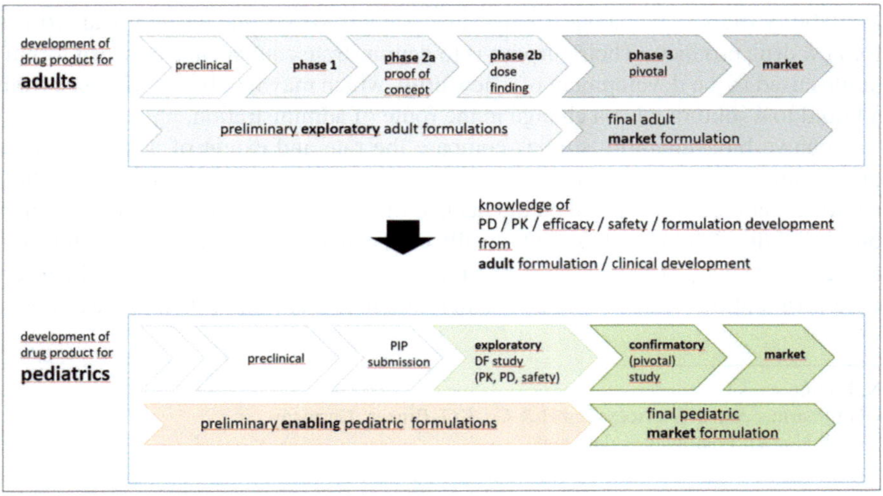

Fig. 19.1 Development of adult and pediatric drug product with an NCE compound

and final pediatric formulations, differing from the adult market formulation(s), and available in suitable dosage strengths to allow dosing of children over the foreseen age/weight range. In the case of complex technical approaches and projects, development of the intended final pediatric market formulations may have to be initiated in parallel to that of the enabling formulation (see Fig. 19.4) [2].

By leveraging all available information from drug development in adults, disease, response to treatment, and the physiological link between adults and children, sponsors may propose a two-step clinical program, composed of a *dose finding exploratory study (exploratory DF study)* followed by a *confirmatory pivotal study* [2] (Fig. 19.2).

Several formulation bridges must be established in order to support clinical extrapolation from:

- *Adult* to *pediatric* patient studies
- From *exploratory* pediatric study to *confirmatory pivotal* pediatric study

The most straightforward pediatric drug product development concept would be to directly develop the final pediatric market formulation *without* using a preliminary enabling pediatric formulation. Following this concept 1, only one bridging study (*RBA 1*, relative bioavailability study 1) is required to compare the pharmacokinetics of the adult reference formulation (adult market formulation) with the final pediatric market formulation in adult volunteers (concept 1, Fig. 19.3).

Often, development of the final pediatric *market* formulation is initiated only when a preliminary *enabling* pediatric formulation had been evaluated in adult

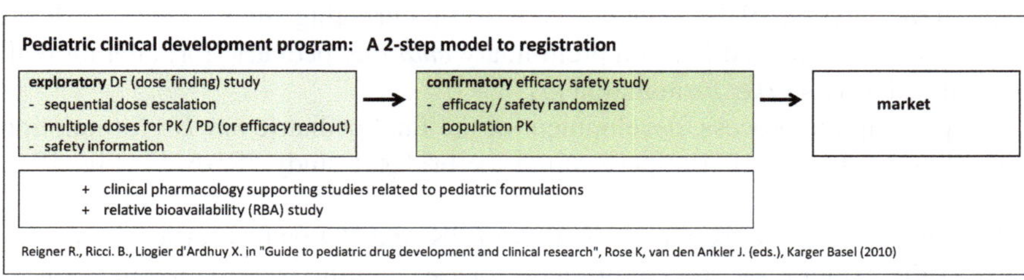

Fig. 19.2 Pediatric drug product clinical development

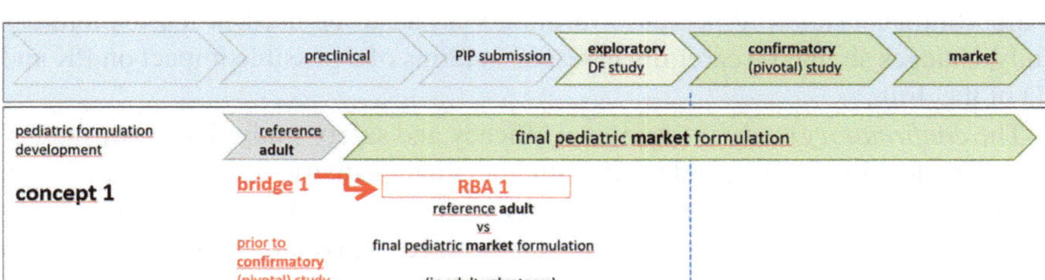

Fig. 19.3 Pediatric formulation development of an NCE compound/*option 1*: Development of final pediatric *market* formulation *without* preliminary enabling pediatric formulation, only 1 RBA bridging study required

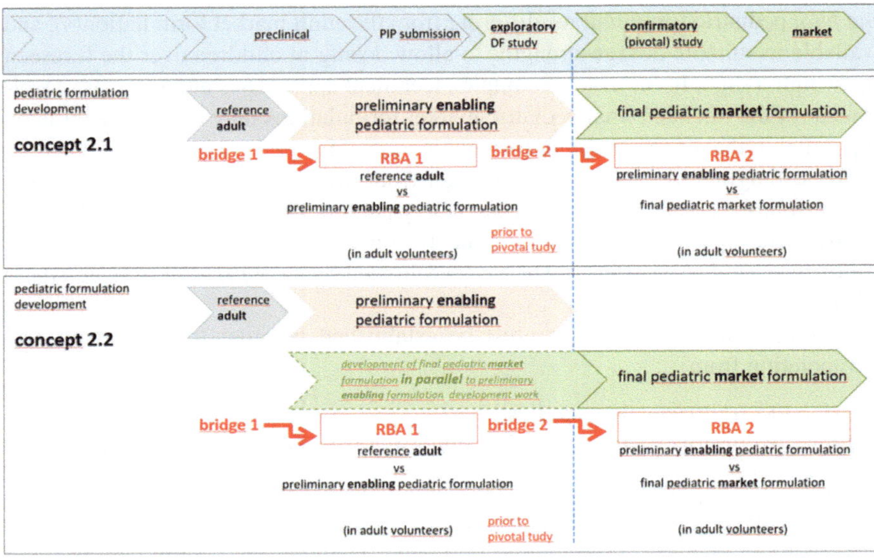

Fig. 19.4 Pediatric formulation development of an NCE compound/*concept 2.1 and 2.2*: Development of final pediatric *market* formulation, *including* a preliminary enabling pediatric formulation; 2 RBA bridging studies required

volunteers successfully: for this purpose, the reference adult (adult market formulation) is compared with a preliminary *enabling* pediatric formulation in a bridge 1 study "RBA 1" (Fig. 19.4, *concept 2.1*). A second bridging study, bridge 2 ("RBA 2") compares the BA and PK of a preliminary *enabling* pediatric formulation with the final pediatric *market* formulation [2].

To speed up the process, development of the final pediatric *market* formulation could be started in parallel to performing the bridge 1 study "RBA 1" (Fig. 19.4, *concept 2.2*).

In order to keep the extent of changes in PK and PD (safety, efficacy) studies between the exploratory and confirmatory phases as little as possible, the *exploratory* dose finding study should—from clinical pharmacology point of view—ideally be conducted with the final pediatric *market* formulation (Fig. 19.3, *concept 1*). If a preliminary *enabling* pediatric formulation had to be used in the exploratory study, changes should be carefully assessed in terms of a possible impact on PK and PD of the drug.

The *confirmatory* study is the pivotal efficacy and safety study. The confirmatory trials should always be conducted with the final pediatric *market* formulation. Similar to adult clinical programs, a change of the formulations used in the confirmatory study and in the market triggers the need for a pivotal BE study–a scenario which should be avoided wherever possible due to potential consequences of failure to establish bioequivalence (BE) [2]. Ideally, the pediatric formulation intended to be commercialized should be used for the complete development program as described in development concept 1 (Fig. 19.3). Similar to clinical programs with

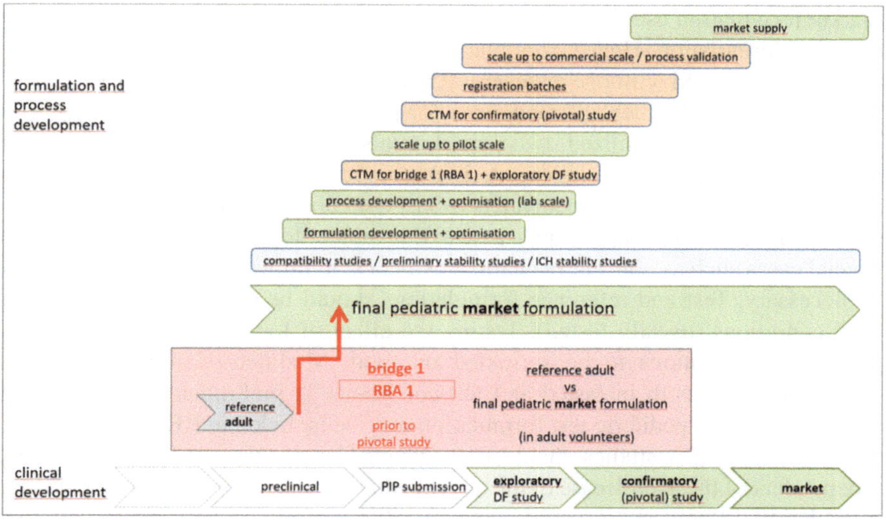

Fig. 19.5 Overview on pediatric drug development of an NCE compound with the "pediatric market formulation concept" (concept 1, see Fig. 19.3)

adults, a change of the formulation during clinical testing should be avoided wherever possible.

Figure 19.5 gives an overview of a pediatric drug product development performed with the final pediatric market formulation from the very beginning following concept 1 (see Fig. 19.3).

19.3 Pediatric Formulation Bridging

Formulation assessments must be performed prior to introduction of the new pediatric formulation in the clinical program. Bridging may be performed using clinical studies, such as relative bioavailability (RBA) studies or bioequivalence (BE) studies, or rely solely on in vitro techniques [2].

19.3.1 Relative Bioavailability (RBA) Studies

The objective of relative BA studies is to compare the new pediatric formulation to the (adult) reference formulation without any formal statistical assessment. RBA studies in pediatric drug development are proposed for comparison between adult and pediatric formulations, in order to derive the drug's pharmacokinetic (PK) parameters in two formulations, and ultimately support dosing selection in pediatric clinical trials. RBA studies might also be applied to compare the final pediatric market formulation used in confirmatory (pivotal) studies to a simpler preliminary enabling formulation. Hence, formulation bridging in pediatric drug development might be performed with one or several RBA studies [2] (Figs. 19.3 and 19.4).

As in the development of medicines for adult use, RBA studies are typically designed using a randomized, two-period, two-sequence crossover comparison of the test and reference formulations. A rigid statistical BE criterion is not required as options to cope with non-bioequivalence exist (i.e., dosage adjustment). The release of the drug from the formulation is assessed in single-dose conditions. However, in special cases such as non-linear pharmacokinetics, multiple-dose RBA studies may be necessary. Test and reference formulations should be compared under similar food conditions (usually fasted). Testing the effect of food may be necessary for pediatric formulations to be sprinkled on food. Modified-release formulations should be tested both in fasted and fed conditions. A preliminary assessment of palatability of the pediatric test formulation may be included in RBA studies [2].

By default, RBA studies should be conducted in *healthy adult volunteers* [3], independent of the age/disease of the target pediatric population to detect significant formulation differences. It would be unethical to perform on children a study that would bring no therapeutic benefit. In some situations, it may be necessary to perform the bridging studies in the patient population rather than with volunteers (e.g., anticancerous agents, RBA study with *adult patients*) studies [2]. The RBA study comparing the adult reference formulation to the first pediatric formulation was conducted prior to the first study in children, in order to assess the performance of the intended pediatric formulation. Adjustment of the formulation/dosage strength may be necessary if the test formulation yields too high exposure (safety concerns), too low exposure (efficacy concerns), or higher variability [2].

19.3.2 Bioequivalence (BE) Studies

In contrast to relative bioavailability (RBA) studies, which aim to optimize formulation development, bioequivalence (BE) studies aim to conclude therapeutic bioequivalence (BE) between test and reference formulations. BE studies are regulatory studies that are needed for the registration of new formulations and therefore have to be conducted in strict accordance with regulatory requirements. Demonstrating BE uses a formal comparative statistical test that has specified criteria for comparison and preset acceptance limits for such criteria, customarily 80–125% for the 90% confidence interval range for the AUC and Cmax parameters [2].

BE studies are needed when there are significant differences between the formulation used in the pivotal study, and the proposed market formulation, or when changes are introduced in the already marketed formulation. Failure to establish BE can have considerable consequences, e.g., delay in registration, or possible need to repeat pivotal efficacy study. Therefore, avoidance of BE studies is strongly advised. Instead, the use of bridging with RBA studies prior to the conduct of the pivotal clinical studies is recommended. When a BE study is unavoidable, it may be extremely useful to precede the BE study by a pilot study in order to generate parameter estimates, assess variability, and optimize sample collection times in preparation for the formal BE study. BE studies should be conducted in the dietary conditions as recommended in the labeling. Health authorities should be consulted in advance for drugs intended to be administered in the fed state, as it may be important to demonstrate BE in both fed and fasted states [2].

19.3.3 Bridging Without Clinical BA and/or BE Studies

Bridging formulations without support from clinical PK studies may be possible, in particular when only very minor formulation or process changes are introduced with the new formulation. In the USA, the FDA's Scale-Up and Post-Approval Change (SUPAC) guidances define the nature of these changes for IR and MR formulations [4, 5]. However, most often these criteria cannot be met, as significant changes in the adult formulation are needed for a child-appropriate formulation.

For drugs that are in the same dosage form and manufactured by the same manufacturing process, but in a different dosage strength than in the adult formulation, a clinical bridging study may be waived based on the in vivo study of the adult formulation. This can be the case when dissolution results are similar under the same dissolution procedures, provided the composition of the adult and pediatric formulations is proportionally similar on both active and inactive ingredients [2]. Further, the drug should exhibit linear elimination pharmacokinetics over the dose range [6].

This approach may be the most appropriate for formulations intended for adolescents and children who can swallow solid dosage forms, and for whom the market formulation may be the same capsules, sachet, stickpacks, etc. as in adults, but of lower dosage strength [2].

For BCS class 1 drugs (high solubility and permeability properties), formulation changes for oral immediate release formulations should not affect in vivo absorption. A BCS class 1 biowaiver approach could be explored provided both the reference formulation (e.g., adult formulation, or enabling pediatric formulation) and the new pediatric formulation are pharmaceutical equivalents exhibiting rapid and similar dissolution profiles. Other conditions should be met, e.g., the excipients should not affect the extent and rate of absorption [2].

19.4 Conclusion on Bridging Studies

Pediatric formulation bridging is typically performed via single-dose, cross-over relative bioavailability (RBA) studies in adult healthy volunteers. Other study designs and study population are to be considered based on drug properties and formulation. The pediatric formulation strategy must be an integral part of the pediatric development plan.

As bioequivalence (BE) studies carry a high risk for a development program, it is important to plan early and aim to bridge formulations with RBA studies prior to pivotal pediatric studies [2].

19.5 Case Studies with Generic Drug Compounds: Bridging/Bioequivalence Studies

Development of pediatric products with generic drug compounds (example for a high-dosed compound: clarithromycin, example for a low-dosed compounds: hydrocortisone) is more straightforward than with NCE compounds (Fig. 19.6).

19.5.1 Clarithromycin 125 mg/5 ml and 250 mg/5 ml, Granules for Oral Suspension

Clarithromycin is an antibiotic drug indicated for different acute and chronic infections in adults, adolescents and children, 6 months to 12 years, when caused by clarithromycin susceptible organisms [7]. In order to mask the extremely bitter taste of clarithromycin, taste-masked micropellets were developed [8]. A 125 and 250 mg dose is presented in an oral liquid. The particle size of the high drug-loaded micropellets is <500 µm. Taste masking must be provided for at least 14 days in an aqueous suspension. Nevertheless, an immediate drug release at pH 6.8 is requested.

The basic formulation concept is based on two main steps (Fig. 19.7):

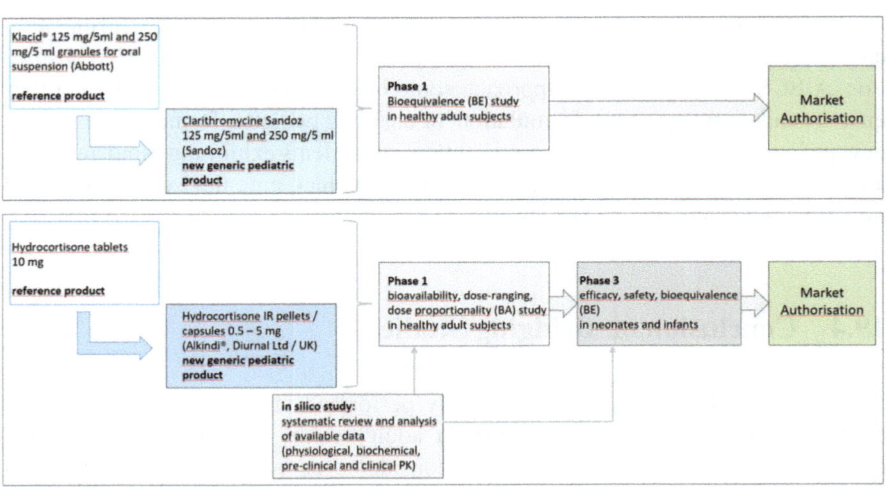

Fig. 19.6 Development of pediatric drug products with generic drug compounds

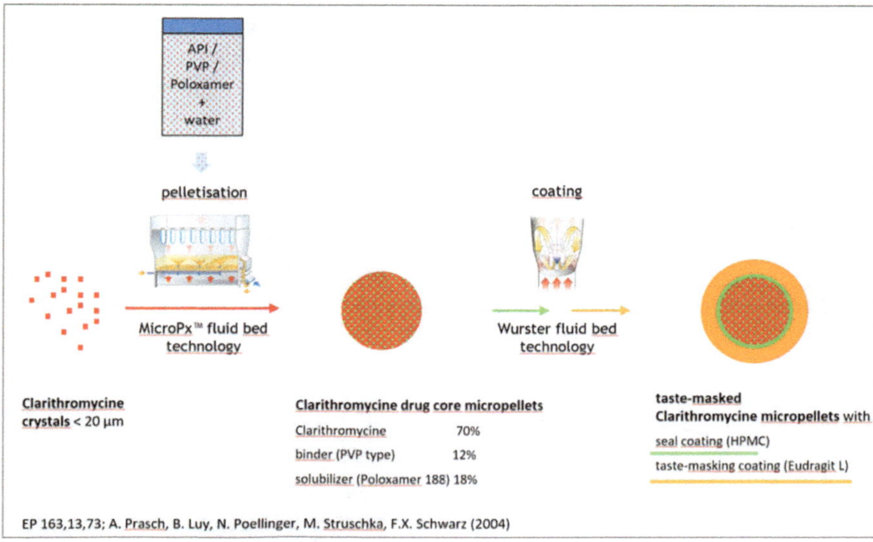

Fig. 19.7 Processing of taste-masked clarithromycin micropellets

1. Core matrix micropellets containing the crystalline micronized API and an emulsifier are produced using continuous MicroPx fluid-bed pelletization technology (particle size of the spherical smooth micropellets: 200–400 µm, process yield: ≥ 95%).
2. Core matrix pellets are coated with a water-soluble HPMC seal-coat and a taste-masking film soluble at pH 6.8 (Eudragit L polymer) using the Wurster (bottom-spray) fluidized-bed technology (particle size of the coated micropellets: <500 µm, process yield: ≥95%).

Clarithromycin micropellets manufactured with continuous MicroPx fluidized bed technology are spherically shaped and in a size range of 200–400 µm. The taste masked micropellets provide an immediate drug release at pH 6.8 (Fig. 19.8).

Clarithromycin Sandoz 125 mg/5 ml and 250 mg/5 ml are white to beige granules, resulting in a white to beige suspension after reconstitution with water. The finished dosage form (dry suspension) consists of coated clarithromycin micropellets ("granules"), which are embedded in a suspension base with powdered sucrose as the main component. For the 125 mg/5 ml form half, the amount of clarithromycin micropellets is used, which is compensated by the carrier maltodextrin [7].

For this generic application, the Marketing Authorization Holder (MAH) has submitted two bioequivalence studies in which the pharmacokinetic profile of the test products clarithromycin Sandoz 125 mg/5 ml and 250 mg/5 ml (Sandoz B.V., NL) is compared with the pharmacokinetic profile of the reference product Klacid® 125 mg/5 ml and 250 mg/ 5 ml granules for oral suspension (Abbott B.V., NL). The formula and preparation of the bioequivalence batches are identical to the formula

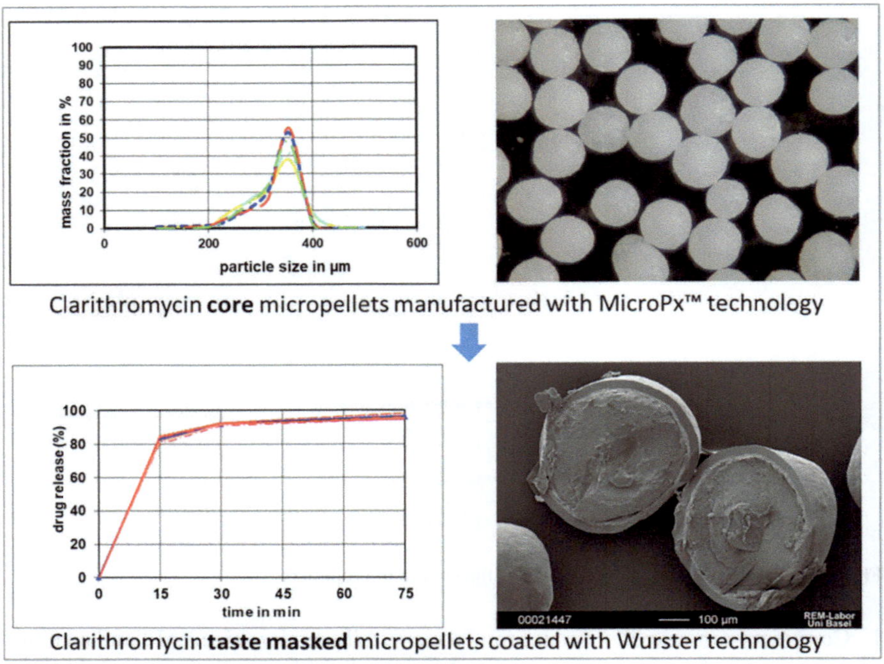

Fig. 19.8 Clarithromycin taste-masked micropellets

proposed for marketing [7]. A preliminary enabling formulation concept was not applied for the development of the generic product.

19.5.1.1 Bioequivalence (BE) Study I: 125 mg/5 ml (Fig. 19.9)

A single-dose, randomized, two-period, two-treatment, two-sequence, crossover BE study was carried out under fasted conditions in 32 healthy subjects (20 males, 12 females), aged 21–43 years. Each subject received a single dose (125 mg/5 ml) of one of the two clarithromycin formulations. The suspension was orally administered with 240 ml water after an overnight fast of at least 10 hours.

19.5.1.2 Bioequivalence (BE) Study II: 250 mg/5 ml (Fig. 19.10)

A single-dose, randomized, two-period, two-treatment, two-sequence, crossover BE study was carried out under fasted conditions in 32 healthy subjects (17 males, 15 females), aged 21–44 years. Each subject received a single dose (250 mg/5 ml) of one of the two clarithromycin formulations. The suspension was orally administered with 240 ml water after an overnight fast of at least 10 hours.

x

Treatment n = 32	AUC 0-t µg.h/ml	AUC 0 - ∞ µg.h/ml	C max µg/ml	T max h	T 1/2 h
Test	6.95 ± 2.21	7.09 ± 2.26	1.10 ± 0.30	2.10 ± 0.88	3.6 ± 0.8
Reference	7.61 ± 2.10	7.74 ± 2.14	1.22 ± 0.30	2.99 ± 0.77	3.8 ± 0.7
*ratio (90% CI)	0.91 (0.87 - 0.94)	0.90 (0.85 - 0.95)	0.88 (0.83 - 0.95)	--	--
CV (%)	12.6	12.5	14.1	--	--

AUC 0-t	area under the plasma concentration-time curve from time zero to t hours
AUC 0 - ∞	area under the plasma concentration-time curve from time zero to infinity
C max	maximum plasma concentration
T max	time for maximum concentration
T 1/2	half-life
*	in-transformed values

Fig. 19.9 Pharmacokinetic parameters of clarithromycin under fasted conditions (125 mg/5 ml)

Treatment n = 32	AUC 0-t µg.h/ml	AUC 0 - ∞ µg.h/ml	C max µg/ml	T max h	T 1/2 h
Test	17.4 ± 4.8	17.6 ± 4.8	2.19 ± 0.61	2.3 ± 0.73	4.8 ± 1.0
Reference	17.5 ± 4.5	17.7 ± 4.5	2.42 ± 0.59	3.38 ± 0.95	4.5 ± 1.0
*ratio (90% CI)	0.99 (0.95 - 1.03)	0.99 (0.95 - 103)	0.90 (0.86 - 0.94)	--	--
CV (%)	9.9	9.8	10.6	--	--

AUC 0-t	area under the plasma concentration-time curve from time zero to t hours
AUC 0 - ∞	area under the plasma concentration-time curve from time zero to infinity
C max	maximum plasma concentration
T max	time for maximum concentration
T 1/2	half-life
*	in-transformed values

Fig. 19.10 Pharmacokinetic parameters of clarithromycin under fasted conditions (250 mg/5 ml)

The 90% confidence intervals calculated for AUC 0-t, AUC 0-∞ and Cmax are within the bioequivalence acceptance range of 0.80–1.25. Based on the pharmacokinetic parameters of clarithromycin under fasted conditions, it can be concluded that clarithromycin Sandoz 125 mg/5 ml and Klacid® 125 mg/5 ml granules for oral suspension and clarithromycin Sandoz 250 mg/5 ml and Klacid 250 mg/5 ml granules for oral suspension are bioequivalent with respect to rate and extent of absorption. Clarithromycin may be taken without reference to food intake. From the literature, it is known that food does not interact with the absorption of clarithromycin. Therefore, a food interaction study is not deemed necessary [7].

19.5.2 *Hydrocortisone 0.5 mg, 1 mg, 2 mg, and 5 mg Granules in Capsules for Opening (Alkindi®, Diurnal Ltd)*

Adrenal failure occurs in approximately 1/10,000 of the adult population and 1/16,000 of infants. It is apparent that dosing regimens for the treatment of children, adults, and elderly adults will vary depending on a number of parameters such as developmental stage and physiological state. The treatment of pediatric adrenal insufficiency has particular problems and requires pharmaceutical formulations that address the pharmacokinetic and pharmacodynamic problems of dosing infants.

Current preparations of hydrocortisone cannot adequately replace cortisol deficiency, especially in the pediatric population because that formulations used do not allow the flexibility of low-dose hydrocortisone to reproduce physiological levels of cortisol. For example, after diagnosis with adrenal insufficiency, usually at birth, a common dose of hydrocortisone prescribed in the United Kingdom is 7.5 mg per day divided into three equal doses (i.e., 3 × 2.5 mg / day). Even lower doses may be required for adequate treatment of neonates (0–28 days old), infants (1–24 months old), and young children (2–6 years old). Halving or quartering tablets for adults do not enable accurate dosing nor low dose titration of hydrocortisone [9]. In a medication safety study, individually compounded capsules were investigated in terms of uniformity of mass, uniformity of drug content and accuracy of the drug content. In total, 18% of the 62 analyzed capsule batches showed insufficient uniformity, and 48% of the batches were inacceptable over all investigations [10].

In the end, suboptimal treatment may result, questioning the efficacy and ethics of such practice, particularly in the most vulnerable patients of all, neonates and infants. Also the quality of individually compounded capsules is not always adequate and needs to be improved. Furthermore, if a subject receiving the medication is able to taste the active ingredient upon ingestion they may refuse to comply with the prescribed dosage regimen.

Hydrocortisone is on EMA unmet medical need priority list (2011) for the use in pediatric population for children suffering from all forms of adrenal insufficiency. As part of an European Commission Framework 7 Grant no. 281654 (work package 3), Glatt Pharmaceutical Services GmbH & Co.KG in cooperation with Diurnal Ltd. (UK) developed in the TAIN (Treatment of Adrenal Insufficiency in Neonates and Infants) project an immediate release formulation of hydrocortisone based on micropellets. Due to the bitter taste of the API, a taste masking coating is necessary [11].

A hydrocortisone drug layer is applied on 350–500 μm cellulose starter beads, seal-coated with a HPMC (hydroxypropylmethylcellulose) based layer and taste-masked with an EC (ethylcellulose)/HPMC = 4/1 polymer coat to nullify the extremely bitter taste of hydrocortisone. In total, only natural or semi-natural polymer compounds are used (Fig. 19.11).

The taste-masked hydrocortisone pellets are manufactured with a three-step fluidized-bed Wurster (bottomspray) process and finally filled into HPMC capsule shells, which serve as packaging material for the pellets but must not be swallowed (Figs. 19.11 and 19.12).

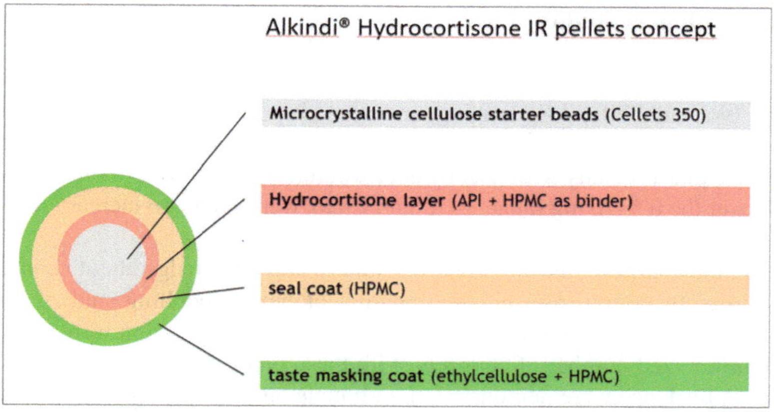

Fig. 19.11 Formulation concept of taste-masked immediate release hydrocortisone pellets. (Alkindi®, Diurnal Ltd., UK)

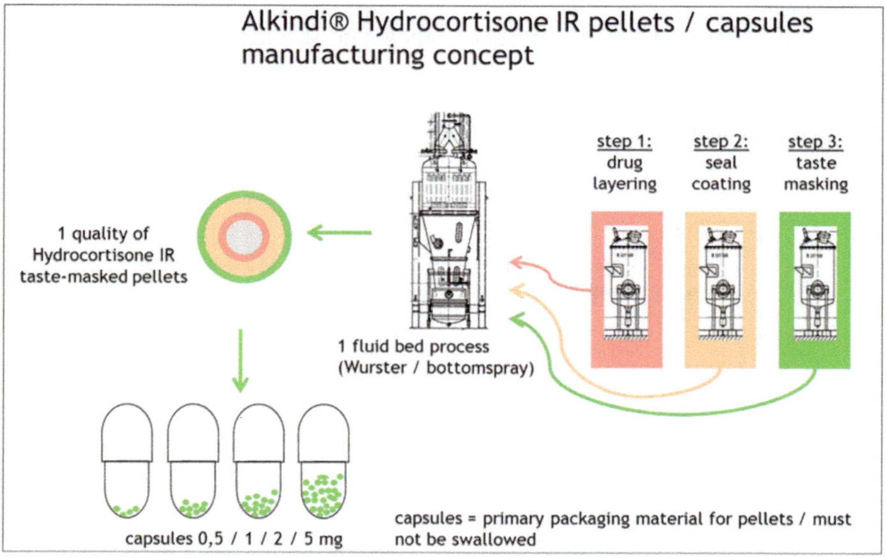

Fig. 19.12 Alkindi® hydrocortisone IR pellets / capsules manufacturing concept

Alkindi capsules with 0.5/1/2/5 mg of hydrocortisone are commercially available. In particular, the low doses of 0.5 mg and 1 mg will give clinicians the flexibility to effectively treat the different patient groups including even premature neonates. The preparation has the advantage of appropriate and easily titratable unit doses, stability for extended shelf-life, compatibility in common food/drink matrices such as yoghurt, milk and water [9]. The content of the capsules (the multiparticulates / pellets = "granules")) may be administered either directly into the patient's mouth or placed on a spoon and then administered into the patient's mouth [12]. Moreover, hydrocortisone pellets can be used for administration via nasogastric tube [13].

19.5.2.1 Clinical Studies

Prior to evaluation of the new pediatric drug product in the pediatric population, the relative bioavailability, dose-ranging, and dose proportionality of Alkindi® pellets were compared with that of a marketed immediate-release 10 mg hydrocortisone tablets in a phase 1 study with healthy adult volunteers. As shown in Figs. 19.13 and 19.14, over the 0.5–10 mg range, AUC and Cmax were shown to increase in a linear fashion [12].

Using the data from the adult phase 1 studies and knowledge about neonate and infant drug absorption from in silico pharmacokinetic modelling studies, phase 3 studies to obtain the information required for a PUMA submission were performed with a minimized number of children. With these phase 3 studies, efficacy, safety, and bioequivalence were demonstrated with male and female neonates and infants [12] (Fig. 19.15).

In total, the following clinical studies were performed:

- Phase 1 PK, safety and dose ranging hydrocortisone study in human adult volunteers
- Pivotal phase 3 hydrocortisone study in neonates (efficacy, safety, bioequivalence)
- Pivotal phase 3 hydrocortisone study in infants (efficacy, safety, bioequivalence)

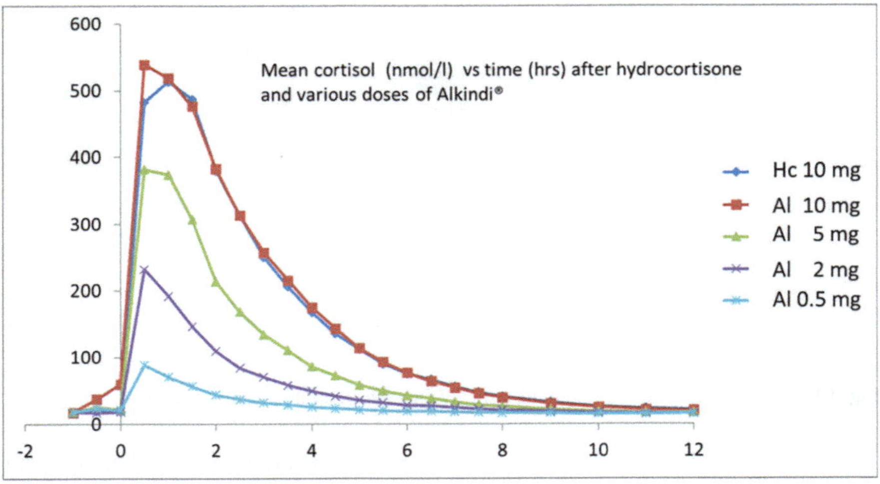

Fig. 19.13 Hydrocortisone pellets (Alkindi®) and hydrocortisone tablets overlap [12]

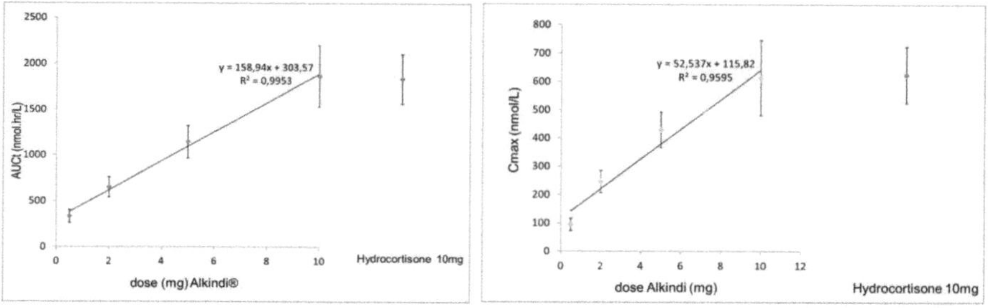

Fig. 19.14 Dose proportionality (AUC, Cmax) from 0.5 to 10 mg Alkindi® [12]

	C max (nmol/l)	AUC 0 - ∞ (h*nmol/l)	tmax (h)
Alkindi 10mg (Geomean)	566	1602	0.75
Hydrocortisone tablet 10 mg (Geomean)	598	1576	1.00
ratio Alkindi to hydrocortisone % (90% CI)	95 (84 - 107)	101 (96 - 107)	0.0 (-0.5 - 0.3)*

AUC 0 - ∞	area under the plasma concentration-time curve from time zero to infinity
C max	maximum plasma concentration
T max	time for maximum concentration
*	mean difference (90% CI)

Fig. 19.15 Summary of statistical analysis of bioequivalence between 10 mg hydrocortisone immediate-release pellets and 10 mg immediate-release tablet reference formulation [12].

19.6 Conclusions and Perspectives

Generally, pediatric drug development starts after development of the drug products for adult patients. This approach extends development time and may necessitate an entirely different dosage form and/or formulation for the pediatric patients which will necessitate pharmaceutical bridging such that the existing clinical knowledge can be applied to the new formulation.

However, if the drug product may have a potentially acceptable therapeutic benefit for the pediatric patients, AND the adults, early development strategy discussions could help to develop "age-friendly" and/or "flexible" dosage forms that are both flexible and available for use by a more diverse patient population including pediatrics and adults. This would lead to an integrated program, reducing the requirement for pharmaceutical bridging studies, instead of a sequential drug development program. The concept will facilitate access to different dosage forms, suitable for the whole patient population from newborn to elderly and a broad dosing range. (Fig. 19.16).

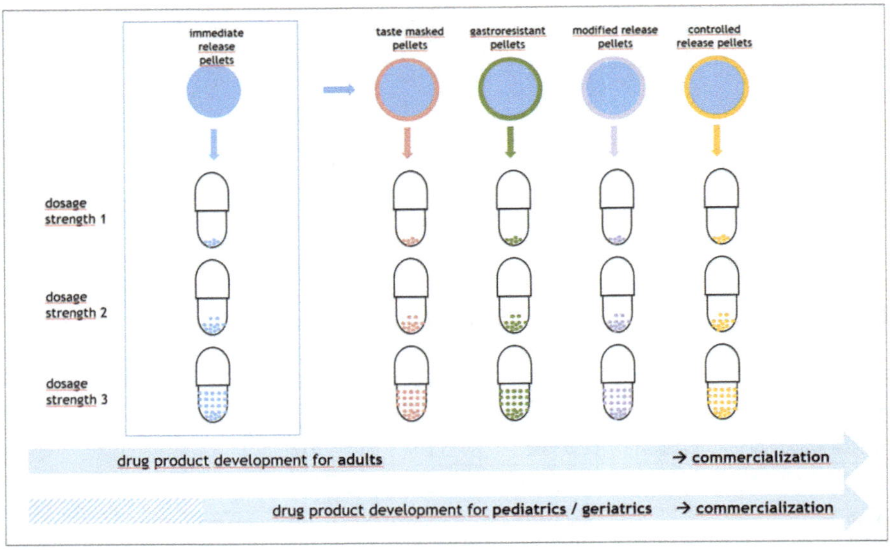

Fig. 19.16 Integrated drug product development concept with (micro)pellets

References

1. EMA. EMA/CHMP/QWP/805880/2012 Rev. 1. Draft guideline on pharmaceutical development of medicines for paediatric use = aus B. Ricci. 2013a.
2. Ricci B. Bridging studies in support of oral pediatric formulation development. Int J Pharm. 2013;457:323–6.
3. EMA. EMA/618604/2008 Rev.7. Questions and answers: positions on specific questions addressed to the pharmacokinetic s working party = aus Ricci B. 2013b.
4. FDA Center for Drug Evaluation and Research (CDER). Guidance for industry immediate release solid oral dosage forms scale-up and postapproval changes: chemistry, manufacturing, and controls, in vitro dissolution testing, and in vivo bioequivalence documentation = from Ricci B. 1995.
5. FDA Center for Drug Evaluation and Research (CDER). Guidance for the industry SUPAC-MR: modified release solid oral dosage forms scale-up and postapproval changes: chemistry, manufacturing, and controls: in vitro dissolution testing and in vivo bioequivalence documentation = from Ricci B. 1997.
6. EMA. CPMP/EWP/QWP/1401/98 Rev.1/Corr. Guideline on the investigation of bioequivalence = lit from Ricci B. 2010.
7. Public assessment report of the Medicines Evaluation Board in the Netherlands. EU-procedure number: NL/H/2099/001–002/DC. Registration number in the Netherlands: RVG 108098–108099 (02.05.2021).
8. Prasch A, Luy B, Poellinger N, Struschka M, Schwarz FX. Micropellets, method for the production thereof, and use thereof. EP. 2004;1:631 373 B1.
9. Wollmer E, Karkossa F, Freerks L, Hetberg AL, Neal G, Porter J, Whitaker M, Margetson D, Klein S. A biopredictive in vitro approach for assessing compatibility with common pediatric dosing vehicles. Pharm Res. 2020;37:203.
10. Kauzor D. Are pediatric CAH patients sufficiently treated with individually compounded capsules? 6th EuPFI conference, Athens 2014.

11. Grave A, Poellinger N, Whitaker M. Taste masked micropellets for pediatric use. 9th world meeting on pharmaceutics, biopharmaceutics and pharmaceutical technology, Lisbon. 2014.
12. Huatan H, Ross R, Whitaker M, Poellinger N, Grave A. Composition comprising hydrocortisone. WO 2014/184525. 2014.
13. Daniel E, Whitaker M, Keevil B, Waler J, Ross R. Accuracy of hydrocortisone dose administration via nasogastric tube. Clin Endocrinol. 2019;90:66–73.

Chapter 20
Modeling and Simulation for Pediatric Formulation Development

Hannah Batchelor

Abstract Pharmacokinetic (PK) modeling is essential in pediatric drug development, offering insights into drug behavior in children. Two primary approaches, population pharmacokinetics (PopPK) and physiologically based pharmacokinetic (PBPK) models, serve distinct purposes.

PopPK models use statistical methods to analyze drug data from various individuals, focusing on population-level variability and factors like age and weight. These models guide dosing by predicting drug concentrations and minimizing adverse effects, especially important in vulnerable populations such as children.

PBPK models, on the other hand, are mechanistic and simulate physiological processes affecting drug absorption, distribution, metabolism, and excretion. They integrate physiological and compound-specific data to predict drug behavior under different conditions, supporting dose scaling and formulation decisions.

Both approaches contribute significantly to pediatric drug development. PopPK models help optimize dosing regimens and reduce clinical trial invasiveness, while PBPK models aid in understanding age-related pharmacokinetic changes and support regulatory submissions. Together, they enhance the development of safe, effective, and age-appropriate therapies, improving pediatric patient outcomes.

Keywords Pharmacokinetics (PK) · Population pharmacokinetics (poppk) · Physiologically-based pharmacokinetic (PBPK) models · Pediatric drug development · Dosing regimens

---

H. Batchelor (✉)
Strathclyde Institute of Pharmacy and Biomedical Sciences, Glasgow, UK
e-mail: Hannah.batchelor@strath.ac.uk

H. Batchelor, K. Rose (eds.), *Pediatric Formulations*, AAPS Advances in the Pharmaceutical Sciences Series 58, https://doi.org/10.1007/978-3-031-77240-5_20

20.1 Introduction

Pharmacokinetic (PK) modeling plays a crucial role in pediatric drug product design by providing insights into how drugs behave in the bodies of children.

Population pharmacokinetics (PopPK) and physiologically based pharmacokinetic (PBPK) models are both valuable tools used in pharmacokinetic analysis, but they differ in their approaches, applications, and levels of complexity. This chapter reports on each individually with their value in pediatric formulation development.

20.2 Population Pharmacokinetic (PopPK) Models

PopPK models are statistical models that analyze pharmacokinetic data obtained from a population of individuals to characterize the variability in drug disposition parameters. These models focus on describing how drug concentrations vary among individuals within a population and aim to identify factors (e.g., age, weight, genetics) that influence drug pharmacokinetics.

PopPK models typically involve nonlinear mixed-effects modeling techniques to estimate population parameters and individual-specific parameters simultaneously.

PopPK models are primarily used to characterize the pharmacokinetic variability in patient populations and to guide dosing recommendations. They are valuable for optimizing dosing regimens, especially in vulnerable populations like pediatrics and geriatrics, where inter-individual variability is significant. PopPK models are frequently employed in clinical pharmacology research, pharmacokinetic-pharmacodynamic (PK-PD) modeling, and drug development studies to inform dose selection and regimen individualization.

PopPK models are typically less complex than PBPK models and focus on characterizing population-level variability in pharmacokinetic parameters. They rely on empirical relationships derived from observed pharmacokinetic data and may incorporate covariates to explain inter-individual variability. PopPK models are computationally less intensive and require fewer physiological assumptions compared to PBPK models.

The value of PopPK models for pediatric populations lies in the value for dose optimization. Pediatric patients often require individualized dosing regimens due to differences in physiology, metabolism, and growth rates compared to adults. Pharmacokinetic modeling allows researchers to predict drug concentrations in pediatric populations based on age, weight, and other relevant factors. By simulating various dosing scenarios, PK modeling helps optimize dosing regimens to achieve desired therapeutic outcomes while minimizing the risk of adverse effects.

Conducting clinical trials in pediatric populations raises ethical considerations due to vulnerability and the need for informed consent. Pharmacokinetic modeling can help minimize the need for invasive procedures and reduce the exposure of pediatric patients to experimental treatments by optimizing study design and dosing strategies. By leveraging existing pharmacokinetic data and simulation techniques, PK modeling contributes to the ethical conduct of pediatric clinical research. As PopPK models enable researchers to predict drug concentrations at unobserved time points using mathematical modeling, they can strategically select sampling times based on model predictions, which can reduce the number of samples required while still obtaining sufficient data to characterize the drug's pharmacokinetic profile accurately. Furthermore, PopPK models account for variability in drug pharmacokinetics due to factors such as age, weight, and genetics. By quantifying and incorporating this variability into the analysis, researchers can reduce the overall variability in pharmacokinetic parameters, which may lead to more precise estimates and consequently reduce the number of samples needed to achieve desired statistical power. In addition, PopPK models can account for inter-individual variability in drug pharmacokinetics, allowing researchers to tailor sampling schedules to individual patients' characteristics. By focusing sampling efforts on patients most likely to provide informative data, researchers can optimize resource utilization and minimize unnecessary sampling in the overall study population.

20.3 Physiologically Based Pharmacokinetic (PBPK) Models

PBPK models are mechanistic models that simulate the physiological processes underlying drug absorption, distribution, metabolism, and excretion (ADME) within the body. These models incorporate physiological parameters such as organ volumes, blood flows, tissue permeabilities, and enzyme kinetics to predict drug concentrations in various tissues and compartments over time. PBPK models simulate drug behavior based on known physiological principles and utilize mathematical equations to describe drug movement within the body.

PBPK models are employed to simulate drug pharmacokinetics under various physiological conditions, such as changes in organ function, disease states, and drug–drug interactions. They are used in drug development to predict human pharmacokinetics based on preclinical data, facilitate dose scaling between species, and support drug formulation and route of administration decisions. PBPK models are also utilized in risk assessment, toxicology studies, and regulatory submissions to evaluate the potential impact of drugs on human health and safety. Pediatric drug products often need to be formulated into age-appropriate dosage forms, such as liquids, suspensions, or chewable tablets, to facilitate administration to children. Pharmacokinetic modeling aids in understanding drug absorption, distribution, and elimination characteristics, which are essential for designing formulations with appropriate bioavailability and dosing flexibility. For instance, modeling can inform decisions regarding particle size, excipients, and dosage strengths to ensure optimal drug delivery and therapeutic efficacy.

Today, PBPK models are routinely employed in drug discovery and development. They play a crucial role in tasks such as selecting initial doses for first-in-human trials, managing drug interactions during clinical studies, and studying the impact of food on drug dissolution and absorption. This approach has seen increasing use in supporting pediatric dosing for recently approved drugs. Consequently, pediatric PBPK (p-PBPK) simulations extend their use beyond drug development by providing insights into off-label use in pediatric clinical care. For instance, PBPK simulation outcomes can contribute to establishing dosing recommendations or enhancing confidence in existing recommendations for pediatric disease treatments.

Pediatric PBPK models have gained prominence over the past decade, primarily because there is a growing availability of physiological parameters for constructing these models, and regulatory guidelines now require pediatric studies as a part of drug development. Thanks to collaborative efforts involving academia, PBPK model developers, pharmaceutical companies, and regulatory agencies, there are now instances where PBPK models have either substituted for or significantly informed clinical studies in children. Nonetheless, the quantity of pediatric PBPK models and their predictive accuracy still falls short when compared to their adult model counterparts.

20.3.1 Principles of PBPK

Physiologically based pharmacokinetic modeling takes a mechanistic approach to predict pharmacokinetics in pediatric populations. Typically, it relies on drug-specific data obtained from adult populations and physiological information relevant to various pediatric age groups. These physiologically based pharmacokinetic (PBPK) models encompass two distinct sets of data: physiological data and compound-specific data. In principle, both compound and population models can be separately developed and subsequently linked, facilitating the reuse of existing compound and population models. This integration of compound models with population models allows for the prediction of pharmacokinetics for any drug within a specific population of interest.

Creating a compound model from the ground up can be a time-intensive process. However, for a substantial number of compounds, model data or model files can be readily accessed in the public domain. This information can be found in sources such as peer-reviewed journals, regulatory assessment documents, and model repositories maintained by PBPK software developers. Frequently, these models were designed for simulating pharmacokinetics in adult populations and were rarely explored for their potential in aiding pediatric drug dosing decisions. A more efficient and practical approach to p-PBPK modeling involves amalgamating existing compound models with pediatric physiological models. This approach proves pragmatic and viable for forecasting pharmacokinetics and offering guidance on drug dosing within the realm of pediatric clinical care. Models can be customized for various populations by modifying system parameters that depend on the specific population in question. This adaptability has been effectively demonstrated in multiple scenarios, such as translating models from animals to humans [1], from one ethnic group to another [2], and from healthy individuals to those with medical conditions [3]. Likewise, when extrapolating from adults to children, the process involves adjusting adult physiological parameters that are age-dependent to values that are suitable for children [4]. This approach is frequently utilized in the development of pediatric models, where the initial evaluation involves key pharmacokinetic processes and confirmation of drug-related parameters unrelated to age within an adult population. Subsequently, age-specific pediatric physiological parameters are incorporated, which may have more limited available data [5]. In practice, a hybrid approach is commonly employed, utilizing a combination of in vitro data and in vivo-derived drug concentrations for the parameterization of models. In situations where essential physiological or drug-related parameters for PBPK models are unavailable, these parameters must be adjusted or tailored by scaling or fitting them based on drug concentration data obtained through clinical measurements.

20.3.1.1 Existing PBPK Models

A tutorial for a general workflow to develop a pediatric PBPK model was described by Maharaj et al., which provides a comprehensive overview [6]. Physiology and compound models for PBPK can typically be sourced from three separate origins. These model types are commonly accessible through specialized modeling software, model repositories, or they may be documented in peer-reviewed publications. Numerous compound models of varying quality, ranging from high-quality to lower-quality models, have been created and made available. These models have been developed using a variety of PBPK software platforms, including both commercial and open-source options.

A range of PBPK software tools are available, including general-purpose modeling tools, such as R, MATLAB, ACSL, as well as bespoke PBPK software platforms, such as GastroPlus (www.simulations-plus.com), PK-Sim (www.open-systems-pharmacology.org), and Simcyp (www.simcyp.com). A comparison of the development of these models is presented in a detailed review by Verscheijden et al. (2020), and the reader is directed there for additional information [7].

Simcyp is a comprehensive tool that incorporates a full physiologically based pharmacokinetic (PBPK) model and extensive libraries pertaining to demography, developmental physiology, and the ontogeny of drug elimination pathways. In addition, Simcyp has developed specialized modules to address specific needs, including the pediatric simulator for pediatric applications. The Simcyp Pediatric module allows users to simulate pharmacokinetic behavior in newborns, infants, and children. This information proves highly beneficial for making dosing decisions, conducting drug–drug interaction (DDI) analyses, addressing safety concerns, optimizing medication design and formulations for children, and designing pediatric clinical studies with a reduced need for subjects. Notably, the software features a built-in Nordic Caucasus (NEC) child population model, facilitating straightforward pediatric modeling. It enables the determination and optimization of appropriate dosage selection for children, spanning from neonates to those aged 2 years, 2–6 years, 6–12 years, and adolescents. Moreover, the software can simulate pharmacokinetic changes in any pediatric age group by adjusting relevant physiological parameters. Details on the development of a pediatric oral absorption model for SimCYP were described in a detailed paper by Johnson et al. 2018 [8].

GastroPlus was originally developed using the mechanical absorption model detailed in the Advanced Compartmental and Transit (ACAT) literature to simulate the absorption of oral drug formulations in the gastrointestinal tract. Over time, it has evolved to encompass whole-body ADME (Absorption, Distribution, Metabolism, and Excretion) modeling for simulating various aspects of drug behavior in both humans and animals. For pediatric modeling, GastroPlus provides the GastroPlus PEAR module, which generates organ weight, volume, and blood perfusion rate estimates that are weight-matched, thereby reflecting demographic data related to age-related physiology. In GastroPlus version 9.6, Chinese population data has been integrated, covering both adults and children.

The PK-Sim software includes a database of population parameters that offers valuable information regarding the relationships between anatomic and physiological parameters like organ sizes, blood flow rates, and tissue composition. These relationships are analyzed with respect to age, gender, body weight, and body mass index, and the data is compiled from a comprehensive literature review. The database encompasses age-specific physiological parameters and population-based parameters. When working on pediatric modeling, specific parameters relevant to a particular age, weight, and height are updated based on the information stored in this database. The primary adjustments typically involve modifying organ volumes, data concerning drug distribution, and applying scaling factors to adjust organ sizes. Notably, the database also includes a subgroup dedicated to "premature babies." The validity of the PKSim model was reported by Yun and Edginton (2019) [9].

Custom PBPK modeling platforms offer a variety of preconfigured population physiology models. Within these platforms, pediatric physiology models encompass adjustments related to age, such as variations in organ blood flows, organ sizes, plasma protein levels, and drug metabolism capabilities, as well as the expression of drug transporters.

20.3.1.2 Verification of Models for Pediatric and Preterm Populations

When it comes to pediatric clinical pharmacokinetic (PK) data, particularly for term and preterm neonates, it is relatively scarce in comparison to data available for adults. Consequently, assessing model performance within the specific pediatric age group and for the desired indication may not always be feasible. Nevertheless, if there is a high level of confidence in the model's performance for children in a different age range or for a different indication, and if the model is sufficiently mechanistically parameterized with a strong assumption of PK similarity among indications (e.g., minimal impact of the new indication's disease on drug absorption and distribution), then the model may still be considered "fit-for-purpose."

20.3.1.3 Regulatory Stance on PBPK Models

As a result of promising reports of using PBPK modeling in pediatrics by several groups [12, 13, 14, 15], the Office of Clinical Pharmacology at the FDA undertook a program of evaluating the effectiveness of PBPK modeling in dose finding for pediatric patients based on submissions made during 2009–2011 [10].

Currently, of all PBPK models used in drug approval applications submitted to the FDA, about 15% serve a pediatric purpose, which is more than seen for other "special populations" [11].

FDA guidance[1] on pediatric formulation development states that, "Modeling and simulation, when appropriate, should be used to inform dose selection and/or trial design." Additional FDA guidance on the content of pediatric study plans states that the "existing of planned modelling and simulation to support dose selection and/or study design, data analysis, and interpretation for planned pediatric studies" should be discussed within the PSP.

The "learn-and-confirm" strategy employed in regulatory submissions with the use of PBPK modeling holds the potential to accelerate pediatric drug development, particularly for the youngest members of the pediatric population. Within this segment of pediatric patients, the existing practice of conducting multiple-dose pharmacokinetic (PK) studies in each age group necessitates a considerable number of pediatric subjects. When considering practicality and ethical considerations, it is essential to minimize the number of pediatric patients involved in clinical trials. A "learn-and-confirm" approach, which involves a reduced number of PK studies across different age groups, followed by reassessment through PBPK modeling, offers a rational means to optimize age-specific pediatric dosing. However, this approach depends on a comprehensive understanding of the desired target concentrations and expected pharmacodynamic responses within each specific age group.

20.3.1.4 Challenges in the Use of Physiologically Based Pharmacokinetic Modeling in Pediatric Drug Development

While physiologically based pharmacokinetic (PBPK) modeling is highly regarded in the context of drug development and regulatory reviews, there are several limitations associated with its theoretical underpinnings and practical application.

To begin with, the development of PBPK models relies heavily on the confidence placed in assumed physiological and pharmacological processes, as well as their mathematical descriptions. The PBPK model's output represents the integrated outcome of a series of simulation processes, including but not limited to protein binding, tissue distribution, human body composition, blood flow rates, and enzymatic reactions within metabolic tissues. Each of these processes is a simplification of the complex realities and, as a result, inherently carries errors when compared to

---

[1] https://www.fda.gov/media/168202/download (accessed March 2024).

real-world situations. These errors may accumulate and magnify in the model's predictions, potentially leading to unreliable outcomes. Presently, major PBPK modeling software packages often employ published or internally validated formulas to construct models, and end-users are typically given the flexibility to adjust confidence levels (such as specifying coefficient of variation values) for simulations. While sensitivity tests can be used to assess the impact of each parameter, there is no universally accepted guideline for selecting suitable ranges of variability for model parameters. End-users either make assumptions about the variability of a physiological parameter or adjust the variability of physiological parameters to better align the PBPK model predictions with clinical data. These practices lack experimental or theoretical validation, a critical issue when developing PBPK models for the pediatric population, whose physiological characteristics are dynamic and highly variable. Therefore, more theoretical and experimental work is needed to validate the variability of relevant physiological parameters in specific populations to enhance the reproducibility of PBPK simulation results.

Currently, the mechanisms underlying age-dependent changes in physiological processes that govern drug ADME are not completely understood. Therefore, the design of initial PK studies in children has relied on PK characteristics in adults and in children in older age groups, using techniques such as allometry. There is a critical need to refine pediatric PBPK models.

Models that depict alternative routes of drug absorption, such as dermal, pulmonary, and ocular routes, have previously been developed for adult populations as well as for rodents and rabbits [12]. Nevertheless, when it comes to children, these models are relatively limited in availability.

The extensive use of PBPK modeling in pediatric drug development and individualized dosing hinges significantly on the quality and accuracy of these models in predicting drug exposure. The variation in claims made by different authors regarding the reliability of model predictions can be attributed to discrepancies in the quality and level of confidence in the estimates of model parameters. Ideally, external datasets are needed to confirm the validity of the scaled parameter in a learning-confirming cycle. As a minimum requirement, biological plausibility of the scaled parameter could be evaluated.

One critical factor that could enhance the development of pediatric PBPK models is addressing the gap in both the quantity and quality of systems data. Given the limited number of clinical studies conducted in children, the advancement of pediatric PBPK modeling will necessitate the collaboration and pooling of data, resources, and expertise. This cooperative effort would involve the amalgamation of tissue samples, plasma, and body fluid data from various institutions and biobanks to encompass the entire age spectrum of pediatric development.

20.4 Conclusions

While both PopPK and PBPK models contribute to pharmacokinetic analysis and drug development, they differ in their approaches, applications, and levels of complexity. PopPK models focus on characterizing population-level variability in pharmacokinetic parameters, while PBPK models simulate drug behavior based on mechanistic principles and physiological processes. Each approach offers unique advantages and is selected based on the specific research or development needs.

Population pharmacokinetics is a powerful tool in supporting the development of safe and effective medicines for pediatric populations. By accounting for inter-individual variability and age-related changes in drug pharmacokinetics, PopPK enables personalized dosing regimens tailored to pediatric patients. Despite challenges, ongoing research efforts and collaborative initiatives hold promise for advancing population pharmacokinetics and improving pediatric medicine development in the future.

PBPK pharmacokinetic modeling is invaluable for pediatric drug product design as it enables the optimization of dosing regimens, formulation development, prediction of age-related pharmacokinetic changes, and support for regulatory submissions. By integrating physiological data, disease-specific parameters, and pharmacokinetic principles, PK modeling facilitates the development of safe, effective, and age-appropriate therapies for pediatric patients, ultimately improving their health outcomes and quality of life.

References

1. Bi Y, et al. A whole-body physiologically based pharmacokinetic model of gefitinib in mice and scale-up to humans. AAPS J. 2016;18:228–38.
2. Matsumoto Y, et al. Application of physiologically based pharmacokinetic modeling to predict pharmacokinetics in healthy Japanese subjects. Clinical Pharmacology & Therapeutics. 2019;105(4):1018–30.
3. Rasool MF, Khalil F, Läer S. A physiologically based pharmacokinetic drug–disease model to predict carvedilol exposure in adult and Paediatric heart failure patients by incorporating pathophysiological changes in hepatic and renal blood flows. Clin Pharmacokinet. 2015;54:943–62.
4. Parrott N, et al. Development of a physiologically based model for oseltamivir and simulation of pharmacokinetics in neonates and infants. Clin Pharmacokinet. 2011;50:613–23.
5. Maharaj A, Edginton A. Physiologically based pharmacokinetic modeling and simulation in pediatric drug development. CPT Pharmacometrics Syst Pharmacol. 2014;3(11):1–13.
6. Maharaj A, Barrett J, Edginton A. A workflow example of PBPK modeling to support pediatric research and development: case study with lorazepam. AAPS J. 2013;15:455–64.
7. Verscheijden LFM, et al. Physiologically-based pharmacokinetic models for children: starting to reach maturation? Pharmacol Ther. 2020;211:107541.
8. Johnson TN, et al. Development and applications of a physiologically-based model of paediatric oral drug absorption. Eur J Pharm Sci. 2018;115:57–67.
9. Yun YE, Edginton AN. Model qualification of the PK-Sim® pediatric module for pediatric exposure assessment of CYP450 metabolized compounds. J Toxic Environ Health A. 2019;82(14):789–814.

10. Leong R, et al. Regulatory experience with physiologically based pharmacokinetic modeling for pediatric drug trials. Clinical Pharmacology & Therapeutics. 2012;91(5):926–31.
11. Grimstein M, et al. Physiologically based pharmacokinetic modeling in regulatory science: an update from the U.S. Food and Drug Administration's Office of Clinical Pharmacology. J Pharm Sci. 2019;108(1):21–5.
12. Le Merdy M, et al. Application of mechanistic ocular absorption modeling and simulation to understand the impact of formulation properties on ophthalmic bioavailability in rabbits: a case study using dexamethasone suspension. AAPS J. 2019;21:1–11.
13. Johnson TN, Small BG, Rowland Yeo K. Increasing application of pediatric physiologically based pharmacokinetic models across academic and industry organizations. CPT Pharmacometrics Syst Pharmacol. 2022 Mar;11(3):373–383. https://doi.org/10.1002/psp4.12764. Epub 2022 Feb 17. PMID: 35174656; PMCID: PMC8923731.
14. Tanaka R, Irie K, & Mizuno T. Physiologically Based Pharmacokinetic Modeling of Antibiotics in Children: Perspectives on Model-Informed Precision Dosing. Antibiotics, 2025;14(6):541. https://doi.org/10.3390/antibiotics14060541.
15. Freriksen JJM, van der Heijden JEM, de Hoop-Sommen MA et al. Physiologically Based Pharmacokinetic (PBPK) Model-Informed Dosing Guidelines for Pediatric Clinical Care: A Pragmatic Approach for a Special Population. Pediatr Drugs 2023;25:5–11. https://doi.org/10.1007/s40272-022-00535-w.

Chapter 21
Paediatric Pharmaceutical Legislation and Its Impact on Adult and Paediatric Drug Development: The EU Regulatory View

Siri Wang and Karl-Heinz Huemer

Abstract The Paediatric Regulation in the European Union sets up a framework governing mandatory requirements and incentives for industry in the development of new medicinal products as well as a series of accompanying measures to facilitate approaches to gain access and improve the exchange of relevant information on medicinal products in paediatric use. This chapter first presents an overview of the evolvement of the regulatory framework in the European Union and the requirements that have to be met by industry. It describes the role of the European Medicines Agency (EMA) and the main points to consider in discussing Paediatric Investigation Plans (PIP). The second part focuses on the specific implications of this regulatory framework on the development of drug formulations for paediatric populations. Essential elements to ensure suitably adapted formulations are reviewed, focusing on dosing appropriateness, acceptability and safety for all relevant age groups.

21.1 Introduction and Historical Background

The use of unlicensed and off-label medicines in children, and the general lack of information and appropriate pharmaceutical formulations to support the administration of many medicines in children, has been a concern for decades. Following similar initiatives in the United States that led to the FDAMA (FDA Modernization Act) in 1997, the discussion in Europe to specifically address requirements for drug use in the paediatric population dates back at least to the 1990s, mainly initiated by

S. Wang (✉)
Norwegian Medical Products Agency (NOMA), Oslo, Norway
e-mail: siri.wang@noma.no; siri.wang@dmp.no

K.-H. Huemer
KHH Consult, Vienna, Austria
e-mail: karl-heinz@khhuemer.com

H. Batchelor, K. Rose (eds.), *Pediatric Formulations*, AAPS Advances in the Pharmaceutical Sciences Series 58, https://doi.org/10.1007/978-3-031-77240-5_21

paediatric academic and scientific societies. The basic idea that there is a need to establish a legislative framework with regard to paediatric medicines was one of the outcomes of an expert round table in 1997 [1].

The European Commission also initiated discussions on the performance of clinical trials in children, mainly in collaboration with the International Conference on Harmonization (ICH), finally resulting in the ICH guideline: "Note for guidance on clinical investigation of medicinal products in the paediatric population" (ICH Topic E11) which entered into force in 2002, with a revision in 2017 [2].

In parallel, at the EMA the Committee for Proprietary Medicinal Products (CPMP) founded an ad hoc Paediatric Expert Group (PEG). With the implementation of Regulation EC no.726/2004, when CPMP was reorganized to the Committee for Medicinal Products for Human Use (CHMP), the PEG was transformed into one of its temporary working parties comprising experts from different areas and also establishing links to other CHMP working parties and the Committee for Orphan Medicinal Products (COMP). The mandate of the PEG was coordinating activities at the EMA and advising its scientific bodies.

The Council of Health Ministers adopted a resolution addressed to the European Commission in 2000 raising the desire for a legislative proposal on the topic of paediatric medicines with high public health priority. Starting with an extended impact assessment (comprising also economic and social consequences), a first draft was proposed in 2004, and the final version of the regulation was adopted by the European Parliament, published in December 2006 and entered into force on 26 Jan 2007. It comprises Regulation No.1901/2006 and amending Regulation No.1902/2006 [3, 4].

With the introduction of the Paediatric Regulation, quite some additional obligations have been introduced for industry. Hence, it should also be considered what was accomplished by these requirements. Due to the long time between agreeing on a Paediatric Investigation Plan (PIP, see below) and the actual marketing (in many cases several years), it took almost a decade until such impact could be observed. The EU Commission has presented such impact outcomes in a 10-year report [5], which clearly shows the increase in the number of clinical trials in children but also the increase in available paediatric medicines and paediatric indications. The regulation has triggered a higher awareness of the problems to generate adequate information on medicines for use in children.

The EU Commission in the last years initiated a broad discussion to evaluate the paediatric and orphan regulatory framework including public consultations to inform possible amendments [6]. Currently the plan is to integrate the requirements into a broader pharmaceutical regulation [7]. Final agreement and implementation are to be expected in the coming years but will likely not have an essential impact on the content of PIPs.

21.2 The Main Points in the EU Paediatric Regulation

The Paediatric Regulation sets up a framework governing mandatory requirements and incentives for industry in the development of new medicinal products as well as a series of accompanying measures to facilitate approaches to gain access and improve the exchange of relevant information on medicinal products in paediatric use. By such means it aims to stimulate the development and accessibility of medicinal products for use in the paediatric population and to ensure that medicinal products used in children are subject to ethical research of high standards and are appropriately authorized. In addition, the information available on the use of medicinal products in the various paediatric populations should be improved.

It is also stated that this should be achieved without subjecting the paediatric population to unnecessary clinical trials and without delaying the authorization of medicinal products for older age groups; to the latter aim, measures in children can be deferred, which means that studies in children can be initiated and/or completed after the application for marketing authorization in adults is submitted. Clearly, to avoid delays, compliance with the required early submission of development plans is critical (see below).

One of the cornerstones of the Paediatric Regulation was establishing a Paediatric Committee (PDCO) within the EMA, composed of members from each EU/EEA member state as well as patient and healthcare professional representatives. This structure might change with the upcoming pharmaceutical legislation, but the tasks related to the regulation will still remain in the mandate of the European Medicines Agency.

The regulation has put into force the requirement for an agreement on the measures that have to be performed to potentially allow approval of a medicinal product in any subgroups of the paediatric population in need of such a treatment. This Paediatric Investigation Plan (PIP) is adopted as a document—the PIP opinion—by EMA and is legally binding. Opinions on agreed PIPs must contain measures to allow assessment of the quality, safety and efficacy of a medicinal product in all concerned paediatric subsets. They also need to include timelines and measures to adapt the formulation of the medicinal product to make its use more acceptable, easier to use, safer or more effective for relevant subsets of the paediatric population. PIP opinions can include measures on generating data in trials but could also contain e.g. extrapolation or modelling approaches, meta-analyses or any other means of generating relevant evidence.

Implementation of all measures (unless deferred, compare below) is obligatory prior to an application for marketing authorization (MAA) for any new medicinal product. For all agreed measures, a compliance check prior to MAA submission has to be performed by EMA (for non-centralized application routes, in principle also by a National Competent Authority (NCA), but this is mostly also delegated to EMA). A positive outcome is required for submitting a valid MAA. On the other hand, a PIP opinion can also contain a product-specific waiver or a class waiver, limiting or obviating the obligations to conduct certain paediatric studies.

The need for agreement on a PIP is also given in case of already authorized medicinal products which are protected either by a supplementary protection certificate (SPC) or by a patent qualifying for such SPC. For these products a PIP is needed if an MAA is planned for a new indication, a new pharmaceutical form or a new route of administration.

Several products are exempted from the need for a PIP, including those submitted via the route of a generic, homoeopathic, herbal, hybrid or well-established use application.

Waivers can be granted for part or all of the paediatric population. This has to be based on evidence either that a product is ineffective or unsafe, or that a condition for which the product is intended does not occur, or that the specific product would not represent a significant therapeutic benefit over existing treatments, as specified in the Paediatric Regulation.

This clearly implies that lack of such evidence would not be a reason for waiving a development. Such waivers can be very specifically limited to one or more paediatric population subsets and/or conditions. The ground for a waiver has to be in line with the options included in the Paediatric Regulation and is granted as part of the opinion, and this information is also published.

Waivers would normally not be accepted based on claimed inappropriateness of the adult formulation, and the requirements for suitable formulations for all relevant age groups would in principle be required. Any request for waivers based on formulation characteristics must be supported by significant evidence that the formulation challenges could not be overcome and that an age-appropriate formulation cannot be developed.

One of the challenges of this regulation in the EU is the early point in time when such a proposal should be submitted. Unless duly justified, this should be no later than upon availability of the human pharmacokinetic data in adults. This should not be misinterpreted as a need to start the paediatric development so early but should rather ensure that there is sufficient time to integrate the paediatric plan appropriately into the overall development of a product. The actual timing of paediatric trials to be performed would then be agreed by also granting deferrals for the planned measures if a delayed initiation or completion for collecting some data seems appropriate. Such early discussion can, depending on the development planned, for example align end points and biomarkers between adult and paediatric development to facilitate extrapolation approaches or safeguard sufficient time in elaborating on age-appropriate formulation efforts without generating delays.

Deferrals can be agreed for initiation or completion of any measures that are included in a paediatric development plan, if scientifically or technically justifiable on grounds related to public health. In practice, this very often will imply that adult data are available prior to initiating paediatric trials. However, also other reasons could be valid justifications for deferring measures, e.g. longer recruitment time. Linking deferral timelines to regulatory milestones (e.g. after approved marketing authorization in adults) rather than scientific reasons would not be considered an appropriate justification. A recurring issue in this context is that agreement on key elements for later stages of development can often not be sufficiently informed by data available at the time of this early discussion. EMA is increasingly flexible to allow postponing final decision on such key elements to a later modification procedure ("stepwise PIP").

Modification of adopted PIPs is possible at any time, e.g. if new data, scientific, clinical or regulatory aspects become available. However, such modifications should be discussed prior to changing a development plan and not—as unfortunately done rather often—in a post hoc attempt to find agreements on already performed measures.

The Paediatric Regulation also foresees a specific voluntary procedure for products not covered by the obligatory requirements as described above. For already marketed off-patent products, a paediatric indication can be claimed by submitting a dossier including documents establishing quality, safety and efficacy in the paediatric population, including an age-appropriate formulation. Such a Paediatric Use Marketing Authorisation (PUMA) would qualify for an 8-year data protection and a 10-year market protection if performed in compliance with a prior agreed PIP. Of note, while in the obligatory PIPs, all paediatric age ranges must be covered unless waived; for a voluntary PIP related to a PUMA, it is possible to cover only some age subsets.

The number of PUMA-related PIPs granted during the first 10 years was very limited, despite some funding from the FP7 programme. Of the 20 PIPs agreed with a view to submit a PUMA, only a handful succeeded. Although the number of PUMAs has increased during the last years, the challenges seem to remain, partly due to limited revenues foreseen for these products, continued off-label use of existing non-adapted products and the lack of appreciation of the added value of these new formulations by downstream decision makers, e.g. from national price and reimbursement perspectives.

21.3 Implications

What does all this mean in practical terms? A company that has the intention to develop a new drug product or is planning new pharmaceutical forms/routes or new indications for an approved medicinal product still under patent protection will have to consider whether this product can fulfil a paediatric unmet need, including any subsets/age ranges up to 18 years. And it will be required to submit this plan to EMA to discuss the planned development or argumentation for waiving or deferring measures in this paediatric development plan.

The paediatric needs that exist in the planned condition will have to be considered. Such unmet medical needs are determined based on the occurrence of a condition and the lack of or limitations in current therapeutic options.

It should be kept in mind that this regulation was introduced to counteract the fact that industry only rarely proceeded in this paediatric product development direction voluntarily. In practice, drug developers still tend to delay PIP submissions, not having decided yet what this development will look like or still looking for arguments to support a desired waiver. For example, companies are unsure about the appropriate design of the paediatric study (e.g. what endpoint could be feasible or how many children could be recruited). Furthermore, the need to develop a formulation in smaller children might depend on the fact that this drug later will be used in a specific (lower) age range. But again: actually, it would facilitate planning to have at least a cursory overview of possible later requirements if they are discussed as early as possible. Should later development generate evidence that paediatric needs or agreed PIP measures are no longer applicable, there is the option to propose changes via a procedure for modification of an agreed PIP at any time and as often as needed.

The risks of delayed submission have become evident in many instances, most notably delays of the planned marketing authorization date due to not having agreed a PIP in time. This is partly due to insufficient understanding of the implications of the Paediatric Regulation, especially in drug developers from outside the EU. Often it also relates to misunderstandings of how paediatric needs are related to a planned condition, which often is artificially narrowed to focus on the population with the highest marketing potential. Such narrowing would not be supported by the Paediatric Regulation, as very often there are differences in conditions in the paediatric population as compared to adults. Hence, this would compromise the rationale behind the regulation, whose main aim is to increase knowledge and availability of drugs in children based on generated evidence, even if conditions are presenting differently from the adult form.

Other difficulties often encountered in late submissions are that data already generated (outside an agreed PIP) are insufficient or that opportunities to cover some points are missed. This can result in seemingly redundant requests, which would be against the intention to prevent unnecessary trials. However, a trial with design flaws (e.g. non-valid endpoint and sample size) cannot be considered as sufficient evidence that would justify not repeating an additional trial with more appropriate design. Here rather the initial trial might be considered unethical, as being insufficiently planned. Often companies are hesitant to include e.g. adolescents in an adult Phase 2 or 3 development. This could result in a request to include them in a separate trial, which will delay both the adolescent data generation and the proceeding to younger age groups. Therefore, these decisions should be discussed well in advance to prevent such situations.

It should also be kept in mind that extrapolation of generated data can be used in some cases to supplement the paediatric development. This can, dependent on the medical setting, affect any data from preclinical, efficacy and safety to dosing and pharmacokinetics. It should also be considered that often in paediatric settings, fully powered comparative trials might not be feasible. But there are many innovative options for other approaches, which in such cases should be proposed and discussed to find a satisfactory agreement. ICH has published a specific guideline that focuses on paediatric extrapolation. ICH E11A [8], which gives a detailed overview of how to include extrapolation concept and methodological implementations in a paediatric development to support safety, efficacy and dose finding. Also, the EMA CHMP has issued more general guidance (not limited to paediatric) on the use of extrapolation [9]. In addition, ICH recently also has published a draft for principles for model-informed drug development (ICH M15) [10] that can support how such key elements can be proposed in a PIP opinion.

In conclusion, early involvement of the regulatory bodies will facilitate further planning. Importantly, in case newly generated evidence would indicate the need for changes in such a programme, this is not restricted by earlier PIP agreements.

21.4 Paediatric Formulations

The Paediatric Regulation highlights in its preamble the problems resulting from the absence of suitably adapted medicinal products and specifically mentions the non-availability of suitable formulations and routes of administration, as well as the use of magistral or officinal formulations of potentially poor quality. These challenges are best known by the paediatric patients themselves, their parents, and the healthcare professionals in their daily struggle to adapt and modify the existing medicinal products in an attempt to benefit from the therapeutic advantage of the product. Crushing, splitting, diluting or dissolving may significantly affect factors like dosing accuracy, PK profile or acceptability, and medication errors may occur when doses are prepared or calculated.

Consequently, the regulation clearly states that the PIP must include a thorough description of any measures to adapt the formulation of the medicinal product so as to make its use more acceptable, easier, safer or more effective for different subsets of the paediatric population. Guidelines have been developed to aid companies in their development strategy related to formulations targeted for children [11].

There are basically three major factors that could decide whether a formulation is suitable for the target age group: the formulation must ensure the right dose, it should be acceptable for the patient, and it must be safe. These factors should be considered and fulfilled for all relevant age groups for which the product is developed. It is therefore expected that the PIP application includes a thorough discussion on the proposed formulations and their suitability for the target age groups, covering aspects of dosing accuracy, acceptability and safety.

At the early phase of product development, it is most often not clear whether the *dose* will be critical in terms of dose-response or whether the need for dose titration is foreseen for other reasons. Such aspects could most likely also influence the choice of formulation. It should be emphasized that for proper dose finding in the paediatric population a certain degree of flexibility in dosing is normally needed, and wide dose banding (often triggered by existing adult formulations that to a limited degree would allow dosing flexibility) could compromise the results of the paediatric clinical trials. Therefore, depending on the lower age cut-off for the development plan and/or the properties of the active substance, a preliminary formulation for use in clinical trials that allows dosing flexibility might be needed. Obviously, in addition to dosing accuracy, both acceptability and safety of such formulations should also be carefully considered. The modification of the preliminary formulation into the product intended to be marketed might necessitate bridging studies depending on the active substance and the formulations proposed.

From a clinical perspective, it is important that the pharmaceutical forms that are being developed have a certain degree of robustness in terms of practical handling and ease of administration to make them useful and safe for both in-hospital and homecare settings, if applicable. This is particularly relevant for medicines for children, since several different caregivers are often involved in addition to the child itself. Dosing device, presentation and proper instructions are vital factors to increase adherence and to reduce the risk of medication errors and should be adapted to the target patient group.

Acceptability implies that a formulation can be easily administered to the relevant age groups and is crucial for optimal adherence and intended effect (for additional details please see Chap. 4). It covers a range of aspects like taste, size, volume, complexity of manipulation, local tolerance and pain. One of the major hurdles for children taking oral medicines is the tablet and capsule size, large sizes being maybe appropriate for adults but definitely less suitable for children. Interestingly, information about size of existing products is often lacking in the initial PIP applications, which could indicate insufficient awareness of this aspect. Size is particularly important if the tablet or capsule is developed to be swallowed whole, due to, e.g., film coating or modified release design. For such products, crushing and scoring may significantly affect taste, bioavailability, or PK, and therefore size is crucial for overall usefulness of the product. Generally, increased attention towards potentially more child-friendly oral solid formulations like 'mini-tablets' and dispersible and/or dispersible tablets is highly needed and welcomed, but is to date not a frequent approach seen in the PIP applications.

Bad taste of oral liquids is a well-known factor that affects medication adherence in children and can be caused by both the active substance and any of the excipients. Companies are encouraged to explore avenues to avoid poor patient acceptability by optimizing taste masking. This could include standard approaches like adding sweeteners or flavours, or more sophisticated methods like microencapsulation. Careful consideration is needed to balance any taste issue with the strength and volume required.

Although significant clinical experience indicates that many children face major challenges in swallowing tablets and oral liquids, more precise knowledge is still limited about taste preferences and which tablet/capsule sizes are appropriate for which age groups. Consequently, to ensure that children in the relevant age group are actually able to take the medicine as intended, data to support and confirm acceptability and palatability of the product is often requested in the PIPs, e.g. as a part of the paediatric clinical studies that will be performed. However, it should be emphasized that evaluation of the patient acceptability of a paediatric preparation should be an *integral* part of the pharmaceutical and clinical development, and *early* focus on the appropriateness of the formulation is important. This would not only ensure optimal product development but also reduce the risk that poorly acceptable formulations might affect the outcome of the paediatric trials .

The importance of acceptability has been flagged in relevant guidance documents [2, 11]; however, no specific recommendations have been given on exactly how to assess acceptability during product development. Supportive data can be gathered from literature data, from adult taste panels, or from non-human sensory studies (e.g. 'electronic tongue', rodent BATA models), but such data may not always be fully relevant, and confirmatory acceptability data in the target population is still important. The choice of method to collect such data is currently left to the applicant. A range of approaches could be taken: From thoroughly assessing compliance during the clinical trial, specifically focusing on challenges related to the formulation, via questionnaires focusing on e.g. preferences or overall acceptability, to more sophisticated methodology covering the full range of multifaceted factors of acceptability (patient factors, product factors and disease factors). The overall goal should be kept in mind: To ensure that treatment with this medicine is actually feasible in the target patient population.

In all cases where a more suitable formulation is not feasible and there is doubt about whether the formulation (and in particular its size) would be appropriate for the patient age group, it is important that alternative approaches are explored and outlined. Recommendations regarding opening capsules, dissolving, dispersing or crushing tablets, or mixing with food can be very useful for patients, parents or healthcare professionals but would require sufficient consideration of the potential impact on the performance of the drug.

The *safety* of the formulation alludes to aspects like the risk of medication errors, appropriate composition and choice of excipients. The strength of any liquid form will decide the volume to be withdrawn from a container and given to the patient. Dosing small children using strengths suitable for adults will often imply too small volumes to accurately administer the dose to the child. Dilution steps to solve the problem with small volumes have been shown to significantly increase the risk of calculation and administration errors and should be avoided if possible. Again, suitable devices to enable accurate dosing are essential. The risk of accidental overdose in the neonatal setting is mentioned below.

Osmolality of the administered drug should be appropriate for the target age group. Depending on the route of administration high-osmolality drugs will have to be diluted to reduce the risk of, e.g., pain, irritation, necrosis and necrotizing enterocolitis, elements that can be age-dependent. However, whenever dilution is necessary, careful consideration must be put on its impact on the stability of the product, on proper instructions for dilution and on the volume load.

Although excipients are traditionally thought to be inert and safe, several cases have shown that this is not always true, particularly for very young children, where a continuously developing organism may give rise to different excipient safety profiles compared to adults. Significant discussions have been ongoing for solvents like ethanol and propylene glycol, preservatives like thiomersal, benzyl alcohol and parabens, and solubilizers and colourants. The regulation's call for adapted formulations implies that products intended for use in children must have an acceptable safety profile also in terms of excipients. The PIP should therefore have a thorough presentation of the safety data available for the proposed excipients, justifying the excipient exposure based on the foreseen doses and taking into account the target age group, route of administration and duration of treatment. There is, however, sparse data available specifically relevant to children and in particular to neonates, and it is generally recommended to avoid excipients with potential safety concerns when developing paediatric formulations, if possible. Additional safety data, e.g., through juvenile animal studies or additional safety monitoring in clinical trials might be requested by regulators whenever the excipient safety profile is not fully reassuring.

21.4.1 Specific Considerations on Formulations for Neonates

Neonates, and even more so preterm newborns, are the patient groups where medicinal products are most often used off-label [12]. The need for PK, efficacy and safety data in this population frequently implies separate clinical trials with careful considerations of sampling scheme, appropriate endpoints and disease characteristics. Particular attention should, however, also be put on whether the formulation is suitable for this patient group, ensuring accurate dosing and safe administration.

In many cases, intravenous administration may be the only feasible route of administration to neonates. Appropriate strength will be vital to ensure sufficient dosing accuracy, often depending on suitable administration and dosing devices such as pumps, but also to consider the limitation in volumes acceptable for neonates with fluid restrictions. In clinical practice, data on compatibility with other commonly administered parenteral drugs will often be needed for treatment to be feasible within the available time and volumes. Therefore, where relevant, such data should be included in the development plan. Ten times dosing error is more often seen in neonatal units due to the fact that the doses are very small compared to the total dose in the vial or bag [13]. Consequently, a separate presentation would often be considered needed.

Intravenous administration may not always be optimal, possible or physicochemically feasible, e.g. due to solubility or stability issues; therefore, in some settings, oral administration is considered needed and appropriate also for neonates. In such cases administration through feeding tubes may be necessary and sometimes the only possible way to administer the drug product. Consequently, factors like adherence to tube material, particle sizes, viscosity and rinsing volumes are essential for safe and accurate dosing, and the PIP would have to include data on dose recovery and feasibility of administration through the relevant tubes. Indeed, such elements may also be relevant for older children (and even adults) whenever tube feeding is likely in the target patient group [14].

In summary, any adult presentation will rarely be entirely suitable for smaller children and especially neonates, and a specific formulation or presentation will often be needed to ensure correct dosing and safe use in these lowest age groups.

21.5 Collaboration on Paediatric Formulations

Other European and worldwide initiatives have been taken during this period, also related to formulations. WHO's campaign 'Make medicines child size' [15] and the European Paediatric Formulation Initiative (EuPFI) [16] are two important examples. EMA and national EU experts have regular contact with FDA and with WHO when relevant, and dialogue is established with EuPFI, where EMA has observer status.

As part of its focus on paediatric formulations, EMA has established an expert group (Paediatric Formulation Operational Expert Group, PFOEG). Attention is put on combined quality, safety and clinical aspects of formulations. The group consists of national quality experts, clinical pharmacy experts, clinicians and academic experts and collaborates with EuPFI and FDA. A systematic approach to PIP quality aspects aims at a broader and more consistent 'cross-product assessment'.

21.6 Achievements of the EU Paediatric Regulation with Regard to Formulations for Children

The focus that has been put on the need for appropriate formulations through the paediatric legislative initiatives has resulted in increased awareness among all stakeholders.

Importantly, a total of 43 *new pharmaceutical forms, strengths or routes* of paediatric relevance were listed in the Annex to the 10-year report (centrally and non-centrally authorised, as line extensions to existing marketing authorisations) [17]. Some of these were very specific lower-age paediatric-targeted formulations/strengths; others were relevant for all age groups. Of note, the approval and availability of additional, paediatric-specific, formulations will most often come at later stages of development, and adequately measuring these achievements therefore probably also requires prolonged time.

The actual availability of these new formulations and strengths throughout Europe is an issue of repeated discussion, in particular for smaller countries [18]. The decision to launch a formulation in the specific countries in Europe rests in the hands of the individual company, and regulatory requirements (e.g. national language packages) and expected limited return on investments could most likely contribute to this 'real-life unavailability' scenario. Although many of these factors concern downstream decision-making at national level and, as such, is outside the scope of the EU law, discussion is needed on how to ensure access to these new formulations and strengths when actually developed and authorised.

21.7 Conclusions

The regulatory paediatric incentives in Europe aim at better medicines, including more suitable formulations, for children in Europe. The focus on medicines properly adapted for the paediatric population will continue.

The importance of early submission of the PIP is valid also for formulation aspects. The final agreed age groups for which the product is intended will inevitably impact the decision on whether the formulation strategy is optimal, and early agreement on the lower age cut-off will be essential for a rational formulation development.

Companies are indeed encouraged to consider new technology and innovative approaches to meet the need for paediatric-specific formulations. In this context it is also important to remind drug developers that the need for flexible dosage forms, both in terms of dosing adjustment and flexibility in mode of administration, is significant also in other patient populations, for example geriatric patients, patients with feeding tubes and intensive care patients. Some of the apparently paediatric-specific factors will be valid also for these settings, and such dosage forms could thereby meet the needs in a bigger market. Therefore, the focus on paediatric formulation development should be an early and integrated part of the overall drug development programme.

References

1. EMA: The European paediatric initiative: History of the Paediatric Regulation (EMEA/17967/04 Rev 1). 2007. https://www.ema.europa.eu/en/documents/other/european-paediatric-initiative-history-paediatric-regulation_en.pdf.
2. ICH: Clinical investigation of medicinal products in the pediatric population E11 (R1). https://database.ich.org/sites/default/files/E11_R1_Addendum.pdf.
3. Regulation (EC) No 1901/2006 on medicinal products for paediatric use. https://eur-lex.europa.eu/legal-content/EN/TXT/PDF/?uri=CELEX:32006R1901.
4. Regulation (EC) No 1902/2006 amending Regulation 1901/2006 on medicinal products for paediatric use. https://eur-lex.europa.eu/legal-content/EN/TXT/?uri=CELEX%3A32006R1902&qid=1621344480633.
5. EC: State of Paediatric Medicines in the EU - 10 years of the EU Paediatric Regulation (COM (2017) 626). https://health.ec.europa.eu/system/files/2017-11/2017_childrensmedicines_report_en_0.pdf.
6. EC: Evaluation of the medicines for rare diseases and children legislation. 2020. https://health.ec.europa.eu/medicinal-products/medicines-children/evaluation-medicines-rare-diseases-and-children-legislation_en.
7. EC: Reform of the EU pharmaceutical legislation. 2023. https://health.ec.europa.eu/medicinal-products/pharmaceutical-strategy-europe/reform-eu-pharmaceutical-legislation_en.
8. ICH/EMA: ICH E11A Guideline on pediatric extrapolation. EMA/CHMP/ICH/205218/2022. 2024. https://www.ema.europa.eu/en/documents/scientific-guideline/ich-guideline-e11a-pediatric-extrapolation-step-5_en.pdf.
9. EMA: Structured guidance on the use of extrapolation. 2022. https://www.ema.europa.eu/en/documents/scientific-guideline/structured-guidance-use-extrapolation_en.pdf.

10. ICH: General principles for model-informed drug development M15. 2024. https://database.ich.org/sites/default/files/ICH_M15_EWG_Step2_DraftGuideline_2024_1031.pdf.
11. EMA: Pharmaceutical development of medicines for paediatric use. 2013. https://www.ema.europa.eu/en/pharmaceutical-development-medicines-paediatric-use-scientific-guideline.
12. EMA: Report on the survey of all paediatric uses of medicinal products in Europe (EMA/794083/2009). 2010. https://www.ema.europa.eu/en/documents/report/report-survey-all-paediatric-uses-medicinal-products-europe_en.pdf.
13. Simpson JH, Lynch R, Grant J, Alroomi L. Reducing medication errors in the neonatal intensive care unit. Arch Dis Child Fetal Neonatal Ed. 2004;89(6):F480–2.
14. EMA: Administration of oral immediate release medicinal products through enteral feeding tubes. 2018. https://www.ema.europa.eu/en/human-regulatory/research-development/scientific-guidelines/qa-quality/quality-medicines-questions-answers-part-2#administration-of-oral-immediate-release-medicinal-products-through-enteral-feeding-tubes-new-december-2018-sec.
15. WHO: https://www.who.int/news/item/05-12-2007-efforts-on-child-appropriate-medicines-intensify.
16. European Paediatric Formulation Initiative (EuPFI). http://www.eupfi.org/.
17. EC: 10-year Report to the European Commission (EMA/231225/2015). 2017. General report on the experience acquired as a result of the application of the Paediatric Regulation https://health.ec.europa.eu/system/files/2020-06/paediatrics_10_years_ema_technical_report_0.pdf.
18. Lepola P, Wang S, Tötterman AM, et al. Does the EU's Paediatric Regulation work for new medicines for children in Denmark, Finland, Norway and Sweden? A cross-sectional study. BMJ Paediatr Open. 2020;4:e000880. https://doi.org/10.1136/bmjpo-2020-000880.

Chapter 22
Global Pediatric Legislation for Pediatric Medicines in LMIC Regions

Hannah Batchelor

Abstract Access to safe and effective pediatric medicines is crucial but remains challenging, especially in low- and middle-income countries (LMICs). Barriers such as inadequate regulatory frameworks, market disincentives, and limited investment hinder access to essential pediatric medications. Global pediatric legislation aims to address these challenges by promoting drug development tailored for children, enhancing regulatory oversight, and improving access to medicines.

Key components of such legislation include incentives for pharmaceutical companies, such as market exclusivity extensions and priority review vouchers, to encourage pediatric formulations and clinical trials. Despite these efforts, LMICs face hurdles in implementation due to financial and infrastructural limitations. Strengthening regulatory capacity and ensuring ethical research practices are vital.

Global initiatives, including those by the WHO and the Gates Foundation, play significant roles in supporting pediatric medicine development and addressing the health needs of children in LMICs. Collaboration among stakeholders is essential to overcome disparities and improve public health outcomes, ensuring better access to medicines for children worldwide.

Keywords Pediatric medicines · Regulatory frameworks · Low- and middle-income countries (LMICs) · Global health · Access to essential medicines

22.1 Introduction

Access to safe and effective pediatric medicines is a global health priority, yet it remains a significant challenge, particularly in low- and middle-income countries (LMICs). Children in these regions often face limited access to essential

H. Batchelor (✉)
Strathclyde Institute of Pharmacy and Biomedical Sciences, Glasgow, Scotland, UK
e-mail: Hannah.batchelor@strath.ac.uk

H. Batchelor, K. Rose (eds.), *Pediatric Formulations*, AAPS Advances in the Pharmaceutical Sciences Series 58, https://doi.org/10.1007/978-3-031-77240-5_22

medications due to various factors, including regulatory barriers, market forces, and insufficient investment in pediatric drug development. To address these disparities, global efforts have been made to establish legislation and regulatory frameworks specifically focused on pediatric medicines. This chapter explores the importance of global pediatric legislation in LMICs, its key components, challenges, and the impact on pediatric medicine access and public health outcomes.

22.2 The Need for Global Pediatric Legislation in LMICs

Children in LMICs bear a disproportionate burden of disease, including infectious diseases, malnutrition, and non-communicable diseases (NCDs). Despite this, the availability of pediatric formulations and dosage forms tailored to their needs remains limited. Several factors contribute to this disparity:

Regulatory Challenges: LMICs often lack robust regulatory frameworks for pediatric medicines, leading to delays in drug approvals and limited availability of essential pediatric formulations. Regulatory agencies may face capacity constraints, inadequate resources, and limited expertise in evaluating pediatric drug safety and efficacy.

Market Forces: Pharmaceutical companies may prioritize investments in drugs targeting adult populations, as they perceive pediatric markets as less profitable due to lower purchasing power and smaller patient populations. Consequently, there is a limited incentive for companies to conduct pediatric drug development or produce pediatric formulations suitable for LMICs.

Ethical Considerations: Conducting clinical trials in pediatric populations raises ethical concerns regarding informed consent, vulnerability, and the risk-benefit balance. These concerns may further deter investment in pediatric drug development, particularly in resource-constrained settings where regulatory oversight and research infrastructure are limited.

Health System Challenges: Weak health systems, inadequate healthcare infrastructure, and limited access to essential medicines exacerbate the challenges of pediatric medicine access in LMICs. This results in disparities in healthcare delivery and contributes to poor health outcomes among children.

In response to these challenges, global pediatric legislation has emerged as a critical tool to promote pediatric drug development, ensure regulatory oversight, and improve access to essential pediatric medicines in LMICs.

22.3 Key Components of Global Pediatric Legislation

Global pediatric legislation encompasses a range of policies, regulations, and initiatives aimed at promoting pediatric drug development, enhancing regulatory oversight, and facilitating access to pediatric medicines.

Many countries have implemented legislative incentives to encourage pharmaceutical companies to conduct pediatric clinical trials and invest in pediatric drug development. Legislation may include provisions to encourage the development of age-appropriate pediatric formulations, such as liquids, suspensions, or dispersible tablets. These formulations address the unique needs of pediatric patients and improve medication adherence and dosing accuracy. Legislative incentives may include extended market exclusivity, patent extensions, or priority review vouchers for pediatric indications. Regulatory agencies, such as the US Food and Drug Administration (FDA) and the European Medicines Agency (EMA), require pharmaceutical companies to submit PIPs as part of the drug approval process. PIPs outline the pediatric studies necessary to support the approval of new drugs or the extension of existing indications to pediatric populations.

Global initiatives, such as the International Council for Harmonisation of Technical Requirements for Pharmaceuticals for Human Use (ICH), promote regulatory harmonization and convergence of standards for pediatric drug development. Harmonized guidelines streamline regulatory processes and facilitate global drug registration, thereby improving access to pediatric medicines. Achieving regulatory harmonization across countries and regions presents challenges due to differences in regulatory frameworks, legal systems, and cultural norms. However, harmonizing pediatric requirements and standards requires collaborative efforts, political commitment, and alignment with global initiatives to address disparities in pediatric medicine access.

22.4 Challenges in Implementing Pediatric Legislation in LMICs

Despite the importance of global pediatric legislation, several challenges hinder its effective implementation in LMICs. For example, LMICs may lack the financial resources, technical expertise, and infrastructure needed to develop and enforce pediatric legislation effectively. Limited funding for regulatory agencies, insufficient staffing, and inadequate training programs impede regulatory capacity to oversee pediatric drug development and ensure compliance with pediatric requirements.

Strengthening regulatory capacity and technical expertise is essential for LMICs to implement pediatric legislation successfully. Capacity-building initiatives, training programs, and knowledge-sharing platforms are needed to enhance regulatory agencies' capabilities in evaluating pediatric medicines and conducting oversight activities.

While legislation may incentivize pharmaceutical companies to develop pediatric formulations, challenges remain in ensuring the availability and affordability of these formulations in LMICs. Supply chain issues, pricing strategies, and market dynamics can affect pediatric medicine access, particularly in resource-constrained settings.

LMICs may face unique ethical challenges in conducting pediatric clinical trials, including issues related to informed consent, vulnerability, and community engagement. Ethical review processes, safeguards for child participants, and adherence to international ethical standards are critical to ensuring the ethical conduct of pediatric research in LMICs.

22.5 Impact of Global Pediatric Legislation on Pediatric Medicine Access in LMICs

Despite these challenges, global pediatric legislation has the potential to significantly improve pediatric medicine access and public health outcomes in LMICs. By providing incentives and regulatory requirements for pediatric studies, legislation stimulates investment in pediatric drug development and expands the availability of pediatric medicines tailored to LMICs' needs. This contributes to the development of new treatments for childhood diseases and reduces reliance on off-label use or adult formulations in pediatric populations.

Legislation that encourages pediatric formulation development and post-marketing surveillance enhances access to essential pediatric medicines in LMICs. Age-appropriate formulations, improved dosing accuracy, and comprehensive pharmacovigilance systems contribute to better health outcomes for children and reduce the burden of childhood diseases in LMICs.

Global pediatric legislation fosters collaboration among regulatory agencies, pharmaceutical companies, healthcare providers, and civil society organizations to address pediatric medicine access challenges in LMICs. Knowledge-sharing, capacity-building initiatives, and public-private partnerships support sustainable solutions and strengthen health systems' resilience.

22.6 WHO Activities on Pediatric Medicine Development for LMIC

The World Health Organization (WHO) plays a central role in promoting pediatric medicine development and improving access to essential medicines for children in low- and middle-income countries (LMICs). WHO maintains the Model List of Essential Medicines (EML), which includes medicines deemed essential for addressing priority health needs globally. The EML serves as a guide for countries in selecting and prioritizing medicines for inclusion in their national essential medicines lists (NEMLs), ensuring access to essential pediatric medicines in LMICs. WHO promotes research and development (R&D) for pediatric medicines, particularly for diseases that disproportionately affect children in LMICs. It supports initiatives to develop new treatments, diagnostics, and vaccines for pediatric conditions, including neglected tropical diseases, malaria, tuberculosis, and HIV/AIDS. WHO collaborates with stakeholders to prioritize R&D investments and accelerate the availability of pediatric medicines.

In addition, WHO develops evidence-based guidelines and technical standards for the use of pediatric medicines in LMICs. These guidelines cover various aspects of pediatric healthcare, including drug selection, dosage, administration, safety monitoring, and treatment protocols. They provide recommendations for healthcare providers and policymakers to ensure the rational use of medicines and optimize pediatric healthcare delivery.

WHO supports efforts to strengthen regulatory capacity and quality assurance systems for pediatric medicines in LMICs. It provides technical assistance to regulatory authorities in assessing the quality, safety, and efficacy of medicines, including pediatric formulations. WHO prequalification programs help ensure that essential medicines, including pediatric formulations, meet international quality standards and are suitable for procurement by international agencies and donors.

Furthermore, WHO leads global efforts to expand access to vaccines for preventable childhood diseases in LMICs. Through initiatives like the Expanded Programme on Immunization (EPI) and the Global Vaccine Action Plan (GVAP), WHO works with countries to increase vaccine coverage, improve immunization systems, and introduce new vaccines. These efforts contribute to reducing childhood morbidity and mortality from vaccine-preventable diseases.

In terms of capacity building, WHO provides technical assistance, training, and capacity-building support to healthcare providers, regulatory authorities, and other stakeholders involved in pediatric medicine development and delivery in LMICs. It offers training programs, workshops, and online resources on pediatric pharmacology, clinical management, rational drug use, and pharmacovigilance to strengthen healthcare systems and improve pediatric healthcare practices.

WHO collaborates with a wide range of partners, including UN agencies, governments, academia, pharmaceutical companies, and civil society organizations, to advance pediatric medicine development and improve child health outcomes in LMICs. Through partnerships, WHO leverages resources, expertise, and innovation to address the multifaceted challenges of pediatric healthcare and ensure access to quality medicines for children worldwide.

22.7 Gates Foundation Activities on Pediatric Medicine Development for LMIC

The Bill and Melinda Gates Foundation, established by Bill Gates and Melinda French Gates, is one of the world's largest private philanthropic organizations. It has been actively involved in improving healthcare in low- and middle-income countries (LMICs), including efforts related to pediatric medicines.

The Gates Foundation provides substantial funding for research and development (R&D) efforts aimed at developing new pediatric medicines and vaccines. It supports innovative approaches to address diseases that disproportionately affect children in LMICs, such as malaria, pneumonia, diarrheal diseases, and neglected tropical diseases. The Gates Foundation collaborates with governments, non-profit organizations, pharmaceutical companies, and research institutions to accelerate the development and delivery of pediatric medicines in LMICs. It supports global health partnerships, such as the Global Alliance for Vaccines and Immunization (GAVI), the Global Fund to Fight AIDS, Tuberculosis and Malaria, and the Coalition for Epidemic Preparedness Innovations (CEPI), which work to improve access to essential medicines and vaccines for children.

The Gates Foundation invests in drug discovery and innovation initiatives focused on pediatric medicines. It supports research efforts to identify new drug targets, develop novel therapeutics, and repurpose existing drugs for pediatric indications. These efforts aim to address the specific needs of children in LMICs, including the development of age-appropriate formulations and dosage forms.

22.8 UNICEF Activities on Pediatric Medicine Development for LMIC

UNICEF (United Nations Children's Fund) plays a significant role in advocating for pediatric medicine legislation and supporting initiatives aimed at improving children's access to essential medicines worldwide. While UNICEF itself does not enact legislation, it works closely with governments, international organizations, civil society, and other stakeholders to advance policies and programs that prioritize children's health needs. UNICEF provides technical assistance to countries in drafting, revising, and implementing legislation related to pediatric medicines. This assistance includes supporting the development of regulatory guidelines, standards, and procedures for the registration, quality assurance, and pharmacovigilance of pediatric drugs.

UNICEF works to improve children's access to essential medicines, including vaccines, antibiotics, antimalarials, and treatments for HIV/AIDS and other pediatric diseases. It collaborates with partners to procure and distribute quality-assured pediatric formulations, promote rational medicine use, and strengthen supply chains to ensure the availability of medicines in underserved communities.

In humanitarian crises and emergencies, UNICEF supports rapid access to essential medicines and medical supplies for children affected by conflict, natural disasters, and epidemics. It works to ensure the timely delivery of pediatric formulations, including therapeutic foods, oral rehydration salts, and medicines for acute and chronic conditions, to affected populations.

22.9 Additional Agencies That Advocate for Pediatric Medicines in LMIC

22.9.1 Coalition for Epidemic Preparedness Innovations (CEPI)

CEPI is a global partnership that aims to accelerate the development of vaccines against emerging infectious diseases, including those that disproportionately affect children in LMICs. By investing in vaccine research and development, CEPI contributes to improving pandemic preparedness and ensuring equitable access to vaccines for children worldwide.

22.9.2 Global Alliance for Vaccines and Immunization (GAVI)

GAVI is an international public-private partnership that supports immunization programs in LMICs by providing access to vaccines, strengthening health systems, and mobilizing resources. GAVI focuses on increasing vaccine coverage and reducing childhood mortality from vaccine-preventable diseases through innovative financing mechanisms and strategic partnerships.

22.9.3 Médecins Sans Frontières (MSF)/Doctors Without Borders

MSF is an international humanitarian organization that provides medical care and assistance to populations affected by conflict, disasters, and epidemics in LMICs. MSF advocates for improved access to essential medicines, including pediatric formulations, and supports efforts to address health disparities and humanitarian crises affecting children in LMICs.

22.9.4 Partnership for Maternal, Newborn, and Child Health (PMNCH)

PMNCH is a global alliance of organizations, governments, and civil society groups working to improve maternal, newborn, and child health outcomes worldwide. PMNCH advocates for increased investments in pediatric health, promotes evidence-based interventions, and fosters collaboration among stakeholders to advance child health agendas in LMICs.

22.9.5 Access to Medicine Foundation

The Access to Medicine Foundation assesses pharmaceutical companies' efforts to improve access to medicines, including pediatric formulations, in LMICs. Through its Access to Medicine Index and other initiatives, the foundation evaluates companies' policies, practices, and contributions to pediatric medicine access and advocates for industry-wide improvements in addressing global health challenges.

22.9.6 International Pediatric Association (IPA)

IPA is a global organization representing pediatric healthcare professionals and advocating for child health worldwide. IPA supports pediatric medicine development, research, and training initiatives, promotes evidence-based pediatric healthcare practices, and collaborates with partners to address pediatric health disparities in LMICs.

22.10 Conclusion

Global pediatric legislation plays a critical role in promoting pediatric drug development, ensuring regulatory oversight, and improving access to essential medicines in LMICs. By incentivizing research and development, encouraging formulation innovation, and strengthening regulatory capacity, legislation addresses

Chapter 23
Impact of the US and EU Pediatric Pharmaceutical Legislation

Klaus Rose

Abstract US and EU pediatric legislation was introduced to improve pediatric healthcare. FDA and EMA reports claim to have triggered such improvements, but their reports use semantic coding that equalizes regulatory parameters with clinical improvement, suggesting that more pediatric studies and labels automatically translate into better child healthcare. For example, the EMA's claim to make medicines "available" for children. "Available" means only that now they have a pediatric label. The real impact of pediatric legislation included triggering many studies in minors; "pediatric" careers in research, regulatory authorities, and companies; and publications of "obvious" results such as that antibiotics work also before the 18th birthday. Positive impacts of pediatric legislation include that it triggered technological advances in physiologically based pharmacokinetics (PBPK) in minors, modeling and simulation (M&S) in minors, and technical formulation development beyond pills and tablets.

Keywords Pediatric drug development (PDD) · Pediatric legislation · On-label/off-label

23.1 Introduction

Pediatric legislation is justified by the claim that it will improve pediatric healthcare [1]. In 2001, the FDA made the tentative prediction that superior drug treatment information would "permit quicker recoveries from childhood illnesses, with fewer attendant hospital stays, physician visits and parental work days lost." [2] Such endpoints would be reasonable. However, reducing the length of hospital stays can

K. Rose (✉)
klausrose Consulting, Pediatric Drug Development and More, Riehen, Switzerland
e-mail: klaus.rose@klausrose.net; http://www.klausrose.net/

H. Batchelor, K. Rose (eds.), *Pediatric Formulations*, AAPS Advances in the Pharmaceutical Sciences Series 58, https://doi.org/10.1007/978-3-031-77240-5_23

also be achieved by improved surgical techniques, different policies of discharge, and/or other technical advances.

The most expensive part of "pediatric drug development" are clinical studies in minors. Companies pay for these studies. Initially, in the USA, these expenses were offset by 6-month patent extension that delayed the introduction of generic copies of the respective medicine [1]. Generic companies lobbied against this. The EU introduced its legislation 10 years later. It followed the US precedence with some key differences. It relies more on coercion, with negligible rewards. Companies must submit a pediatric investigation plan (PIP) years before the marketing approval application. Drugs for orphan diseases, biologics, and vaccines were excluded from the US legislation, while the EU includes them [3, 4].

Pediatric legislation has triggered many studies in minors. The key question is if these studies contributed to advancing child healthcare. We also analyze further impacts.

23.2 Label Changes, Number of Clinical and Other Studies

FDA and EMA documents show that the pediatric legislation established a gigantic approval machine for drugs for minors, including label changes, "pediatric" studies, and many other administrative steps [5, 6]. We need to keep in mind that many study participants are only administratively still children, but physiologically already mature. Specifically, EMA reports need to be read with adequate decoding. "Availability" does not mean that medicines were made available. It means that now they have a "pediatric" label, children defined as <18 years old.

The 2017 EU commission report lists 949 agreed PIPs and 1275 agreed PIP modifications by 1016 [6]. Since then, these numbers have continued to rise.

Pediatric formulations are a relatively small part of "pediatric drug development," but one of those few elements that advance technical development, including development of formulations other than tablets and pills. But also people in formulation development should have an idea of the entire "pediatric drug development".

Olski et al. report a retrospective analysis of PIPs submitted to the EMA between 2007 and 2009 and conclude that "The key objective of the Paediatric Regulation, namely, the availability of medicines with age-appropriate information, is going to be achieved" [7]. See above for decoding the term "available."

23.3 Awareness of the Need for Research with Children and the Removal of Mental Barriers

Without pediatric legislation, this book would not have been published, an undeniable impact. Under the threat of non-registration, awareness of "pediatric drug development" has spread very fast into the pharmaceutical industry.

Criminal experiments performed during the Second World War by German and Japanese researchers had a considerable influence on public perception of experiments with humans. It took decades to develop a framework to balance the wish to learn against the need to protect the patient. The mainstream attitude was for a long time that children needed protection against research. But then, experimental treatment of childhood cancer saved many lives. It took decades to fine-tune their use in pediatric malignancies [8]. Pediatric cancer treatment was revolutionized by oncologists that systematically tested available drugs in international clinical trials [9–11]. In this revolution, the regulatory authorities did not play any major role [12, 13]. Today, most institutional review boards (IRBs)/ethics committees (ECs) accept the need for clinical research in children [14]. Parents' consent depends largely on the severity of the disease. If the child's life is at stake, most do not hesitate. Otherwise, they hesitate more. The awareness of the need for clinical research in children has increased. Pediatric legislation has contributed to this movement.

But not every experimental treatment of child cancer is reasonable. In 2001, the FDA justified a pemetrexed study in refractory or relapsed pediatric malignancies by arguing that experimental treatment in children was state of the art [15]. Comparable pediatric written requests were issued for cytarabine, anastrozole, carboplatin, clofarabine, and more [16, 17]. But in the early 2000s, young patients with a relapsed or refractory malignancy after standard chemotherapy had failed did not have any chance of an additional follow-up round with a single cytotoxic. All key players in healthcare profited from these studies in terms of money and/or reputation: the FDA; the researching clinicians (paid for this trial; publications; networking; career advancement); and the company that received a patent extension. The only ones who did not benefit from these studies were the patients and their families. The situation is comparable in the EMA-coordinated attempt to repeat the successful switch of targets in early pediatric oncology treatment with cytotoxic agents from adult to pediatric cancer. For every newly developed targeted anticancer drug, the EMA demands "pediatric" studies, often in patients that physiologically are already mature. Chemotherapy agents poison the entire body in the hope to kill enough malignant cells so that the body will deal with those left over. The switching of targets of chemotherapeutic agents from adults to young patients worked in early pediatric oncology. But targeted anticancer drugs are not chemotherapeutic agents. EMA-enforced pediatric trials with targeted anticancer drugs have not resulted in any clinical success [18]. The US RACE for Children Act of 2015 adopted this strategy from the EU precedence, supported in this by many cancer research centers [17–23]. It is psychologically understandable that "something" should be done, but if this does not make sense, it raises false hopes and is a waste of time and resources.

What minors really need is not pseudo-research that serves formal regulatory targets, but true innovation. Such innovation is taking place also in pediatrics, including the survival of Emily Whitehead by involving the immune system in fighting malignant cells [24–27]; multiple new innovative drugs against cystic fibrosis and other diseases that start in children [17]; and the distancing of the international academic world of epileptology from pointless "pediatric" studies that prove that antiseizure medications work also before the 18th birthday [28].

23.4 Pediatric Research Networks

Before the US legislation, a network of pediatric pharmacology research units (PPRUs) had been established [29], later expanded to the RPDP (Research in Pediatric Developmental Pharmacology [30]. Today's European Society of Developmental Perinatal and Pediatric Pharmacology (ESDPPP) was founded in 1988 by pediatricians and pharmacologists [31]. Furthermore, the EMA has established a European network of pediatric research networks (European Network of Paediatric Research at the European Medicines Agency, Enpr-EMA) [32]. This network of networks aims at coordinating the activities of European pediatric research networks. In 2014, it had 38 registered networks or centers [33]. Enpr-EMA does not directly distribute funds, but it is a strong and large network where pediatricians can find support for their research activities.

23.5 Child Health

In its report to Congress 2001, the FDA gave a few tentative parameters how better medicines for children might have measurable health outcomes: "Superior drug treatment information is expected to permit quicker recoveries from childhood illnesses, with fewer attendant hospital stays, physician visits and parental work days lost" [2]. So far, no convincing reports have been brought forward that would elucidate the impact of US or EU pediatric legislation in this dimension.

When the US legislation was introduced, there was palpable enthusiasm among pediatricians, regulatory authorities, and patient organizations. A fast improvement in pediatric pharmaceutical treatment was expected. These hopes have made way for a more conservative assessment. In the long run, it might sometimes benefit minors that more information on side effects, adequate dosing, counterindications, and more are available in drug labels. However, when the concept of "pediatric drug development" emerged, the labels were often the only immediately available information at hand. But technology has evolved. The "pediatric" labels in administratively/legally defined "children" represent a movement that was born out of the pre-internet era. Pediatric oncology emerged in these times without licensing the drugs used. The hope for a "pediatric" switch of targeted anticancer drugs is based on the hope of being able to force innovation with a mechanical formula. It has not worked so far and never will.

23.6 Public Expenses

At the EMA, roughly 30 positions were created to overlook the PIP procedure. In the larger EU countries, one or several employees of the national regulatory authorities work exclusively to support the respective national PDCO representative. In

2013, the EU commission estimated the amount for the period 2007–2012 as more than EUR 39 million, plus costs covered by the regulatory authorities of the EU Member States [34].

23.7 Administrative and Financial Burden for Industry

The PIP procedure between company and the EMA's pediatric committee (PDCO) takes about one year. PIPs are designed as plans that, once approved, cover the entire pediatric development, which can take several years or even decades. With limited clinical data and without efficacy data at the time of the PIP negotiation, little of what is planned will later be executed. The negotiation procedure is exhausting for both sides. At its the end, the PIP contains commitments that often represent EMA's enthusiastic assumptions about the numbers of existing patients that in reality do not exist [17, 35].

The EMA and the EU Commission estimate that per PIP 3–5 modifications of the original PIP have to be negotiated [6, 36], not including PIPs that are withdrawn after negotiating for a year when the PDCO comes up with new ideas in the last moment, the company disagrees, and withdraws the PIP to avoid a negative opinion on the EMA website.

Large companies have internal pediatric matrixes replenished by new hires with PIP experience in other companies. It is different for small- and medium-sized companies. PIP consultants work either in regulatory consultancies and/or in specialized companies. Service organizations offer complete regulatory packages that include PIP negotiations.

The EU Commission estimates the average R&D costs per PIP at 18.9 million, and the total costs for the whole industry at EUR 2.1 billion per year. On top of this, companies incur overhead costs of around EUR 720,000 in relation to filing the initial submission of a PIP and for subsequent modifications [36].

23.8 Patients

Already for clinical studies triggered by the US pediatric legislation, an increasing number of studies were performed outside of the USA [37]. The execution of clinical trials has largely switched from in-house to outsourcing to clinical research organizations (CROs).

Recruitment of minor patients in frequent diseases is less of a problem. But as innovative drug development is moving more toward rare diseases and there is competition between different companies, recruitment of minors into PIP-triggered studies is often challenging or even impossible. Entire PIPs that demanded up to 20 studies have been modified into complete waivers [17].

23.9 Modeling and Simulation (M&S)/Physiologically Based Pharmacokinetics (PBBK) in Minors

Modeling and simulation (M&S) uses physical, physiological, mathematical, or other models to simulate processes supported by computation power. It allows avoiding real-world experimentation, which is usually costly and time-consuming. M&S and more so the advanced physiologically based pharmacokinetics (PBBK) modeling have gained an increasing role in estimating drug doses for minors in recent years [38–40]. The literature and the regulatory agencies continue to use these terms in the context of "pediatric drug development" and continue to uncritically use the term "extrapolation" from adults to children [41], which is flawed in adolescents that have already a mature body [17, 42]. Nevertheless, the willingness of the authorities to refrain from physical studies with minors in some cases, [43, 44] and the increasing demand for additional M&S studies in PIPs show how reality has slowly entered the dogmatic world of "pediatric drug development."

23.10 Conclusions

Modern drugs save lives and/or improve the quality of life. Millions of people live only due to their availability. Many diseases do no longer spread terror as they are prevented by vaccination.

Myocardial infarction and stroke were in the past among the most targeted diseases by drug development. We know the underlying causes, including hypertension, dyslipidemia, obesity, suboptimal nutrition, and "modern lifestyle," including smoking, TV, computer gaming, recreational drugs, and lack of physical activities. Not everybody with such a lifestyle will have a stroke or a myocardial infarction. But to prevent this in one person many must take daily pills. As long as further beta-blockers were prescribed, more were developed. We already have several generations of antihypertensives. Some companies dream of yet another one. But effective drugs exist already, the original products today inexpensive generics, and companies are looking for areas where the potential return of investment is high, e.g., rare diseases [45].

The consideration of children in the drug development process claims to be driven by public health. Drug development is market-driven but is also strongly connected to academic science. BioNTech developed the first mRNA vaccine against COVID-19. It was developing mRNA technology for therapeutic vaccination against cancer. When it was surprised by the pandemic, it switched targets, partnered with Pfizer, and won the race [46]. This shows two factors: scientific know-how, and economic competition. Long term, not the big players win the competition, but the fastest and most flexible ones.

References

1. Hirschfeld S, Saint-Raymond A. Pediatric regulatory initiatives. Handb Exp Pharmacol. 2011;205:245–68.
2. FDA. The pediatric exclusivity provision – January 2001. Status report to Congress. 2001. http://www.fda.gov/downloads/drugs/developmentapprovalprocess/developmentresources/ucm049915.pdf.
3. Ramet J, Lehman B, Rose K. Europe and the path to better medicines for children. In: Rose, van den Anker, editors. Guide to paediatric drug development & clinical research. Basel: Karger; 2010.
4. Wright EA, Rose K. European Union paediatric regulation theory and practice. In: Rose, van den Anker, editors. Guide to paediatric drug development & clinical research. Basel: Karger; 2010.
5. FDA. BPCA and PREA status report to Congress July 9, 2012–June 30, 2016. 2016. https://www.fda.gov/media/99184/download.
6. EMA. 10-year report to the European Commission general report on the experience acquired as a result of the application of the Paediatric Regulation. 2017. https://health.ec.europa.eu/system/files/2020-06/paediatrics_10_years_ema_technical_report_0.pdf.
7. Olski TM, Lampus SF, Gherarducci G, Saint-Raymond A. Three years of paediatric regulation in the European Union. Eur J Clin Pharmacol. 2011;67:245–52.
8. Adamson PC. Improving the outcome for children with cancer: development of targeted new agents. CA Cancer J Cli. 2015;65(3):212–20.
9. Pritchard-Jones K, Pieters R, Reaman GH, et al. Improving cancer care for children and young people 1 – sustaining innovation and improvement in the treatment of childhood cancer: lessons from high-income countries. Lancet Oncol. 2013;14(3):e95–e103. http://press.thelancet.com/childhoodcancer1.pdf.
10. Magrath I, Steliarova-Foucher E, Epelman S, et al. Improving cancer care for children and young people 2 – paediatric cancer in low-income and middle-income countries. Lancet Oncol. 2013;14(3):e104–16. http://press.thelancet.com/childhoodcancer2.pdf.
11. Boklan J. Little patients, losing patience: pediatric cancer drug. Mol Cancer Ther. 2006;5:1905–8. http://mct.aacrjournals.org/content/5/8/1905.full.pdf.
12. Sullivan R, Kowalczyk JR, Agarwal B, et al. Improving cancer care for children and young people 4 – new policies to address the global burden of childhood cancers. Lancet Oncol. 2013;14(3):e125–35. http://press.thelancet.com/childhoodcancer4.pdf.
13. Vassal G, Zwaan CM, Ashley D. Improving cancer care for children and young people 3 – new drugs for children and adolescents with cancer: the need for novel development pathways. Lancet Oncol. 2013;14(3):e117–24. http://press.thelancet.com/childhoodcancer3.pdf.
14. Ethical considerations for paediatric trials – how can ethics committees in the European Member States and the Paediatric Committee at the European Medicines Agency work together? EMA Workshop 2011. Documentation on: http://www.ema.europa.eu/ema/index.jsp?curl=pages/news_and_events/events/2011/07/event_detail_000443.jsp&mid=WC0b01ac058004d5c3.
15. FDA. FDA Pemetrexed pediatric written request 2001. 2001. https://www.fda.gov/media/80043/download.
16. FDA. Written requests issued. 2022. https://www.fda.gov/drugs/development-resources/written-requests-issued.
17. Rose K. Considering the patient in pediatric drug development. How good intentions turned into harm. Elsevier, London, 2020.
18. Rose K. Pediatric oncology at the crossroads: a call for change. Pharmaceut Med. 2020;34(5):297–300. https://rdcu.be/b8drm.
19. H.R.1231. RACE for Children Act 115th Congress (2017–2018) https://www.congress.gov/bill/115th-congress/house-bill/1231.

20. Reaman GH. FDARA 2017 and the RACE for Children Act: implications for pediatric cancer drug development. Available online: https://www.accelerate-platform.eu/wp-content/uploads/sites/4/2018/02/1640-REAMAN.pdf.
21. Curesearch of children's cancer. The RACE for Children Act: a game changer for children's cancer. https://curesearch.org/The-RACE-for-Children-Act-A-Game-Changer-for-Childrens-Cancer.
22. Pearson ADJ, Karres D, Reaman G, et al. The RACE to accelerate drug development for children with cancer. Lancet Child Adolesc Health. 2020;4(10):714–6.
23. Endorsements of RACE Act. The RACE for Children Act coalition. https://www.kidsvcancer.org/race-for-children-act/endorsements-for-race-act/.
24. Maude SL, Laetsch TW, Buechner J, et al. Tisagenlecleucel in children and young adults with B-cell lymphoblastic leukemia. N Engl J Med. 2018;378:439–48.
25. First child to receive CAR T-Cell therapy is 10 years cancer-free. https://www.youtube.com/watch?v=0ei1yRxbdfU.
26. 7 year old girl beats acute lympoblastic leukemia: Emily Whitehead's immunotherapy story. https://www.youtube.com/watch?v=Sz11e0r2L_s.
27. New cancer cure: world's first child treated with immunotherapy is cancer-free 10 years later. https://www.youtube.com/watch?v=DP7fFoKPWcM.
28. Rose K, Ettienne EB, Grant-Kels JM, Striano P, Neubauer D, Oishi T. Neurology's vital role in preventing unnecessary and potentially harmful pediatric studies. Expert Rev Neurother. 2022;22(3):209–19. https://doi.org/10.1080/14737175.2022.2045953.
29. U.S. Department of Health & Human Services (DHHS): Obestric and Pediatric Pharmacology Branch Report to the NACHHD Council, 2008 https://www.nichd.nih.gov/sites/default/files/publications/pubs/documents/oppb_council_2008_historical.pdf.
30. US National Institute of Health (NIH): specialized centers in Research in Pediatric Developmental Pharmacology (RPDP) 2011. http://www.nichd.nih.gov/research/supported/Pages/scrpdp.aspx#overview.
31. European Society of Developmental Perinatal and Pediatric Pharmacology (ESDPPP). https://www.esdppp.org/about-esdppp/.
32. European Network of Paediatric Research at the European Medicines Agency (Enpr-EMA). http://www.ema.europa.eu/ema/index.jsp?curl=pages/partners_and_networks/general/general_content_000303.jsp&mid=WC0b01ac05801df74a.
33. Pelle B, Eichler I, Turner M. PO-0930 Enpr-ema, a platform for disseminating good practices about paediatric medicines research across Europe and with international partners. Arch Dis Child. 2014;99(Suppl. 2) https://adc.bmj.com/content/99/Suppl_2/A554.2.
34. EU Commission 2013: report from the Commission to the European Parliament and the Council. Better medicines for children—from concept to reality. General report on experience acquired as a result of the application of Regulation (EC) No 1901/2006 on medicinal products for paediatric use. http://ec.europa.eu/health/files/paediatrics/2013_com443/paediatric_report-com(2013)443_en.pdf.
35. Hwang TJ, Tomasi PA, Bourgeois FT. Delays in completion and results reporting of clinical trials under the paediatric regulation in the European Union: a cohort study. PLoS Med. 2018;15(3):e1002520. https://www.ncbi.nlm.nih.gov/pmc/articles/PMC5832187/pdf/pmed.1002520.pdf.
36. EU Commission 2017: report from the Commission to the European Parliament and the Council. State of Paediatric Medicines in the EU – 10 years of the EU Paediatric Regulation. https://www.europarl.europa.eu/RegData/docs_autres_institutions/commission_europeenne/com/2017/0626/COM_COM(2017)0626_EN.pdf.
37. Pasquali SK, Burstein DS, Benjamin DK Jr, et al. Globalization of pediatric research: analysis of clinical trials completed for pediatric exclusivity. Pediatrics. 2010;126:e687–92. http://pediatrics.aappublications.org/content/126/3/e687.full.pdf+html.

38. Vinks AA, Emoto C, Fukuda T. Modeling and simulation in pediatric drug therapy: application of pharmacometrics to define the right dose for children. Clin Pharmacol Ther. 2015;98(3):298–308.
39. Verscheijden LFM, Koenderink JB, Johnson TN, et al. Physiologically-based pharmacokinetic models for children: starting to reach maturation? Pharmacol Ther. 2020;211:107541. https://www.sciencedirect.com/science/article/pii/S0163725820300693?via%3Dihub.
40. Maharaj AT, Edginton AN. Physiologically based pharmacokinetic modeling and simulation in pediatric drug development. CPT Pharmacometrics Syst Pharmacol. 2014;3:e148. https://ascpt.onlinelibrary.wiley.com/doi/pdfdirect/10.1038/psp.2014.45.
41. ICH Harmonized Guideline. Pediatric Extrapolation. Draft version, April 2022. https://database.ich.org/sites/default/files/ICH_E11A_Document_Step2_Guideline_2022_0404_0.pdf.
42. Momper JD, Mulugeta Y, Green DJ, et al. Adolescent dosing and labeling since the Food and Drug Administration Amendments Act of 2007. JAMA Pediatr. 2013;167(10):926–32.
43. FDA. Pediatric trial design and modeling: moving into the next decade. 2017. https://www.fda.gov/drugs/news-events-human-drugs/pediatric-trial-design-and-modeling-moving-next-decade.
44. Peck C. Modeling & simulation in pediatric drug development and regulation. EMEA, workshop on modelling in paediatric medicines, London, April 14–15, 2008. https://www.ema.europa.eu/en/documents/presentation/presentation-modelling-simulation-paediatric-drug-development-regulation_en.pdf.
45. Thorat C, et al. What the orphan drug act has done lately for children with rare diseases: a 10-year analysis. Pediatrics. 2012;129:516–21. http://pediatrics.aappublications.org/content/129/3/516.long.
46. Rose K. The COVID-19 pandemic: a global high-tech challenge at the interface of science, politics, and illusions. London: Elsevier; 2022.

Chapter 24
The Dangerous Business of Predicting the Future

Daniel Bar-Shalom, Hannah Batchelor, Linda F. McElhiney, and Klaus Rose

Abstract New drugs do not fall from the sky. Their development has become a complex, expensive, and multidimensional process. There is the past without which we cannot understand the present; the present which this book discusses; and the future. We can extrapolate some technical developments, such as the expanding technique of microencapsulation. The more complex other dimensions are, the more difficult it is to make predictions, specifically where there are interfaces between politics, weltanschauung, and technical developments. Adolescents are *legally* still children, but no longer *physiologically*. Adolescents and prepubertal older school children rarely need age-appropriate pediatric formulations. Pointless and/or exaggerated "pediatric" clinical studies in this age group need to be challenged by those working in pharmaceutical development. Formulation development for the benefit of young and very young children and babies is probably the most valuable aspect of pediatric legislation.

Keywords Pediatric drug development (PDD) · Pediatric legislation · On-label/off-label

D. Bar-Shalom
Department of Pharmacy, University of Copenhagen, Copenhagen, Denmark

H. Batchelor
Strathclyde Institute of Pharmacy and Biomedical Sciences, Glasgow, UK
e-mail: hannah.batchelor@strath.ac.uk

L. F. McElhiney
Indiana University Health, University Boulevard Adult Outpatient Center (AOC), Indianapolis, IN, USA

K. Rose (✉)
klausrose Consulting, Pediatric Drug Development and More, Riehen, Switzerland
e-mail: klaus.rose@klausrose.net

H. Batchelor, K. Rose (eds.), *Pediatric Formulations*, AAPS Advances in the Pharmaceutical Sciences Series 58, https://doi.org/10.1007/978-3-031-77240-5_24

24.1 The Dangerous Business of Predicting the Future

On a purely technical level, it is relatively safe to predict which innovations will be available for the compounding of pediatric formulations in the near future. Multiparticulate formulations seem to be one path where a lot of research will be invested in the near future, seeing the promising results of research performed so far.

Another question are challenges and opportunities of automatic compounding. To integrate automatic, personalized compounding into the drug development process might e.g. result in no longer registering e.g. tablets or pre-packed liquid formulations. It would allow medication to be compounded in a personalized way in the hospital and maybe even in specialized community pharmacies and then given to the individual patient. This would require a rather radical re-thinking about the relationship of drugs during clinical trials and in later medical use, both on the side of regulatory authorities and developing companies.

Reviews of the US and EU pediatric legislation praise it, as expected [1–4]. But the scientific foundations of pediatric legislation have also been critically addressed [5–7]. It is unpredictable when science, public opinion, and politics follow suit.

The recipients of pediatric medicines are changing. Preterm newborns that survive today with or without long-lasting health problems would not have survived half a century ago. Neonatology and the drugs used will certainly further evolve. Despite the need for more pediatric formulations and dosage forms to meet the medical needs of pediatric patients, there are still not enough clinical studies done with new drugs and medications in this patient population.

During the COVID-19 pandemic, Russia and China developed their own vaccines soon, hoping to make quick political gains from it. Both failed, and the West's life science industry showed its strengths in developing effective vaccines. Start-up companies are partially replacing the rather ineffective discovery and development machinery of large pharmaceutical companies. The first mRNA vaccine against COVID-19 was developed by a company that did not exist 15 years ago. It partnered with an old-style life science industry behemoth, a partnership from which both companies profited [8].

To what degree will innovative and breakthrough medicines be facilitated and reimbursed, even if they cost a lot of money? When drugs for rare diseases were rare, the money for a few patients with very high daily treatment costs could usually be found somewhere. Now we have a growing pipeline of orphan drugs. The fight for limited resources will continue and will certainly become harsher.

Will the principle of a standard dose remain as the routine way of registration, or will individualized and/or personalized doses become part of the registration process, specifically in neonatology?

Increasingly, medicines are paired with diagnosing kits toward a personalized and individualized treatment. For now, companies continue to print patient information. It is imaginable that eventually posting the full version on the Internet will suffice, with only short summaries printed.

Which new pediatric excipients will be developed? Which food ingredients will be used as pharmaceutical, pediatric excipients? Will novel taste masking/taste concealing/taste suppressant strategies be developed? Many taste-testing studies are done with adult participants; however, pediatric patients' taste-perception is completely different and more sensitive. A taste or flavor that may be tolerated or acceptable in an adult may not be acceptable at all in a child. Pediatric patients are more susceptible to sweet-tasting substances and sensitive to sour or bitter-tasting substances.

Specifically in serious conditions, polypharmacy will play an even stronger role. How will pharmaceutical and medical training keep pace with this?

With compounding tragedies in the USA [9], standardizing and regulating compounding has gained momentum. Private organizations, such as the United States Pharmacopeia, are continuously evaluating their compendial standards for compounding to possibly provide global standards for compounding so that the same compound may be prepared in the same manner, whether it is in Europe, Asia, or the Americas.

The need for compounding will always exist as long as there are patients who need specialized dosage forms that are not commercially available. The key issue will always be how do pharmacists develop and prepare these to provide high-quality care.

We have listed those unknowns that we are aware of. Others will evolve. Life is always challenging.

We hope you enjoyed the book.

References

1. FDA. The pediatric exclusivity provision. January 2001. Status report to Congress. 2001. https://www.fda.gov/downloads/Drugs/DevelopmentApprovalProcess/DevelopmentResources/UCM049915.pdf.
2. FDA. BPCA and PREA status report to Congress July 9, 2012—June 30, 2016. 2016. https://www.fda.gov/media/99184/download.
3. FDA. BPCA and PREA status report to Congress July 1, 2015—June 30, 2020. 2020. https://www.fda.gov/media/157840/download.
4. EMA. 10-year Report to the European Commission General report on the experience acquired as a result of the application of the Paediatric Regulation. 2017. https://health.ec.europa.eu/system/files/2020-06/paediatrics_10_years_ema_technical_report_0.pdf.
5. Rose K. Considering the patient in pediatric drug development. How good intentions turned into harm. London: Elsevier; 2020.
6. Rose K. Blind trust – how parents with a sick child can escape the labyrinth of lies, hypocrisy and false promises of researchers and regulatory authorities. London: Hammersmith; 2022.
7. Rose K, Tanjinatus O, Grant-Kels JM, et al. Minors and a dawning paradigm shift in "pediatric" drug development. J Clin Pharmacol. 2020; https://accp1.onlinelibrary.wiley.com/doi/epdf/10.1002/jcph.1806
8. Rose K. The COVID-19 pandemic: a global high-tech challenge at the interface of science, politics, and illusions. London: Elsevier; 2022.
9. Kastango ES. Lessons learned from compounding tragedies. Can J Hosp Pharm. 2013;66(3):152–3.

Index

A

Acceptability, 23, 38, 39, 48–63, 72, 89, 92, 135, 136, 151, 153–154, 156, 158, 161–164, 169–173, 181, 182, 184, 211, 217, 220, 229, 233, 248–250, 252, 261, 274, 275, 305, 306, 313, 314, 337–339, 352, 356, 429–431
Access to essential medicines, 438, 440, 442–444
Adhesiveness, 219
Adolescents, 6–7, 17, 18, 20, 24, 38, 39, 51, 52, 124, 151, 154, 158, 164, 168, 180, 183, 212, 240, 267, 283, 292, 296, 354, 386–389, 401, 402, 417, 428, 450
Aerosol clouds, 283, 288–290, 294, 295, 303
Aqueous humour, 307, 308, 311, 312

B

Bioequivalence studies, 35, 395, 396, 403
Buccal tablets, 233

C

Children, 1, 9, 33, 48, 86, 123, 154, 180, 212, 228, 240, 260, 282, 301, 331, 344, 388, 397, 420, 423, 437, 446
Coatings, 81, 87–100, 103, 104, 106, 107, 112, 136, 154, 164, 165, 168, 181, 185, 188, 194, 201–204, 206, 207, 234, 235, 313, 319, 334, 358, 359, 361, 406, 430
Cochlear implants, 324
Cohesiveness, 217, 219

D

Definition, 1–2, 4, 5, 71–72, 138, 142, 165, 387
Deposition fractions, 284, 285, 287, 296
Diabetic macular oedema, 319
(oro)dispersible solids, 163, 430
Dosage forms, 6, 15, 17, 35, 36, 38–45, 62, 70–82, 86, 89, 92, 100, 103, 105, 109, 110, 114, 115, 123–126, 132, 137, 145, 150–159, 161–165, 168, 169, 173, 180–194, 203, 205, 208, 212–217, 223, 228, 229, 234, 235, 249, 251, 253, 268, 287, 333, 335, 338, 339, 345, 346, 350, 352–355, 401, 403, 409, 415, 434, 438, 442, 456, 457
Dose selection, 10–14, 36, 414, 419
Dosing regimens, 5, 34, 43, 44, 49, 180, 208, 406, 414, 421
Drug development, 34, 35, 48, 54, 153, 384, 388, 397, 409, 414–416, 419, 429, 449, 450
Drug elimination, 13, 14, 26–27, 417

E

Enabling formulations, 396, 397, 400, 404
European Medicines Agency, 2, 48

H. Batchelor, K. Rose (eds.), *Pediatric Formulations*, AAPS Advances in the Pharmaceutical Sciences Series 58, https://doi.org/10.1007/978-3-031-77240-5

Evaluations, 6, 59, 61–63, 74, 75, 77–78, 81, 110–115, 203, 204, 235, 236, 344, 350–352, 356, 375, 408, 417, 430
Excipients, 40, 42, 70, 92, 103, 104, 106, 108, 115, 125, 126, 128, 134–138, 140, 145, 151, 153–156, 159, 165, 166, 180, 184–186, 190, 196, 198, 201, 216, 220–223, 233, 234, 247, 248, 250, 251, 253, 254, 262, 267–270, 275, 287, 289, 295, 302, 312, 314, 323, 336, 338, 344–375, 401, 415, 430–432, 457

F
Fentanyl citrate, 233, 235
Flexible solid dosage forms, 184, 409, 434
Fluticasone propionate, 290
Formulation, 4, 9, 33, 48, 123, 151, 154, 180, 213, 229, 240, 260, 282, 303, 334, 352, 383, 396, 415, 423, 444, 446
Formulation development, 4, 36, 48, 78, 126, 128, 138, 168, 181, 182, 241, 253, 267, 384, 390, 397, 398, 400, 414–421, 434, 440, 446

G
Gastric functionality, 16–18
Global health, 155, 184, 437, 442, 444
Gummies, 162, 212–216, 223

H
Human panels, 49, 77–78, 110, 111

I
Inactive ingredients, 314, 345, 350, 401
Inertial impaction, 283–287, 303
Infants, 4–6, 11, 14–17, 19–26, 38, 39, 73, 75, 88, 123, 125, 126, 134, 143–145, 154, 159–161, 164, 169, 180, 182, 184, 231, 236, 240, 244, 248, 252, 254, 263–268, 270–274, 282, 287, 294, 296, 307–310, 320, 321, 331, 333, 334, 344, 346, 347, 354, 356, 360, 363, 368, 372, 386, 387, 406, 408, 417
Integrated development concept with micropellets, 409, 410
Interactions, 35, 42, 50, 54, 55, 70, 73–77, 86, 89, 92, 107, 112, 163, 167, 208, 218, 219, 263, 269, 285, 290, 291, 296, 305, 345, 353–354, 389, 405, 415, 417
In vitro/in vivo testing, 78–81, 110, 144, 306

J
Jellies, 171, 212–223

L
Low- and middle-income countries (LMICs), 10, 42, 156, 335, 437–444
Lung deposition, 283–285, 288–292, 295–297

M
Medicinal products, 1, 4–6, 10, 48, 71, 77, 139, 142, 153–155, 173, 263, 287, 319, 337, 348, 349, 351–353, 355, 387, 424–427, 429, 432
Micropellet manufacturing technologies, 180–208
Micropellets and minitablets, 180–208, 220, 410
Mouthfeel, 49, 72, 78, 81, 89, 101, 154, 164, 169, 217, 219
Multiparticulates, 72, 78, 92, 96–100, 154, 155, 157, 158, 163–165, 181–184, 187, 188, 201, 205, 207, 208, 213, 217, 218, 220, 407

N
National Competent Authority (NCA), 425
Neonates, 4–5, 10, 15–17, 19–27, 39, 75, 88, 125, 134, 144, 151, 154, 155, 161, 168, 171, 180, 181, 183, 228, 240, 260–272, 274, 307–309, 312, 318, 320, 332–334, 344, 346, 356, 360, 369, 370, 372, 374, 385, 406–408, 417, 418, 432

O
Ocular surface, 309, 312, 314, 315, 317
On-label/off-label, 382–390, 445–450, 456–457
Optimized API solubility and bioavailability, 187, 190
Oral absorption, 5, 6, 10, 14–23, 418
Oral liquids, 39, 89, 123–145, 151, 156, 181, 185–186, 335, 336, 352, 354, 375, 402, 430

P

Paediatric, 154, 228, 241, 260, 283, 303, 332, 352, 418, 423
Paediatric formulations, 4, 88, 150, 151, 153, 156, 164, 168, 169, 173, 301–325, 345–346, 356, 429–434
Paediatric patients, 1–7, 10, 16, 89, 94, 97–99, 110, 151, 157, 233, 235, 236, 245, 260, 264, 271, 272, 282, 290, 293, 303, 308–311, 314, 316–319, 321, 333, 334, 336, 429
Paediatric regulations, 150, 424–429, 433, 446
Palatability, 48, 50, 59–61, 70–82, 89, 126, 128, 135, 153, 154, 159, 161, 162, 164, 170–172, 180, 181, 184, 188, 219, 220, 234, 236, 354, 355, 400, 430
Parenteral, 23–24, 144, 228, 260, 261, 263, 264, 267–270, 272, 274, 275, 345, 347, 359, 360, 368–370, 372–374, 432
Pass metabolism, 228
Pediatric, 1, 10, 34, 48, 88, 123, 180, 211, 385, 396, 414, 438, 445
Pediatric drug development (PDD), 37, 382–390, 399, 400, 409, 419–420, 438–440, 444–450, 456–457
Pediatric formulation bridging, 401
Pediatric formulations, 33–45, 124, 125, 127, 134, 138, 145, 180, 181, 183, 211, 352, 355–375, 387, 388, 396–402, 414–421, 438–444, 446, 456
Pediatric legislation, 382–390, 437–450, 456–457
Pediatric medicines, 70, 180, 181, 188, 201, 350, 355, 375, 384–385, 421, 437–444, 456
Pellets and micropellets, 180–208
Personalized medicines, 187
Pharmaceutical development, 34, 36, 49, 58, 60, 63, 71, 150, 151, 155, 158, 161–163, 165–169, 173, 183, 187, 338, 352, 355
Pharmacokinetics (PK), 4, 10, 27, 39, 41, 44, 88, 138, 228, 261, 265, 308, 345, 346, 397–401, 408, 414–416, 418–421, 428–430, 432, 450
Physiologically-based pharmacokinetic (PBPK) models, 414–420
Population pharmacokinetics (PopPK), 414–415, 421
Posterior segments, 308, 312–313, 318

Q

Quality target product profile (QTPP), 34–45, 49, 352

R

Regulatory frameworks, 142, 348–353, 375, 424, 438, 439
Relative bioavailability studies, 215, 395–409

S

Safety, 2, 23, 34, 35, 40, 44, 48, 50, 52, 78, 88, 104, 115, 125, 126, 134–136, 138, 142, 145, 150, 151, 153–155, 158, 173, 192–194, 208, 223, 228, 236, 262, 266, 267, 273, 274, 319, 333, 338, 344, 345, 347–356, 367, 375, 383, 384, 389, 396, 398, 400, 406, 408, 415, 417, 425, 427–429, 431–433, 438, 441
Skin, 14, 24, 52, 70, 73, 239–254, 260, 265–267, 274, 275, 346, 355, 371, 373, 374
Small intestinal functionality, 19–21
Smell, 59, 70–72, 74, 154, 161, 162, 219, 223
Soft-pudding, 213–216
Solid dosage forms, 16, 18, 36, 40, 42–44, 114, 123–125, 132, 137, 151, 181, 208, 229, 233, 262, 345, 353, 354, 356, 401
Solid formulations, 16, 78, 150–173, 182–184, 430
Solutions, 21, 52, 53, 75, 78–80, 88, 90, 91, 96, 104–108, 110, 111, 115, 123–128, 130–132, 136–138, 140, 145, 151, 156, 159, 168, 192, 195, 198, 203, 218, 229, 231, 234, 236, 242, 246, 248, 250, 251, 262, 264, 268–270, 273, 274, 283, 287–290, 293–295, 302, 304, 308–312, 314, 315, 317, 322–324, 333, 335–337, 354, 355, 369, 373, 440
Suspensions, 38, 52, 90, 91, 98–100, 104, 105, 123–125, 128–132, 135–140, 144, 151, 154, 156, 170, 185, 186, 198, 203, 218, 219, 229, 250, 287–289, 293–295, 304, 312, 315, 333, 335, 336, 354, 355, 403–405, 415, 439
Swallowability, 48–50, 59–61, 74, 81, 92, 125, 135, 153, 157, 161, 169–172, 213, 217, 220, 223

Syrups, 78, 88, 124–127, 130, 135, 145, 151, 156, 160–162, 170, 171, 213, 214, 216, 235, 348, 354, 361, 367, 375

T

Target product profile (TPP), 34–38, 45, 138
Taste masking, 78, 86–89, 92, 94, 97, 102–105, 108–116, 126, 135–136, 151, 154, 161, 164, 181, 185, 186, 188, 201, 203, 204, 219, 234, 305, 331, 355, 356, 403, 406, 430, 457
Tastes, 39, 43, 48, 49, 51–53, 59, 61, 62, 70–81, 86–116, 125, 126, 128, 135–137, 153, 154, 160–164, 181, 185, 187, 188, 203–204, 216, 217, 219, 223, 234, 235, 305, 314, 345, 354, 356, 365, 402, 403, 406, 430, 431, 457
Textures, 9, 48, 71–74, 77, 154, 163, 168, 213, 214, 216, 217, 219, 234
Topical, 14, 24, 229, 236, 239–254, 273, 274, 283, 303, 310, 311, 313, 314, 316, 317, 321–322, 346, 355, 357, 358, 363, 368, 370–375
Toxicity, 5, 24, 104, 155, 192, 222, 229, 239, 248, 253, 254, 261, 268, 269, 287, 303, 314, 344, 346, 347, 349–351, 355, 356, 369, 373, 375, 385
Transdermal, 5, 233, 239–254

MIX
Papier aus verantwortungsvollen Quellen
Paper from responsible sources
FSC® C105338

If you have any concerns about our products,
you can contact us on
ProductSafety@springernature.com

In case Publisher is established outside the EU,
the EU authorized representative is:
Springer Nature Customer Service Center GmbH
Europaplatz 3, 69115 Heidelberg, Germany

Printed by Libri Plureos GmbH
in Hamburg, Germany